de Gruyter Series in Nonlinear Analysis and Applications 12

de Gruyter Series in Nonlinear Analysis and Applications

1 Multivalued Differential Equations, *Klaus Deimling, 1992*

2 Approximation Procedures in Nonlinear Oscillation Theory, *Nikolai A. Bobylev, Yuri M. Burman and Sergey K. Korovin, 1994*

3 Sobolev Spaces of Fractional Order, Nemytskij Operators, and Nonlinear Partial Differential Equations, *Thomas Runst and Winfried Sickel, 1996*

4 Relaxation in Optimization Theory and Variational Calculus, *Tomás Roubícek, 1997*

5 Quasilinear Elliptic Equations with Degenerations and Singularities, *Pavel Drábek, Alois Kufner and Francesco Nicolosi, 1997*

6 Introduction to Neural Dynamics and Signal Transmission Delay, *Jianhong Wu, 2001*

7 Condensing Multivalued Maps and Semilinear Differential Inclusions in Banach Spaces, *Mikhail Kamenskii, Valeri Obukhovskii and Pietro Zecca, 2001*

8 Equivariant Degree Theory, *Jorge Ize and Alfonso Vignoli, 2003*

9 Nonlinear Integral Operators and Applications, *Carlo Bardaro, Julian Musielak and Gianluca Vinti, 2003*

10 Nonlinear Spectral Theory, *Jürgen Appell, Espedito De Pascale and Alfonso Vignoli, 2004*

11 Homotopy of Extremal Problems, *Stanislav V. Emelyanov, Sergey K. Korovin, Nikolai A. Bobylev and Alexander V. Bulatov, 2007*

Victor G. Zvyagin
Dmitry A. Vorotnikov

Topological Approximation Methods for Evolutionary Problems of Nonlinear Hydrodynamics

Walter de Gruyter · Berlin · New York

Authors

Victor G. Zvyagin
Research Institute of Mathematics
Voronezh State University
Universitetskaya pl. 1
394006 Voronezh
Russia

Dmitry A. Vorotnikov
Research Institute of Mathematics
Voronezh State University
Universitetskaya pl. 1
394006 Voronezh
Russia

Keywords
Fluid dynamics, evolutionary equations, rheology, viscoelastic media, non-Newtonian fluids, attractors, weak solutions, Jeffreys' model, topological degree

Mathematics Subject Classification 2000
Primary 76-XX, 76-02, 76A05, 76A10; secondary 47Hxx, 35B41

ISBN 978-3-11-020222-9
ISSN 0941-813X

Bibliographic information published by the Deutsche Nationalbibliothek

The Deutsche Nationalbibliothek lists this publication in the Deutsche Nationalbibliografie; detailed bibliographic data are available in the Internet at http://dnb.d-nb.de.

Printed in Germany.
Cover design: Thomas Bonnie, Hamburg
Typeset using the authors' LaTeX files: Matthias Pfizenmaier, Berlin
Printing and binding: Hubert & Co. GmbH & Co. KG, Göttingen

Preface

The well-known methods of investigation for evolutionary problems of fluid dynamics are the Faedo–Galerkin method [36, 37, 61], the iteration method [2, 3], the method of evolution equations in Banach spaces [51, 78] and some others. In the present book we describe one more method of study of such problems, which is especially appropriate for the research of weak solvability for initial-boundary value problems arising in nonlinear hydrodynamics. Here we use this method for the investigation of some models for motion of viscoelastic media, its employment for other models can be found e.g. in works [29, 64, 71]. The outline of application of this method is the following one.

One begins with the interpretation of an initial-boundary value problem as an operator equation in the function space which naturally corresponds to the considered problem. As a rule, the maps involved in this equation do not possess good operator properties, so at once it is not possible to apply any principles of nonlinear analysis for the proof of the solvability of the problem. Therefore one finds an approximation of this equation (which consists in smoothing of nonlinear terms, or in adding terms of higher order with a small parameter, or in some other operations) and studies the solvability of this approximating equation in the spaces with more suitable topological properties. For this purpose, one applies the technique of the Leray–Schauder topological degree or its generalizations. It is important to point out that in this situation the approximating equation has natural equation properties, and small variations of the right-hand side and of the initial data imply small variation of the solution set for this equation. In particular, this gives opportunity to apply various approximate methods for the analysis of this equation, and the convergence of the approximate solutions to the solutions of the approximating equation is guaranteed. The last step of the method is the passage to the limit in the approximating equation as the approximation parameters tend to zero, and here the solutions of the approximating equation converge to a solution of the original equation (usually in a topology which is much weaker than the one of the spaces where the approximating equation was studied).

In particular, this method turned out to be useful in those problems of non-Newtonian hydrodynamics where it is hard or impossible to express the deviatoric stress tensor via the velocity vector function explicitly (and, consequently, one has to examine systems of two basic variables: the velocity vector and the stress tensor).

The book contains preliminary material from rheology, which is required for understanding the models under consideration. This material is written from the mathematician's point of view. Besides, for the sake of relative completeness of the research, we give some results on the existence of strong solutions for the initial-boundary value problems describing the motion of the media which are modeled by the considered models. Let us turn to more detailed description of the book's contents.

The **first** chapter of the book is the just mentioned introduction to the rheology of viscoelastic and nonlinear-viscous media. Let us point out that we do not try to describe all existing models, and we mainly restrict ourselves to the ones which are the objects of our research in the subsequent chapters. Section 1.1 is a summary of the basic laws which are independent of particular media. Section 1.2 is devoted to the brief analysis of one-dimensional viscoelastic models, which are generalized to higher dimensions in Section 1.3. The nonlinear-viscous models are handled in Section 1.4. In Section 1.5, we combine the viscoelastic and nonlinear-viscous approaches.

The **second** chapter is concerned with the function spaces which are required for the mathematical study of the hydrodynamical equations and with the interrelations between these spaces (embedding theorems etc.). Section 2.1 deals with the spaces of functions defined on subsets of $\mathbb{R}^n$ and with values in finite-dimensional spaces. In Section 2.2, we give the necessary results which involve the vector functions with values in Banach spaces.

The **third** chapter contains miscellaneous facts on linear (Section 3.1) and nonlinear (Section 3.2) non-evolutionary and evolutionary operator equations in Banach spaces, which we need for the investigation of equations of hydrodynamics (in particular, the notion of Leray–Schauder degree).

The **fourth** chapter is a rather general self-contained theory of attractors for evolutionary equations in Banach spaces. It is used in Chapter 6 for construction of attractors for the weak solutions to the initial-boundary value problems from the dynamics of viscoelastic media. Section 4.1 contains the classical issues of attractor theory; its generalizations for the evolution equations without uniqueness of solutions, without invariance of the trajectory space etc. are given in Section 4.2 (autonomous case) and Section 4.3 (non-autonomous case).

The goal of the **fifth** chapter is to present some results on the strong solvability and some solution properties for the initial-boundary value and Cauchy problems describing the motion of viscoelastic media. Unfortunately, as in the classical dynamics of Newtonian fluids in three-dimensional domains, for the problems under consideration one cannot prove global in time existence of strong solutions for arbitrary data even in two dimensions (although there is no counterexample). Section 5.1 is devoted to a local existence result for the initial-boundary value problem for the system of equations of motion for the Jeffreys viscoelastic medium. In the subsequent sections of the chapter we study the initial-value problem for the system of motion equations for a more general combined class of models for nonlinear viscoelastic media in the whole space $\mathbb{R}^2$ or $\mathbb{R}^3$. In Section 5.2, we formulate the problem and the main (existence and uniqueness) result and introduce additional notations. In Section 5.3, the operator treatment of the considered problem is realized. In Section 5.4, an auxiliary problem depending on a parameter is introduced and investigated. The existence of solutions of this problem and a uniform a priori estimate are proved. In Section 5.5, the passage to the limit as the parameter tends to zero is carried out, and the global strong solution of the original problem is obtained (for small initial data and body force). In Section 5.6, we study the continuous dependence of solutions on the problem data.

As we already mentioned, the problem of existence of global strong solutions to the initial-boundary value problems for the equations of viscoelastic fluid mechanics is open in the general case. A possible way to break this deadlock is to investigate weak solutions to these problems. This is realized in the **sixth** chapter. In Section 6.1, we give a set-theoretic scheme for weak formulation of problems and prove some important equalities. In Section 6.2, we introduce the concept of a weak solution to the initial-boundary value problem for the system of motion equations for the Jeffreys viscoelastic medium, and then we formulate the main existence theorem for it. In Section 6.3, existence of solutions of an auxiliary problem, depending on several parameters, is proved via obtaining a priori estimates and application of the Leray–Schauder degree theory. In Section 6.4, the passage to the limit as one of these parameters tends to zero is carried out. With the help of the obtained result, in Section 6.5 we prove existence of a weak solution of the initial-boundary value problem for the Jeffreys model (here we also touch on the existence of pressure).

The uniqueness of weak solutions for the majority of equations of hydrodynamics is still an open problem. For example, for the Navier–Stokes equations in the two-dimensional case a weak solution is unique, and in the three dimensions there are only conditional results. In Section 6.6 we present two results of the latter kind for the Jeffreys model. Section 6.7 (autonomous case) and Section 6.8 (non-autonomous case) are devoted to the study of attractors of weak solutions for the system of motion equations for the Jeffreys viscoelastic medium. In Section 6.9, we investigate stationary (independent of time) solutions of this system.

The **seventh** chapter represents another approach to the study of the equations of viscoelastic fluid mechanics. It is based on regularization ideas, and the problems arising here contain less unknown functions. In Section 7.1 we describe the regularization procedure for the equations of motion of the Jeffreys viscoelastic medium. We give the weak statement of the regularized initial-boundary value problem and formulate some existence results. In Sections 7.2 – 7.4, we apply the approximating-topological procedure for the proof of these theorems. In Section 7.5, we suggest another weak formulation of the regularized problem and compare it with the first one. This helps to establish (in Section 7.6) the convergence of the solutions of the regularized problems to the solutions of the original problem in some generalized sense. Section 7.7 is a sort of appendix to Chapter 7. Here we describe necessary constructions of regularization operators.

Voronezh, November 2007

Victor G. Zvyagin
Dmitry A. Vorotnikov

Contents

Chapter 1

Non-Newtonian flows

1.1 Principles of flow description

1.1.1 The basic characteristics of a flow

For mathematical description of behaviour of the real fluids and of the media close to fluids, in hydrodynamics it is usually supposed that the particles of a medium are infinitesimal, and they are situated at the points of an open set (domain) of the n-dimensional space. The domain corresponds to a vessel filled with this medium. In the course of time, the particles move and describe some trajectories in the space. The "vessel" can also move with the course of time; it can have elastic boundary, which changes its shape under the influence of the flow; it can have holes, through which the medium flows in or out, and so on.

Remark 1.1.1. The most typical situation is $n = 3$; the case $n = 2$ corresponds to the so called plane-parallel flows, but from the cognitive point of view one can also investigate other cases.

The problem of description of motion for a medium may be reduced to the description of motion for each point particle of the medium. Consider such a particle. Its position is described by a function $x(t)$ of time t with values in the points of the space.

Assume that in the space an origin and an orthonormal basis are fixed. In this case the coordinates of a point (or of a vector) are denoted as $x = (x_1, x_2, \dots, x_n)$. Thus the space is identified with the arithmetical n-dimensional space $\mathbb{R}^n$.

The velocity of the particle with a trajectory $x(t)$ at the moment t is

$$v(t) = x'(t). \qquad (1.1.1)$$

Remark 1.1.2. Here and below in this chapter we consider all functions to be sufficiently smooth, so that all required derivatives exist.

Definition 1.1.1. A linear operator transforming (n-dimensional) vectors to vectors will be briefly called *tensor*.

Remark 1.1.3. Provided a basis is chosen, a tensor may be identified with a matrix $n \times n$. The elements of this matrix will be called components of the tensor.

Remark 1.1.4. The tensor is called orthogonal, symmetric etc., if the corresponding operator (or matrix) is orthogonal, symmetric and so on.

Denote by $v(t,x)$ the velocity of the particle which at the moment of time t is situated at the spatial point x. Consider the gradient of this function, i.e. the tensor with the components

$$(\nabla v)_{ij}(t,x) = \frac{\partial v_i(t,x)}{\partial x_j}. \tag{1.1.2}$$

Its symmetrical part

$$\mathcal{E} = \frac{1}{2}(\nabla v + \nabla v^{\top}) \tag{1.1.3}$$

is called the *strain velocity tensor*, and the skew-symmetrical one

$$W = \frac{1}{2}(\nabla v - \nabla v^{\top}) \tag{1.1.4}$$

is called the *vorticity tensor*.

Consider a particle inside the medium. Assume that it is surrounded by an imaginary surface of an arbitrary shape. The remaining part of the medium effects the particle through this surface. Consider a small flat part of this surface of area ΔS with the exterior normal vector $\vec{n}$. Denote by $P_{\vec{n}}$ the force of action of the outer part of the medium onto the particle through the considered area ΔS.

The vector

$$p_{\vec{n}} = \lim_{\Delta S \to 0} \frac{P_{\vec{n}}}{\Delta S}$$

is called *stress*.

The fundamental Cauchy theorem [63] says that there exists a symmetric tensor $\mathrm{T}(t,x)$ such that the stress at the moment t at the point x in the direction $\vec{n}$ is expressed by the formula:

$$p_{\vec{n}}(t,x) = \mathrm{T}(t,x)\vec{n}.$$

This tensor is called the *stress tensor*.

The tensor

$$\sigma = \mathrm{T} - \frac{1}{n}\operatorname{Tr}\mathrm{T}\, I, \tag{1.1.5}$$

where I is the unit tensor, is called the *deviatoric stress tensor*. Roughly speaking, it characterizes the forces of interior friction in a medium.

1.1.2 Newtonian fluid

The science that treats deformation and flow of materials is called *rheology*. The *rheological* behavior of a particular material depends on the relations between stresses, strains, stretchings in it. The most important of such relations is the *constitutive equation* (it is also called *constitutive law, constitutive relation*), which gives the connection between the deviatoric stress tensor σ and various characteristics of deformation.

The most simple relation of this kind, describing a fluid, looks like

$$\sigma = 2\eta(\mathcal{E} - \frac{1}{n}\operatorname{Tr}\mathcal{E}\, I), \tag{1.1.6}$$

where η is a scalar parameter called *viscosity*.

When $\eta = 0$, the fluid is called *ideal*. For $\eta > 0$ we have the classical *Newtonian fluid*. It is the basic object of the classical hydrodynamics. In this book, however, our prime interest is in the models different from the Newtonian one.

1.1.3 Equation of motion

Assume that the velocity field $v(t, x)$ is given at every geometrical point x of a spatial domain (where the medium is moving) and at every moment t in some time interval. Then, to describe the motion of the particle which at the initial moment t_0 is at a point x_0, it suffices to solve the Cauchy problem

$$x'(t) = v(t, x(t)), \tag{1.1.7}$$

$$x(t_0) = x_0. \tag{1.1.8}$$

If the velocity field v is regular enough, this problem has a unique solution.

Hence, the crucial problem for description of the medium motion is to find the velocity field $v(t, x)$. The basic tool to realize this is to use the following relation between the velocity and the stress tensor, which is called the *equation of motion* [10, 23]:

$$\rho\frac{\partial v}{\partial t} + \rho\sum_{i=1}^{n} v_i \frac{\partial v}{\partial x_i} - \operatorname{Div}\mathrm{T} = \rho f. \tag{1.1.9}$$

Here $f(t, x)$ is the *body force* (i.e. the mass density of exterior forces effecting the particles of the material, for example, gravitational forces), $\rho(t, x)$ is the density of the medium, while the divergence of the tensor $\operatorname{Div}\mathrm{T}(t, x)$ is the vector

$$\left(\sum_{j=1}^{n}\frac{\partial \mathrm{T}_{1j}(t,x)}{\partial x_j}, \sum_{j=1}^{n}\frac{\partial \mathrm{T}_{2j}(t,x)}{\partial x_j}, \dots, \sum_{j=1}^{n}\frac{\partial \mathrm{T}_{nj}(t,x)}{\partial x_j}\right)$$

$$= \sum_{j=1}^{n}\left(\frac{\partial \mathrm{T}_{1j}(t,x)}{\partial x_j}, \frac{\partial \mathrm{T}_{2j}(t,x)}{\partial x_j}, \dots, \frac{\partial \mathrm{T}_{nj}(t,x)}{\partial x_j}\right).$$

For a *homogeneous incompressible* medium, one has more information. Firstly, the divergence of velocity is zero:

$$\operatorname{div} v(t,x) = \sum_{j=1}^{n} \frac{\partial v_j(t,x)}{\partial x_j} = 0. \tag{1.1.10}$$

Secondly, because of the constraints generated by the incompressibility, the trace of the stress tensor has to be considered as completely independent of the deformation characteristics, so one has to introduce the unknown scalar function $p(t,x) = -\frac{1}{n}\operatorname{Tr}\mathrm{T}$, called the *hydrostatic pressure* (see [63], Chapter IV, §7). Formula (1.1.5) implies

$$\mathrm{T} = -pI + \sigma. \tag{1.1.11}$$

Thirdly, the density is constant, and it is possible to consider it equal to one.

Note that $\operatorname{grad} p = (\frac{\partial p}{\partial x_1}, \dots, \frac{\partial p}{\partial x_n}) = \operatorname{Div}(pI)$. Therefore, (1.1.9) and (1.1.5) yield

$$\frac{\partial v}{\partial t} + \sum_{i=1}^{n} v_i \frac{\partial v}{\partial x_i} - \operatorname{Div}\sigma + \operatorname{grad} p = f. \tag{1.1.12}$$

This is the equation of motion for homogeneous incompressible media.

We have now

$$\begin{aligned} 2\operatorname{Div}\mathcal{E} &= 2\sum_{j=1}^{n}\left(\frac{\partial \mathcal{E}_{1j}}{\partial x_j}, \frac{\partial \mathcal{E}_{2j}}{\partial x_j}, \dots, \frac{\partial \mathcal{E}_{nj}}{\partial x_j}\right) \\ &= \sum_{j=1}^{n}\left(\frac{\partial^2 v_1}{\partial x_j^2}, \frac{\partial^2 v_2}{\partial x_j^2}, \dots, \frac{\partial^2 v_n}{\partial x_j^2}\right) + \sum_{j=1}^{n}\left(\frac{\partial^2 v_j}{\partial x_j \partial x_1}, \frac{\partial^2 v_j}{\partial x_j \partial x_2}, \dots, \frac{\partial^2 v_j}{\partial x_j \partial x_n}\right). \end{aligned}$$

Condition (1.1.10) implies that the second sum is equal to zero. Therefore

$$2\operatorname{Div}\mathcal{E} = \Delta v, \tag{1.1.13}$$

where Δ is the Laplacian $\sum_{j=1}^{n} \frac{\partial^2}{\partial x_j^2}$.

Then from Newton's constitutive relation (1.1.6) and equation of motion (1.1.12) we get the equation of motion for the Newtonian fluid:

$$\frac{\partial v}{\partial t} + \sum_{i=1}^{n} v_i \frac{\partial v}{\partial x_i} - \eta\Delta v + \operatorname{grad} p = f, \tag{1.1.14}$$

which is usually called the Navier–Stokes equation.

1.1.4 No-slip condition

Assume that the medium is contained in a motionless vessel which may be identified with an open set (domain) Ω in the n-dimensional space $\mathbb{R}^n$ (see Remark 1.1.1). The *no-slip condition* means that the velocity of the medium vanishes on the boundary of Ω:

$$v(t,x) = 0, \ x \in \partial\Omega. \tag{1.1.15}$$

A justification for the no-slip condition may be found if we remember that a real medium consists of molecules. When a medium is stationary, its molecules still move with a random velocity, but the mean velocity of this motion is zero. When the medium begins to move, there appears a non-zero mean velocity, sometimes called the bulk velocity, which is superimposed on the random velocity of the stationary state. Just this bulk velocity corresponds to the velocity (1.1.1) at the idealized hydrodynamical approach described in Section 1.1.1. Near the interface between the medium and the solid boundary of the vessel there exists an attraction between the molecules of the medium and the molecules of the solid boundary. The attractive force is usually so strong that the bulk velocity of the medium vanishes at this interface.

The no-slip condition can be illustrated by the following folklore example. If you like Italian or French cuisine, your dishes may be sometimes stained with cheese. When you put them into your dishwasher, you realize that they are not washed up well. This is not surprising, since, due to the no-slip condition, the velocity vanishes at the interface between the water and the cheese, so the stream itself is not able to carry away the cheese. A possible way out is to scrub the cheese by yourself.

1.2 One-dimensional models of viscoelastic media

1.2.1 Method of mechanical models

In rheology, for construction of constitutive relations describing materials with more complicated properties than the Newtonian fluid, the method of mechanical models is often used. Let us describe the essence of this heuristic method.

The basic properties for rheology are *elasticity, viscosity* and *plasticity*. For representation of the elasticity a spiral *spring* is used. It satisfies Hooke's law: the length variation of the spring is in direct proportion to the force applied to its ends. This model is denoted by the letter H. For representation of viscosity one uses a viscous damper (*dashpot*). The velocity of the dashpot (i.e. the rate of change of its length) is in direct proportion to the applied force. This model is denoted by N. The model for plasticity is not used in this book.

Assume that investigated media consist of microcomplexes of small springs and dashpots. Within these complexes they are connected in parallel (denoted by $|$) or in series (denoted by $-$). For the parallel connection the loadings perceived by each element are summarized and the velocities of each element are the same. For the connection in series the velocities are summarized and the loadings are the same.

However for rheological relations one does not need the forces and the velocities but the stresses and strain velocities. The stress (in a spring or dashpot) may be understood as the ratio of the resistance force (which is equal in absolute value to the applied force) to the cross-section area (of the spring or dashpot). The strain velocity ε may be understood as a half of the ratio of the velocity of the spring or dashpot (i.e. the rate of change of the length) to the average longitudinal length of the spring or dashpot.

Now one has to write relations between the stress and the strain velocity for the spring and the dashpot. For the dashpot it is very simple. If the force is in direct proportion to the velocity, then the stress is in direct proportion to the strain velocity:

$$\sigma_N = 2\eta\varepsilon_N. \tag{1.2.1}$$

Remark 1.2.1. Note that (1.2.1) resembles (1.1.6), so η has the physical sense of viscosity.

Now, differentiating (with respect to time) Hooke's law for the spring, we conclude that the derivative of the force is in direct proportion to the velocity of variation of length of the spring. Therefore the derivative of the stress is in direct proportion to the strain velocity:

$$\dot{\sigma}_H = 2\mu\varepsilon_H. \tag{1.2.2}$$

From here (but only in Section 1.2) the time derivative is denoted by a point. The constants of proportionality μ and η are usually positive.

1.2.2 The Maxwell body

The simplest structure of a model for a viscoelastic medium is the Maxwell body with the symbolical notation $M = H - N$, i.e. a spring and a dashpot connected in series. Let us deduce the constitutive equation for this body. For this purpose, recall that for the connection in series the stress is constant

$$\sigma_M = \sigma_N = \sigma_H, \tag{1.2.3}$$

and the strain velocities are summarized:

$$\varepsilon_M = \varepsilon_H + \varepsilon_N. \tag{1.2.4}$$

From equalities (1.2.1) – (1.2.4) it follows that

$$\varepsilon_M = \frac{\dot{\sigma}_M}{2\mu} + \frac{\sigma_M}{2\eta}. \tag{1.2.5}$$

This is the constitutive relation for the Maxwell body. It looks like a linear differential equation for σ_M. Multiplying both parts of this equation by $e^{\frac{\mu}{\eta}t}$, we get:

$$e^{\frac{\mu}{\eta}t}\mathcal{E}_M = e^{\frac{\mu}{\eta}t}\frac{\dot{\sigma_M}}{2\mu} + e^{\frac{\mu}{\eta}t}\frac{\sigma_M}{2\eta},$$

$$e^{\frac{\mu}{\eta}t}\mathcal{E}_M = \frac{(e^{\frac{\mu}{\eta}t}\sigma_M)^{\cdot}}{2\mu}.$$

It implies

$$e^{\frac{\mu}{\eta}t}\sigma_M = \sigma_{M_0} + 2\mu\int_0^t e^{\frac{\mu}{\eta}s}\mathcal{E}_M(s)\,ds,$$

σ_{M_0} is the stress at the initial moment (here and below it is the moment $t = 0$). We conclude that the solution of equation (1.2.5) has the following form:

$$\sigma_M = e^{-\frac{\mu}{\eta}t}\Big(\sigma_{M_0} + 2\mu\int_0^t e^{\frac{\mu}{\eta}s}\mathcal{E}_M(s)\,ds\Big). \tag{1.2.6}$$

1.2.3 The Jeffreys body

The most typical model of a viscoelastic medium is the Jeffreys model J. It consists of a Maxwell body and one more dashpot, connected in parallel, i.e. $J = M|N$. The viscosities of mediums in two different dashpots may differ; denote them by η_1 and η_2. Thus, for the Maxwell constituent of the Jeffreys body we have:

$$\sigma_M = e^{-\frac{\mu}{\eta_1}t}\Big(\sigma_{M_0} + 2\mu\int_0^t e^{\frac{\mu}{\eta_1}s}\mathcal{E}_M(s)\,ds\Big), \tag{1.2.7}$$

and for the Newtonian constituent

$$\sigma_N = 2\eta_2\mathcal{E}_N. \tag{1.2.8}$$

But for the parallel connection one has:

$$\sigma_J = \sigma_M + \sigma_N, \tag{1.2.9}$$

$$\mathcal{E}_J = \mathcal{E}_M = \mathcal{E}_N. \tag{1.2.10}$$

Equalities (1.2.7) – (1.2.10) yield

$$\sigma_J = e^{-\frac{\mu}{\eta_1}t}\Big(\sigma_{M_0} + 2\mu\int_0^t e^{\frac{\mu}{\eta_1}s}\mathcal{E}_J(s)\,ds\Big) + 2\eta_2\mathcal{E}_J. \tag{1.2.11}$$

Let us denote the stress and the strain velocity of the Jeffreys body at the initial moment of time by σ_{J_0} and $\mathcal{E}_{J_0}$. Taking $t = 0$ in (1.2.11), we get the equality:

$$\sigma_{J_0} = \sigma_{M_0} + 2\eta_2\mathcal{E}_{J_0}. \tag{1.2.12}$$

Then (1.2.11) and (1.2.12) imply

$$\sigma_J = e^{-\frac{\mu}{\eta_1}t}\left(\sigma_{J_0} - 2\eta_2 \varepsilon_{J_0} + 2\mu \int_0^t \varepsilon_J(s) e^{\frac{\mu}{\eta_1}s}\, ds\right) + 2\eta_2 \varepsilon_J. \qquad (1.2.13)$$

This is the rheological relation for the Jeffreys body with explicitly expressed stress. Another form of the constitutive law for the Jeffreys body, which we deduce just below, is more traditional.

Differentiate expression (1.2.11) with respect to t:

$$\dot{\sigma}_J = e^{-\frac{\mu}{\eta_1}t} 2\mu e^{\frac{\mu}{\eta_1}t} \varepsilon_J(t) - \frac{\mu}{\eta_1} e^{-\frac{\mu}{\eta_1}t}\left(\sigma_{M_0} + 2\mu \int_0^t e^{\frac{\mu}{\eta_1}s} \varepsilon_J(s)\, ds\right) + 2\eta_2 \dot{\varepsilon}_J. \qquad (1.2.14)$$

On the other hand, (1.2.11) implies:

$$e^{-\frac{\mu}{\eta_1}t}\left(\sigma_{M_0} + 2\mu \int_0^t e^{\frac{\mu}{\eta_1}s} \varepsilon_J(s)\, ds\right) = \sigma_J - 2\eta_2 \varepsilon_J. \qquad (1.2.15)$$

Equalities (1.2.14) and (1.2.15) yield

$$\dot{\sigma}_J = 2\mu \varepsilon_J - \frac{\mu}{\eta_1}(\sigma_J - 2\eta_2 \varepsilon_J) + 2\eta_2 \dot{\varepsilon}_J.$$

Multiply this by $\frac{\eta_1}{\mu}$ and move the term σ_J to the left-hand side:

$$\sigma_J + \frac{\eta_1}{\mu}\dot{\sigma}_J = 2(\eta_1 + \eta_2)\varepsilon_J + 2\frac{\eta_1 \eta_2}{\mu}\dot{\varepsilon}_J. \qquad (1.2.16)$$

Denote $\lambda_1 = \frac{\eta_1}{\mu}$, $\lambda_2 = \frac{\eta_1 \eta_2}{\mu(\eta_1 + \eta_2)}$, $\eta_J = \eta_1 + \eta_2$. Then (1.2.16) may be rewritten as

$$\sigma_J + \lambda_1 \dot{\sigma}_J = 2\eta_J(\varepsilon_J + \lambda_2 \dot{\varepsilon}_J). \qquad (1.2.17)$$

That is how another form of the constitutive equation for the Jeffreys body looks like. The parameter η_J is called the viscosity of the Jeffreys body. The parameter λ_1 is called the *relaxation time*, and the parameter λ_2 is the *retardation time*. Since the viscosities are positive and $\lambda_2 = \lambda_1 \frac{\eta_2}{\eta_1 + \eta_2}$, one has

$$\lambda_2 < \lambda_1.$$

From the physical point of view, the parameters $\lambda_1, \lambda_2, \eta_J$ are preferable with respect to the parameters μ_1, η_1, η_2 because the first ones are not only parameters of a mechanical model, but can be measured for real media possessing viscoelastic properties of the Jeffreys model.

Assume that the strain velocity is constant and equal to ε_{J_0}. Then $\dot{\varepsilon}_J = 0$, and equation (1.2.17) may be transformed as follows:

$$\sigma_J - 2\eta_J \varepsilon_{J_0} + \lambda_1(\sigma_J - 2\eta_J \varepsilon_{J_0})' = 0.$$

Solving this differential equation, we obtain

$$\sigma_J - 2\eta_J \varepsilon_{J_0} = (\sigma_{J_0} - 2\eta_J \varepsilon_{J_0}) e^{-\frac{t}{\lambda_1}}.$$

We have the following expression for the stress:

$$\sigma_J = 2\eta_J \varepsilon_{J_0} + (\sigma_{J_0} - 2\eta_J \varepsilon_{J_0}) e^{-\frac{t}{\lambda_1}}. \tag{1.2.18}$$

If $\sigma_{J_0} = 2\eta_J \varepsilon_{J_0}$, then $\sigma_J = 2\eta_J \varepsilon_{J_0}$, and the stress does not depend on time. If deformation in a body remains constant, the strain velocity ε_{J_0} is equal to zero, and the stress decreases according to the exponential law:

$$\sigma_J = \sigma_{J_0} e^{-\frac{t}{\lambda_1}}.$$

Thus, the relaxation time λ_1 is the time, during which, for constant deformation, the stress is reduced to $1/e$-th part.

If there is no stress ($\sigma_J = 0$), then (1.2.17) implies

$$\varepsilon_J + \lambda_2 \dot{\varepsilon}_J = 0,$$

and

$$\varepsilon_J = e^{-\frac{t}{\lambda_2}} \varepsilon_{J_0}, \tag{1.2.19}$$

i.e. the strain velocity decreases according to the exponential law, and during the retardation time λ_2 the strain velocity is reduced to $1/e$-th part.

Let us express the "mechanical" parameters as functions of the viscosity of the Jeffreys body and the times of relaxation and retardation:

$$\eta_2 = \eta_J \frac{\lambda_2}{\lambda_1}, \quad \eta_1 = \eta_J \Big(1 - \frac{\lambda_2}{\lambda_1}\Big), \quad \mu = \eta_J \frac{\lambda_1 - \lambda_2}{\lambda_1^2}.$$

Thus, (1.2.13) turns into

$$\sigma_J = e^{-\frac{t}{\lambda_1}} \Big(\sigma_{J_0} - 2\eta_J \frac{\lambda_2}{\lambda_1} \varepsilon_{J_0} + 2\eta_J \frac{\lambda_1 - \lambda_2}{\lambda_1^2} \int_0^t e^{\frac{s}{\lambda_1}} \varepsilon_J(s)\, ds\Big) + 2\eta_J \frac{\lambda_2}{\lambda_1} \varepsilon_J. \tag{1.2.20}$$

This is one more form of the constitutive relation for the Jeffreys body.

Remark 1.2.2. The equations (1.2.17) and (1.2.20) are also rheological relations for another viscoelastic body. It is the Lethersich body [49] with symbolical notation $(N|H) - N$, where the viscosities of two dashpots N may differ.

Remark 1.2.3. For $\lambda_2 = \eta_2 = 0$ the Jeffreys body turns to be equivalent to the Maxwell one: it suffices to compare (1.2.11) and (1.2.6).

1.3 Multidimensional models of viscoelastic media

1.3.1 Passage to multidimensional models

The employment of the models obtained in the previous section for description of mediums in the "real" n-dimensional space (see Remark 1.1.1) requires generalization of the equations (first of all, of the constitutive laws) to the n-dimensional case. It is the last step of the method of mechanical models.

The problem is how to understand the components of the obtained constitutive relations, for example, of (1.2.17), in the n-dimensional case. Here the answer is that the times of relaxation and retardation of the Jeffreys body remain scalars, as well as the viscosity, so it is possible to measure them for particular materials. The stress and the strain velocity become the deviatoric stress tensor and the strain velocity tensor, respectively (see however Remarks 1.3.1 and 1.3.5). The only problem is how to understand the time derivative denoted by a point. There is no direct answer to this question, there are several variants (almost all of them, however, may be criticized), and the choice of a variant determines the obtained model. The following subsections are devoted to the analysis of the principal variants.

Remark 1.3.1. We will describe all the variants by the example of the Jeffreys model. Henceforth, the indices $_J$ at the symbols of viscosity, stress tensor and strain velocity tensor will be omitted for brevity. In Section 1.3 it is implicitely assumed that we deal with *incompressible medium*, i.e.

$$\operatorname{div} v = \operatorname{Tr} \mathcal{E} = 0.$$

However, the results remain valid for *compressible medium* with or without minor changes (such as to replace $\mathcal{E}$ with $\widetilde{\mathcal{E}} = \mathcal{E} - \frac{1}{n} \operatorname{Tr} \mathcal{E}\, I$ or to stop treating σ as a deviatoric stress tensor, cf. Remark 1.3.5 below).

1.3.2 Partial derivative

The simplest way out, which causes minor mathematical difficulties, is the choice of the partial derivative with respect to time t as a substitute for the derivative denoted by a point.

The constitutive relation (1.2.17) becomes

$$\sigma + \lambda_1 \frac{\partial \sigma}{\partial t} = 2\eta\big(\mathcal{E} + \lambda_2 \frac{\partial \mathcal{E}}{\partial t}\big), \tag{1.3.1}$$

and the equivalent constitutive relation (1.2.20) becomes

$$\sigma = e^{-\frac{t}{\lambda_1}} \left(\sigma_0 - 2\eta \frac{\lambda_2}{\lambda_1} \mathcal{E}_0 + 2\eta \frac{\lambda_1 - \lambda_2}{\lambda_1^2} \int_0^t e^{\frac{s}{\lambda_1}} \mathcal{E}(s)\, ds \right) + 2\eta \frac{\lambda_2}{\lambda_1} \mathcal{E}. \tag{1.3.2}$$

In the case of a homogeneous incompressible medium this and (1.1.13) imply:

$$\operatorname{Div} \sigma = e^{-\frac{t}{\lambda_1}} \left(\operatorname{Div} \sigma_0 - \eta \frac{\lambda_2}{\lambda_1} \Delta v|_{t=0} + \eta \frac{\lambda_1 - \lambda_2}{\lambda_1^2} \int_0^t e^{\frac{s}{\lambda_1}} \Delta v(s)\, ds \right) + \eta \frac{\lambda_2}{\lambda_1} \Delta v.$$

Substituting it into the equation of motion (1.1.12), we get the following equation of motion for a homogeneous incompressible viscoelastic medium

$$\begin{aligned} \frac{\partial v}{\partial t} + \sum_{i=1}^n v_i \frac{\partial v}{\partial x_i} - \eta \frac{\lambda_2}{\lambda_1} \Delta v - \eta \frac{\lambda_1 - \lambda_2}{\lambda_1^2} \int_0^t e^{\frac{s-t}{\lambda_1}} \Delta v(s)\, ds + \operatorname{grad} p \\ = f + e^{-\frac{t}{\lambda_1}} \Big(\operatorname{Div} \sigma_0 - \eta \frac{\lambda_2}{\lambda_1} \Delta v|_{t=0} \Big). \end{aligned} \tag{1.3.3}$$

1.3.3 Substantial derivative

Equation (1.3.1) describes the relation between stresses, strains and their partial derivatives with respect to time, i.e. with the rates of their change at each fixed geometrical point of space. It is more natural from different points of view to use the approach when the constitutive relation connects stresses and strains with the rates of their change for each particle. For this purpose, introduce an auxiliary function z. Let t and τ be two arbitrary moments of time and let x be a spatial point. Consider the particle which at the moment τ is at the spatial point x. Then $z(t, \tau, x)$ expresses the spatial position of this particle at the moment t.

From this definition we can see that

$$\frac{\partial z(t, \tau, x)}{\partial t} = v(t, z(t, \tau, x)), \tag{1.3.4}$$

$$z(t, t, x) = x, \tag{1.3.5}$$

$$z(s, t, z(t, \tau, x)) = z(s, \tau, x). \tag{1.3.6}$$

Let $A(t, x)$ be a sufficiently smooth function of time and space. The rate of change of A for a fixed particle is equal to

$$\begin{aligned} \lim_{\Delta t \to 0} \frac{A(t + \Delta t, z(t + \Delta t, t, x)) - A(t, x)}{\Delta t} &= \frac{\partial}{\partial \tau} A(\tau, z(\tau, t, x)) \Big|_{\tau = t} \\ &= \frac{\partial A(\tau, x)}{\partial \tau} \Big|_{\tau = t} + \sum_{i=1}^n \frac{\partial A(\tau, x)}{\partial x_i} \frac{\partial z_i(\tau, t, x)}{\partial \tau} \Big|_{\tau = t} \\ &= \frac{\partial A(t, x)}{\partial t} + \sum_{i=1}^n v_i(t, x) \frac{\partial A(t, x)}{\partial x_i}. \end{aligned} \tag{1.3.7}$$

This function is called the substantial derivative of A and is denoted by $\frac{dA}{dt}$. Observe that

$$\frac{dx}{dt} = \frac{\partial x}{\partial t} + \sum_{i=1}^{n} v_i \frac{\partial x}{\partial x_i} = \sum_{i=1}^{n} v_i e_i = v,$$

where e_i, $i = 1, 2, \dots, n$, are the basis vectors of the space. Thus, we have the following property of the substantial derivative:

$$v(t, x) = \frac{dx}{dt}.$$

To obtain an analogue of (1.2.20), we have to use the inverse operation for the new derivation $\frac{d}{dt}$. This inverse operation is the integration along the trajectories of particles. Let us study this problem more formally.

We have to find the solution to the following Cauchy problem

$$\begin{aligned} \frac{d}{dt} U(t, x) &= A(t, x), \\ U(0, x) &= U_0(x), \end{aligned} \tag{1.3.8}$$

where A and U_0 are arbitrary sufficiently smooth known functions.

Let us show that

$$\frac{d}{dt} \int_0^t A(s, z(s, t, x))\, ds = A(t, x). \tag{1.3.9}$$

Really, the continued equality (1.3.7) gives

$$\frac{dA}{dt}(t, x) = \frac{\partial}{\partial \tau} A(\tau, z(\tau, t, x))\Big|_{\tau=t}. \tag{1.3.10}$$

Taking into account (1.3.6) and (1.3.10), we obtain:

$$\begin{aligned} \frac{d}{dt} \int_0^t A(s, z(s, t, x))\, ds &= \frac{\partial}{\partial \tau} \int_0^\tau A(s, z(s, \tau, z(\tau, t, x)))\, ds\Big|_{\tau=t} \\ &= \frac{\partial}{\partial \tau} \int_0^\tau A(s, z(s, t, x))\, ds\Big|_{\tau=t} = A(\tau, z(\tau, t, x))\Big|_{\tau=t} = A(t, x). \end{aligned}$$

Observe that the function $U_0(z(0, t, x))$ possesses the value $U_0(x)$ at $t = 0$. Besides, using (1.3.6) and (1.3.10), we obtain

$$\frac{d}{dt} U_0(z(0, t, x)) = \frac{\partial}{\partial \tau} U_0(z(0, \tau, z(\tau, t, x)))\Big|_{\tau=t} = \frac{\partial}{\partial \tau} U_0(z(0, t, x))\Big|_{\tau=t} = 0.$$

Thus, the solution of problem (1.3.8) has the following representation:

$$U_0(z(0, t, x)) + \int_0^t A(s, z(s, t, x))\, ds. \tag{1.3.11}$$

But this problem cannot have more than one solution. Really, let U_1 and U_2 be two solutions of problem (1.3.8), and $w = U_1 - U_2$. Then we have:

$$\frac{d}{dt} w(t,x) = 0, \quad w(0,x) = 0.$$

Equality (1.3.10) yields

$$\frac{\partial}{\partial \tau} w(\tau, z(\tau,t,x))\Big|_{\tau=t} = 0.$$

Substituting here $z(t,0,y)$ for x, where y is an arbitrary point, we get:

$$0 = \frac{\partial}{\partial \tau} w(\tau, z(\tau,t,z(t,0,y)))\Big|_{\tau=t} = \frac{\partial}{\partial \tau} w(\tau, z(\tau,0,y))\Big|_{\tau=t} = \frac{\partial}{\partial t} w(t, z(t,0,y)).$$

Let $w_1(t,y) = w(t,z(t,0,y))$. Then

$$\frac{\partial}{\partial t} w_1(t,y) = 0, \quad w_1(0,y) = 0.$$

Hence, $w_1 \equiv 0$. Then

$$0 = w_1(t, z(0,t,x)) = w(t, z(t,0,z(0,t,x))) = w(t,x)$$

for any x and t. Thus, $w \equiv 0$, and uniqueness of the solution for problem (1.3.8) is proven.

Now we can return to deriving of constitutive relations. Using the substantial derivative, rewrite the constitutive equation (1.2.17) as

$$\sigma + \lambda_1 \frac{d}{dt}\sigma = 2\eta(\mathcal{E} + \lambda_2 \frac{d}{dt}\mathcal{E}). \tag{1.3.12}$$

We are going to derive an explicit expression for σ, i.e. an analogue of (1.2.20). Let us carry out a number of transformations of equation (1.3.12):

$$\sigma - 2\eta\frac{\lambda_2}{\lambda_1}\mathcal{E} + \lambda_1 \frac{d}{dt}(\sigma - 2\eta\frac{\lambda_2}{\lambda_1}\mathcal{E}) = 2\eta(\frac{\lambda_1 - \lambda_2}{\lambda_1}\mathcal{E}),$$

$$\frac{e^{\frac{t}{\lambda_1}}}{\lambda_1}(\sigma - 2\eta\frac{\lambda_2}{\lambda_1}\mathcal{E}) + e^{\frac{t}{\lambda_1}} \frac{d}{dt}(\sigma - 2\eta\frac{\lambda_2}{\lambda_1}\mathcal{E}) = 2\eta e^{\frac{t}{\lambda_1}}(\frac{\lambda_1 - \lambda_2}{\lambda_1^2}\mathcal{E}),$$

$$\frac{d}{dt}(e^{\frac{t}{\lambda_1}}(\sigma - 2\eta\frac{\lambda_2}{\lambda_1}\mathcal{E})) = 2\eta e^{\frac{t}{\lambda_1}}(\frac{\lambda_1 - \lambda_2}{\lambda_1^2}\mathcal{E}).$$

In accordance with (1.3.11), the solution for this equation has the following form:

$$e^{\frac{t}{\lambda_1}}(\sigma - 2\eta\frac{\lambda_2}{\lambda_1}\mathcal{E}) = \sigma_0(z(0,t,x)) - 2\eta\frac{\lambda_2}{\lambda_1}\mathcal{E}_0(z(0,t,x)) + 2\eta\frac{\lambda_1 - \lambda_2}{\lambda_1^2}\int_0^t e^{\frac{s}{\lambda_1}}\mathcal{E}(s, z(s,t,x))\,ds.$$

It implies an expression for σ:

$$\sigma(t,x) = e^{-\frac{t}{\lambda_1}}\Big[\sigma_0(z(0,t,x)) - 2\eta\frac{\lambda_2}{\lambda_1}\mathcal{E}_0(z(0,t,x)) + 2\eta\frac{\lambda_1-\lambda_2}{\lambda_1^2}\int_0^t e^{\frac{s}{\lambda_1}}\mathcal{E}(s,z(s,t,x))\,ds\Big] + 2\eta\frac{\lambda_2}{\lambda_1}\mathcal{E}(t,x). \tag{1.3.13}$$

In the case of a homogeneous incompressible medium, (1.3.13) and (1.1.13) yield

$$\operatorname{Div}\sigma(t,x) = e^{-\frac{t}{\lambda_1}}\operatorname{Div}\Big[\sigma_0(z(0,t,x)) - 2\eta\frac{\lambda_2}{\lambda_1}\mathcal{E}_0(z(0,t,x))\Big] + 2\eta\frac{\lambda_1-\lambda_2}{\lambda_1^2}\int_0^t e^{\frac{s-t}{\lambda_1}}\operatorname{Div}[\mathcal{E}(s,z(s,t,x))]\,ds + \eta\frac{\lambda_2}{\lambda_1}\Delta v.$$

To get the equation of motion of the homogeneous incompressible Jeffreys' viscoelastic medium, substitute it into (1.1.12):

$$\frac{\partial v}{\partial t} + \sum_{i=1}^n v_i\frac{\partial v}{\partial x_i} - \eta\frac{\lambda_2}{\lambda_1}\Delta v - 2\eta\frac{\lambda_1-\lambda_2}{\lambda_1^2}\int_0^t e^{\frac{s-t}{\lambda_1}}\operatorname{Div}[\mathcal{E}(s,z(s,t,x))]\,ds + \operatorname{grad} p = f + e^{-\frac{t}{\lambda_1}}\operatorname{Div}\Big[\sigma_0(z(0,t,x)) - 2\eta\frac{\lambda_2}{\lambda_1}\mathcal{E}_0(z(0,t,x))\Big]. \tag{1.3.14}$$

1.3.4 Principle of material frame-indifference. Frame-indifferent functions

Before passing to the description of other objects able to replace in the n-dimensional situation the time derivative denoted by a point, it is necessary to say a few words about the principle of material frame-indifference. It is one of the main principles of rational mechanics, which expresses the fact that the properties of a material do not depend on the choice of observer.

The observer in rational mechanics is identified with a frame, i.e. a certain correspondence between the spatial points and the elements x of the space $\mathbb{R}^n$, as well as between the moments of time and the elements t of the scalar axis $\mathbb{R}$. It is considered that when the observer is changed, the metrics in $\mathbb{R}^n$ and in the scalar axis are conserved, the time direction is also conserved. Then the most general change of coordinates for each point looks like [63]:

$$t^* = t + a, \tag{1.3.15}$$

$$x^* = x_0^*(t) + Q(t)(x - x_0), \tag{1.3.16}$$

where a is a time value, x_0 is a spatial point, $x_0^*(t)$ is a function of time with the values at the spatial points, Q is a time dependent orthogonal tensor.

A change of observer induces some transformation of vectors and tensors. However, the *principle of material frame-indifference* [63] states that such a transformation (i.e. a change of coordinates (1.3.15) – (1.3.16)) should not affect the structure of the formulas expressing physical properties of a medium and containing the time variable t, the spatial variable x and their functions.

Remark 1.3.2. In spite of these statements, it should be pointed out that the form of the equations of motion (1.1.9) and (1.1.12) depends on the choice of observer [63]. The point here is that, generally speaking, the equation of motion has form (1.1.9) only in *inertial* [63] reference frame. If a change of variables of the form (1.3.15) – (1.3.16) transforms inertial frames to inertial ones (it means, in particular, that Q is independent of time t), the equation of motion does not change its form.

Take an arbitrary vector, which is *geometrical*, i.e. it is a directed segment existing in the space irrespectively of the observer. Assume that in the initial frame of reference it has the form $w = \overrightarrow{x_1 x_2}$. Then in the new frame of reference $w^* = \overrightarrow{x_1^* x_2^*} = \overrightarrow{x_0^*(t) + Q(t)(x_1 - x_0),\, x_0^*(t) + Q(t)(x_2 - x_0)} = Q(t)\overrightarrow{x_1 x_2} = Q(t)w$.

Let some tensor $\mathcal{T}$ transform geometrical vectors to geometrical ones. Then in the initial frame of reference we have:

$$w_1 = \mathcal{T} w_2,$$

where w_1 and w_2 are two trial geometrical vectors. Q is an orthogonal tensor, so $Q(t)^\top Q(t) = I$ and

$$w_1^* = Q(t)w_1 = Q(t)\mathcal{T} Q(t)^\top Q(t)w_2 = \mathcal{T}^* w_2^*,$$

where $\mathcal{T}^* = Q(t)\mathcal{T} Q(t)^\top$.

Based on this, a vector-valued function $w(t, x)$ of time and space is called *frame-indifferent* if under the change of frame (1.3.15) – (1.3.16) its coordinates are transformed as follows:

$$w^*(t^*, x^*) = Q(t)w(t, x); \tag{1.3.17}$$

a tensor-valued function $\mathcal{T}(t, x)$ is called *frame-indifferent* if under the similar change one has:

$$\mathcal{T}^*(t^*, x^*) = Q(t)\mathcal{T}(t, x)Q(t)^\top. \tag{1.3.18}$$

The stress tensor T is an example of a frame-indifferent function [63].

Moreover, a scalar function $A(t, x)$ is called *frame-indifferent* if under the same change of frame:

$$A^*(t^*, x^*) = A(t, x). \tag{1.3.19}$$

The density $\rho(t, x)$ is an example of such a function.

1.3.5 The Zaremba–Żórawski theorem

For checking frame-indifference of constitutive relations we are going to use the following statement on the transformation of the strain velocity and vorticity tensors under a change of observer; it is called the Zaremba–Żórawski theorem.

Theorem 1.3.1. *Under the frame change (1.3.15) – (1.3.16), the following transformations occur:*

$$\mathcal{E}^*(t^*, x^*) = Q(t)\mathcal{E}(t, x)Q(t)^\top, \tag{1.3.20}$$

i.e. the strain velocity tensor is frame-indifferent;

$$W^*(t^*, x^*) = Q(t)W(t, x)Q(t)^\top + Q'(t)Q(t)^\top, \tag{1.3.21}$$

i.e. the vorticity tensor is not frame-indifferent.

Proof. By definition, $z(t, \tau, x)$, for fixed arguments, is a point in the space. Therefore, (1.3.16) gives:

$$z^*(t^*, \tau^*, x^*) = x_0^*(t) + Q(t)(z(t, \tau, x) - x_0). \tag{1.3.22}$$

This yields

$$\begin{aligned}
\frac{\partial z^*(t^*, \tau^*, x^*)}{\partial t^*} &= \frac{\partial[x_0^*(t) + Q(t)(z(t, \tau, x) - x_0)]}{\partial(t + a)} \\
&= x_0^{*\prime}(t) + Q'(t)(z(t, \tau, x) - x_0) + Q(t)\frac{\partial z(t, \tau, x)}{\partial t} \\
&= x_0^{*\prime}(t) + Q'(t)Q(t)^\top(z^*(t^*, \tau^*, x^*) - x_0^*(t)) + Q(t)\frac{\partial z(t, \tau, x)}{\partial t}.
\end{aligned}$$

Substituting $t = \tau, t^* = \tau^*$ into this equality and using (1.3.4) and (1.3.5), we obtain:

$$\begin{aligned}
v^*(t^*, x^*) &= \left.\frac{\partial z^*(t^*, \tau^*, x^*)}{\partial t^*}\right|_{\tau^*=t^*} \\
&= x_0^{*\prime}(t) + Q'(t)Q(t)^\top(x^* - x_0^*(t)) + Q(t)v(t, x).
\end{aligned} \tag{1.3.23}$$

Therefore

$$\begin{aligned}
(\nabla v)^* &= \frac{\partial v^*(x^*, t^*)}{\partial x^*} \\
&= \frac{\partial(x_0^{*\prime}(t) + Q'(t)Q(t)^\top(x^* - x_0^*(t)) + Q(t)v(t, x))}{\partial x^*} \\
&= Q'(t)Q(t)^\top + \frac{\partial[Q(t)v(t, x)]}{\partial x}\frac{\partial x}{\partial x^*} \\
&= Q'(t)Q(t)^\top + Q(t)(\nabla v)\frac{\partial(x_0 + Q(t)^\top(x^* - x_0^*(t)))}{\partial x^*} \\
&= Q'(t)Q(t)^\top + Q(t)(\nabla v)Q(t)^\top.
\end{aligned}$$

Thus,

$$(\nabla v)^* = Q(t)(\nabla v)Q(t)^\top + Q'(t)Q(t)^\top. \tag{1.3.24}$$

Transpose this equality:

$$[(\nabla v)^*]^\top = Q(t)(\nabla v)^\top Q(t)^\top + Q(t)Q'(t)^\top. \tag{1.3.25}$$

Differentiating the identity $Q(t)Q(t)^\top = I$, we get

$$Q'(t)Q(t)^\top + Q(t)Q'(t)^\top = 0. \tag{1.3.26}$$

Taking into account (1.3.26), calculate the half-sum of equalities (1.3.24) and (1.3.25):

$$\begin{aligned}\mathcal{E}^* &= \frac{1}{2}((\nabla v)^* + [(\nabla v)^*]^\top)\\ &= \frac{1}{2}[Q(t)(\nabla v)Q(t)^\top + Q'(t)Q(t)^\top + Q(t)(\nabla v)^\top Q(t)^\top + Q(t)Q'(t)^\top]\\ &= Q(t)\mathcal{E}Q(t)^\top.\end{aligned}$$

Now consider the half-difference of (1.3.24) and (1.3.25):

$$\begin{aligned}W^* &= \frac{1}{2}((\nabla v)^* - [(\nabla v)^*]^\top)\\ &= \frac{1}{2}[Q(t)(\nabla v)Q(t)^\top + Q'(t)Q(t)^\top - Q(t)(\nabla v)^\top Q(t)^\top - Q(t)Q'(t)^\top]\\ &= Q(t)WQ(t)^\top + Q'(t)Q(t)^\top.\end{aligned}$$

□

1.3.6 Objective derivatives

According to the principle of material frame-indifference, a constitutive equation, as an expression of physical properties of a medium, should not change its form with a change of frame. Let us check whether this statement is valid for constitutive relations (1.3.1) and (1.3.12).

As we have already mentioned, the stress tensor T is frame-indifferent. Let us show that the deviatoric stress tensor is also frame-indifferent. We have:

$$\begin{aligned}\sigma^* &= \mathrm{T}^* - \frac{1}{n}\mathrm{Tr}\,\mathrm{T}^* I = Q(t)\mathrm{T}Q(t)^\top - \frac{1}{n}\mathrm{Tr}(Q(t)\mathrm{T}Q(t)^\top)I\\ &= Q(t)\mathrm{T}Q(t)^\top - \frac{1}{n}\sum_{i,j,k=1}^{n} q_{ij}\mathrm{T}_{jk}q_{ik}I\\ &= Q(t)\mathrm{T}Q(t)^\top - \frac{1}{n}\sum_{j,k=1}^{n} [Q(t)^\top Q(t)]_{jk}\mathrm{T}_{jk}I\end{aligned}$$

$$= Q(t)\mathrm{T}Q(t)^\top - \frac{1}{n}\sum_{j,k=1}^{n} I_{jk}\mathrm{T}_{jk}I$$

$$= Q(t)\mathrm{T}Q(t)^\top - \frac{1}{n}\mathrm{Tr}\,\mathrm{T}I = Q(t)\sigma Q(t)^\top. \tag{1.3.27}$$

This yields, in particular,

$$\sigma = Q(t)^\top \sigma^* Q(t); \tag{1.3.28}$$

and (1.3.20) implies:

$$\mathcal{E} = Q(t)^\top \mathcal{E}^* Q(t). \tag{1.3.29}$$

Substitute these representations into (1.3.1):

$$Q(t)^\top\sigma^* Q(t) + \lambda_1 \frac{\partial Q(t)^\top \sigma^* Q(t)}{\partial t} = 2\eta\Big(Q(t)^\top \mathcal{E}^* Q(t) + \lambda_2 \frac{\partial Q(t)^\top \mathcal{E}^* Q(t)}{\partial t}\Big).$$

Having applied $Q(t)$ from the left, and $Q(t)^\top$ from the right, we arrive at

$$\sigma^* + \lambda_1 \frac{\partial \sigma^*}{\partial t^*} + \lambda_1 Q(t)Q'(t)^\top \sigma^* Q(t)Q(t)^\top + \lambda_1 Q(t)Q(t)^\top \sigma^* Q'(t)Q(t)^\top$$
$$= 2\eta\big(\mathcal{E}^* + \lambda_2 \frac{\partial \mathcal{E}^*}{\partial t^*} + \lambda_2 Q(t)Q'(t)^\top \mathcal{E}^* Q(t)Q(t)^\top$$
$$+ \lambda_2 Q(t)Q(t)^\top \mathcal{E}^* Q'(t)Q(t)^\top\big).$$

Now, use (1.3.26) and (1.3.15):

$$\sigma^* + \lambda_1 \frac{\partial \sigma^*}{\partial t^*} + \lambda_1 Q(t^*-a)Q'(t^*-a)^\top \sigma^* - \lambda_1 \sigma^* Q(t^*-a)Q'(t^*-a)^\top$$
$$= 2\eta\Big(\mathcal{E}^* + \lambda_2 \frac{\partial \mathcal{E}^*}{\partial t^*} + \lambda_2 Q(t^*-a)Q'(t^*-a)^\top \mathcal{E}^* - \lambda_2 \mathcal{E}^* Q(t^*-a)Q'(t^*-a)^\top\Big).$$

Thus, after a change of observer additional terms appear in the constitutive relation (1.3.1). Similarly one shows that the same additional terms emerge in (1.3.12).

Remark 1.3.3. If a change of variables of the form (1.3.15) – (1.3.16) transforms inertial frames to inertial ones (see Remark 1.3.2), then $Q'(t)$ is identically zero, so (1.3.1) and (1.3.12) do not change their form.

Nevertheless, the appearing terms add up to null function provided the object which we substitute for the time derivative denoted by a point becomes more complicated.

Definition 1.3.1. An operator of the form

$$\frac{D\mathcal{T}(t,x)}{Dt} = \frac{d\mathcal{T}(t,x)}{dt} + G(\nabla v(t,x), \mathcal{T}(t,x)), \tag{1.3.30}$$

where G is a symmetric tensor-valued function of two tensor arguments, $\mathcal{T}(t,x)$ is a symmetric tensor-valued function, is called an *objective derivative*, if, for any change of frame (1.3.15) – (1.3.16), the equality

$$\frac{D^*\mathcal{T}^*}{Dt^*} = Q(t)\frac{D\mathcal{T}}{Dt}Q(t)^\top \tag{1.3.31}$$

holds for all frame-indifferent symmetric tensor-valued functions $\mathcal{T}(t,x)$.

Remark 1.3.4. The symbol $\frac{D^*}{Dt^*}$ denotes the representation of the operator $\frac{D}{Dt}$ in the new frame, i.e. the expression of the form

$$\frac{d^*\mathcal{T}^*}{dt^*} + G((\nabla v)^*, \mathcal{T}^*) = \frac{\partial \mathcal{T}^*}{\partial t^*} + \sum_{i=1}^{n} v_i^* \frac{\partial \mathcal{T}^*}{\partial x_i^*} + G((\nabla v)^*, \mathcal{T}^*).$$

The choice of the function G is realized according to various mechanical and experimental reasons.

Given an objective derivative $\frac{D}{Dt}$, we can pass from (1.2.17) to the constitutive relation (see Remark 1.3.5)

$$\sigma + \lambda_1 \frac{D\sigma}{Dt} = 2\eta\big(\mathcal{E} + \lambda_2 \frac{D\mathcal{E}}{Dt}\big). \tag{1.3.32}$$

This constitutive relation satisfies the principle of material frame-indifference. Really, substituting representations (1.3.28) and (1.3.29) into (1.3.32) and using property (1.3.31) of objective derivatives, we obtain:

$$Q(t)^\top \sigma^* Q(t) + \lambda_1 Q(t)^\top \frac{D^*\sigma^*}{Dt^*} Q(t) = 2\eta\Big(Q(t)^\top \mathcal{E}^* Q(t) + \lambda_2 Q(t)^\top \frac{D^*\mathcal{E}^*}{Dt^*} Q(t)\Big),$$

whence

$$\sigma^* + \lambda_1 \frac{D^*\sigma^*}{Dt^*} = 2\eta\left(\mathcal{E}^* + \lambda_2 \frac{D^*\mathcal{E}^*}{Dt^*}\right). \tag{1.3.33}$$

For various objective derivatives it can turn out to be very hard or impossible to express the tensor σ in terms of the strain characteristics explicitly from (1.3.32). Therefore, for description of motion, for example, of a homogeneous incompressible viscoelastic medium, it is necessary to consider system (1.1.10), (1.1.12), (1.3.32) with a large number of unknown functions.

1.3.7 Examples of objective derivatives

The simplest example of an objective derivative of a tensor is *Jaumann's derivative*:

$$\frac{D_0\mathcal{T}(t,x)}{Dt} = \frac{d\mathcal{T}(t,x)}{dt} + \mathcal{T}(t,x)W(t,x) - W(t,x)\mathcal{T}(t,x), \tag{1.3.34}$$

which is also called *corotational*. Let us show that it satisfies condition (1.3.31). We have, using (1.3.10), (1.3.21) and (1.3.26):

$$\begin{aligned}
\frac{D_0^*\mathcal{T}^*}{Dt^*} &= \frac{\partial}{\partial\tau^*}\mathcal{T}^*(\tau^*, z^*(\tau^*, t^*, x^*))\Big|_{\tau^*=t^*} + \mathcal{T}^*W^* - W^*\mathcal{T}^* \\
&= \frac{\partial}{\partial\tau}\big[Q(\tau)\mathcal{T}(\tau, z(\tau, t, x))Q(\tau)^\top\big]\Big|_{\tau=t} + Q(t)\mathcal{T}Q(t)^\top\big(Q(t)WQ(t)^\top \\
&\quad + Q'(t)Q(t)^\top\big) - \big(Q(t)WQ(t)^\top + Q'(t)Q(t)^\top\big)Q(t)\mathcal{T}Q(t)^\top \\
&= Q'(t)\mathcal{T}Q(t)^\top + Q(t)\frac{d\mathcal{T}}{dt}Q(t)^\top + Q(t)\mathcal{T}Q'(t)^\top + Q(t)\mathcal{T}WQ(t)^\top \\
&\quad + Q(t)\mathcal{T}Q^\top(t)Q'(t)Q(t)^\top - Q(t)W\mathcal{T}Q(t)^\top - Q'(t)Q(t)^\top Q(t)\mathcal{T}Q(t)^\top \\
&= Q(t)\frac{D_0\mathcal{T}}{Dt}Q(t)^\top + Q'(t)\mathcal{T}Q(t)^\top + Q(t)\mathcal{T}Q'(t)^\top \\
&\quad - Q(t)\mathcal{T}Q'(t)^\top Q(t)Q(t)^\top - Q'(t)Q(t)^\top Q(t)\mathcal{T}Q(t)^\top = Q(t)\frac{D_0\mathcal{T}}{Dt}Q(t)^\top.
\end{aligned}$$

Below we show (see Remark 1.4.2) that every objective derivative may be represented as

$$\frac{D\mathcal{T}(t,x)}{Dt} = \frac{D_0\mathcal{T}(t,x)}{Dt} + G_1(\mathcal{E}(t,x), \mathcal{T}(t,x)), \tag{1.3.35}$$

where G_1 is a symmetric tensor-valued function of two symmetric tensor arguments. The sense of this representation is that every objective derivative is the sum of Jaumann's derivative and some expression which is independent of the vorticity tensor W. One can also give some information on the structure of G_1 (Corollary 1.4.2).

An elementary generalization of Jaumann's derivative is *Oldroyd's derivative*:

$$\frac{D_{\mathbf{a}}\mathcal{T}}{Dt} = \frac{D_0\mathcal{T}}{Dt} - \mathbf{a}(\mathcal{E}\mathcal{T} + \mathcal{T}\mathcal{E}). \tag{1.3.36}$$

This construction depends on a parameter $\mathbf{a} \in [-1, 1]$. For $\mathbf{a} = 0$ it becomes Jaumann's derivative.

For $\mathbf{a} = 1$, (1.3.36) is called the *upper-convected Maxwell derivative* (UCM), and, for $\mathbf{a} = -1$, the *lower-convected Maxwell derivative* (LCM).

A more general variant of objective derivative is due to Spriggs [7] and depends on three parameters $\mathbf{a}, \mathbf{b}, \mathbf{c}$:

$$\frac{D_{\mathbf{abc}}\mathcal{T}}{Dt} = \frac{D_{\mathbf{a}}\mathcal{T}}{Dt} + \mathbf{b}\,\mathrm{Tr}(\mathcal{T}\mathcal{E})I + \mathbf{c}\,\mathrm{Tr}(\mathcal{T})\mathcal{E}, \tag{1.3.37}$$

where I is the unit tensor.

Remark 1.3.5. For generalizations of Jaumann's derivative, (1.3.32) may contradict (1.1.5) for the latter implies $\mathrm{Tr}\,\sigma = 0$. In this case, in (1.3.32), σ should not be understood as the deviatoric stress tensor but as a certain *extra-stress tensor*, which still satisfies (1.1.11) with some other p.

1.4 Nonlinear effects in viscous media

1.4.1 Nonlinear viscosity and viscoelasticity

The conception of viscoelastic medium is not unique for explanation and description of non-Newtonian behaviour of real fluids and the media close to fluids. There is another class of models: nonlinear-viscous media. Here it is supposed that the stress tensor at a moment t at a point x is a function of other characteristics of a medium, taken at the same point and at the same moment:

$$\mathrm{T}(t,x) = g_1(t,x,v(t,x),\nabla v(t,x),\rho(t,x)). \tag{1.4.1}$$

Then (1.1.5) implies that σ is also a function of these characteristics:

$$\sigma(t,x) = g_2(t,x,v(t,x),\nabla v(t,x),\rho(t,x)). \tag{1.4.2}$$

Remark 1.4.1. Obviously, constitutive relations (1.3.2), (1.3.13) and, all the more, (1.3.32) cannot be reduced to (1.4.2). The deviatoric stress tensor $\sigma(t,x)$ in these equations turns out to depend on characteristics of a medium at other points and at other moments. This property is a feature of *viscoelastic* models. In this sense, viscoelastic media and the models reducible to (1.4.2) may be considered as two opposite classes of models. On the other hand, nonlinear viscosity and viscoelasticity can be combined in one model (see Section 1.5).

1.4.2 Noll's theorem and the Stokes conjecture.

Relation (1.4.2) is the constitutive relation for the nonlinear-viscous medium. According to the principle of material frame-indifference, its structure should not depend on observer. This gives opportunity to simplify and specify the form of the constitutive relation (1.4.2). The first step in this direction is Noll's theorem.

Theorem 1.4.1. *If relation (1.4.2) satisfies the principle of material frame-indifference, then g_2 depends only on $\mathcal{E}$ and ρ:*

$$g_2(t,x,v,\nabla v,\rho) = g_3(\mathcal{E},\rho). \tag{1.4.3}$$

Proof. By virtue of the principle of material frame-indifference, for any change of observer (1.3.15) – (1.3.16) the form of the constitutive relation (1.4.2) remains constant. Note that the density ρ is frame-indifferent. Then, in view of representations (1.3.20), (1.3.21), (1.3.23) and (1.3.28), we get

$$\begin{aligned} g_2(t,x,v,\nabla v,\rho) &= Q^\top g_2(t^*,x^*,v^*,(\nabla v)^*,\rho^*)Q \\ &= Q^\top g_2(t+a,x_0^*+Q(x-x_0),x_0^{*\prime}+Q'Q^\top(x^*-x_0^*) \\ &\quad + Qv, Q\mathcal{E}Q^\top + QWQ^\top + Q'Q^\top,\rho)Q. \end{aligned} \tag{1.4.4}$$

Let x_1 and t_1 be a fixed spatial point and a moment of time. Consider the tensor-valued function

$$Q(t) = e^{(t_1 - t)W(t_1, x_1)}, \tag{1.4.5}$$

where $W(t_1, x_1)$ is the vorticity tensor at (t_1, x_1).

Let us show that $Q(t)$ are orthogonal tensors. By a property of the exponential function, $Q'(t) = -Q(t)W(t_1, x_1)$. We have:

$$\begin{aligned} Q(t_1)Q(t_1)^\top &= I, \\ (Q(t)Q(t)^\top)' &= Q'(t)Q(t)^\top + Q(t)Q'(t)^\top \\ &= -Q(t)WQ(t)^\top - Q(t)W^\top Q(t)^\top = 0 \end{aligned}$$

since W is skew-symmetric.

This yields

$$Q(t)Q(t)^\top \equiv I.$$

Observe that $Q(t_1) = I, Q'(t_1) = -W(t_1, x_1)$.

Put in (1.4.4): $t = t_1, x = x_1$, Q as in (1.4.5), and

$$x_0 = 0, \tag{1.4.6}$$

$$x_0^*(t) = -x_1 - (t - t_1)(v(t_1, x_1) - W(t_1, x_1)x_1), \tag{1.4.7}$$

$$a = -t_1. \tag{1.4.8}$$

We obtain:

$$\begin{aligned} &g_2\big(t_1, x_1, v(t_1, x_1), \nabla v(t_1, x_1), \rho(t_1, x_1)\big) \\ &\quad = g_2\big(-a + a, -x_1 + x_1, -\big(v(t_1, x_1) - W(t_1, x_1)x_1\big) - W(t_1, x_1)x_1 \\ &\qquad\quad + v(t_1, x_1), \mathcal{E}(t_1, x_1) + W(t_1, x_1) - W(t_1, x_1), \rho(t_1, x_1)\big) \\ &\quad = g_2(0, 0, 0, \mathcal{E}(t_1, x_1), \rho(t_1, x_1)). \end{aligned}$$

Since t_1 and x_1 have been chosen arbitrarily, this continued equality implies the statement of the theorem. □

Remark 1.4.2. Let $\frac{D}{Dt}$ be an arbitrary objective derivative. Then, by (1.3.31), $\frac{D\mathcal{T}}{Dt}$ is frame-indifferent for any frame-indifferent tensor $\mathcal{T}(t, x)$. Since Jaumann's derivative is objective, $\frac{D\mathcal{T}}{Dt} - \frac{D_0\mathcal{T}}{Dt}$ is also frame-indifferent. But by (1.3.30) and (1.3.34) this expression is equal to $G(\nabla v, \mathcal{T}) - \mathcal{T}W + W\mathcal{T}$. Denote it by $G_1(\nabla v, \mathcal{T})$. Take arbitrary t_1 and x_1, and define Q by formula (1.4.5). Since $G_1(\nabla v, \mathcal{T})$ is frame-indifferent, just as in the proof of Noll's theorem we obtain:

$$\begin{aligned} G_1(\nabla v, \mathcal{T}) &= Q^\top G_1\big((\nabla v)^*, \mathcal{T}^*\big)Q \\ &= Q^\top G_1\big(Q\mathcal{E}Q^\top + QWQ^\top + Q'Q^\top, Q\mathcal{T}Q^\top\big)Q. \end{aligned} \tag{1.4.9}$$

At the moment t_1 and at the point x_1 one has:

$$G_1\big(\nabla v(t_1, x_1), \mathcal{T}(t_1, x_1)\big) = G_1\big(\mathcal{E}(t_1, x_1), \mathcal{T}(t_1, x_1)\big),$$

This implies representation (1.3.35).

Now, by Noll's theorem, relation (1.4.2) can be rewritten as:

$$\sigma(t, x) = g_3\big(\mathcal{E}(t, x), \rho(t, x)\big).$$

In the case when the deviatoric stress tensor does not depend on density (for example, the density is constant), we simply have:

$$\sigma(t, x) = g_3\big(\mathcal{E}(t, x)\big). \tag{1.4.10}$$

This relation is called the Stokes conjecture.

1.4.3 The Wang and Rivlin–Ericksen theorems

The presence of frame-indifference gives an opportunity to get some information on the function g_3 from (1.4.10) as well as on the function G_1 from (1.3.35). The main tool here is the following theorem, which is a particular case of Wang's theorem ([80], p. 215, see also [79], p. 197).

Theorem 1.4.2. *Let $n = 3$ and let a symmetric tensor-valued function $\zeta(\mathcal{T}_1, \mathcal{T}_2)$ of two symmetric tensor arguments satisfy the condition*

$$\zeta(\mathcal{T}_1, \mathcal{T}_2) = Q^\top \zeta(Q\mathcal{T}_1 Q^\top, Q\mathcal{T}_2 Q^\top) Q, \tag{1.4.11}$$

for each orthogonal tensor Q and all symmetric tensors $\mathcal{T}_1, \mathcal{T}_2$. Then ζ can be represented as follows:

$$\begin{aligned}\zeta(\mathcal{T}_1, \mathcal{T}_2) = \alpha_0 I &+ \alpha_1 \mathcal{T}_1 + \alpha_2 \mathcal{T}_1^2 + \alpha_3 \mathcal{T}_2 + \alpha_4 \mathcal{T}_2^2 \\ &+ \alpha_5(\mathcal{T}_1\mathcal{T}_2 + \mathcal{T}_2\mathcal{T}_1) + \alpha_6(\mathcal{T}_1^2\mathcal{T}_2 + \mathcal{T}_2\mathcal{T}_1^2) \\ &+ \alpha_7(\mathcal{T}_1\mathcal{T}_2^2 + \mathcal{T}_2^2\mathcal{T}_1) + \alpha_8(\mathcal{T}_1^2\mathcal{T}_2^2 + \mathcal{T}_2^2\mathcal{T}_1^2)\end{aligned} \tag{1.4.12}$$

where each α_j is a scalar function of ten scalar arguments:

$$\begin{aligned}\alpha_j = \alpha_j\big(&\operatorname{Tr}\mathcal{T}_1, \operatorname{Tr}(\mathcal{T}_1^2), \operatorname{Tr}(\mathcal{T}_1^3), \operatorname{Tr}(\mathcal{T}_2), \operatorname{Tr}(\mathcal{T}_2^2), \operatorname{Tr}(\mathcal{T}_2^3), \operatorname{Tr}(\mathcal{T}_1\mathcal{T}_2), \\ &\operatorname{Tr}(\mathcal{T}_1^2\mathcal{T}_2), \operatorname{Tr}(\mathcal{T}_1\mathcal{T}_2^2), \operatorname{Tr}(\mathcal{T}_1^2\mathcal{T}_2^2)\big).\end{aligned}$$

Remark 1.4.3. This particular case of Wang's theorem has no concern with the controversy related to the proof of his representation theorems (see [57], [81]).

The following statement is a special case of the well-known Rivlin–Ericksen theorem.

Corollary 1.4.1. *For $n = 3$, the function g_3 from (1.4.10) may be represented as follows:*

$$g_3(\mathcal{E}) = \varphi_0(I_1, I_2, I_3)I + \varphi_1(I_1, I_2, I_3)\mathcal{E} + \varphi_2(I_1, I_2, I_3)\mathcal{E}^2, \tag{1.4.13}$$

where

$$I_1 = \operatorname{Tr}\mathcal{E} = \operatorname{div} v,\ I_2 = \operatorname{Tr}\mathcal{E}^2 = \sum_{i,j=1}^{n} \mathcal{E}_{ij}^2,\ I_3 = \det\mathcal{E},$$

and $\varphi_0, \varphi_1, \varphi_2$ are scalar functions of three scalar arguments.

Proof. Observe that (1.4.3) and (1.4.4) yield:

$$g_3(\mathcal{E}) = Q^\top g_3(\mathcal{E}^*)Q,$$

and (1.3.20) implies

$$g_3(\mathcal{E}) = Q^\top g_3(Q\mathcal{E}Q^\top)Q, \tag{1.4.14}$$

for any orthogonal tensor Q.

Without loss of generality we may assume that $g_3(\mathcal{T}_1)$ is defined for all symmetric tensors $\mathcal{T}_1$ (for example, put $g_3(\mathcal{T}_1) = 0$ for all $\mathcal{T}_1$ which cannot be represented as $Q\mathcal{E}Q^\top$).

It is easy to check that

$$\operatorname{Tr}(\mathcal{T}_1^3) = -\frac{1}{2}(\operatorname{Tr}\mathcal{T}_1)^3 + \frac{3}{2}\operatorname{Tr}\mathcal{T}_1\ \operatorname{Tr}(\mathcal{T}_1^2) + 3\det\mathcal{T}_1. \tag{1.4.15}$$

It remains to apply Theorem 1.4.2 with $\zeta(\mathcal{T}_1, \mathcal{T}_2) = g_3(\mathcal{T}_1)$ and to use representation (1.4.12) with $\mathcal{T}_2 = I$. □

For a homogeneous incompressible medium one has $I_1 = \operatorname{div} v = 0$. Therefore, in this case, (1.4.13) has a simpler form:

$$g_3(\mathcal{E}) = \varphi_0(I_2, I_3)I + \varphi_1(I_2, I_3)\mathcal{E} + \varphi_2(I_2, I_3)\mathcal{E}^2. \tag{1.4.16}$$

It yields

$$\operatorname{Div}\sigma = \operatorname{Div}(\varphi_0 I + \varphi_1\mathcal{E} + \varphi_2\mathcal{E}^2) = \nabla\varphi_0 + \operatorname{Div}(\varphi_1\mathcal{E} + \varphi_2\mathcal{E}^2).$$

To obtain the equation of motion, substitute it into (1.1.12):

$$\frac{\partial v}{\partial t} + \sum_{i=1}^{n} v_i \frac{\partial v}{\partial x_i} - \nabla\varphi_0 - \operatorname{Div}(\varphi_1\mathcal{E} + \varphi_2\mathcal{E}^2) + \operatorname{grad} p = f. \tag{1.4.17}$$

Let $\widetilde{p}(t,x) = p(t,x) - \varphi_0(I_2(t,x), I_3(t,x))$. We have obtained the general equation of motion for a homogeneous incompressible nonlinear-viscous medium:

$$\frac{\partial v}{\partial t} + \sum_{i=1}^{n} v_i \frac{\partial v}{\partial x_i} - \operatorname{Div}\left(\varphi_1(I_2, I_3)\mathcal{E} + \varphi_2(I_2, I_3)\mathcal{E}^2\right) + \nabla\widetilde{p} = f. \tag{1.4.18}$$

Remark 1.4.4. This shows that $\widetilde{p}$ has the same "essence" as the hydrostatic pressure p, so without loss of generality of the model one can assume that $\varphi_0 \equiv 0$. However, it can be applied to the equation of motion, which does not contain σ, and it cannot be done directly for (1.4.13), because it may violate the assertion $\operatorname{Tr}\sigma = 0$, which follows from (1.1.5).

Due to representation (1.4.9), Theorem 1.4.2 is applicable to the function G_1 from (1.3.35). Thus, we have

Corollary 1.4.2. *For $n = 3$, every objective derivative may be represented as*

$$\frac{D\mathcal{T}(t,x)}{Dt} = \frac{D_0\mathcal{T}(t,x)}{Dt} + \alpha_0 I + \alpha_1\mathcal{E} + \alpha_2\mathcal{E}^2 + \alpha_3\mathcal{T} + \alpha_4\mathcal{T}^2 + \alpha_5(\mathcal{E}\mathcal{T} + \mathcal{T}\mathcal{E})$$
$$+\alpha_6(\mathcal{E}^2\mathcal{T} + \mathcal{T}\mathcal{E}^2) + \alpha_7(\mathcal{E}\mathcal{T}^2 + \mathcal{T}^2\mathcal{E}) + \alpha_8(\mathcal{E}^2\mathcal{T}^2 + \mathcal{T}^2\mathcal{E}^2) \tag{1.4.19}$$

where each α_j is a scalar function of ten scalar arguments:

$$\begin{aligned}\alpha_j = \alpha_j(&\operatorname{Tr}\mathcal{E}, \operatorname{Tr}(\mathcal{E}^2), \operatorname{Tr}(\mathcal{E}^3), \operatorname{Tr}(\mathcal{T}), \operatorname{Tr}(\mathcal{T}^2), \operatorname{Tr}(\mathcal{T}^3),\\ &\operatorname{Tr}(\mathcal{E}\mathcal{T}), \operatorname{Tr}(\mathcal{E}^2\mathcal{T}), \operatorname{Tr}(\mathcal{E}\mathcal{T}^2), \operatorname{Tr}(\mathcal{E}^2\mathcal{T}^2)).\end{aligned} \tag{1.4.20}$$

Remark 1.4.5. The results of this subsection are also true for $n = 2$. Their derivation from the given three-dimensional results is a good exercise in matrix theory.

1.4.4 Oldroyd's method. Models of Prandtl and Eyring

The functions $\varphi_0, \varphi_1, \varphi_2$ in (1.4.13) are to be determined experimentally. In order to simplify the construction of these functions, one may apply some reduction to a one-dimensional model (as in the case of the method of mechanical models). For example, Oldroyd [44] suggested the following procedure. Using simple flows, one finds a one-dimensional connection between the stress $\sigma_1 = \sigma\vec{n}$ and the strain velocity $\mathcal{E}_1 = \mathcal{E}\vec{n}$ in a certain direction $\vec{n}$ (for example, in the longitudinal direction, for the pipe flow):

$$\sigma_1 = \psi(\mathcal{E}_1). \tag{1.4.21}$$

It is assumed that $\psi(0) = 0$.

Let $\psi_1(\mathcal{E}_1) = \frac{\psi(\mathcal{E}_1)}{\mathcal{E}_1}$ $(\mathcal{E}_1 \neq 0)$. We have:

$$\sigma_1 = \psi_1(\mathcal{E}_1)\mathcal{E}_1 \quad (\mathcal{E}_1 \neq 0). \tag{1.4.22}$$

But since $\psi(0) = 0$,

$$\sigma_1 = 0 \quad (\mathcal{E}_1 = 0). \tag{1.4.23}$$

In order to pass to the n-dimensional situation, one substitutes σ for σ_1, and $\mathcal{E}$ for $\mathcal{E}_1$. However, since ψ_1 is a function of scalar argument, one has to substitute some scalar describing the strain velocity for this argument. The simplest variant is the Euclidean norm of the strain velocity tensor: $\sqrt{\sum_{i,j=1}^{n} \mathcal{E}_{ij}^2} = \sqrt{I_2}$ (for simplicity, here we assume incompressibility).

We obtain the following constitutive relation:

$$\sigma = \psi_1(\sqrt{I_2})\mathcal{E} \quad (\mathcal{E} \neq 0); \tag{1.4.24}$$

$$\sigma = 0 \quad (\mathcal{E} = 0). \tag{1.4.25}$$

In order to get a constitutive law of the same form as (1.4.13), put

$$\varphi_0 \equiv \varphi_2 \equiv 0, \quad \varphi_1(I_2, I_3) = \psi_1(\sqrt{I_2}) \quad (I_2 \neq 0), \qquad \varphi_1(I_2, I_3) = 0 \quad (I_2 = 0). \tag{1.4.26}$$

Let us consider the application of this method to the models of Prandtl and Eyring.

In this subsection A, B, C mean positive constants, which may be measured for real fluids. The one-dimensional Prandtl's model [83] looks like

$$\sigma_1 = A \arcsin\left(\frac{\mathcal{E}_1}{C}\right), \tag{1.4.27}$$

where σ_1 and $\mathcal{E}_1$ are the stress and the strain velocity. "Extracting viscosity", we get

$$\sigma_1 = \frac{A \arcsin\left(\frac{\mathcal{E}_1}{C}\right)}{\mathcal{E}_1}\mathcal{E}_1. \tag{1.4.28}$$

Passing to the n-dimensional model, we have:

$$\sigma = \frac{A \arcsin\left(\frac{\sqrt{I_2}}{C}\right)}{\sqrt{I_2}}\mathcal{E}, \tag{1.4.29}$$

together with (1.4.25).

Eyring's model [83] has (in the one-dimensional situation) the form:

$$\sigma_1 = \frac{\mathcal{E}_1}{B} + C \sin\left(\frac{\sigma_1}{A}\right). \tag{1.4.30}$$

We have :

$$\mathcal{E}_1 = B\sigma_1 - BC \sin\left(\frac{\sigma_1}{A}\right). \tag{1.4.31}$$

Consider the case $C < A$. Then the function in the right-hand side of (1.4.31) is invertible on a neighborhood of the origin. Denote the inverse function by ς. Evidently, $\varsigma(0) = 0$. Then

$$\sigma_1 = \varsigma(\mathcal{E}_1) \tag{1.4.32}$$

for small strain velocities.

This implies the n-dimensional constitutive law:

$$\sigma = \frac{\varsigma(\sqrt{I_2})}{\sqrt{I_2}}\mathcal{E}, \tag{1.4.33}$$

together with (1.4.25).

Remark 1.4.6. The functions $\frac{A \arcsin\left(\frac{\sqrt{I_2}}{C}\right)}{\sqrt{I_2}}$ and $\frac{\varsigma(\sqrt{I_2})}{\sqrt{I_2}}$ are even (as functions of I_2), so they are in fact smooth functions of I_2.

1.5 Combined models of nonlinear viscoelastic media

1.5.1 Nonlinear differential constitutive relations

In this section we assume that $n = 2$ or 3.

Rewrite the Jeffreys relation (1.3.32) using (1.3.35):

$$\sigma + \lambda_1\Big(\frac{D_0\sigma}{Dt} + G_1(\mathcal{E}, \sigma)\Big) = 2\eta\Big(\mathcal{E} + \lambda_2\Big(\frac{D_0\mathcal{E}}{Dt} + G_1(\mathcal{E}, \mathcal{E})\Big)\Big). \tag{1.5.1}$$

Denote $\mu_1 = \eta\frac{\lambda_2}{\lambda_1}$, $\eta_1 = \eta - \mu_1$, $\tau = \sigma - 2\mu_1\mathcal{E}$. Then (1.5.1) implies

$$\tau + \lambda_1\frac{D_0\tau}{Dt} + \lambda_1 G_1(\mathcal{E}, \tau + 2\mu_1\mathcal{E}) - 2\eta\lambda_2 G_1(\mathcal{E}, \mathcal{E}) = 2\eta_1\mathcal{E}. \tag{1.5.2}$$

Denote $\beta(\tau, \mathcal{E}) = \lambda_1 G_1(\mathcal{E}, \tau + 2\mu_1\mathcal{E}) - 2\eta\lambda_2 G_1(\mathcal{E}, \mathcal{E})$.

We have:

$$\tau + \lambda_1\frac{D_0\tau}{Dt} + \beta(\tau, \mathcal{E}) = 2\eta_1\mathcal{E}. \tag{1.5.3}$$

As in the case with Corollary 1.4.2, Theorem 1.4.2 is applicable to β. Hence, $\beta(\tau, \mathcal{E})$ may be represented as

$$\begin{aligned}&\alpha_0 I + \alpha_1\mathcal{E} + \alpha_2\mathcal{E}^2 + \alpha_3\tau + \alpha_4\tau^2 + \alpha_5(\mathcal{E}\tau + \tau\mathcal{E})\\ &\quad + \alpha_6(\mathcal{E}^2\tau + \tau\mathcal{E}^2) + \alpha_7(\mathcal{E}\tau^2 + \tau^2\mathcal{E}) + \alpha_8(\mathcal{E}^2\tau^2 + \tau^2\mathcal{E}^2)\end{aligned} \tag{1.5.4}$$

where each α_j is a scalar function of ten scalar arguments:

$$\alpha_j = \alpha_j\big(\operatorname{Tr}\mathcal{E}, \operatorname{Tr}(\mathcal{E}^2), \operatorname{Tr}(\mathcal{E}^3), \operatorname{Tr}(\tau), \operatorname{Tr}(\tau^2), \operatorname{Tr}(\tau^3), \operatorname{Tr}(\mathcal{E}\tau), \operatorname{Tr}(\mathcal{E}^2\tau), \operatorname{Tr}(\mathcal{E}\tau^2), \operatorname{Tr}(\mathcal{E}^2\tau^2)\big).$$

Relation (1.5.3) is a widespread form of constitutive equations for viscoelastic medium. Here are some examples of the function $\beta(\tau, \mathcal{E})$ (see [21, 22]).

The Giesekus model has

$$\beta(\tau, \mathcal{E}) = -\lambda_1(\tau\mathcal{E} + \mathcal{E}\tau) + \alpha\tau^2. \tag{1.5.5}$$

The Phan-Thien and Tanner model corresponds to

$$\beta(\tau, \mathcal{E}) = -\lambda_1(\tau\mathcal{E} + \mathcal{E}\tau) + \alpha\tau \operatorname{Tr}\,\tau. \tag{1.5.6}$$

In both cases α is a constant.

Larson's generalization of these two models has

$$\beta(\tau, \mathcal{E}) = -\lambda_1(\tau\mathcal{E} + \mathcal{E}\tau) + \alpha_1\tau^2 + \alpha_2\tau, \tag{1.5.7}$$

where α_1 and α_2 are scalar functions of $\operatorname{Tr}\,\tau$ and $\det\tau$, and $\alpha_2(0,0) = 0$.

The Larson model (another one) corresponds to

$$\beta(\tau, \mathcal{E}) = -\lambda_1(\tau\mathcal{E} + \mathcal{E}\tau) + 2\alpha_3 \operatorname{Tr}(\tau\mathcal{E})(\tau + I) \tag{1.5.8}$$

where α_3 is a scalar function of $\operatorname{Tr}\,\tau$.

The Oldroyd "8 constants" model has

$$\beta(\tau, \mathcal{E}) = -a\lambda_1(\tau\mathcal{E} + \mathcal{E}\tau) + b\operatorname{Tr}(\tau\mathcal{E})I + c\operatorname{Tr}(\tau)\mathcal{E} + c_1\mathcal{E}^2 + c_2 I_2 I, \tag{1.5.9}$$

where a, b, c, c_1, c_2 are constants. For $c_1 = c_2 = 0$ it is equivalent to Jeffreys' model with Spriggs' derivative (1.3.37).

Remark 1.5.1. Using the notation $\tau = \sigma - 2\mu_1\mathcal{E}$ and (1.1.13), one rewrites the equation of motion (1.1.12) as

$$\frac{\partial v}{\partial t} + \sum_{i=1}^{n} v_i \frac{\partial v}{\partial x_i} - \mu_1 \Delta v - \operatorname{Div}\,\tau + \operatorname{grad} p = f. \tag{1.5.10}$$

Remark 1.5.2. There is an interesting generalization of Jeffreys' model. It is assumed that the relaxation time λ_1 and the viscosity (more precisely, the viscosity of Maxwell's constituent $(1 - \lambda_2/\lambda_1)\eta$) depend on the Euclidean norm I_2 of the strain velocity tensor (compare with (1.4.26)). Such models are called the models of the White–Metzner type [22]. Mathematical investigation of the boundary value problems for equations of motion of these media was carried out by Hakim (see [26]).

1.5.2 Combined models

In order to get wider and more realistic classes of constitutive relations one can use the following simple heuristic argument inspired by the method of mechanical models.

Assume that we have m various models, each one with its own constitutive relation, which gives a connection between the deviatoric stress tensor σ and deformation characteristics. Let M_j, $j = 1, \dots, m$, be the "symbolical representations" (compare with Section 1.2.1) for each of these models. Consider the combined model $M_1|M_2|\dots|M_m$. Roughly speaking, this combined model describes the mixture of given m mediums. Since in the case of parallel connection the stresses perceived by the elements are summarized, and the deformation is common, we have the following constitutive law for the combined model:

$$\sigma = \sum_{k=1}^{m} \sigma_{M_k}. \tag{1.5.11}$$

Let us apply this method to the general models that we have: nonlinear-viscous fluid satisfying the Stokes conjecture (1.4.10) and the Jeffreys model (1.3.32). Observe that combining of several relations like (1.4.10) cannot produce something essentially new. Thus, we combine several Jeffreys' models (1.3.32), each one with its own parameters, with one nonlinear-viscous model (1.4.10), and obtain the constitutive law:

$$\sigma = \sum_{k=0}^{r} \sigma_k, \tag{1.5.12}$$

$$\sigma_0 = \varphi_0(I_1, I_2, I_3)I + \varphi_1(I_1, I_2, I_3)\mathcal{E} + \varphi_2(I_1, I_2, I_3)\mathcal{E}^2, \tag{1.5.13}$$

$$\begin{aligned} \sigma_k + \lambda_k\Big(\frac{D_0\sigma_k}{Dt} + G_k(\mathcal{E}, \sigma_k)\Big) \\ = 2\nu_k\Big(\mathcal{E} + \lambda_{2,k}\Big(\frac{D_0\mathcal{E}}{Dt} + G_k(\mathcal{E}, \mathcal{E})\Big)\Big), \quad k = 1, \dots, r. \end{aligned} \tag{1.5.14}$$

Here ν_k are viscosities, λ_k are relaxation times, and $\lambda_{2,k}$ are retardation times.

Assume that the medium is homogeneous and incompressible. Then the argument I_1 disappears in (1.5.13), as in (1.4.16).

Introduce the notations: $\mu_k = \nu_k \dfrac{\lambda_{2,k}}{\lambda_k}$, $\eta_k = \nu_k - \mu_k$, $\tau^k = \sigma_k - 2\mu_k\mathcal{E}$, $\beta_k(\tau, \mathcal{E}) = \lambda_k G_k(\mathcal{E}, \tau^k + 2\mu_k\mathcal{E}) - 2\nu_k\lambda_{2,k}G_k(\mathcal{E}, \mathcal{E})$, $k = 1, \dots, r$.

Then, as in the previous subsection, we can rewrite (1.5.14) as

$$\tau^k + \lambda_k \frac{D_0\tau^k}{Dt} + \beta_k(\tau^k, \mathcal{E}) = 2\eta_k\mathcal{E} \tag{1.5.15}$$

where

$$\begin{aligned}\beta_k(\tau,\mathcal{E}) &= \alpha_0^k I + \alpha_1\mathcal{E} + \alpha_2^k\mathcal{E}^2 + \alpha_3^k\tau + \alpha_4^k\tau^2 + \alpha_5^k(\mathcal{E}\tau + \tau\mathcal{E}) \\ &\quad + \alpha_6^k(\mathcal{E}^2\tau + \tau\mathcal{E}^2) + \alpha_7^k(\mathcal{E}\tau^2 + \tau^2\mathcal{E}) + \alpha_8^k(\mathcal{E}^2\tau^2 + \tau^2\mathcal{E}^2).\end{aligned} \tag{1.5.16}$$

Each α_j^k is a scalar function of nine scalar arguments:

$$\begin{aligned}\alpha_j^k = \alpha_j^k\big(&\mathrm{Tr}(\mathcal{E}^2), \mathrm{Tr}(\mathcal{E}^3), \mathrm{Tr}(\tau), \mathrm{Tr}(\tau^2), \\ &\mathrm{Tr}(\tau^3), \mathrm{Tr}(\mathcal{E}\tau), \mathrm{Tr}(\mathcal{E}^2\tau), \mathrm{Tr}(\mathcal{E}\tau^2), \mathrm{Tr}(\mathcal{E}^2\tau^2)\big).\end{aligned} \tag{1.5.17}$$

Let $\widetilde{p} = p - \varphi_0(I_2, I_3), \widetilde{\varphi}_1 = \varphi_1 + 2\sum_{j=1}^{r} \mu_k$ and

$$\tau^0 = \Psi(\mathcal{E}(u)) = \widetilde{\varphi}_1(I_2, I_3)\mathcal{E} + \varphi_2(I_2, I_3)\mathcal{E}^2. \tag{1.5.18}$$

Then (1.1.11) and (1.5.12) yield:

$$\mathrm{T} = -\widetilde{p}I + \sum_{k=0}^{r} \tau^k. \tag{1.5.19}$$

The system (1.5.15) – (1.5.19) is the general combined constitutive law for homogeneous incompressible nonlinear viscoelastic medium. For description of motion, it must be considered together with (1.1.9) and (1.1.10).

Chapter 2

Basic function spaces. Embedding and compactness theorems

2.1 Function spaces and embeddings

2.1.1 Lebesgue and Sobolev spaces

Let $n \in \mathbb{N}$ and $\mathbb{R}^n$ be the arithmetical n-dimensional space. In this book open sets in $\mathbb{R}^n$ are often called *domains*.

Let $\Omega \subset \mathbb{R}^n$ be a domain. Denote by $L_p(\Omega)$, $p \in \mathbb{R}$, $1 \leq p < \infty$, the space of all Lebesgue measurable functions $u : \Omega \to \mathbb{R}$ for which the function $|u(x)|^p$ is Lebesgue integrable. Of course, if two measurable or integrable functions coincide almost everywhere in Ω, they are considered to be equal. The norm in the space $L_p(\Omega)$, $1 \leq p < \infty$, is defined as

$$\|u\|_{L_p(\Omega)} = \left(\int_\Omega |u(x)|^p \, dx \right)^{\frac{1}{p}}.$$

In the case $p = \infty$ the space $L_p(\Omega)$ is defined as the space of all Lebesgue measurable functions $u : \Omega \to \mathbb{R}$ which are bounded by a constant almost everywhere on Ω. The norm in this space is given by the formula

$$\|u\|_{L_\infty(\Omega)} = \operatorname*{ess\,sup}_{x \in \Omega} |u(x)|.$$

A well-known property of L_p-norms is Hölder's inequality: for

$$1 \leq p, p_1, p_2 \leq \infty, \quad \frac{1}{p} = \frac{1}{p_1} + \frac{1}{p_2}, \quad \psi_1 \in L_{p_1}(\Omega), \quad \psi_2 \in L_{p_2}(\Omega),$$

one has that the pointwise product $\psi_1 \psi_2 \in L_p(\Omega)$, and

$$\|\psi_1 \psi_2\|_{L_p(\Omega)} \leq \|\psi_1\|_{L_{p_1}(\Omega)} \|\psi_2\|_{L_{p_2}(\Omega)}. \tag{2.1.1}$$

Here it is supposed $\frac{1}{\infty} = 0$.

For $u, v \in L_1(\Omega)$ such that $uv \in L_1(\Omega)$, introduce the bilinear form

$$(u, v) = \int_\Omega u(x) v(x) \, dx.$$

Hereafter we write simply $\|u\|$ for $\|u\|_{L_2(\Omega)}$. This norm becomes Euclidean if we define the scalar product in $L_2(\Omega)$ by the formula

$$(u, v)_{L_2(\Omega)} = (u, v).$$

Denote by $C(\overline{\Omega})$ the set of continuous functions $u : \overline{\Omega} \to \mathbb{R}$ (here $\overline{\Omega}$ is the closure of Ω). If Ω is bounded, $C(\overline{\Omega})$ is a Banach space with the norm

$$\|u\|_{C(\overline{\Omega})} = \max_{x \in \overline{\Omega}} |u(x)|.$$

For $u \in C(\overline{\Omega})$,

$$\|u\|_{C(\overline{\Omega})} = \|u\|_{L_\infty(\Omega)}.$$

Denote by $C^1(\overline{\Omega})$ the set of continuously differentiable functions $u : \overline{\Omega} \to \mathbb{R}$ (more precisely, they must be defined in a neighbourhood of Ω in order to deal with the derivatives on the boundary). If Ω is bounded, it is a Banach space with the norm

$$\|u\|_{C^1(\overline{\Omega})} = \|u\|_{C(\overline{\Omega})} + \sum_{i=1}^{n} \|\frac{\partial u}{\partial x_i}\|_{C(\overline{\Omega})}.$$

If α is a *multi-index*, i.e. $\alpha = (\alpha_1, \alpha_2, \ldots, \alpha_k)$, where $k, \alpha_1, \alpha_2, \ldots, \alpha_k \in \mathbb{N} \cup \{0\}$, and if $u \in L_p(\Omega)$, then $D^\alpha u$ stands for the generalized (Sobolev) partial derivative of the function $u(x)$, $x \in \Omega$:

$$D^\alpha u = \frac{\partial^{|\alpha|} u}{\partial x_1^{\alpha_1} \partial x_2^{\alpha_2} \ldots \partial x_k^{\alpha_k}}.$$

Here $|\alpha| = \sum_{j=1}^{k} \alpha_j$.

Denote by $W_p^r(\Omega)$, $1 \le p \le \infty, r \in \mathbb{N}$ or $r = 0$, the Sobolev space of such functions $u \in L_p(\Omega)$ that $D^\alpha u \in L_p(\Omega)$ for $|\alpha| \le r$. The norm in this space is

$$\|u\|_{W_p^r(\Omega)} = \Big(\sum_{|\alpha| \le r} \|D^\alpha u\|^p_{L_p(\Omega)} \Big)^{\frac{1}{p}}, \quad p < \infty;$$

$$\|u\|_{W_p^r(\Omega)} = \max_{|\alpha| \le r} \|D^\alpha u\|_{L_\infty(\Omega)}, \quad p = \infty.$$

The norm in $W_2^r(\Omega)$ is Euclidean, and the corresponding scalar product can be defined as

$$(u, v)_{W_2^r(\Omega)} = \sum_{|\alpha| \le r} (D^\alpha u, D^\alpha v).$$

Hereafter we often write $(u, v)_r$ for $(u, v)_{W_2^r(\Omega)}$ and $\|u\|_r$ for $\|u\|_{W_2^r(\Omega)}$. Note that $(u, v)_0 = (u, v)$ and $\|u\|_0 = \|u\|$.

Denote by $C_0^\infty(\Omega)$ the set of smooth scalar functions with compact support in Ω. Denote by $\overset{\circ}{W}{}_p^r(\Omega)$ the subspace of $W_p^r(\Omega)$ which is the closure of $C_0^\infty(\Omega)$.

Now we can define $W_p^{-r}(\Omega)$, $1 < p \leq \infty, r \in \mathbb{N}$, as the dual space of $\overset{\circ}{W}{}_q^r(\Omega)$ where $\frac{1}{p} + \frac{1}{q} = 1$.

Remark 2.1.1. We recall that the dual space X^* of a real normed space X consists of all continuous linear functionals $\varphi : X \to \mathbb{R}$. In this book we denote the value of a functional $\varphi \in X^*$ on an element $w \in X$ by

$$\langle \varphi, w \rangle_{X^* \times X}$$

or simply by

$$\langle \varphi, w \rangle$$

(it is called the "bra–ket" notation). The norm in the space X^* is defined as

$$\|\varphi\|_{X^*} = \sup_{\|w\|_X = 1} |\langle \varphi, w \rangle|.$$

The $*$-*weak* topology on X^* may be set as follows: $\varphi_m \to \varphi_0$ $*$-weakly if $\langle \varphi_m, w \rangle \to \langle \varphi_0, w \rangle$ for all $w \in X$. If X is separable, any bounded set in X^* is relatively compact in this $*$-weak topology.

We recall also that X is called *reflexive* if every linear continuous functional on X^* may be represented in the form

$$\langle \cdot, w \rangle,$$

where w is some element of X. Every bounded set in a reflexive space X is relatively compact in the weak topology of X.

The spaces $W_2^r(\Omega)$ and $\overset{\circ}{W}{}_2^r(\Omega)$ are also denoted $H^r(\Omega)$ and $H_0^r(\Omega)$, respectively. We shall use the generalization of $H^r(\Omega)$, the Sobolev–Slobodetskii space $H^s(\Omega)$ (or $W_2^s(\Omega)$), where $s \in \mathbb{R} \backslash \mathbb{N}$, $s > 0$. Let the brackets $[\cdot]$ and $\{\cdot\}$ denote the integer part and the fractional part of a number, respectively. Then $H^s(\Omega)$ consists of the functions $u \in H^{[s]}(\Omega)$ for which the norm

$$\|u\|_{H^s(\Omega)} = \Big(\|u\|_{[s]}^2 + \sum_{|\alpha|=[s]} \int_\Omega \int_\Omega \frac{|D^\alpha u(x) - D^\alpha u(y)|^2}{\|x-y\|_{\mathbb{R}^n}^{n+2\{s\}}} \, dxdy \Big)^{1/2}$$

is finite.

Remark 2.1.2. Another approach to the definition of $H^s(\Omega)$ in the case $\Omega = \mathbb{R}^n$ will be described in Section 5.2.2.

Denote by $H_0^s(\Omega)$ the subspace of $H^s(\Omega)$ which is the closure of $C_0^\infty(\Omega)$.

The Sobolev and Sobolev–Slobodetskii spaces are Banach spaces. They are reflexive for $1 < p < \infty$ and separable for $p \neq \infty$.

Following L. Schwartz, denote by $\mathcal{D}(\Omega)$ the set $C_0^\infty(\Omega)$ equipped with the following topology: a sequence $\psi_k \in \mathcal{D}(\Omega)$ converges to $\psi_0 \in \mathcal{D}(\Omega)$ if

i) there exists a compact set $\omega \subset \Omega$ such that $\psi_k(x) = \psi_0(x)$ for every natural k and any $x \in \Omega \backslash \omega$,

ii) for every multi-index α

$$\sup_{x\in\Omega} |D^\alpha(\psi_k(x) - \psi_0(x))| \to 0.$$

Let $\mathcal{D}'(\Omega)$ be the set of continuous linear functionals on the space $\mathcal{D}(\Omega)$. The elements of the space $\mathcal{D}'(\Omega)$ are called *generalized functions, scalar distributions* or simply *distributions*. The value of a functional $\varphi \in \mathcal{D}'(\Omega)$ on an element $\psi \in \mathcal{D}(\Omega)$ is denoted as $\langle\varphi, \psi\rangle$ (the "bra–ket" notation).

The topology on $\mathcal{D}'(\Omega)$ can be set as follows: a sequence $u_k \in \mathcal{D}'(\Omega)$ converges to $u_0 \in \mathcal{D}'(\Omega)$ if for any $\psi \in \mathcal{D}(\Omega)$ one has

$$\langle u_k, \psi\rangle \to \langle u_0, \psi\rangle.$$

We recall the following chain of inclusions:

$$\mathcal{D}(\Omega) \subset W_p^r(\Omega) \subset L_p(\Omega) \subset W_p^{-r}(\Omega) \subset \mathcal{D}'(\Omega) \tag{2.1.2}$$

where $1 < p \leq \infty, r \in \mathbb{N}$, and the corresponding embeddings are continuous.

Remark 2.1.3. An embedding of two topological vector spaces $A \subset B$ is called *continuous* if the topology of B induces on A a topology which is weaker than the topology of A, i.e. the intersections of the open sets of B with A are open in A. If A and B are normed spaces, then the embedding $A \subset B$ is continuous if and only if there exists a constant $C > 0$ such that for any $w \in A$ one has $\|w\|_B \leq C\|w\|_A$. We recall also that the embedding $A \subset B$ is called *compact* if any bounded in A set is relatively compact in B.

Remark 2.1.4. The third embedding in (2.1.2) is understood in a special sense. For any $u \in L_p(\Omega)$ one can consider the linear functional $\varphi(u) \in W_p^{-r}(\Omega) = \overset{\circ}{W}{}_q^r(\Omega)^*$, $\frac{1}{p} + \frac{1}{q} = 1$, determined by the formula

$$\langle\varphi(u), \psi\rangle = \int_\Omega \psi(x)u(x)\,dx, \quad \psi \in \overset{\circ}{W}{}_q^r(\Omega).$$

The functional $\varphi(u)$ and the function u are usually identified. Thus, the forth embedding in (2.1.2) and the third one (for $p = 2$) imply, in particular,

$$\langle u, \psi \rangle = (u, \psi), \quad u \in L_2(\Omega), \quad \psi \in \mathcal{D}(\Omega).$$

More information concerning interrelations of the Sobolev and Sobolev–Slobodetskii spaces for various parameters p, r and s is given by the Sobolev embedding theorems and Rellich–Kondrashov compactness theorems (see e.g. [1, 11, 43, 59]). Here we recall the embeddings which will be used in this book.

Theorem 2.1.1. a) *Let Ω be an arbitrary domain in $\mathbb{R}^n$, and let $1 < p_1 \le p_2 < \infty$, $r_1 \in \mathbb{N} \cup \{0\}$, $r_2 \in \mathbb{Z}$. If*

$$\frac{n}{p_1} - r_1 \le \frac{n}{p_2} - r_2, \tag{2.1.3}$$

then

$$\overset{\circ}{W}{}^{r_1}_{p_1}(\Omega) \subset W^{r_2}_{p_2}(\Omega), \tag{2.1.4}$$

and the embedding is continuous. If Ω is sufficiently regular (see Remark 2.1.5), then (2.1.3) implies

$$W^{r_1}_{p_1}(\Omega) \subset W^{r_2}_{p_2}(\Omega) \tag{2.1.5}$$

with continuous embedding.

b) *Let $s \in \mathbb{R}$, $r \in \mathbb{Z}$, $s \ge r \ge 0$, $2 \le p < \infty$. If*

$$\frac{n}{2} - s \le \frac{n}{p} - r, \tag{2.1.6}$$

then

$$H^s_0(\Omega) \subset W^r_p(\Omega), \tag{2.1.7}$$

and the embedding is continuous. If, in addition, Ω is sufficiently regular, then

$$H^s(\Omega) \subset W^r_p(\Omega) \tag{2.1.8}$$

with continuous embedding.

Remark 2.1.5. We call a domain $\Omega \subset \mathbb{R}^n$ *sufficiently regular* if its boundary $\partial\Omega$ is a C^k-smooth manifold for some $k \in \mathbb{N}$ and if Ω is locally located on one side of $\partial\Omega$. In particular, the domain $\Omega = \mathbb{R}^n$ with empty boundary is considered to be sufficiently regular. Of course, embeddings (2.1.5) and (2.1.8) are valid for wider classes of domains (see the references above), but we shall not use such embeddings in this book. Moreover, most of the results will be based on the embeddings like (2.1.4) which do not require any regularity of Ω.

Theorem 2.1.2. *Assume that under the conditions of Theorem 2.1.1 the domain Ω is bounded and inequalities (2.1.3) and (2.1.6) are strict. Then embeddings (2.1.4), (2.1.5), (2.1.7) and (2.1.8) are compact.*

Theorem 2.1.3. *Let Ω be an arbitrary domain in $\mathbb{R}^n$.*

a) *Let $1 < p < \infty$, $r \in \mathbb{N}$. If*

$$\frac{n}{p} < r, \tag{2.1.9}$$

then

$$\overset{\circ}{W}{}^r_p(\Omega) \subset L_\infty(\Omega), \tag{2.1.10}$$

and any function from $\overset{\circ}{W}{}^r_p(\Omega)$ is almost everywhere equal to a continuous on Ω function. If Ω is sufficiently regular, then

$$W^r_p(\Omega) \subset L_\infty(\Omega), \tag{2.1.11}$$

and any function from $W^r_p(\Omega)$ is almost everywhere equal to a continuous on Ω function.

b) *If*

$$s \in \mathbb{R}, \quad n < 2s, \tag{2.1.12}$$

then

$$H^s_0(\Omega) \subset L_\infty(\Omega), \tag{2.1.13}$$

and any function from $H^s_0(\Omega)$ is almost everywhere equal to a continuous on Ω function. If Ω is sufficiently regular, then

$$H^s(\Omega) \subset L_\infty(\Omega), \tag{2.1.14}$$

and any function from $H^s(\Omega)$ is almost everywhere equal to a continuous on Ω function.

c) *Embeddings (2.1.10), (2.1.11), (2.1.13), (2.1.14) are continuous. If, in addition, Ω is bounded, then they are also compact.*

In particular, for $n = 2, 3$ one has

$$H^1_0(\Omega) \subset L_4(\Omega), \tag{2.1.15}$$

$$H^s_0(\Omega) \subset L_\infty(\Omega), \quad s > \frac{3}{2}, \tag{2.1.16}$$

and for sufficiently regular Ω:

$$H^1(\Omega) \subset L_4(\Omega), \tag{2.1.17}$$

$$H^s(\Omega) \subset L_\infty(\Omega), \quad s > \frac{3}{2}. \tag{2.1.18}$$

If Ω is bounded, embeddings (2.1.15) – (2.1.18) are compact.

Corollary 2.1.1. *If Ω is sufficiently regular, $n = 2, 3$, then the following inequalities are valid:*

$$\|uv\| \le C\|u\|_2\|v\|, \quad u \in W_2^2(\Omega), \quad v \in L_2(\Omega), \tag{2.1.19}$$

$$\|uv\| \le C\|u\|_1\|v\|_1, \quad u, v \in W_2^1(\Omega), \tag{2.1.20}$$

$$\|uv\|_1 \le C\|u\|_1\|v\|_2, \quad u \in W_2^1(\Omega), \quad v \in W_2^2(\Omega), \tag{2.1.21}$$

$$\|uv\|_2 \le C\|u\|_2\|v\|_2, \quad u, v \in W_2^2(\Omega). \tag{2.1.22}$$

Remark 2.1.6. Hereafter in this section C stands for various constants independent of u, v.

Proof. To obtain (2.1.19), we apply Hölder's inequality (2.1.1) with $p = p_2 = 2$, $p_1 = \infty$ and use (2.1.18) with $s = 2$:

$$\|uv\| \le \|u\|_{L_\infty(\Omega)}\|v\| \le C\|u\|_2\|v\|.$$

To get (2.1.20), we use (2.1.1) with $p_1 = p_2 = 4$, $p = 2$ and embedding (2.1.17):

$$\|uv\| \le \|u\|_{L_4(\Omega)}\|v\|_{L_4(\Omega)} \le C\|u\|_1\|v\|_1.$$

Now, we have

$$\|uv\|_1 = \Big(\|uv\|^2 + \sum_{i=1}^n \|\frac{\partial(uv)}{\partial x_i}\|^2\Big)^{1/2} \le \Big(\|uv\|^2 + \sum_{i=1}^n (\|\frac{\partial u}{\partial x_i}v\| + \|u\frac{\partial v}{\partial x_i}\|)^2\Big)^{1/2}.$$

Due to (2.1.19) and (2.1.20), this does not exceed

$$C\Big(\|u\|_1^2\|v\|_1^2 + \sum_{i=1}^n (\|\frac{\partial u}{\partial x_i}\|\|v\|_2 + \|u\|_1\|\frac{\partial v}{\partial x_i}\|_1)^2\Big)^{1/2} \le C\|u\|_1\|v\|_2.$$

Finally, applying (2.1.19) – (2.1.21), we have

$$\begin{aligned}
\|uv\|_2 &= \Big(\|uv\|_1^2 + \sum_{i,j=1}^n \|\frac{\partial^2(uv)}{\partial x_i \partial x_j}\|^2\Big)^{1/2} \\
&\le \Big(\|uv\|_1^2 + \sum_{i,j=1}^n (\|\frac{\partial^2 u}{\partial x_i \partial x_j}v\| + \|\frac{\partial u}{\partial x_i}\frac{\partial v}{\partial x_j}\| + \|\frac{\partial u}{\partial x_j}\frac{\partial v}{\partial x_i}\| + \|u\frac{\partial^2 v}{\partial x_i \partial x_j}\|)^2\Big)^{1/2} \\
&\le \Big(\|u\|_1^2\|v\|_2^2 + \sum_{i,j=1}^n (\|\frac{\partial^2 u}{\partial x_i \partial x_j}\|\|v\|_2 + \|\frac{\partial u}{\partial x_i}\|_1\|\frac{\partial v}{\partial x_j}\|_1 \\
&\quad + \|\frac{\partial u}{\partial x_j}\|_1\|\frac{\partial v}{\partial x_i}\|_1 + \|u\|_2\|\frac{\partial^2 v}{\partial x_i \partial x_j}\|)^2\Big)^{1/2} \\
&\le C\|u\|_2\|v\|_2,
\end{aligned}$$

and (2.1.22) is proven. □

Let Z denote any of the classes introduced above (L_p, H^s etc.) We shall use the notations $Z(\Omega)^n$ and $Z(\Omega, \mathbb{R}^n)$ for the Cartesian product of n spaces $Z(\Omega)$. The norm in this space is defined as

$$\|u\|_{Z(\Omega)^n} = \Big(\sum_{i=1}^{n} \|u_i\|_{Z(\Omega)}^2\Big)^{1/2},$$

where

$$u = (u_1, u_2, \dots, u_n);$$

the scalar product (if applicable) may be given as

$$(u, v)_{Z(\Omega)^n} = \sum_{i=1}^{n} (u_i, v_i)_{Z(\Omega)};$$

the "bra–ket" product (for $Z = \overset{\circ}{W}{}_q^r(\Omega), \mathcal{D}$) is

$$\langle u, v\rangle_{Z^*(\Omega)^n \times Z(\Omega)^n} = \sum_{i=1}^{n} \langle u_i, v_i\rangle_{Z^*(\Omega)\times Z(\Omega)}.$$

A useful (but simple) inequality is

$$\|u\|_{Z(\Omega)^n} \le \sum_{i=1}^{n} \|u_i\|_{Z(\Omega)}.$$

We recall Ladyzhenskaya's inequalities:

$$\|u\|_{L_4(\Omega)} \le 2^{1/4} \|u\|^{1/2} \|\operatorname{grad} u\|_{L_2(\Omega)^n}^{1/2}, \quad n = 2, \tag{2.1.23}$$

$$\|u\|_{L_4(\Omega)} \le 2^{1/2} \|u\|^{1/4} \|\operatorname{grad} u\|_{L_2(\Omega)^n}^{3/4}, \quad n = 3, \tag{2.1.24}$$

for any domain $\Omega \subset \mathbb{R}^n$ and $u \in H_0^1(\Omega)$. Here

$$(\operatorname{grad} u)_j = \frac{\partial u}{\partial x_j}.$$

See e.g. [61], Lemmas III.3.3 and III.3.5, for the proofs of these inequalities. Note that (2.1.23) and (2.1.24) imply

$$\|u\|_{L_4(\Omega)} \le 2^{1/2} \|u\|^{1/4} \|u\|_1^{3/4}, \quad n = 2, 3. \tag{2.1.25}$$

Let us mention also Friedrichs' inequality. For any bounded domain Ω:

$$\|v\| \le K_0(\Omega) \|\operatorname{grad} v\|_{L_2(\Omega)^n}, \quad v \in H_0^1(\Omega), \tag{2.1.26}$$

where the constant K_0 is independent of v (see e.g. [20]).

2.1.2 The spaces used in hydrodynamics

In this subsection we are going to introduce the function spaces suitable for description of velocities and stress tensors.

Let

$$\mathcal{V} = \{u \in C_0^\infty(\Omega, \mathbb{R}^n),\ \operatorname{div} u = 0\}.$$

The symbols $H = H(\Omega)$, $V = V(\Omega)$, $V_\delta = V_\delta(\Omega)$ ($\delta \in (0, 1]$) denote the closures of $\mathcal{V}$ in $L_2(\Omega, \mathbb{R}^n)$, $H^1(\Omega, \mathbb{R}^n)$, $H^\delta(\Omega, \mathbb{R}^n)$, respectively.

The space V is especially important in hydrodynamics since the assertion $v \in V$ expresses conditions (1.1.10) and (1.1.15).

As a rule, using the Riesz representation theorem, one identifies H with H^*, and, hence, has (see Section 2.2.8)

$$V \subset H \equiv H^* \subset V^*. \tag{2.1.27}$$

Denote by $\mathbb{R}^{n\times n}$ the space of matrices of the order $n \times n$ with the following scalar product: for $A = (A_{ij})$, $B = (B_{ij})$

$$(A, B)_{\mathbb{R}^{n\times n}} = \sum_{i,j=1}^{n} A_{ij} B_{ij}$$

and by $\mathbb{R}_S^{n\times n}$ its subspace of symmetric matrices.

Denote by $\mathbb{R}^{n\times n\times n}$ the space of ordered collections of n matrices of the order $n \times n$ with the following scalar product: for $A = (A_1, \dots, A_n)$, $B = (B_1, \dots, B_n)$

$$(A, B)_{\mathbb{R}^{n\times n\times n}} = \sum_{i=1}^{n} (A_i, B_i)_{\mathbb{R}^{n\times n}}.$$

The symbol ∇u stands for the Jacobi matrix of a vector function $u : \Omega \subset \mathbb{R}^n \to \mathbb{R}^n$ (cf. (1.1.2)). The symbol $\nabla\tau$ denotes the ordered collection of the Jacobi matrices of the columns of a matrix function $\tau : \Omega \subset \mathbb{R}^n \to \mathbb{R}^{n\times n}$.

In this book the Euclidean norms in $\mathbb{R}^n$, $\mathbb{R}^{n\times n}$, $\mathbb{R}_S^{n\times n}$ are denoted as $|\cdot|$.

Let Z denote any of the classes introduced in the previous subsection. By analogy with the case described in the end of that subsection, we use the notations $Z(\Omega, \mathbb{R}^{n\times n})$, $Z(\Omega, \mathbb{R}_S^{n\times n})$, $Z(\Omega, \mathbb{R}^{n\times n\times n})$ for the Cartesian product of the corresponding number of spaces $Z(\Omega)$. The norms, the scalar and "bra-ket" products in these spaces may be defined analogously. The space $H^s(\Omega, \mathbb{R}_S^{n\times n})$ will also be denoted as $H_M^s(\Omega)$.

As in the case of scalar functions, we shall write $\|u\|$ for $\|u\|_{L_2(\Omega,F)}$, $(u, v)_r$ for $(u, v)_{W_2^r(\Omega,F)}$ and $\|u\|_r$ for $\|u\|_{W_2^r(\Omega,F)}$, (u, v) for $\int_\Omega (u(x), v(x))_F\, dx$. Here F stands for any of the spaces $\mathbb{R}^n, \mathbb{R}^{n\times n}, \mathbb{R}_S^{n\times n}, \mathbb{R}^{n\times n\times n}$.

Sometimes (mostly in Chapter 6) we shall write simply L_p, H^s, W_p^r etc. instead of $L_p(\Omega, F)$, $H^s(\Omega, F)$, $W_p^r(\Omega, F)$ when it is clear from the context which domain Ω and space F are used.

It is clear that for these Cartesian product spaces one has embedding theorems similar to the ones described in Section 2.1.1.

Denote by $Y(\Omega)$ the completion of $\mathcal{V}$ with respect to the Euclidean norm

$$\|u\|_{Y(\Omega)} = \|\nabla u\|,$$

which corresponds to the scalar product

$$(u, v)_{Y(\Omega)} = (\nabla u, \nabla v).$$

If Ω is bounded, (2.1.26) implies

$$\|u\|_1^2 \leq (1 + K_0^2(\Omega))\|\nabla u\|^2, u \in V, \tag{2.1.28}$$

so $Y(\Omega)$ coincides with V (up to equivalent norm).

2.2 Spaces of vector functions

2.2.1 Preliminaries

Let E be a Banach space and let $J = (\alpha, \beta)$ be an interval of the real axis $\mathbb{R}$, $-\infty \leq \alpha < \beta \leq +\infty$.

A function $u : J \to E$ is called *simple* if there exist a finite number of mutually disjoint Lebesgue measurable subsets $B_i \subset J$, $i = 1, 2, \ldots, m$, such that the function u is identically a constant x_i in each B_i and vanishes in $J \setminus \bigcup_{i=1}^{m} B_i$.

The *Bochner integral* of a simple function u is defined by the formula

$$\int_\alpha^\beta u(t)\,dt = \sum_{i=1}^{m} \operatorname{mes}(B_i) x_i .$$

A function $u : J \to E$ is called *Bochner measurable* if there exists a sequence of simple functions $\{u_k\}$ such that for almost all $t \in J$:

$$u_k(t) \xrightarrow[k\to\infty]{} u(t) \text{ in } E.$$

If the condition

$$\lim_{k\to\infty} \int_\alpha^\beta \|u(t) - u_k(t)\|_E\,dt = 0$$

holds for this sequence, then the function u is called *Bochner integrable*. The Bochner integral of such a function is

$$\int_\alpha^\beta u(t)\,dt = \lim_{k\to\infty} \int_\alpha^\beta u_k(t)\,dt.$$

As well as in the case of scalar functions, if two (Bochner) measurable functions differ on a null set, they are considered to be equal.

Denote by $L_p(\alpha,\beta;E)$, $1 \le p \le \infty$, the sets of such measurable functions $u : J \to E$ that

$$\|u\|_{L_p(\alpha,\beta;E)} = \left(\int_\alpha^\beta \|u(t)\|_E^p \, dt\right)^{\frac{1}{p}} < +\infty \quad (1 \le p < \infty);$$

or

$$\|u\|_{L_p(\alpha,\beta;E)} = \operatorname*{ess\,sup}_{t\in J} \|u(t)\|_E < +\infty \quad (p = \infty).$$

They are Banach spaces with the indicated norms. They are separable provided $p < \infty$ and E is separable, and are reflexive for $1 < p < \infty$ and reflexive E. We recall also that

$$L_p(\alpha,\beta;E)^* = L_{\frac{1}{1-1/p}}(\alpha,\beta;E^*), \quad 1 < p < \infty,$$
$$L_1(\alpha,\beta;E)^* = L_\infty(\alpha,\beta;E^*).$$

It is easy to see that a measurable function $u(t)$ belongs to the space $L_p(\alpha,\beta;E)$ if and only if the scalar function $\|u(t)\|_E$ belongs to $L_p(\alpha,\beta) = L_p(\alpha,\beta;\mathbb{R})$, whereas

$$\|u\|_{L_p(\alpha,\beta;E)} = \| \|u(t)\|_E \|_{L_p(\alpha,\beta)}.$$

A function $u : J \to E$ is Bochner integrable if and only if $u \in L_1(\alpha,\beta;E)$.

Remark 2.2.1. In this chapter, when dealing with bounded intervals J, for convenience we use the interval $J = (0,T)$, $0 < T < \infty$. However, each definition or statement using $J = (0,T)$ is good for any J with $\alpha,\beta \ne \pm\infty$ as well.

Lemma 2.2.1. a) *Let* $1 \le p \le q \le \infty$, *and* $u \in L_q(0,T;E)$. *Then* $u \in L_p(0,T;E)$, *and*

$$\|u\|_{L_p(0,T;E)} \le T^{\frac{1}{p}-\frac{1}{q}} \|u\|_{L_q(0,T;E)}.$$

b) *Let*

$$p, p_1, p_2, q, q_1, q_2 \ge 1, \quad \frac{1}{p} = \frac{1}{p_1} + \frac{1}{p_2}, \; \frac{1}{q} = \frac{1}{q_1} + \frac{1}{q_2},$$
$$u \in L_{p_1}(\alpha,\beta;L_{q_1}(\Omega)), \quad v \in L_{p_2}(\alpha,\beta;L_{q_2}(\Omega)).$$

Then the pointwise product uv,

$$uv(t)(x) = u(t)(x)v(t)(x),$$

belongs to the space $L_p(\alpha,\beta;L_q(\Omega))$, *and*

$$\|uv\|_{L_p(\alpha,\beta;L_q(\Omega))} \le \|u\|_{L_{p_1}(\alpha,\beta;L_{q_1}(\Omega))} \|v\|_{L_{p_2}(\alpha,\beta;L_{q_2}(\Omega))}. \tag{2.2.1}$$

Proof. a) Hölder's inequality (2.1.1) with $\Omega = (0,T)$, $\psi_1 = 1$, $\psi_2 = \|u\|_E$, $\frac{1}{p_1} = \frac{1}{p} - \frac{1}{q}$, $p_2 = q$ implies

$$\|\|u\|_E\|_{L_p(0,T)} \leq \|1\|_{L_{p_1}(0,T)}\|\|u\|_E\|_{L_q(0,T)},$$

so

$$\|u\|_{L_p(0,T;E)} \leq \|1\|_{L_{p_1}(0,T)}\|u\|_{L_q(0,T;E)}.$$

It remains to observe that

$$\|1\|_{L_{p_1}(0,T)} = \left(\int_0^T dt\right)^{\frac{1}{p_1}} = T^{\frac{1}{p_1}} = T^{\frac{1}{p}-\frac{1}{q}}.$$

b) It suffices to apply Hölder's inequality (2.1.1) twice:

$$\begin{aligned}\|uv\|_{L_p(\alpha,\beta;L_q(\Omega))} &= \|\|uv\|_{L_q(\Omega)}\|_{L_p(\alpha,\beta)}\\ &\leq \|\|u\|_{L_{q_1}(\Omega)}\|v\|_{L_{q_2}(\Omega)}\|_{L_p(\alpha,\beta)}\\ &\leq \|\|u\|_{L_{q_1}(\Omega)}\|_{L_{p_1}(\alpha,\beta)}\|\|v\|_{L_{q_2}(\Omega)}\|_{L_{p_2}(\alpha,\beta)}\\ &= \|u\|_{L_{p_1}(\alpha,\beta;L_{q_1}(\Omega))}\|v\|_{L_{p_2}(\alpha,\beta;L_{q_2}(\Omega))}.\end{aligned}$$

□

Lemma 2.2.1, a), implies that for $1 \leq p \leq q \leq \infty$ the space $L_q(0,T;E)$ is continuously embedded into $L_p(0,T;E)$.

Denote by $L_{p,\mathrm{loc}}(\alpha,\beta;E)$ the set of measurable functions $u : J \to E$ which belong to $L_p(t_1,t_2;E)$ for any $t_1,t_2 \in \mathbb{R}, \alpha < t_1 < t_2 < \beta$.

Denote by $C_w(\overline{J};E)$ the set of the functions $u : \overline{J} \to E$ which are *weakly continuous*, i.e. for each $g \in E^*$ the function $\langle g, u(\cdot)\rangle : \overline{J} \to \mathbb{R}$ is continuous. Here $\overline{J}$ is the closure of J.

Denote by $C(\overline{J};E)$ the set of continuous functions $u : \overline{J} \to E$.

Note that $C([0,T];E)$ is a Banach space for the norm

$$\|u\|_{C([0,T],E)} = \max_{t\in[0,T]} \|u(t)\|_E.$$

It is easy to see that the space $C([0,T];E)$ is continuously embedded into $L_p(0,T;E)$. At $1 \leq p < \infty$ the embedding is dense; at $p = \infty$ for any $u \in C([0,T];E)$ one has

$$\|u\|_{C([0,T];E)} = \|u\|_{L_\infty(0,T;E)}.$$

Remark 2.2.2. Let i_p stand for the natural embedding operator from $C([0,T];E)$ into $L_p(0,T;E)$. Besides, $C_w([0,T];E) \subset L_p(0,T;E)$ (see [20], Theorem 1.9). Let j_p be the natural embedding. Hereafter we shall often identify $C([0,T];E)$ and $i_p(C([0,T];E))$, as well as $C_w([0,T];E)$ and $j_p(C_w([0,T];E))$.

Note also that $C([0,+\infty);E)$ is a Fréchet space for the pre-norm

$$\|v\|_{C([0,+\infty);E)} = \sum_{i=0}^{+\infty} 2^{-i} \frac{\|v\|_{C([0,i];E)}}{1+\|v\|_{C([0,i];E)}},$$

and $C((-\infty,+\infty);E)$ is a Fréchet space for the pre-norm

$$\|v\|_{C((-\infty,+\infty);E)} = \sum_{i=0}^{+\infty} 2^{-i} \frac{\|v\|_{C([-i,i];E)}}{1+\|v\|_{C([-i,i];E)}}.$$

For more details on Bochner integrable and measurable functions see, for example, [84].

Denote by $\mathcal{D}'(0,T;E)$ the set of linear continuous maps from the space $\mathcal{D}(0,T) = \mathcal{D}((0,T))$ into the space E considered with the weak topology. The topology on $\mathcal{D}'(0,T;E)$ can be set as follows: a sequence $u_k \in \mathcal{D}'(0,T;E)$ converges to $u_0 \in \mathcal{D}'(0,T;E)$ if for any $\psi \in \mathcal{D}(0,T)$ and any linear continuous functional φ on E one has

$$\varphi(u_k(\psi)) \to \varphi(u_0(\psi)).$$

The elements of $\mathcal{D}'(0,T;E)$ are called distributions on $(0,T)$ with values in E.

Definition 2.2.1. The *generalized (distributional) derivative* of a distribution $u \in \mathcal{D}'(0,T;E)$ is the distribution determined by the formula

$$u'(\psi) = -u(\psi'), \quad \forall\, \psi \in \mathcal{D}(0,T).$$

The operator of differentiation is a linear continuous operator on $\mathcal{D}'(0,T;E)$.

As in the scalar case (Remark 2.1.4), it turns out that every element $u \in L_{1,\mathrm{loc}}(0,T;E)$ can be identified with a distribution from $\mathcal{D}'(0,T;E)$ (i.e. with a weakly continuous linear map from $\mathcal{D}(0,T)$ into E) according to the following formula:

$$u(\psi) = \int_0^T \psi(t)u(t)\,dt, \quad \psi \in \mathcal{D}(0,T).$$

In this sense it is possible to consider $L_{1,\mathrm{loc}}(0,T;E) \subset \mathcal{D}'(0,T;E)$.

Lemma 2.2.2. *Let $u \in \mathcal{D}'(0,T;E)$ and $u' = 0$. Then there exists an element $b \in E$ such that $u \equiv b$, i.e. for any $\psi \in \mathcal{D}(0,T)$:*

$$u(\psi) = b\int_0^T \psi(t)\,dt.$$

Proof. Fix a function $\varphi \in \mathcal{D}(0,T)$ such that $\int_0^T \varphi(t)\,dt = 1$. Let $b = u(\varphi)$. Take an arbitrary function $\psi \in \mathcal{D}(0,T)$. The function $g(s) = \psi(s) - \varphi(s)\int_0^T \psi(t)\,dt$ belongs to $\mathcal{D}(0,T)$. Then the function $\eta(\xi) = \int_0^\xi g(s)\,ds$ also belongs to $\mathcal{D}(0,T)$. Really, it is smooth and vanishes at small ξ. For ξ close to T, we have:

$$\begin{aligned}\eta(\xi) &= \int_0^\xi \psi(s)\,ds - \int_0^\xi \varphi(s)\int_0^T \psi(t)\,dt\,ds \\ &= \int_0^T \psi(s)\,ds - \int_0^T \varphi(s)\int_0^T \psi(t)\,dt\,ds \\ &= \int_0^T \psi(s)\,ds - \int_0^T \psi(t)\,dt = 0.\end{aligned}$$

Now we have

$$\begin{aligned}u(\psi) - b\int_0^T \psi(t)\,dt &= u(\psi) - u(\varphi)\int_0^T \psi(t)\,dt \\ &= u\Big(\psi - \varphi\int_0^T \psi(t)\,dt\Big) = u(\eta') = -u'(\eta) = 0. \qquad \square\end{aligned}$$

Lemma 2.2.3. *Let $u \in \mathcal{D}'(0,T;E)$ and $u' \in L_p(0,T;E)$, $1 \le p \le \infty$. Then $u \in C([0,T];E)$, and the following representation takes place:*

$$u(t) = u(0) + \int_0^t u'(s)\,ds. \tag{2.2.2}$$

Proof. Note first that $u' \in L_p(0,T;E) \subset L_1(0,T;E)$. Since $C([0,T];E)$ is dense in $L_1(0,T;E)$, there exists a sequence $g_n \in C([0,T];E)$ such that $g_n \to u'$ in $L_1(0,T;E)$. By the Newton – Leibnitz formula we have for an arbitrary $\psi \in \mathcal{D}(0,T)$:

$$\int_0^T \Big(\int_0^t g_n(s)\,ds \cdot \psi(t)\Big)'\,dt = \Big(\int_0^t g_n(s)\,ds \cdot \psi(t)\Big)\Big|_0^T = 0.$$

But

$$\Big(\int_0^t g_n(s)\,ds \cdot \psi(t)\Big)' = g_n(t)\psi(t) + \int_0^t g_n(s)\,ds \cdot \psi'(t).$$

Therefore

$$\int_0^T g_n(t)\psi(t)\,dt + \int_0^T \int_0^t g_n(s)\,ds \cdot \psi'(t)\,dt = 0.$$

Passing to the limit as $n \to \infty$, we obtain

$$\int_0^T u'(t)\psi(t)\,dt + \int_0^T \int_0^t u'(s)\,ds \cdot \psi'(t)\,dt = 0.$$

Using the definition of generalized derivative, we rewrite this as

$$u'(\psi) - \left(\int_0^t u'(s)\, ds\right)'(\psi) = 0.$$

Since the function $\psi \in \mathcal{D}(0, T)$ was taken arbitrary, this yields

$$\left(u - \int_0^t u'(s)\, ds\right)' = 0.$$

By Lemma 2.2.2 there exists b such that $u - \int_0^t u'(s)\, ds = b$. Thus,

$$u = b + \int_0^t u'(s)\, ds.$$

Hence, $u \in C([0, T]; E)$. It remains to observe that at $t = 0$ the left-hand side of the obtained equality is $u(0)$, and the right-hand one is b, therefore this equality implies the statement of the lemma. □

In conclusion, let us define some other spaces of vector functions.

Denote by $C^r(0, T; E), r \in \mathbb{N}$, the space of the functions $u \in C(0, T; E)$ which have r continuous derivatives with respect to t. The norm in this space is

$$\|u\|_{C^r(0,T;E)} = \sum_{m=0}^{r} \|u^{(m)}(t)\|_{C(0,T;E)}.$$

Denote by $W_p^r(0, T; E)$, $1 \le p \le \infty, r \in \mathbb{N}$ or $r = 0$, the Sobolev space of the functions $u \in L_p(0, T; E)$ for which the norm

$$\|u\|_{W_p^r(0,T;E)} = \sum_{m=0}^{r} \|u^{(m)}(t)\|_{L_p(0,T;E)}$$

is finite. These spaces are Banach spaces. They are separable provided $p < \infty$ and E is separable, and are reflexive for $1 < p < \infty$ and reflexive E.

Corollary 2.2.1. *One has*

$$W_1^1(0, T; E) \subset C([0, T]; E),$$

and the embedding is continuous.

Proof. The inclusion follows immediately from Lemma 2.2.3. The continuity is due to the inequality

$$\begin{aligned}
\max_{t\in[0,T]} \|u(t)\|_E &\le \frac{1}{T}\int_0^T \|u(s)\|_E\, ds + \max_{t,s\in[0,T]} \|u(t) - u(s)\|_E \\
&\le \frac{1}{T}\int_0^T \|u(s)\|_E\, ds + \int_0^T \|u'(s)\|_E\, ds \\
&\le C\,\|u\|_{W_1^1(0,T;E)}.
\end{aligned}$$

□

Denote by $C_0^\infty(0,T;E)$ the set of smooth functions with values in E which have a compact support in $(0,T)$. Denote by $\overset{\circ}{W}{}_p^r(0,T;E)$ the subspace of $W_p^r(0,T;E)$ which is the closure of $C_0^\infty(0,T;E)$.

Denote by $W_p^{-r}(0,T;E)$ the dual space for the space $\overset{\circ}{W}{}_q^r(0,T;E^*)$, $\frac{1}{p}+\frac{1}{q}=1$. Here it is assumed that E is reflexive and $q<\infty$ (we do not need here a definition for the general case, which becomes more complicated).

Every element $u \in W_p^{-r}(0,T;E) = (\overset{\circ}{W}{}_q^r(0,T;E^*))^*$ can be identified with a distribution from $\mathcal{D}'(0,T;E)$ according to the following formula:

$$\langle \xi, u(\psi)\rangle_{E^*\times E} = \langle u, \xi\psi\rangle_{W_p^{-r}(0,T;E)\times \overset{\circ}{W}{}_q^r(0,T;E^*)}, \quad \psi \in \mathcal{D}(0,T), \quad \xi \in E^*.$$

2.2.2 Classical criteria of compactness

A subset K of a Banach space E is called *compact* if one can select a finite subcovering from every open covering of this set; $K \subset E$ is called *relatively compact* if its closure $\overline{K}$ is compact.

Let us recall the well-known Hausdorff and Arzela–Ascoli criteria of relative compactness.

Theorem 2.2.1. *A set $K \subset E$ is relatively compact if and only if for every $\varepsilon > 0$ there exists a finite set $e_i \subset K$, $i = 1,\dots,j$, such that for every $x \in K$ there is an element e_i such that $\|e_i - x\|_E \le \varepsilon$. This set $\{e_i, i=1,\dots,j\}$ is called ε-net of the set K.*

Corollary 2.2.2. *If K is a uniform limit of relatively compact sets (i.e. for any $\varepsilon > 0$ there exists a relatively compact set K_ε such that for every $x \in K$ there is $y \in K_\varepsilon$ such that $\|x-y\|_E \le \varepsilon$), then K is relatively compact.*

Proof. Fix $\varepsilon > 0$. Choose a relatively compact set $K_{\frac{\varepsilon}{2}}$. By Theorem 2.2.1 it possesses an $\frac{\varepsilon}{2}$-net $\{e_i,\ i=1,\dots,j_\varepsilon\}$. Then $\{e_i,\ i=1,\dots,j_\varepsilon\}$ is an ε-net for K. Really, for any $x \in K$ there is $y \in K_{\frac{\varepsilon}{2}}$ such that $\|x-y\|_E \le \frac{\varepsilon}{2}$, and for this y there exists e_i such that $\|y-e_i\|_E \le \frac{\varepsilon}{2}$. By the triangle inequality $\|x-e_i\|_E \le \varepsilon$. Since ε has been arbitrary, by Theorem 2.2.1 K is a relatively compact set. □

Theorem 2.2.2. *A set $F \subset C([0,T];E)$ is relatively compact if and only if*

$$F(t) = \{f(t) \mid f \in F\} \text{ is relatively compact in } E\ \forall t \in [0,T], \tag{2.2.3}$$

$$\text{and } F \text{ is equicontinuous.} \tag{2.2.4}$$

The second condition means that for any $\varepsilon > 0$ there is $\delta > 0$ such that for all $t_1, t_2 \in [0,T]$, $|\,t_1 - t_2\,| \le \delta$, and every $f \in F$ one has $\|f(t_2) - f(t_1)\|_E \le \varepsilon$.

2.2.3 Compactness in $L_p(0, T; E)$

For a function $f : [0, T] \to E$ and a number $h > 0$ put $(\tau_h f)(t) = f(t + h)$. Then we have a function $\tau_h f : [-h, T - h] \to E$.

Theorem 2.2.3 ([54]). *Let E be a Banach space. A set $F \subset L_p(0, T; E)$, $1 \le p < \infty$, is relatively compact if and only if*

$$\text{the set } \left\{ \int_{t_1}^{t_2} f(t)\, dt \mid f \in F \right\} \text{ is relatively compact in } E \tag{2.2.5}$$

for every $t_1, t_2 \in (0, T)$, $t_1 < t_2$, and

$$\|\tau_h f - f\|_{L_p(0,T-h;E)} \underset{h\to 0}{\longrightarrow} 0 \text{ uniformly with respect to } f \in F, \tag{2.2.6}$$

i.e. for any $\varepsilon > 0$ there is $\delta > 0$ such that for every $f \in F$ and $h < \delta$ one has $\|\tau_h f - f\|_{L_p(0,T-h;E)} \le \varepsilon$.

Proof. The first stage. Let F be a relatively compact subset of $L_p(0, T; E)$.

Let us show that the linear map $l : f \mapsto \int_{t_1}^{t_2} f(t)\, dt$, $t_1, t_2 \in (0, T)$, $t_1 < t_2$, from $L_p(0, T; E)$ into E, is bounded. Really, using Lemma 2.2.1, a), we get

$$\left\| \int_{t_1}^{t_2} f(t)\, dt \right\|_E \le \int_{t_1}^{t_2} \|f(t)\|_E\, dt \le \|f\|_{L_1(0,T;E)} \le T^{1-\frac{1}{p}} \|f\|_{L_p(0,T;E)}.$$

Therefore the map l is continuous, so it transforms relatively compact sets to relatively compact ones. Thus, (2.2.5) is fulfilled.

Fix $\varepsilon > 0$. By Theorem 2.2.1 there is a finite $\frac{\varepsilon}{5}$-net $\{f_i, i = 1, \dots, j\}$ in the space $L_p(0, T; E)$ for the set F. Since $C([0, T]; E)$ is dense in $L_p(0, T; E)$, there exist $e_i \in C([0, T]; E)$, $\|e_i - f_i\|_{L_p(0,T;E)} \le \frac{\varepsilon}{5}$, $i = 1, \dots, j$. For continuous functions e_i we have $\|\tau_h e_i - e_i\|_{C([0,T-h];E)} \underset{h\to 0}{\longrightarrow} 0$. Therefore $\|\tau_h e_i - e_i\|_{L_p(0,T-h;E)} \underset{h\to 0}{\longrightarrow} 0$, and for δ small enough:

$$\|\tau_h e_i - e_i\|_{L_p(0,T-h;E)} \le \frac{\varepsilon}{5} \quad \text{for} \quad h < \delta.$$

For any element $f \in F$ there is an element f_i from the $\frac{\varepsilon}{5}$-net such that $\|f - f_i\|_{L_p(0,T;E)} \le \frac{\varepsilon}{5}$. By the triangle inequality:

$$\begin{aligned}
&\|\tau_h f - f\|_{L_p(0,T-h;E)} \\
&\le \|\tau_h(f - f_i)\|_{L_p(0,T-h;E)} + \|\tau_h(f_i - e_i)\|_{L_p(0,T-h;E)} \\
&\qquad + \|\tau_h e_i - e_i\|_{L_p(0,T-h;E)} + \|e_i - f_i\|_{L_p(0,T-h;E)} + \|f_i - f\|_{L_p(0,T-h;E)} \\
&= \|f - f_i\|_{L_p(h,T;E)} + \|f_i - e_i\|_{L_p(h,T;E)} + \|\tau_h e_i - e_i\|_{L_p(0,T-h;E)} \\
&\qquad + \|e_i - f_i\|_{L_p(0,T-h;E)} + \|f_i - f\|_{L_p(0,T-h;E)} \\
&\le 5 \cdot \frac{\varepsilon}{5} = \varepsilon.
\end{aligned}$$

We have proved that condition (2.2.6) also holds.

The second stage. Assume, on the contrary, that conditions (2.2.5) and (2.2.6) hold. Put for $f \in F$ and $0 < a < T$:

$$(M_a f)(t) = \frac{1}{a} \int_t^{t+a} f(s)\, ds. \tag{2.2.7}$$

We have for $t_1, t_2 \in [0, T-a],\ t_1 < t_2$:

$$\begin{aligned}\|(M_a f)(t_2) - (M_a f)(t_1)\|_E &= \left\| \frac{1}{a} \left(\int_{t_2}^{t_2+a} f(s)\, ds - \int_{t_1}^{t_1+a} f(s)\, ds \right) \right\|_E \\ &= \frac{1}{a} \left\| \int_{t_1}^{t_1+a} (\tau_{t_2-t_1} f - f)(s)\, ds \right\|_E \le \frac{1}{a} \|\tau_{t_2-t_1} f - f\|_{L_1(0,T-(t_2-t_1);E)} \\ &\le \frac{T^{1-1/p}}{a} \|\tau_{t_2-t_1} f - f\|_{L_p(0,T-(t_2-t_1);E)}.\end{aligned}$$

Condition (2.2.6) yields that the right-hand side tends to zero uniformly with respect to f as $|t_2 - t_1| \to 0$. Hence, the set $M_a F = \{M_a f \,|\, f \in F\}$ is contained in $C([0, T-a]; E)$ and is equicontinuous on $[0, T-a]$.

The condition (2.2.5) with $t_1 = t,\ t_2 = t + a,\ t \in (0, T-a)$ gives that the set $(M_a F)(t) = \{\frac{1}{a} \int_t^{t+a} f(s)\, ds | f \in F\}$ is relatively compact in E for all $t \in (0, T-a)$. The set $(M_a F)(0)$ is the uniform limit of the relatively compact sets $(M_a F)(\varepsilon)$ as $\varepsilon \to 0$. In fact, equicontinuity of $M_a F$ implies that for any $\varepsilon > 0$ there is $\delta(\varepsilon) > 0$ such that for any $f \in F$ one has $\|M_a f(\delta) - M_a f(0)\| \le \varepsilon$.

Therefore the set $(M_a F)(0)$ is also relatively compact. Similarly, $(M_a F)(T-a)$ is relatively compact as the uniform limit of the sets $(M_a F)(T - a - \delta)$ as $\delta \to 0$.

We have shown that on the segment $[0, T-a]$ the set $M_a F$ satisfies conditions (2.2.3) and (2.2.4) of Theorem 2.2.2. Hence, $M_a F$ is relatively compact in $C([0, T-a], E)$ and therefore in $L_p(0, T-a; E)$ for any $a \in (0, T)$.

We have for $t \in [0, T-a]$:

$$\begin{aligned}M_a f(t) - f(t) &= \frac{1}{a} \int_t^{t+a} f(s)\, ds - f(t) \\ &= \frac{1}{a} \int_0^a f(t+h)\, dh - \frac{1}{a} \int_0^a f(t)\, dh = \frac{1}{a} \int_0^a (\tau_h f - f)(t)\, dh.\end{aligned} \tag{2.2.8}$$

Hence,

$$\|M_a f - f\|_{L_p(0,T-a;E)} \le \sup_{h \in [0,a]} \|\tau_h f - f\|_{L_p(0,T-a;E)}. \tag{2.2.9}$$

Let us show that a set of restrictions $F|_{[0,T_1]} = \{f|_{[0,T_1]} \,\big|\, f \in F\}$ is the uniform limit of the sets $M_a F$ in $L_p(0, T_1; E)$ for every $T_1 \in (0, T)$ as $a \to 0$ (here and below we omit the symbol of restriction). Really, fix $\varepsilon > 0$. For $a < T - T_1$, (2.2.9) yields:

$$\|M_a f - f\|_{L_p(0,T_1;E)} \leq \sup_{h\in[0,a]} \|\tau_h f - f\|_{L_p(0,T-h;E)}. \tag{2.2.10}$$

By condition (2.2.6) there is a such that the right-hand side and, consequently, the left-hand side of (2.2.10) does not exceed ε for any $f \in F$. Thus, F is relatively compact in $L_p(0, T_1; E)$ as the uniform limit of the relatively compact sets $M_a F$.

Let us show that F is relatively compact in $L_p(T - T_1, T; E)$. Consider the set

$$\widetilde{F} = \{\widetilde{f}(t) = f(T-t) | f \in F\}.$$

For $\widetilde{F}$ condition (2.2.5) holds. Let us show that condition (2.2.6) also holds for it. We have for $\widetilde{f} \in \widetilde{F}$ that $\widetilde{f}(t) = f(T-t)$, $f \in F$. Then

$$\|\tau_h \widetilde{f} - \widetilde{f}\|_{L_p(0,T-h;E)} = \left(\int_0^{T-h} \left\|\widetilde{f}(t+h) - \widetilde{f}(t)\right\|_E^p dt\right)^{\frac{1}{p}}$$

$$= \left(\int_0^{T-h} \left\|f(T-t-h) - f(T-t)\right\|_E^p dt\right)^{\frac{1}{p}}$$

$$= \left(-\int_0^{T-h} \left\|f(T-t-h) - f(T-t)\right\|_E^p d(T-t-h)\right)^{\frac{1}{p}}$$

$$= \left(\int_0^{T-h} \left\|f(s) - f(s+h)\right\|_E^p ds\right)^{\frac{1}{p}} = \|f - \tau_h f\|_{L_p(0,T-h;E)}.$$

The last expression tends to zero as $h \to 0$ uniformly with respect to $f \in F$ by condition (2.2.6) for F.

Thus, for $\widetilde{F}$ conditions (2.2.5) and (2.2.6) are fulfilled. Therefore, just as it was shown above for F, $\widetilde{F}$ is relatively compact in $L_p(0, T_1; E)$. It implies that F is relatively compact in $L_p(T - T_1, T; E)$. Taking $T_1 \geq \frac{T}{2}$ we conclude that F is relatively compact in $L_p(0, T; E)$. The theorem is completely proved. □

Remark 2.2.3. For $p = \infty$, the statement of Theorem 2.2.3 as it is formulated above is incorrect. Really, consider the set F consisting of only one function $f(t) = \operatorname{sign}(t - \frac{T}{2})u$, where $u \in E$, $u \neq 0$.

Obviously, it is compact in $L_\infty(0, T; E)$. We have: $\tau_h f(t) - f(t) = u[\operatorname{sign}(t - \frac{T}{2} + h) - \operatorname{sign}(t - \frac{T}{2})]$. Hence, $\tau_h f - f$ is identically equal to $2u$ in the interval $(\frac{T}{2} - h, \frac{T}{2})$. Therefore $\|\tau_h f - f\|_{L_\infty(0,T-h;E)}$ does not converge to zero as $h \to 0$, i.e. condition (2.2.6) is not fulfilled.

Remark 2.2.4. However conditions (2.2.5) and (2.2.6) for $p = \infty$ are close to the conditions of Theorem 2.2.2 on compactness in $C([0,T];E)$. Condition (2.2.6), as we shall see, is equivalent to the equicontinuity (2.2.4), and condition (2.2.5) resembles condition (2.2.3). As the following theorem shows, it is not an accidental coincidence.

Theorem 2.2.4 ([54]). *A set $F \subset L_\infty(0,T;E)$ satisfies conditions* (2.2.5) *and* (2.2.6) *with $p = \infty$ if and only if F is contained in $C([0,T];E)$ and is relatively compact in this space.*

Proof. The first stage. Let F be a relatively compact subset of $C([0,T];E)$. Then F is relatively compact in $L_p(0,T;E)$, and by Theorem 2.2.3 F satisfies condition (2.2.5). By Theorem 2.2.2 F satisfies condition (2.2.4). It remains to observe that (2.2.4) yields (2.2.6) with $p = \infty$. In fact, for $f \in F \subset C([0,T];E)$:

$$\|\tau_h f - f\|_{L_\infty(0,T-h;E)} = \|\tau_h f - f\|_{C([0,T-h];E)} = \max_{t\in[0,T-h]} \|f(x+h)-f(x)\|_E, \tag{2.2.11}$$

which in view of the equicontinuity (2.2.4) tends to zero uniformly with respect to f as $h \to 0$.

The second stage. Assume, on the contrary, that conditions (2.2.5) and (2.2.6) with $p = \infty$ hold. As in the second stage of the proof of Theorem 2.2.3, it is possible to consider the set $M_a F$ and to verify that it is relatively compact in $C([0,T-a];E)$ for every $a \in (0,T)$. Then (2.2.8) yields (2.2.9) and (2.2.10) for $p = \infty$. As in the proof of Theorem 2.2.3, condition (2.2.6) implies that $M_a f \xrightarrow[a\to 0]{} f$ in $L_\infty(0,T_1;E)$, $T_1 \in (0,T)$, uniformly with respect to $f \in F$. Then for any sequence $a_n \to 0$ the sequence $M_{a_n} f$ is fundamental:

$$\|M_{a_n} f - M_{a_m} f\|_{L_\infty(0,T_1;E)} = \|M_{a_n} f - M_{a_m} f\|_{C([0,T_1];E)} \xrightarrow[\max(n,m)\to\infty]{} 0.$$

Since the space $C([0,T_1];E)$ is complete, $M_a f \xrightarrow[a\to 0]{} f$ in $C([0,T_1];E)$ uniformly with respect to $f \in F$, i.e. F is the uniform limit of the relatively compact sets $M_a F$ in $C([0,T_1];E)$. Therefore F is relatively compact in $C([0,T_1];E)$. As well as in the proof of Theorem 2.2.3, having considered the set $\widetilde{F} = \{f(T-t) | f \in F\}$, which satisfies conditions (2.2.5) and (2.2.6) and therefore is relatively compact in $C([0,T_1];E)$, we conclude that F is relatively compact in $C([T-T_1,T];E)$. Taking $T_1 > \frac{T}{2}$ we get that F is relatively compact in $C([0,T];E)$. □

Remark 2.2.5. At the first stage of the proof of Theorem 2.2.4 we have shown that for $F \subset L_\infty(0,T;E)$ the condition of equicontinuity (2.2.4) implies (2.2.6) with $p = \infty$; at the second stage we have got that condition (2.2.6) with $p = \infty$ yields that, for any $f \in F$, $M_a f \xrightarrow[a\to 0]{} f$ in $C([0,T_1];E)$. Hence, $f \in C([0,T_1];E)$ for $f \in F$. But if F satisfies (2.2.6), then $\widetilde{F}$ satisfies (2.2.6), so for $\widetilde{f} \in F$ one has

$M_a \widetilde{f} \xrightarrow[a\to 0]{} \widetilde{f}$ in $C([0,T_1];E)$, which gives $f \in C([T-T_1,T];E)$ for all $f \in F$. Taking $T_1 > \frac{T}{2}$, we get $F \subset C([0,T];E)$. Then (2.2.6) and (2.2.11) yield (2.2.4). Thus, for $F \subset L_\infty(0,T;E)$ conditions (2.2.4) and (2.2.6) are equivalent.

2.2.4 Compactness of sets of vector functions with values in an "intermediate" space

Let $X \subset E \subset Y$ be Banach spaces where the embedding $X \subset E$ is compact, and $E \subset Y$ continuously.

Lemma 2.2.4 ([37]). *For any $\eta > 0$ there exists a natural number N such that for all $u \in X$:*

$$\|u\|_E \le \eta\|u\|_X + N\|u\|_Y. \tag{2.2.12}$$

Proof. For every natural n, denote by V_n the set $\{u \in E \,|\, \|u\|_E - \eta - n\|u\|_Y < 0\}$. Since E is embedded into Y continuously, the function $\|u\|_E - \eta - n\|u\|_Y : E \to \mathbb{R}$ is continuous. Therefore the sets V_n are open. As n increases, the sequence of sets V_n extends and their unit covers the whole space E. Since the embedding $X \subset E$ is compact, the unit sphere S of the space X is compact in E. Therefore from the open covering $\{V_n\}$ of the set S one can select a finite subcovering $\{V_{n_1}, V_{n_2}, \dots, V_{n_j}\}$. Let $N = \max(n_1, n_2, \dots, n_j)$. Then $S \subset V_N$, i.e. for any $u \in S$:

$$\|u\|_E - \eta - N\|u\|_Y < 0. \tag{2.2.13}$$

Now, if $u \neq 0$, the element $\frac{u}{\|u\|_X} \in S$ satisfies estimate (2.2.13), i.e.

$$\left\|\frac{u}{\|u\|_X}\right\|_E - \eta - N\left\|\frac{u}{\|u\|_X}\right\|_Y < 0.$$

Multiplying by $\|u\|_X$, we obtain estimate (2.2.12). If $u = 0$, estimate (2.2.12) is obvious. □

On application of the compactness theorems obtained in this section, difficulties usually arise at the check of condition (2.2.6). The use of the space triples $X \subset E \subset Y$ gives opportunity to weaken this condition in exchange for a strengthening of condition (2.2.5).

Theorem 2.2.5 ([54]). *Let $X \subset E \subset Y$ be Banach spaces where the embedding $X \subset E$ is compact, and $E \subset Y$ continuously. Let $F \subset L_p(0,T;X)$, $1 \le p \le \infty$. If*

$$F \text{ is bounded in } L_p(0,T;X), \tag{2.2.14}$$

$$\|\tau_h f - f\|_{L_p(0,T-h;Y)} \xrightarrow[h\to 0]{} 0 \text{ uniformly with respect to } f \in F, \tag{2.2.15}$$

then F is relatively compact in $L_p(0,T;E)$ (at $p = \infty$ the set F is contained in $C([0,T];E)$ and is relatively compact in this space).

Proof. It suffices to check that conditions (2.2.5) and (2.2.6) hold: in this case the statement of the theorem follows from Theorem 2.2.3 (or Theorem 2.2.4, for $p = \infty$).

As we have already observed, the linear map $f \mapsto \int_{t_1}^{t_2} f(t)\,dt,\ t_1, t_2 \in (0,T)$, $t_1 < t_2$, is continuous from $L_p(0,T;X)$ into X. Therefore, since F is bounded in $L_p(0,T;X)$, the set $\{\int_{t_1}^{t_2} f(t)\,dt \mid f \in F\}$ is bounded in X. Hence, this set is relatively compact in E, i.e. the condition (2.2.5) holds.

Let us show now that the condition (2.2.6) is fulfilled. Fix $\varepsilon > 0$. Since F is bounded in $L_p(0,T;X)$, there is a number $R > 0$ such that $\|f\|_{L_p(0,T;X)} < R,\ f \in F$. Applying Lemma 2.2.4 with $\eta = \frac{\varepsilon}{4R}$, we get that for some N and almost all $t \in (0,T)$:

$$\|\tau_h f(t) - f(t)\|_E \le \frac{\varepsilon}{4R}\|\tau_h f(t) - f(t)\|_X + N\|\tau_h f(t) - f(t)\|_Y.$$

Therefore

$$\|\tau_h f - f\|_{L_p(0,T-h;E)} \le \frac{\varepsilon}{4R}\|\tau_h f - f\|_{L_p(0,T-h;X)} + N\|\tau_h f - f\|_{L_p(0,T-h;Y)}. \tag{2.2.16}$$

By condition (2.2.15) the second term in the right-hand side of (2.2.16) uniformly with respect to f tends to zero as $h \to 0$ and for h small enough does not exceed $\frac{\varepsilon}{2}$. On the other hand,

$$\begin{aligned}\frac{\varepsilon}{4R}\|\tau_h f - f\|_{L_p(0,T-h;X)} &\le \frac{\varepsilon}{4R}\left(\|\tau_h f\|_{L_p(0,T-h;X)} + \|f\|_{L_p(0,T-h;X)}\right)\\ &= \frac{\varepsilon}{4R}\left(\|f\|_{L_p(h,T;X)} + \|f\|_{L_p(0,T-h;X)}\right) \le \frac{\varepsilon}{4R}(R+R) = \frac{\varepsilon}{2}.\end{aligned}$$

Now (2.2.16) implies

$$\|\tau_h f - f\|_{L_p(0,T-h;E)} \le \varepsilon,$$

for all $f \in F$. Thus, condition (2.2.6) also holds. □

2.2.5 The Aubin–Simon theorem

The compactness theorems are frequently applied to establish relative compactness of sets of solutions for differential equations. Here it often happens that it is known that the set of solutions is bounded in a comparatively narrow space $L_p(0,T;X)$, and the set of their derivatives with respect to t is bounded in some "weak sense", for example, in the metric of a space $L_r(0,T;Y)$ wide enough. In this case one may often apply the following theorem.

Theorem 2.2.6. *Let $X \subset E \subset Y$ be Banach spaces where the embedding $X \subset E$ is compact, and $E \subset Y$ continuously. Let $F \subset L_p(0,T;X)$, $1 \le p \le \infty$. Assume* *that for any $f \in F$ its generalized derivative f' in the space $\mathcal{D}'(0,T;Y)$ belongs to $L_r(0,T;Y)$, $1 \le r \le \infty$. Let*

$$F \text{ be bounded in } L_p(0,T;X), \tag{2.2.17}$$

$$\{f' | f \in F\} \text{ be bounded in } L_r(0,T;Y). \tag{2.2.18}$$

Then for $p < \infty$ F is relatively compact in $L_p(0,T;E)$; for $p = \infty$ and $r > 1$, F is relatively compact in $C([0,T];E)$.

Remark 2.2.6. The first proof of this statement was given by Aubin [8] for the case when $r > 1$, $1 < p < \infty$, and the spaces X and Y are reflexive. Aubin's result was generalized by several authors (for example, Dubinskii [19] gave a proof for the case $p = 2$, $r = 1$, X, E, Y are Hilbert spaces). In the general case the theorem was formulated and proved by Simon [54].

Remark 2.2.7. In the case $p = \infty$, $r = 1$ the statement of Theorem 2.2.6 is incorrect. Really, fix an arbitrary nonzero element $b \in X$ and an arbitrary smooth function $\varphi : \mathbb{R} \to \mathbb{R}$, $\varphi \not\equiv 0$, vanishing outside of the interval $(0,T)$. Consider the set

$$G = \{g_n(t) = b\varphi(nt) | n \in \mathbb{N}\}.$$

We have:

$$\begin{aligned}
\|\varphi(nt)\|_{L_\infty(0,T)} &= \operatorname*{ess\,sup}_{t\in(0,T)} |\varphi(nt)| = \operatorname*{ess\,sup}_{s\in(0,nT)} |\varphi(s)| \\
&= \|\varphi\|_{L_\infty(0,nT)} = \|\varphi\|_{L_\infty(0,T)}, \\
\|\varphi(nt)\|_{L_1(0,T)} &= \int_0^T |\varphi(nt)|\,dt = \frac{1}{n}\int_0^{nT} |\varphi(nt)|\,d(nt) \\
&= \frac{1}{n}\|\varphi\|_{L_1(0,nT)} = \frac{1}{n}\|\varphi\|_{L_1(0,T)}, \\
\|(\varphi(nt))'\|_{L_1(0,T)} &= \|n\varphi'(nt)\|_{L_1(0,T)} = n \cdot \frac{1}{n}\|\varphi'\|_{L_1(0,T)} = \|\varphi'\|_{L_1(0,T)}.
\end{aligned}$$

Therefore

$$\begin{aligned}
\|g_n\|_{L_\infty(0,T;X)} &= \Big\| \|b\varphi(nt)\|_X \Big\|_{L_\infty(0,T)} \\
&= \|b\|_X \|\varphi(nt)\|_{L_\infty(0,T)} = \|b\|_X \|\varphi\|_{L_\infty(0,T)},
\end{aligned}$$

i.e. the set G is bounded in $L_\infty(0,T;X)$;

$$\begin{aligned}
\|g_n\|_{L_1(0,T;E)} &= \Big\| \|b\varphi(nt)\|_E \Big\|_{L_1(0,T)} \\
&= \|b\|_E \|\varphi(nt)\|_{L_1(0,T)} = \frac{1}{n}\|b\|_E \|\varphi\|_{L_1(0,T)},
\end{aligned}$$

i.e.

$$g_n \xrightarrow[n\to\infty]{} 0 \text{ in } L_1(0,T;E); \tag{2.2.19}$$

$$\|g_n'\|_{L_1(0,T;Y)} = \Big\| \|(b\varphi(nt))'\|_Y \Big\|_{L_1(0,T)}$$

$$= \|b\|_Y \|(\varphi(nt))'\|_{L_1(0,T)} = \|b\|_Y \|\varphi'\|_{L_1(0,T)},$$

i.e. the set $\{g_n'\}$ is bounded in $L_1(0,T;Y)$.

Thus, the set G satisfies conditions of Theorem 2.2.6 with $p=\infty$ and $r=1$. However the set G is not relatively compact either in $C([0,T];E)$ or in $L_\infty(0,T;E)$. Assume the contrary. Then there is a subsequence $\{g_{n_m}\} \subset G$ and an element $g \in L_\infty(0,T;E)$ such that $g_{n_m} \xrightarrow[m\to\infty]{} g$ in $L_\infty(0,T;E)$. It yields $g_{n_m} \xrightarrow[m\to\infty]{} g$ in $L_1(0,T;E)$, and (2.2.19) gives that $g=0$. So

$$g_{n_m} \xrightarrow[m\to\infty]{} 0 \text{ in } L_\infty(0,T;E).$$

On the other hand,

$$\|g_{n_m}\|_{L_\infty(0,T;E)} = \|b\|_E \|\varphi(nt)\|_{L_\infty(0,T)} = \|b\|_E \|\varphi\|_{L_\infty(0,T)}.$$

Passing to the limit as $m\to\infty$, we get

$$0 = \|b\|_E \|\varphi\|_{L_\infty(0,T)},$$

which contradicts the assumptions $b \neq 0$ and $\varphi \not\equiv 0$.

For the proof of Theorem 2.2.6 we need the following

Lemma 2.2.5. *Let $u \in L_r(0,T;Y)$, $r \ge 1$, $h \in (0,T)$. Then for the function*

$$J_h u(t) = \int_t^{t+h} u(s)\,ds = \int_0^h u(t+s)\,ds$$

the following estimates take place

$$\|J_h u\|_{L_p(0,T-h;Y)} \le T^{1-\frac{1}{r}} h^{\frac{1}{p}} \|u\|_{L_r(0,T;Y)}, \quad 1 \le p < \infty, \tag{2.2.20}$$

$$\|J_h u\|_{L_p(0,T-h;Y)} \le h^{1-\frac{1}{r}} \|u\|_{L_r(0,T;Y)}, \quad p=\infty. \tag{2.2.21}$$

Proof. Let k be a natural number such that $kh \ge T$. Define the function $\widetilde{u} : [0,kh] \to Y$ by the rule: $\widetilde{u}(t) = u(t)$ for $t \in [0,T]$, $u(t) = 0$ for $t > T$. We have:

$$\|J_h u\|_{L_p(0,T-h;Y)} \le \|J_h \widetilde{u}\|_{L_p(0,kh-h;Y)}$$

$$= \left(\int_0^{kh-h} \left(\int_0^h \|\widetilde{u}(t+s)\|\,ds \right)^p dt \right)^{\frac{1}{p}} \tag{2.2.22}$$

$$= \left(\sum_{j=0}^{k-2} \int_0^h \left(\int_0^h \|\widetilde{u}(t+jh+s)\|_Y\,ds \right)^p dt \right)^{\frac{1}{p}}.$$

Using the simple scalar inequality

$$\sum_{j=0}^{m} a_j^p \le \left(\sum_{j=0}^{m} a_j\right)^p, \quad a_0,\dots,a_m \ge 0, \ \ p \ge 1,$$

we conclude that the right-hand side of (2.2.22) does not exceed

$$\begin{aligned}
&\left(\int_0^h \left(\sum_{j=0}^{k-2} \int_0^h \|\widetilde{u}(t+jh+s)\|_Y\, ds\right)^p dt\right)^{\frac{1}{p}} \\
&= \left(\int_0^h \left(\int_0^{kh-h} \|\widetilde{u}(t+s)\|_Y\, ds\right)^p dt\right)^{\frac{1}{p}} \\
&= \left(\int_0^h \left(\int_t^{kh-h+t} \|\widetilde{u}(s)\|_Y\, ds\right)^p dt\right)^{\frac{1}{p}} \\
&\le \left(\int_0^h \left(\int_0^{kh} \|\widetilde{u}(s)\|_Y\, ds\right)^p dt\right)^{\frac{1}{p}} = \left(h\left(\int_0^T \|u(s)\|_Y\, ds\right)^p\right)^{\frac{1}{p}} \\
&= h^{\frac{1}{p}} \int_0^T \|u(s)\|_Y\, ds = h^{\frac{1}{p}} \|u\|_{L_1(0,T;Y)},
\end{aligned}$$

which does not exceed $h^{\frac{1}{p}} T^{1-\frac{1}{r}} \|u\|_{L_r(0,T;Y)}$ in view of Lemma 2.2.1, a). Estimate (2.2.20) is proved. We have then:

$$\|J_h u\|_{L_\infty(0,T-h;Y)} = \operatorname*{ess\,sup}_{t\in(0,T-h)} \|J_h u(t)\|_Y \le \operatorname*{ess\,sup}_{t\in(0,T-h)} \int_0^h \|u(t+s)\|_Y\, ds.$$

Lemma 2.2.1, a), gives that the right-hand side does not exceed

$$\begin{aligned}
\operatorname*{ess\,sup}_{t\in(0,T-h)} h^{1-\frac{1}{r}} \left(\int_0^h \|u(t+s)\|_Y^r\right)^{\frac{1}{r}} &\le h^{1-\frac{1}{r}} \left(\int_0^T \|u(s)\|_Y^r\, ds\right)^{\frac{1}{r}} \\
&= h^{1-\frac{1}{r}} \|u\|_{L_r(0,T;Y)}.
\end{aligned}$$

□

Proof of Theorem 2.2.6. It suffices to check the conditions of Theorem 2.2.5. The condition (2.2.14) coincides with (2.2.17).

For $f \in F$ we have $f' \in L_r(0,T;Y)$. Then by formula (2.2.2):

$$\begin{aligned}
(\tau_h f - f)(t) &= f(t+h) - f(t) \\
&= f(0) + \int_0^{t+h} f'(s)\, ds - f(0) - \int_0^t f'(s)\, ds = \int_t^{t+h} f'(s)\, ds.
\end{aligned}$$

So by Lemma 2.2.5:

$$\|\tau_h f - f\|_{L_p(0,T-h;Y)} \le T^{1-\frac{1}{r}} h^{\frac{1}{p}} \|f'\|_{L_r(0,T;Y)}, \quad 1 \le p \le \infty,$$

$$\|\tau_h f - f\|_{L_p(0,T-h;Y)} \le h^{1-\frac{1}{r}} \|f'\|_{L_r(0,T;Y)}, \quad p = \infty.$$

The right-hand sides tend to zero as $h \to 0$ uniformly with respect to $f \in F$, therefore (2.2.15) holds. □

2.2.6 Theorem on "partial" compactness

Using the same technique as has been just described, Simon [54] proved the following interesting result.

Theorem 2.2.7. *Let $X \subset E \subset Y$ be Banach spaces where the embedding $X \subset E$ is compact, and $E \subset Y$ continuously. Let $F \subset L_{1,\mathrm{loc}}(0,T;X) \cap L_p(0,T;E)$, $1 < p \le \infty$. Assume that for any $f \in F$ its generalized derivative f' in the space $\mathfrak{D}'(0,T;Y)$ belongs to $L_{1,\mathrm{loc}}(0,T;Y)$. Let*

$$F \text{ be bounded in } L_p(0,T;E). \tag{2.2.23}$$

Assume that for any $t_1, t_2 \in (0,T)$, $t_1 < t_2$, the set

$$\{f|_{[t_1,t_2]} \mid f \in F\} \text{ is bounded in } L_1(t_1,t_2;X), \tag{2.2.24}$$

while the set

$$\{f'|_{[t_1,t_2]} \mid f \in F\} \text{ is bounded in } L_1(t_1,t_2;Y). \tag{2.2.25}$$

Then for any $q < p$ F is relatively compact in $L_q(0,T;E)$.

2.2.7 Lemma on weak continuity of essentially bounded functions

The following lemma by J.-L. Lions and E. Magenes illustrates the subtle connection between boundedness and continuity.

Lemma 2.2.6. *Let E and E_0 be Banach spaces, $E \subset E_0$ continuously. Let the space E be reflexive. Then*

$$C_w([0,T];E_0) \cap L_\infty(0,T;E) \subset C_w([0,T];E).$$

Remark 2.2.8. See the proof in [61], Chapter III, Lemma 1.4, as well as [55], p. 232 where it is also shown that without the condition of reflexivity of E there may be no such embedding.

Remark 2.2.9. Lemma 2.2.6 shows that the values of functions from $C_w([0,T];E_0) \cap L_\infty(0,T;E)$ belong to E at every $t \in [0,T]$.

2.2.8 Lemma on differentiability of the quadrate of the norm of a vector function

Assume that we have two Hilbert spaces, $X \subset Y$ with continuous embedding operator $i : X \to Y$, and $i(X)$ is dense in Y. The adjoint operator $i^* : Y^* \to X^*$ is continuous and, since $i(X)$ is dense in Y, one-to-one. Since i is one-to-one, $i^*(Y^*)$ is dense in X^*, and one may identify Y^* with a dense subspace of X^*. Due to the Riesz representation theorem, one may also identify Y with Y^*. We arrive at the chain of inclusions:

$$X \subset Y \equiv Y^* \subset X^*. \tag{2.2.26}$$

Both embeddings here are dense and continuous. Observe that in this situation, for $f \in Y, u \in X$, their scalar product in Y coincides with the value of the functional f from X^* on the element $u \in X$:

$$(f, u)_Y = \langle f, u \rangle. \tag{2.2.27}$$

The following lemma is a particular case of a general interpolation theorem by Lions and Magenes.

Lemma 2.2.7. *If a function u belongs to $L_2(0, T; X)$ and its derivative u' belongs to $L_2(0, T; X^*)$, then $u \in C([0, T]; Y)$, and*

$$\frac{d}{dt}\|u\|_Y^2 = 2\langle u', u\rangle \tag{2.2.28}$$

in the sense of scalar distributions on $(0, T)$.

An elementary proof may be found in [61], Lemma 1.2 of Chapter 3.
The spaces V and H from Section 2.1.2 are a typical example of the spaces X and Y.
Lemma 2.2.7 and Corollary 2.2.1 (with $E = \mathbb{R}$) imply

Corollary 2.2.3. *If some set F is bounded in $L_2(0, T; X)$, and the set $\{f' | f \in F\}$ is bounded in $L_2(0, T; X^*)$, then F is bounded in $C([0, T]; Y)$.*

2.2.9 Two lemmas on absolutely continuous vector functions

We finish this chapter with two technical lemmas. The first one is close to Lemma 3.1.1 from [61].

Lemma 2.2.8. *Let X be a Banach space, and let $u, g \in L_1(0, T; X)$. Let Z be an everywhere dense set in the $*$-weak topology of X^*. Then, the following four conditions are equivalent:*

i) *For each* $v \in Z$,

$$\frac{d}{dt}\langle v, u\rangle = \langle v, g\rangle \tag{2.2.29}$$

in the sense of scalar distributions on $(0, T)$;

ii) *for each* $v \in X^*$, *the distributional derivative* $\frac{d}{dt}\langle v, u(t)\rangle$ *is integrable on* $(0, T)$ *and almost everywhere equal to* $\langle v, g(t)\rangle$;

iii) $g = u'$ *in the sense of Definition 2.2.1;*

iv) $u \in C([0, T]; X)$, *and*

$$u(t) = u(0) + \int_0^t g(s)\, ds, \quad t \in [0, T]. \tag{2.2.30}$$

Proof. i) → ii). Let $v \in X^*$ and $v_m \in Z$, $v_m \underset{m\to\infty}{\rightharpoonup} v$ $*$-weakly. Then $\frac{d}{dt}\langle v_m, u\rangle = \langle v_m, g\rangle$. Passing to the limit in the sense of scalar distributions on $(0, T)$, we conclude that $\frac{d}{dt}\langle v, u\rangle = \langle v, g\rangle$. Since the right-hand side is integrable, the left-hand one is also integrable, so the equality holds almost everywhere on $(0, T)$.

ii) → iii). Take $\psi \in \mathcal{D}(0, T)$ and $v \in X^*$. Then

$$\begin{aligned}\langle v, u'(\psi)\rangle &= -\langle v, u(\psi')\rangle = -\Big\langle v, \int_0^T \psi'(t)u(t)\, dt\Big\rangle \\ &= -\int_0^T \langle v, u(t)\rangle\psi'(t)\, dt = \int_0^T \frac{d}{dt}\langle v, u(t)\rangle\psi(t)\, dt \\ &= \int_0^T \langle v, g(t)\rangle\psi(t)\, dt = \Big\langle v, \int_0^T g(t)\psi(t)\, dt\Big\rangle = \langle v, g(\psi)\rangle.\end{aligned}$$

Since $\psi \in \mathcal{D}(0, T)$ and $v \in X^*$ are arbitrary, $u' = g$.

iii) → iv) follows from Lemma 2.2.3.

iv) → ii). We have:

$$\frac{d}{dt}\langle v, u(t)\rangle = \frac{d}{dt}\Big\langle v, u(0) + \int_0^t g(s)\, ds\Big\rangle = \langle v, g(t)\rangle.$$

ii) → i) is clear. □

Lemma 2.2.9. *Let* E *be a reflexive Banach space,* $u, v \in C([0, T]; E)$, *and* $v' \in L_p(0, T; E)$, $1 < p < \infty$. *If there is a constant* C *such that for all* $t, s \in [0, T]$ *one has*

$$\|u(t) - u(s)\|_E \le C\|v(t) - v(s)\|_E, \tag{2.2.31}$$

then $u' \in L_p(0, T; E)$.

Proof. We shall use the notations τ_h and M_a from Section 2.2.3. Denote by u_h, $h > 0$, the function $\frac{1}{h}(\tau_h u - u)$. Fix a number $T_1 \in (\frac{T}{2}, T)$. Let $0 \le t \le T_1$. Then

$$\int_0^t u_h(s)\,ds = \frac{1}{h}\int_t^{t+h} u(s)\,ds - \frac{1}{h}\int_0^h u(s)\,ds \xrightarrow[h\to 0,\ T_1 \le T-h]{} u(t) - u(0) \tag{2.2.32}$$

in E.

The set $\{v'\}$ is compact in $L_p(0, T; E)$. As in the proof of Theorem 2.2.3 one shows that $M_h v' \xrightarrow[h\to 0]{} v'$ in $L_p(0, T_1; E)$. Let $h_k \xrightarrow[k\to\infty]{} 0$. Then the sequence $v_{h_k} = M_{h_k} v' = \frac{1}{h_k}(\tau_{h_k} v - v)$ is bounded in $L_p(0, T_1; E)$. But due to (2.2.31):

$$\|u_{h_k}\|_{L_p(0,T_1;E)} \le C\|v_{h_k}\|_{L_p(0,T_1;E)},$$

so $\{u_{h_k}\}$ is also bounded. Since E is reflexive, $L_p(0, T_1; E)$ is also reflexive. Therefore without loss of generality we may assume that there is $g \in L_p(0, T_1; E)$ such that $u_{h_k} \rightharpoonup g$ weakly in $L_p(0, T_1; E)$. The linear operator $l : w \mapsto \int_0^t w(s)\,ds$, $0 \le t \le T_1$, from $L_p(0, T_1; E)$ to E, is bounded (cf. the proof of Theorem 2.2.3). Hence,

$$\int_0^t u_{h_k}(s)\,ds \xrightarrow[k\to\infty]{} \int_0^t g(s)\,ds$$

weakly in E. But (2.2.32) gives

$$u(t) = u(0) + \int_0^t g(s)\,ds,$$

i.e.

$$u' = g \in L_p(0, T_1; E).$$

Repeating these arguments for the functions $\widetilde{u}(t) = u(T - t)$ and $\widetilde{v}(t) = v(T - t)$, we conclude that $\widetilde{u}' \in L_p(0, T_1; E)$, i.e. $u' \in L_p(T - T_1, T; E)$. Thus, $u' \in L_p(0, T; E)$. □

Chapter 3

Operator equations in Banach spaces

This chapter contains miscellaneous facts on linear and nonlinear non-evolutionary and evolutionary operator equations in Banach spaces, which are used later during the investigation of equations of hydrodynamics.

3.1 Linear equations

3.1.1 The Lax–Milgram theorem

The following projection theorem (see [61], Theorem 2.2 of Chapter 1) is a good tool to prove existence of solutions for linear equations (of *weak* solutions for elliptic equations, especially).

Theorem 3.1.1. *Let W be a separable real Hilbert space and let $b : W \times W \to \mathbb{R}$ be a continuous bilinear form. Let b be coercive, i.e. there is $\alpha > 0$ such that*

$$b(u,u) \geq \alpha \|u\|_W^2, \quad \forall u \in W. \tag{3.1.1}$$

Then, for each $g \in W^*$, there exists unique $u \in W$ such that

$$b(u,v) = \langle g, v \rangle, \quad \forall v \in W. \tag{3.1.2}$$

3.1.2 Characterization of the gradient of a distribution

The following profound result is due to de Rham and J.-L. Lions ([37], p. 67, see also [52], [61], [82], [56]).

Lemma 3.1.1. *Let Ω be an arbitrary domain in $\mathbb{R}^n$, $n \in \mathbb{N}$, and let $g \in \mathcal{D}'(\Omega)^n$. Then,*

$$\langle g, v \rangle = 0 \textit{ for any } v \in C_0^\infty(\Omega)^n \textit{ with } \operatorname{div} v = 0 \tag{3.1.3}$$

if and only if there exists $q \in \mathcal{D}'(\Omega)$ such that

$$\operatorname{grad} q = g. \tag{3.1.4}$$

The basic regularity result for equation (3.1.4) is

Lemma 3.1.2 (see [41], p. 320). *Let Ω be a sufficiently regular bounded domain in $\mathbb{R}^n$. If $q \in \mathcal{D}'(\Omega)$ and* $\operatorname{grad} q \in H^{-1}(\Omega)^n$, *then $q \in L_2(\Omega)$.*

Corollary 3.1.1. *Assume that Ω is the union of a finite number of sufficiently regular bounded domains in $\mathbb{R}^n$. If $q \in \mathcal{D}'(\Omega)$ and* $\operatorname{grad} q \in H^{m-2}(\Omega)^n$, *$m \in \mathbb{N}$, then $q \in H^{m-1}(\Omega)$.*

Proof. Let $\Omega = \bigcup_{k=1}^{l} \Omega_k$ where $\Omega_k, k = 1, \ldots, l$, are sufficiently regular bounded domains. Let $q \in \mathcal{D}'(\Omega)$, and $g = \operatorname{grad} q \in H^{m-2}(\Omega)^n$, $m \in \mathbb{N}$. In particular, g belongs to $H^{-1}(\Omega)^n$, and its restriction $g|_{\Omega_k} \in H^{-1}(\Omega_k)^n$ can be defined as

$$\langle g|_{\Omega_k}, \varphi\rangle_{H^{-1}(\Omega_k)^n \times H_0^1(\Omega_k)^n} = \langle g, \widetilde{\varphi}\rangle_{H^{-1}(\Omega)^n \times H_0^1(\Omega)^n}$$

where $\varphi \in H_0^1(\Omega_k)^n$, and $\widetilde{\varphi} \in H_0^1(\Omega)^n$ coincides with φ almost everywhere on Ω_k and is identically zero outside Ω_k. Observe that $g|_{\Omega_k} = \operatorname{grad}(q|_{\Omega_k})$ where

$$\langle q|_{\Omega_k}, \phi\rangle_{\mathcal{D}'(\Omega_k)\times\mathcal{D}(\Omega_k)} = \langle q, \widetilde{\phi}\rangle_{\mathcal{D}'(\Omega)\times\mathcal{D}(\Omega)}.$$

Here $\phi \in \mathcal{D}(\Omega_k)$, and $\widetilde{\phi} \in \mathcal{D}(\Omega)$ coincides with ϕ on Ω_k and is identically zero outside Ω_k. By Lemma 3.1.2, $q|_{\Omega_k} \in L_2(\Omega_k)$ (for each k). Therefore $q \in L_2(\Omega)$. Since $\operatorname{grad} q \in H^{m-2}(\Omega)^n$, we conclude that $q \in H^{m-1}(\Omega)$. □

Corollary 3.1.2. *Let Ω be as in Corollary 3.1.1. Denote the connected components of Ω by Ω^k, $k = 1, 2, \ldots$. Fix a function $\vartheta \in L_2(\Omega)$ with $\int_{\Omega^k} \vartheta(x)\,dx \neq 0$ for each k. Denote by $E_{\vartheta,\Omega}$ the subspace of $H^{m-1}(\Omega)$ which is the intersection of the kernels of the functionals $(\cdot, \vartheta)_{L_2(\Omega^k)}$. Denote by $H_G^{m-2}(\Omega)^n$ the subspace of $H^{m-2}(\Omega)^n$ which consists of the functions satisfying condition (3.1.3). Then the operator* grad *is an isomorphism of $E_{\vartheta,\Omega}$ onto $H_G^{m-2}(\Omega)^n$.*

Proof. The operator $\operatorname{grad} : E_{\vartheta,\Omega} \subset H^{m-1}(\Omega) \to H_G^{m-2}(\Omega)^n \subset H^{m-2}(\Omega)^n$ is continuous. Given $g \in H_G^{m-2}(\Omega)^n$, by Lemma 3.1.1 and Corollary 3.1.1, there exists $q \in H^{m-1}(\Omega)$ satisfying (3.1.4). Then the function q_* defined as

$$q_*(x) = q(x) - \frac{(q, \vartheta)_{L_2(\Omega^k)}}{\int_{\Omega^k} \vartheta(x)\,dx}, \quad x \in \Omega^k,$$

belongs to $E_{\vartheta,\Omega}$, and $\operatorname{grad} q_* = g$. Thus, the operator $\operatorname{grad} : E_{\vartheta,\Omega} \to H_G^{m-2}(\Omega)^n$ is surjective. It remains to show that its kernel is trivial. Let q belong to the kernel. For each k one has $\operatorname{grad} q(x) = 0$ at almost all $x \in \Omega^k$, so $q(x)$ does not depend on $x \in \Omega^k$ (due to connectedness of Ω^k). Hence, $0 = (q, \vartheta)_{L_2(\Omega^k)} = q|_{\Omega^k} \int_{\Omega^k} \vartheta(x)\,dx$ for each k, so $q \equiv 0$. □

Corollary 3.1.3. *Under the conditions of Corollary 3.1.2, the operator* grad *is an isomorphism of $Z(0, T; E_{\vartheta,\Omega})$ onto $Z(0, T; H_G^{m-2}(\Omega)^n)$ where the symbol Z stands for C, C^r ($r \in \mathbb{N}$) or W_p^r ($r \in \mathbb{Z}, 1 \le p \le \infty$).*

The corollary follows straightforwardly from Corollary 3.1.2.

Now, let Z stand for C, C^r, L_p or W_p^r ($r \in \mathbb{N}, 1 \le p \le \infty$). For any domain $\Omega \subset \mathbb{R}^n$, let the symbol $Z(0, T; H_{\mathrm{loc}}^{m-1}(\Omega))$ ($m \in \mathbb{N}$) denote the set of functions u which

i) are defined almost everywhere on $(0, T)$,

ii) possess values in the set of Lebesgue measurable functions on Ω,

iii) for every open ball B such that the closure of B is contained in Ω, the restrictions $u|_B$ (defined as $u|_B(t) = u(t)|_B$ for almost all $t \in (0, T)$) belong to $Z(0, T; H^{m-1}(B))$.

Corollary 3.1.4. *Let Ω be a connected domain in $\mathbb{R}^n$. Let a function $\vartheta \in L_2(\Omega)$ have a compact support, and $\int_\Omega \vartheta(x)\ dx \neq 0$. Then for any $g \in Z(0, T; H_G^{m-2}(\Omega)^n)$ there exists a unique function $q \in Z(0, T; H_{\mathrm{loc}}^{m-1}(\Omega))$ such that $g(t) = \operatorname{grad} q(t)$ and $(q(t), \vartheta)_{L_2(\Omega)} = 0$ for almost all $t \in (0, T)$.*

Proof. Let $\mathcal{B}$ be the collection of all domains satisfying the properties

i) it may be represented as union of a finite number of sufficiently regular bounded domains;

ii) it is contained in Ω;

iii) it contains the support of ϑ;

iv) it is connected.

Let us show that for any open ball B such that its closure $\overline{B}$ is contained in Ω, there is a set from $\mathcal{B}$ which contains $\overline{B}$. Really, for any point $x \in \overline{B} \cup \operatorname{supp} \vartheta$ there is a ball $B(x) \subset \Omega$. Since $\overline{B} \cup \operatorname{supp} \vartheta$ is a compact set, its covering $\{B(x)\}$ has a finite subcovering $B(x_1), \dots, B(x_k)$. Since Ω is an open connected set, any two balls from this subcovering can be connected with a broken line within Ω. The union of small regular neighbourhoods of these broken lines and of the balls $B(x_1), \dots, B(x_k)$ belongs to $\mathcal{B}$ and contains $\overline{B}$.

Let $g \in Z(0, T; H_G^{m-2}(\Omega)^n)$. By Corollary 3.1.3, for any domain ω from $\mathcal{B}$, there is a unique function $q_\omega \in Z(0, T; H^{m-1}(\omega))$ such that $g(t) = \operatorname{grad} q_\omega(t)$ and $(q_\omega(t), \vartheta)_{L_2(\omega)} = 0$ for almost all $t \in (0, T)$. Then one can define the required function q by the formula $q(t)(x) = q_\omega(t)(x)$ where ω is a domain from $\mathcal{B}$ containing x. This definition is consistent, i.e. $q(t)(x)$ does not depend on the choice of ω. In fact, for any two domains $\omega_1, \omega_2 \in \mathcal{B}$ containing x, one has $\omega_1 \cup \omega_2 \in \mathcal{B}$, and $q_{\omega_1} = q_{\omega_1 \cup \omega_2}\big|_{\omega_1}$, $q_{\omega_2} = q_{\omega_1 \cup \omega_2}\big|_{\omega_2}$. Thus, $q_{\omega_1}(t)(x) = q_{\omega_1 \cup \omega_2}(t)(x) = q_{\omega_2}(t)(x)$. Note that $q|_B \in Z(0, T; H^{m-1}(B))$ for any open ball B such that its closure $\overline{B}$ is contained in Ω, so $q \in Z(0, T; H_{\mathrm{loc}}^{m-1}(\Omega))$. Uniqueness of each q_ω implies the uniqueness of q. □

Corollary 3.1.5. *Let Ω be an arbitrary domain in $\mathbb{R}^n$. Denote the connected components of Ω by Ω^k, $k = 1, 2, \dots$. Let a function $\vartheta \in L_2(\Omega)$ have a compact support, and $\int_{\Omega^k} \vartheta(x)\, dx \neq 0$ for each k. Then for any $g \in Z(0, T; H_G^{m-2}(\Omega)^n)$ there exists a unique function $q \in Z(0, T; H_{\text{loc}}^{m-1}(\Omega))$ such that $g(t) = \operatorname{grad} q(t)$ and $(q(t), \vartheta)_{L_2(\Omega^k)} = 0$ for each k and almost all $t \in (0, T)$.*

It results from application of Corollary 3.1.4 to each connected component of Ω.

3.1.3 An existence lemma

Assume that there are two Hilbert spaces, $X \subset Y$ with continuous embedding operator $i : X \to Y$, and $i(X)$ is dense in Y. Then, as in Subsection 2.2.8, we have (2.2.26) and (2.2.27).

The following lemma is a particular case of Theorem 1.1 from [20], Chapter VI.

Lemma 3.1.3. *Let X be separable and let $A : X \to X^*$ be a continuous linear operator. Assume that there is $\alpha > 0$ such that*

$$\langle Au, u \rangle \geq \alpha \|u\|_X^2, \quad \forall u \in X. \tag{3.1.5}$$

Then, given $a \in Y$ and $f \in L_2(0, T; X^*)$, the Cauchy problem

$$u'(t) + Au(t) = f(t), \quad t \in (0, T), \tag{3.1.6}$$

$$u(0) = a \tag{3.1.7}$$

possesses a unique solution in the class

$$u \in L_2(0, T; X), \quad u' \in L_2(0, T; X^*). \tag{3.1.8}$$

Remark 3.1.1. Due to Lemma 2.2.7, the solution u belongs to $C([0, T]; Y)$, so the initial condition (3.1.7) holds in the usual sense.

3.1.4 Strongly positive operators and parabolic equations

In this subsection we discuss some properties of abstract parabolic equations.

Let E be a Banach space. In the remainder of the chapter, we write $\|\cdot\|$ for the norm in E and for the norm of a bounded linear operator in E.

We consider the autonomous equation of the form:

$$\frac{dv}{dt} + \mathcal{A}(t)v = 0 \tag{3.1.9}$$

Here $v : [0, T] \to E$ is an unknown function; $\mathcal{A}(t) : D(\mathcal{A}) \subset E \to E$ is a linear unbounded operator for each $t \in [0, T]$.

We also consider the non-autonomous equation

$$\frac{dv}{dt} + \mathcal{A}(t)v = f(t) \tag{3.1.10}$$

where $f : [0, T] \to E$ is a given function.

Remark 3.1.2. The space E may be both complex and real. In the latter case, however, for issues concerning eigenvalues, complex numbers etc. one has to consider the complexification of the space E and the corresponding extension of the operator $\mathcal{A}$.

Definition 3.1.1 ([33]). A linear unbounded operator $\mathcal{B}$ with a dense domain in E is called *strongly positive* if for any complex number λ with Re $\lambda \geq 0$ the operator $\mathcal{B} + \lambda I$ has a bounded inverse operator and

$$\|(\mathcal{B} + \lambda I)^{-1}\| \leq \frac{K_1}{1 + |\lambda|}, \tag{3.1.11}$$

where the constant K_1 does not depend on λ.

A strongly positive operator $\mathcal{B}$ generates an analytical semigroup $e^{-\mathcal{B}t}$, $t \geq 0$, and it is possible to define its fractional powers (see [33, 58]):

$$\mathcal{B}^\alpha = \frac{1}{\Gamma(-\alpha)} \int_0^\infty \tau^{-\alpha-1} e^{-\tau\mathcal{B}} d\tau \quad (\alpha < 0), \tag{3.1.12}$$

where Γ is Euler's gamma-function;

$$\mathcal{B}^\alpha = (\mathcal{B}^{-\alpha})^{-1} \quad (\alpha > 0); \tag{3.1.13}$$

it is also assumed that $\mathcal{B}^0 = I$ (the identity operator). At $\alpha > 0$ the operators $\mathcal{B}^\alpha$ are not bounded. Their domains $D(\mathcal{B}^\alpha)$ are dense in E, and $D(\mathcal{B}^\alpha) \subset D(\mathcal{B}^\beta)$ for $\alpha > \beta$.

Theorem 3.1.2 (see [33], Theorem 14.8). *Let $\mathcal{B}_1$, $\mathcal{B}_2$ be two strongly positive operators in E. Assume that*

$$D(\mathcal{B}_1) \supset D(\mathcal{B}_2), \tag{3.1.14}$$

$$\|\mathcal{B}_1 u\| \leq K_2 \|\mathcal{B}_2 u\|, \; u \in D(\mathcal{B}_2), \tag{3.1.15}$$

where K_2 is independent of u. Then for any $0 \leq \varepsilon_1 < \varepsilon_2 \leq 1$,

$$D(\mathcal{B}_1^{\varepsilon_1}) \supset D(\mathcal{B}_2^{\varepsilon_2}), \tag{3.1.16}$$

$$\|\mathcal{B}_1^{\varepsilon_1} u\| \leq K_3(\varepsilon_1, \varepsilon_2) \|\mathcal{B}_2^{\varepsilon_2} u\|, \; u \in D(\mathcal{B}_2^{\varepsilon_2}), \tag{3.1.17}$$

where $K_3(\varepsilon_1, \varepsilon_2)$ is independent of u, $\mathcal{B}_1$, $\mathcal{B}_2$ but depends on K_2 (the notation $K_3(\varepsilon_1, \varepsilon_2)$ certainly means that K_3 depends on ε_1 and ε_2).

Definition 3.1.2. If an operator $\mathcal{A}(t)$ has a dense in E domain D which is independent of t, if it is strongly positive for every t, and if K_1 in (3.1.11) does not depend on t, we call $\mathcal{A}(t)$ *uniformly strongly positive* (with respect to t).

The following four results are due to Sobolevskii (see [58]).

Theorem 3.1.3. *Let an operator $\mathcal{A}(t)$ be uniformly strongly positive. Assume that for some $\varepsilon > 0$ and every $t, \tau, s \in [0, T]$ one has:*

$$\|(\mathcal{A}(t) - \mathcal{A}(\tau))\mathcal{A}^{-1}(s)\| \leq C|t - \tau|^{\varepsilon} \tag{3.1.18}$$

(hereafter in this subsection C stands for various constants which do not depend on t, s, τ).

Then there exists a function $U(t, \tau)$ (the "resolving operator" for $\mathcal{A}$) with values in the space of linear bounded operators in E, which is defined and is strongly continuous on the triangle $0 \leq \tau \leq t \leq T$. This operator function is uniformly differentiable with respect to t for $t > \tau$, and

$$\frac{\partial U(t, \tau)}{\partial t} + \mathcal{A}(t)U(t, \tau) = 0. \tag{3.1.19}$$

The following identities hold for $0 \leq \tau \leq s \leq t \leq T$:

$$U(t, \tau) = U(t, s)U(s, \tau), \tag{3.1.20}$$

$$U(t, t) = I. \tag{3.1.21}$$

For any $v_0 \in E$, the formula

$$v(t) = U(t, 0)v_0 \tag{3.1.22}$$

determines a solution of equation (3.1.9) which is continuous on $[0, T]$ and continuously differentiable for $t > 0$. If $v_0 \in D$, then $v \in C^1([0, T]; E)$.

Theorem 3.1.4. *Under the conditions of Theorem 3.1.3, the following estimates are valid:*

$$\|\mathcal{A}^{\alpha}(t)U(t, \tau)\mathcal{A}^{-\beta}(\tau)\| \leq C|t - \tau|^{\beta-\alpha}, \tag{3.1.23}$$

where $0 \leq \tau \leq t \leq T$, $0 \leq \beta \leq \alpha < 1 + \varepsilon$;

$$\|\mathcal{A}^{\alpha}(0)(U(t + \Delta t, 0) - U(t, 0))\mathcal{A}^{-\beta}(0)\| \leq C\Delta t^{\beta-\alpha}, \tag{3.1.24}$$

where $0 \leq t \leq t + \Delta t \leq T$, $0 \leq \alpha \leq \beta \leq 1$.

Theorem 3.1.5. *Let $\mathcal{A}(t)$ satisfy the conditions of Theorem 3.1.3. Let $f \in C([0, T]; E)$ and let $\delta > 0$ be such that*

$$\|f(t) - f(\tau)\| \leq C|t - \tau|^{\delta} \tag{3.1.25}$$

for all $t, \tau \in [0, T]$.

Given $v_0 \in E$, the function

$$v(t) = U(t,0)v_0 + \int_0^t U(t,s)f(s)ds \tag{3.1.26}$$

is continuous on $[0, T]$ and continuously differentiable for $t > 0$ solution of equation (3.1.10). The function $\mathcal{A}(t)v(t)$ is continuous at $t > 0$.

If $v_0 \in D$, then

$$v \in C^1([0, T]; E), \tag{3.1.27}$$

$$\mathcal{A}(\cdot)v \in C([0, T]; E). \tag{3.1.28}$$

If $v_0 \in D(\mathcal{A}^{1+\alpha}(0))$ where $\alpha < \min(\delta, \varepsilon)$, and $\mathcal{A}^\alpha(0)f \in C([0, T]; E)$, then

$$\mathcal{A}^\alpha(0)\mathcal{A}(t)v(t) \in C([0, T]; E). \tag{3.1.29}$$

Theorem 3.1.6. *Under the conditions of the Theorem 3.1.5, one has the following inequalities*

$$\begin{aligned} &\left\| \mathcal{A}^\alpha(0) \left[\int_\tau^{t+\Delta t} U(t+\Delta t, s) f(s)\,ds - \int_\tau^t U(t,s) f(s)\,ds \right] \right\| \\ &\quad \le K_4(\alpha)\Delta t^{1-\alpha}(|\ln \Delta t| + 1)\|f\|_{C([\tau,t+\Delta t];E)}, \end{aligned} \tag{3.1.30}$$

where $0 \le \tau \le t \le t + \Delta t \le T$, $0 \le \alpha < 1$,

$$\left\| \frac{\partial}{\partial t} \int_\tau^t U(t,s)f(s)ds \right\| \le K_4' \cdot (\|f(t)\| + C), \tag{3.1.31}$$

where $0 \le \tau \le t \le T$ and C is the constant from (3.1.25).

Remark 3.1.3. The conditions of Theorem 3.1.3 may be replaced by the following one: the operator $\mathcal{A}(0)$ is strongly positive and for all $t, \tau \in [0, T]$

$$\|(\mathcal{A}(t) - \mathcal{A}(\tau))\mathcal{A}^{-1}(0)\| \le C|t - \tau|^\varepsilon \tag{3.1.32}$$

provided

$$(1 + K_1)CT^\varepsilon < 1, \tag{3.1.33}$$

where C and K_1 are the constants from (3.1.32) and estimate (3.1.11) for $\mathcal{A}(0)$. Let us show this. Take any complex λ with $\operatorname{Re}\lambda \ge 0$ and $s \in [0, T]$. We have:

$$\begin{aligned} &\|([\mathcal{A}(s) + \lambda I] - [\mathcal{A}(0) + \lambda I])(\mathcal{A}(0) + \lambda I)^{-1}\| \\ &\quad \le \|(\mathcal{A}(s) - \mathcal{A}(0))\mathcal{A}^{-1}(0)\| \|\mathcal{A}(0)(\mathcal{A}(0) + \lambda I)^{-1}\| \\ &\quad \le \|(\mathcal{A}(s) - \mathcal{A}(0))\mathcal{A}^{-1}(0)\| \|I - \lambda(\mathcal{A}(0) + \lambda I)^{-1}\| \\ &\quad \le Cs^\varepsilon\big(1 + \frac{|\lambda|K_1}{1 + |\lambda|}\big) \le (1 + K_1)CT^\varepsilon, \end{aligned}$$

whence

$$\|[\mathcal{A}(s) + \lambda I](\mathcal{A}(0) + \lambda I)^{-1} - I\| \le (1 + K_1)CT^{\varepsilon} < 1.$$

As is well known, this yields that the operator $[\mathcal{A}(s)+\lambda I](\mathcal{A}(0)+\lambda I)^{-1}$ is invertible (and, hence, $\mathcal{A}(s) + \lambda I$ is also invertible), and we have the following bound for the norm of the inverse operator:

$$\|[\mathcal{A}(0) + \lambda I](\mathcal{A}(s) + \lambda I)^{-1}\| \le \frac{1}{1 - (1 + K_1)CT^{\varepsilon}}. \tag{3.1.34}$$

Thus, by estimate (3.1.11) for $\mathcal{A}(0)$,

$$\begin{aligned} \|(\mathcal{A}(s) + \lambda I)^{-1}\| &\le \|[\mathcal{A}(0) + \lambda I]^{-1}\| \|[\mathcal{A}(0) + \lambda I](\mathcal{A}(s) + \lambda I)^{-1}\| \\ &\le \frac{K_1/(1 + |\lambda|)}{1 - (1 + K_1)CT^{\varepsilon}}, \end{aligned}$$

so $\mathcal{A}(s)$ is uniformly strongly positive.

Using (3.1.32) and (3.1.34) for $\lambda = 0$, we obtain

$$\begin{aligned} \|(\mathcal{A}(t) - \mathcal{A}(\tau))\mathcal{A}^{-1}(s)\| &\le \|(\mathcal{A}(t) - \mathcal{A}(\tau))\mathcal{A}^{-1}(0)\| \|\mathcal{A}(0)\mathcal{A}^{-1}(s)\| \\ &\le \frac{C}{1 - (1 + K_1)CT^{\varepsilon}}|t - \tau|^{\varepsilon}. \end{aligned}$$

This estimate is similar to (3.1.18).

Remark 3.1.4. Let us show that (3.1.28) may be rewritten as

$$\mathcal{A}(0)v \in C([0, T]; E). \tag{3.1.35}$$

Really, for $0 \le t, t + \Delta t \le T$, one has

$$\begin{aligned} \|\mathcal{A}(0)(v(t + \Delta t) - v(t))\| &\le \|\mathcal{A}(0)\mathcal{A}^{-1}(t)\| \, \|\mathcal{A}(t)(v(t + \Delta t) - v(t))\| \\ &\le \|\mathcal{A}(0)\mathcal{A}^{-1}(t)\| \, (\|(\mathcal{A}(t) - \mathcal{A}(t + \Delta t))v(t + \Delta t)\| \\ &\quad + \|\mathcal{A}(t + \Delta t)v(t + \Delta t) - \mathcal{A}(t)v(t)\|) \\ &\le C(\|(\mathcal{A}(t) - \mathcal{A}(t + \Delta t))\mathcal{A}^{-1}(t + \Delta t)\| \|\mathcal{A}(t + \Delta t)v(t + \Delta t)\| \\ &\quad + \|\mathcal{A}(t + \Delta t)v(t + \Delta t) - \mathcal{A}(t)v(t)\|) \\ &\le C|\Delta t|^{\varepsilon} + \|\mathcal{A}(t + \Delta t)v(t + \Delta t) - \mathcal{A}(t)v(t)\| \xrightarrow[\Delta t \to 0]{} 0. \end{aligned}$$

Remark 3.1.5. Inequality (3.1.31) immediately implies

$$\begin{aligned}
&\|\mathcal{A}(0)\int_0^t U(t,s)f(s)ds\| \\
&\le \|\mathcal{A}(0)\mathcal{A}^{-1}(t)\|\|\mathcal{A}(t)\int_0^t U(t,s)f(s)ds\| \\
&= \|\mathcal{A}(0)\mathcal{A}^{-1}(t)\|\|\frac{\partial}{\partial t}\int_0^t U(t,s)f(s)ds - f(t)\| \\
&\le C(K_4'\cdot(\|f(t)\|+C)+\|f(t)\|).
\end{aligned} \tag{3.1.36}$$

3.2 Nonlinear equations

3.2.1 An existence theorem

We will need the following existence theorem for nonlinear equations of parabolic type.

Theorem 3.2.1. a) *Let $\mathcal{B} : D(\mathcal{B}) \to E$ be a strongly positive operator in a Banach space E. Let α and R be some numbers, $0 \le \alpha < 1$, $R > 0$. Assume that for every $v \in D(\mathcal{B}^\alpha)$ such that $\|\mathcal{B}^\alpha v\| < R$ there is a linear operator $\mathcal{A}(v)(\cdot) : D(\mathcal{B}) \to E$, and*

$$\|(\mathcal{A}(v) - \mathcal{A}(w))\mathcal{B}^{-1}\| \le K_5\|\mathcal{B}^\alpha(v-w)\|, \tag{3.2.1}$$

where K_5 does not depend on $v, w \in D(\mathcal{B}^\alpha)$ such that $\|\mathcal{B}^\alpha v\|$, $\|\mathcal{B}^\alpha w\| < R$.
Assume that for every v_0 such that

$$v_0 \in D(\mathcal{B}),\ \|\mathcal{B}^\alpha v_0\| < R, \tag{3.2.2}$$

the linear operator $\mathcal{A}_0 = \mathcal{A}(v_0)$ is strongly positive and the constant K_1 in (3.1.11) does not depend on v_0. Assume also that for all $v \in D(\mathcal{B})$:

$$K_6\|\mathcal{A}_0 v\| \le \|\mathcal{B}v\| \le K_7\|\mathcal{A}_0 v\|, \tag{3.2.3}$$

where K_6, K_7 do not depend on v, v_0.
Let $f_1 : [0,T] \times D(\mathcal{B}^\alpha) \to E$ and

$$\|f_1(t,v) - f_1(s,w)\| \le K_8(\|\mathcal{B}^\alpha(v-w)\| + \|t-s\|), \tag{3.2.4}$$

where K_8 does not depend on t, s and $v, w \in D(\mathcal{B}^\alpha)$ such that $\|\mathcal{B}^\alpha v\|$, $\|\mathcal{B}^\alpha w\|$ $< R$.

Then, for any v_0 satisfying (3.2.2), there exists a solution v of the problem

$$\frac{dv}{dt} + \mathcal{A}(v)v = f_1(t, v) \tag{3.2.5}$$

$$v(0) = v_0 \tag{3.2.6}$$

in the class

$$v \in C^1([0, t_0]; E), \ \mathcal{B}v \in C([0, t_0]; E), \tag{3.2.7}$$

where $t_0 = t_0(v_0)$ is some positive number.

b) *One can take $t_0 = T$ provided it is a priori known that the solution v is bounded in the following sense:*

$$\begin{aligned} \|\mathcal{B}^{\beta} v(t)\| &< K_9, \\ \|\mathcal{B}^{\alpha} v(t)\| &< K_9', \end{aligned} \tag{3.2.8}$$

where β is some fixed number, $\alpha < \beta \le 1$, K_9 does not depend on t_0 and on $t \in [0, t_0]$, and $K_9' < R$.

Remark 3.2.1. As a matter of fact, this theorem is very close to a particular case of Theorem 7 from [58]. However, we give here the proof.

Proof. Take v_0 satisfying (3.2.2), and consider the set $Q = Q(t_0, \eta)$ which consists of the functions $v \in C([0, t_0]; E)$ with

$$v(0) = \mathcal{B}^{\alpha} v_0 \tag{3.2.9}$$

such that for any $t, \tau \in [0, t_0]$

$$\|v(t) - v(\tau)\| \le |t - \tau|^{\eta} \tag{3.2.10}$$

(the parameters $t_0, \eta > 0$ will be defined later). The set Q is closed in $C([0, t_0]; E)$.

Since

$$\|v(0)\| = \|\mathcal{B}^{\alpha} v_0\| < R,$$

(3.2.10) implies

$$\|v(t)\| < R \tag{3.2.11}$$

for all $t \in [0, t_0]$, $v \in Q$, provided

$$t_0^{\eta} < R - \|\mathcal{B}^{\alpha} v_0\|.$$

Hence, for $v \in Q$ there is the operator function

$$\mathcal{A}_v(t) = \mathcal{A}(\mathcal{B}^{-\alpha} v(t)) \tag{3.2.12}$$

defined on $D(\mathcal{B})$ for all $t \in [0, t_0]$. Now, (3.2.1), (3.2.3) and (3.2.10) yield

$$\|(\mathcal{A}_v(t) - \mathcal{A}_v(\tau))\mathcal{A}_0^{-1}\| \le \|(\mathcal{A}_v(t) - \mathcal{A}_v(\tau))\mathcal{B}^{-1}\| \, \|\mathcal{B}\mathcal{A}_0^{-1}\| \le K_5 K_7 |t - \tau|^{\eta}. \tag{3.2.13}$$

By (3.2.9)

$$\mathcal{A}_v(0) = \mathcal{A}(v_0) = \mathcal{A}_0, \tag{3.2.14}$$

and $\mathcal{A}_0$ is strongly positive, so (3.2.13) implies (see Remark 3.1.3) that, for t_0 sufficiently small and $t \in [0, t_0]$, the operator $\mathcal{A}_v(t)$ is uniformly strongly positive. Furthermore,

$$\|(\mathcal{A}_v(t) - \mathcal{A}_v(\tau))\mathcal{A}_v^{-1}(s)\| \leq K_{10}|t - \tau|^{\eta} \tag{3.2.15}$$

for all $t, \tau, s \in [0, t_0]$. By Theorem 3.1.3 there is a resolving operator $U_v(t, \tau)$ for $\mathcal{A}_v$. Then (3.1.24) gives

$$\|\mathcal{A}_0^{\alpha_1}(U_v(t + \Delta t, 0) - U_v(t, 0))\mathcal{A}_0^{-\beta_1}\| \leq K_{11}\Delta t^{\beta_1 - \alpha_1} \tag{3.2.16}$$

for $0 \leq t \leq t + \Delta t \leq t_0$, $0 \leq \alpha_1 \leq \beta_1 \leq 1$.

Let $f_v(t) = f_1(t, \mathcal{B}^{-\alpha} v(t))$. Then (3.2.4) and (3.2.10) imply:

$$\|f_v(t) - f_v(\tau)\| \leq K_{12}\|t - \tau\|^{\eta}. \tag{3.2.17}$$

Furthermore, by (3.2.4) and (3.2.11),

$$\begin{aligned} \|f_v(t)\| &\leq \|f_1(t, \mathcal{B}^{-\alpha} v(t)) - f_1(0, 0)\| + \|f_1(0, 0)\| \\ &\leq K_8(R + T) + \|f_1(0, 0)\| = K_{13}. \end{aligned} \tag{3.2.18}$$

Since K_5, K_8, K_{10} do not depend on s, t, τ, v and v_0, the constants K_{11}, K_{12}, K_{13} are also independent of these variables.

By (3.1.30) we have the inequality:

$$\begin{aligned} &\left\| \mathcal{A}_0^{\alpha_1} \left[\int_0^{t+\Delta t} U_v(t + \Delta t, s) f_v(s)\, ds - \int_0^t U_v(t, s) f_v(s)\, ds \right] \right\| \\ &\quad \leq K_{14}\Delta t^{1-\alpha_1}(|\ln \Delta t| + 1)\|f_v\|_{C([0,t_0];E)}, \quad 0 \leq \alpha_1 < 1. \end{aligned} \tag{3.2.19}$$

Introduce the operator $w : Q \to C([0, t_0]; E)$ by the formula:

$$w(v)(t) = \mathcal{B}^{\alpha} U_v(t, 0)v_0 + \mathcal{B}^{\alpha} \int_0^t U_v(t, s) f_v(s)\, ds. \tag{3.2.20}$$

Let $\alpha_1, \beta_1, \eta_1, \eta$ be numbers such that $\alpha < \alpha_1 < \alpha_1 + \eta < \alpha_1 + \eta_1 < \beta_1 < \beta$.

By (3.2.3) and Theorem 3.1.2 we have:

$$\|\mathcal{A}_0^{\gamma_1} u\| \leq K_{15}(\gamma_1, \gamma)\|\mathcal{B}^{\gamma} u\|, \tag{3.2.21}$$

$$\|\mathcal{B}^{\gamma_1} u\| \leq K_{16}(\gamma_1, \gamma)\|\mathcal{A}_0^{\gamma} u\| \tag{3.2.22}$$

for $0 < \gamma_1 < \gamma < 1$.

Using (3.2.16), (3.2.18) – (3.2.22) we get:

$$\begin{aligned}
&\|w(v)(t+\Delta t)-w(v)(t)\| \\
&\quad \le \|\mathcal{B}^{\alpha}\mathcal{A}_0^{-\alpha_1}\|\Big(\|\mathcal{A}_0^{\alpha_1}(U_v(t+\Delta t,0)-U_v(t,0))\mathcal{A}_0^{-\beta_1}\|\|\mathcal{A}_0^{\beta_1}\mathcal{B}^{-\beta}\|\|\mathcal{B}^{\beta}v_0\| \\
&\qquad + \Big\|\mathcal{A}_0^{\alpha_1}\Big(\int_0^{t+\Delta t} U_v(t+\Delta t,s)f_v(s)\,ds-\int_0^t U_v(t,s)f_v(s)\,ds\Big)\Big\|\Big) \\
&\quad \le K_{17}(\Delta t^{\beta_1-\alpha_1}+\Delta t^{1-\alpha_1}(|\ln\Delta t|+1)) \le K_{18}\Delta t^{\eta}t_0^{\eta_1-\eta},
\end{aligned}$$

where K_{18} depends only on $\|\mathcal{B}^{\beta}v_0\|$.

For t_0 small enough,

$$\|w(v)(t+\Delta t)-w(v)(t)\| \le \Delta t^{\eta}. \tag{3.2.23}$$

But

$$w(v)(0)=\mathcal{B}^{\alpha}v_0. \tag{3.2.24}$$

Therefore w transforms Q into itself.

Let us show that the map w is contracting. Let $v_1, v_2 \in Q$, and put $z_1 = \mathcal{B}^{-\alpha}w(v_1), z_2 = \mathcal{B}^{-\alpha}w(v_2)$. It is easy to see that

$$z_1(0)=z_2(0)=v_0. \tag{3.2.25}$$

Theorem 3.1.5 gives that, for $t>0$, z_1, z_2 are continuously differentiable and

$$\frac{dz_i(t)}{dt}+\mathcal{A}_{v_i}(t)z_i(t)=f_{v_i}(t) \quad (i=1,2). \tag{3.2.26}$$

Hence,

$$\frac{d(z_1-z_2)}{dt}+\mathcal{A}_{v_1}(t)(z_1-z_2)=(\mathcal{A}_{v_2}(t)-\mathcal{A}_{v_1}(t))z_2+f_{v_1}(t)-f_{v_2}(t). \tag{3.2.27}$$

Let us show that

$$\|\mathcal{A}_0 z_2(t)\| \le K_{19}t^{\beta_1-1}. \tag{3.2.28}$$

We have:

$$\|\mathcal{A}_0 z_2(t)\| \le \|\mathcal{A}_0 U_{v_2}(t,0)v_0\| + \Big\|\mathcal{A}_0\int_0^t U_{v_2}(t,s)f_{v_2}(s)\,ds\Big\|. \tag{3.2.29}$$

Due to Remark 3.1.5 the second term in the right-hand side of (3.2.29) is simply bounded by a constant $K_{23} \le K_{24}t^{\beta_1-1}$. Putting in (3.2.15) $s=\tau$, $v=v_2$, we get

$$\|\mathcal{A}_{v_2}(t)\mathcal{A}_{v_2}^{-1}(\tau)\| \le \|\mathcal{A}_{v_2}(\tau)\mathcal{A}_{v_2}^{-1}(\tau)\| + K_{10}t_0^{\eta} \le K_{20}. \tag{3.2.30}$$

Hence, by Theorem 3.1.2,

$$\|\mathcal{A}_{v_2}^{\gamma_1}(t)\mathcal{A}_{v_2}^{-\gamma_2}(\tau)\| \le K_{21}, \quad \gamma_2 > \gamma_1. \tag{3.2.31}$$

With the help of (3.1.23), we can estimate now the first term in the right-hand side of (3.2.29):

$$\begin{aligned}\|\mathcal{A}_0 U_{v_2}(t,0)v_0\| &\le \|\mathcal{A}_0\mathcal{A}_{v_2}(t)^{-1}\|\|\mathcal{A}_{v_2}(t)U_{v_2}(t,0)\mathcal{A}_0^{-\beta_1}\|\|\mathcal{A}_0^{\beta_1}\mathcal{B}^{-\beta}\|\|\mathcal{B}^{\beta}v_0\| \\ &\le K_{22}t^{\beta_1-1}.\end{aligned}$$

Using (3.2.27) and (3.2.28) one can show (cf. [58], p. 341) that

$$z_1(t) - z_2(t) = \int_0^t U_{v_1}(t,s)[(\mathcal{A}_{v_2}(s) - \mathcal{A}_{v_1}(s))z_2(s) + (f_{v_1}(s) - f_{v_2}(s))]\,ds. \tag{3.2.32}$$

Applying to both sides the operator $\mathcal{B}^{\alpha}$ and taking into account (3.2.22), (3.1.23), (3.2.1), (3.2.28) and (3.2.4), we obtain:

$$\begin{aligned}&\|w(v_1)(t) - w(v_2)(t)\| \\ &\le \|\mathcal{B}^{\alpha}\mathcal{A}_0^{-\alpha_1}\| \int_0^t \|\mathcal{A}_0^{\alpha_1}U_{v_1}(t,s)\|[\|(\mathcal{A}_{v_2}(s) - \mathcal{A}_{v_1}(s))\mathcal{B}^{-1}\|\|\mathcal{B}\mathcal{A}_0^{-1}\|\|\mathcal{A}_0 z_2(s)\| \\ &\qquad + \|(f_{v_1}(s) - f_{v_2}(s))\|]\,ds \\ &\le K_{25}\int_0^t |t-s|^{-\alpha_1}\left(\|v_1(s) - v_2(s)\|s^{\beta_1-1} + \|v_1(s) - v_2(s)\|\right)ds \\ &\le K_{25}t_0^{\beta_1-\alpha_1}\|v_1 - v_2\|_{C([0,t_0];E)}\int_0^t |t-s|^{-\beta_1}\left(s^{\beta_1-1}+1\right)ds,\end{aligned}$$

whence

$$\|w(v_1) - w(v_2)\|_{C([0,t_0];E)} \le K_{26}t_0^{\beta_1-\alpha_1}\|v_1 - v_2\|_{C([0,t_0];E)}. \tag{3.2.33}$$

For t_0 small enough, $K_{26}t_0^{\beta_1-\alpha_1} < 1$, and w is a contraction. Thus, by the Banach principle, the map w has a fixed point $v_* \in Q$. Then the function

$$v(t) = \mathcal{B}^{-\alpha}v_*(t)$$

is a solution of the equation:

$$v(t) = U_{\mathcal{B}^{\alpha}v}(t,0)v_0 + \int_0^t U_{\mathcal{B}^{\alpha}v}(t,s)f_{\mathcal{B}^{\alpha}v}(s)\,ds. \tag{3.2.34}$$

By Theorem 3.1.5, $v(t)$ is a solution of (3.2.5) – (3.2.6). The linear operator $\mathcal{B}\mathcal{A}_0^{-1}$ is bounded, so (3.1.27) and (3.1.35) imply (3.2.7).

It remains to observe that the constants in the proof of the theorem, including t_0, are either independent of v_0 or depend on $\|\mathcal{B}^{\beta}v_0\|$ and on $R - \|\mathcal{B}^{\alpha}v_0\|$. Therefore, a priori bound (3.2.8) gives an opportunity to construct a solution, step by step, on the whole segment $[0, T]$. □

3.2.2 The Leray–Schauder degree

Here we recall some elementary facts from the Leray–Schauder degree theory [40, 32].

Let E be a normed space and B be the class of open bounded subsets of E. Denote by Ξ the set of triples $(I - k, D, p)$ where I is the identity operator in E, $p \in E$, $D \in B$, $k : \overline{D} \to E$ is a *compact operator* (i.e. k is continuous and its image is a relatively compact set), $p \notin (I - k)(\partial D)$. Here ∂D denotes the boundary of D.

Theorem 3.2.2. *There exists a unique map $d_{LS} : \Xi \to \mathbb{Z}$ satisfying the following four conditions (axioms).*

1) *(Normalization). For any $D \in B$ such that $0 \in D$, one has*

$$d_{LS}(I, D, 0) = 1.$$

2) *(Additivity). For any $(I - k, D, p) \in \Xi$ and all open sets $D_1, D_2 \subset D$ such that $p \notin f(\overline{D} \backslash (D_1 \bigcup D_2))$ one has*

$$d_{LS}((I-k)|_{\overline{D}_1}, D_1, p) + d_{LS}((I-k)|_{\overline{D}_2}, D_2, p) = d_{LS}(I-k, D, p).$$

3) *(Homotopic invariance). Let $D \in B$, $D \neq \varnothing$, and let $h : [0,1] \times \overline{D} \to E$ be a compact operator. Assume that $p \neq x - h(t,x)$ for $t \in [0,T], x \in \partial D$. Such h is called a homotopy. Then $d_{LS}(I - h_t, D, p)$ does not depend on $t \in [0,T]$ where $h_t : \overline{D} \to E$, $h_t(x) = h(t,x)$.*

4) *(Homogeneity). For any $(I - k, D, p) \in \Xi$ such that $D \neq \varnothing$, one has*

$$d_{LS}(f, D, p) = d_{LS}(f - p, D, 0).$$

Remark 3.2.2. Note that the fact that h is continuous and $h(t) : \overline{D} \to E$ is compact for all $t \in [0,T]$ does not imply that h is compact (see [32], p. 129). However, if $h_0, h_1 : \overline{D} \to E$ are compact, then the *linear homotopy* $h(t,x) = (1-t)h_0(x) + th_1(x) : [0,T] \times \overline{D} \to E$ is compact.

The map d_{LS} is called the *Leray–Schauder topological degree*. The most important property of the degree is

Theorem 3.2.3. *Let $(I-k, D, p) \in \Xi$ and $d_{LS}(I-k, D, p) \neq 0$. Then the equation*

$$x - k(x) = p \tag{3.2.35}$$

has a solution $x_0 \in D$.

A well-known consequence of this theorem is the Schauder fixed point principle. We shall use it in the following form:

Theorem 3.2.4. *Let A be a non-empty compact convex set in a normed space E. Then for any continuous map $\varphi : A \to A$ there is a point $x_* \in A$ such that $x_* = \varphi(x_*)$.*

Axioms 1) – 4) imply

Corollary 3.2.1. *For all $D \in B$ and $p \in D$,*

$$d_{LS}(I, D, p) = 1.$$

Chapter 4

Attractors of evolutionary equations in Banach spaces

4.1 Attractors of autonomous equations: classical approach

4.1.1 Attractor of a semigroup

Let E be an arbitrary set.

Definition 4.1.1. A family of mappings $\mathcal{S}_t : E \to E$, $t \geq 0$, is called a *semigroup* if $\mathcal{S}_0$ is the identity map I and

$$\mathcal{S}_t \circ \mathcal{S}_\tau = \mathcal{S}_{t+\tau} \tag{4.1.1}$$

for any $t, \tau \geq 0$.

Hereafter we assume that E is a Banach space.

Definition 4.1.2. A semigroup $\mathcal{S}_t$ is called *bounded* in E if, for any bounded set $B \subset E$, the set $\bigcup_{t \geq 0} \mathcal{S}_t B$ is also bounded in E.

Let F be a topological space such that $E \cap F \neq \varnothing$.

Definition 4.1.3. A set $P \subset F$ is called (E, F)*-attracting* (for the semigroup $\mathcal{S}_t$) if for any bounded set $B \subset E$ and any open neighborhood W of P in F there exists $h \geq 0$ such that $\mathcal{S}_t B \subset W$ for all $t \geq h$.

Definition 4.1.4. A set $P \subset E$ is called *absorbing* (for the semigroup $\mathcal{S}_t$) if for any bounded set $B \subset E$ there is $h \geq 0$ such that for all $t \geq h$ one has $\mathcal{S}_t B \subset P$.

Definition 4.1.5. A set $A \subset E$ is called *invariant* (for the semigroup $\mathcal{S}_t$) if

$$\mathcal{S}_t A = A$$

for any $t \geq 0$.

Definition 4.1.6. A set $\mathcal{A} \subset E \cap F$ is called an (E, F)*-attractor* (of the semigroup $\mathcal{S}_t$) if

i) $\mathcal{A}$ is compact in F and bounded in E;

ii) $\mathcal{A}$ is invariant for the semigroup $\mathcal{S}_t$;

iii) $\mathcal{A}$ is (E, F)-attracting for the semigroup $\mathcal{S}_t$.

Various criteria for the existence of such an attractor may be found, for instance, in [27, 28, 9]. They are mainly based on the assumption that there is some (E, F)-attracting or absorbing set P, and the attractor (under some assumptions on the semigroup $\mathcal{S}_t$) can be found by the formula

$$\mathcal{A} = \bigcap_{s>s_0} [\bigcup_{t\geq s} \mathcal{S}_t P], \tag{4.1.2}$$

where s_0 is large enough and $[\cdot]$ stands for the closure in F.

Lemma 4.1.1. *Let $\mathcal{S}_t : E \to E$ be a semigroup. Let F be a T_1 space (i.e. the one-point sets in F are closed), and $E \cap F \neq \varnothing$. Let $A \subset E$ be bounded and an invariant set for the semigroup $\mathcal{S}_t$, and let $P \subset F$ be an (E, F)-attracting set for the semigroup $\mathcal{S}_t$. Then $A \subset P$.*

Proof. For any open neighborhood W of P in F, there is $h \geq 0$ such that $A = \mathcal{S}_t A \subset W$ for all $t \geq h$. Hence, $A \subset F$. If there is a point $x \in A$ such that $x \notin P$, then $W_x = F \backslash \{x\}$ is an open neighborhood of P in F. Therefore $A \subset W_x$, and we arrive at a contradiction. □

Corollary 4.1.1. *Let $\mathcal{S}_t : E \to E$ be a semigroup. Let F be a T_1 space, $E \cap F \neq \varnothing$. If there exists an (E, F)-attractor of the semigroup $\mathcal{S}_t$, it is unique.*

Proof. If there are two (E, F)-attractors $\mathcal{A}_1$ and $\mathcal{A}_2$, then, by Lemma 4.1.1, $\mathcal{A}_1 \subset \mathcal{A}_2$ and $\mathcal{A}_2 \subset \mathcal{A}_1$. □

4.1.2 Global (E, E_0)-attractors of evolutionary equations

Let E and E_0 be Banach spaces, $E \subset E_0$. Consider an abstract differential equation

$$u'(t) = A(u(t)), \tag{4.1.3}$$
$$u(t) \in E, \ A : D(A) \subset E \to R(A).$$

The symbol "=" may be understood in any appropriate sense (e.g. in the sense of some topological space containing both E and $R(A)$). The derivative "′" may also be considered in any generalized sense. The nonlinear operator A is arbitrary (it may even be multi-valued, but in this case the symbol "=" must be replaced by "$\subset$"), but here we consider A to be independent of t (this means that the equation is "autonomous").

We shall investigate attractors of solutions of this equation which belong to the space $C([0,+\infty); E_0) \cap L_\infty(0,+\infty; E)$.

Hereafter it is supposed that the space E is reflexive. Then, by Lemma 2.2.6, $C([0,+\infty); E_0) \cap L_\infty(0,+\infty; E) \subset C_w([0,+\infty); E)$. Hence, the values of functions from $C([0,+\infty); E_0) \cap L_\infty(0,+\infty; E)$ belong to E at every $t \geq 0$.

Consider the translation (shift) operators $T(h)$,

$$T(h)(u)(t) = u(t+h),$$

$h \geq 0$ for $u \in C([0,+\infty); E_0), L_\infty(0,+\infty; E)$, or $h \in \mathbb{R}$ for $C((-\infty,+\infty); E_0)$, $L_\infty(-\infty,+\infty; E)$.

For any fixed $h \geq 0$ the operators $T(h)$ are continuous bounded mappings of the spaces $C([0,+\infty); E_0)$ and $L_\infty(0,+\infty; E)$ into themselves.

Assume that for any $b \in E$ equation (4.1.3) possesses a unique solution u_b in a certain class $\mathcal{F} \subset C([0,+\infty); E_0) \cap L_\infty(0,+\infty; E)$, satisfying the initial condition

$$u_b(0) = b. \tag{4.1.4}$$

Define the mapping $S_t : E \to E, t \geq 0$, by the formula

$$S_t(b) = u_b(t). \tag{4.1.5}$$

If

$$T(h)\mathcal{F} \subset \mathcal{F} \tag{4.1.6}$$

for all $h \geq 0$, then

$$T(h)u_b = u_{S_h(b)}. \tag{4.1.7}$$

This implies that S_t is a semigroup.

Definition 4.1.7. The (E, E_0)-attractor of the semigroup S_t is called the *global* (E, E_0)-*attractor* of the evolutionary equation (4.1.3).

4.2 Attractors of autonomous problems without uniqueness of the solution

In this section we describe a more general approach to attractors of equation (4.1.3), which does not require uniqueness of the solution to the Cauchy problem (4.1.3) – (4.1.4) and existence of the semigroup S_t.

4.2.1 Basic definitions

Let some set

$$\mathcal{H}^+ \subset C([0,+\infty); E_0) \cap L_\infty(0,+\infty; E)$$

of solutions (strong, weak, etc.) for equation (4.1.3) on the positive axis be fixed. The set $\mathcal{H}^+$ will be called the *trajectory space* and its elements will be called *trajectories*.

Remark 4.2.1. We do not assume that $T(h)\mathcal{H}^+ \subset \mathcal{H}^+$ for $h \geq 0$ (see also Remarks 4.2.5, 4.2.6).

Remark 4.2.2. Usually an appropriate trajectory space must be such that for every $a \in E$ there exists (but is not necessarily unique) a trajectory u satisfying the initial condition $u(0) = a$.

Remark 4.2.3. Below the concrete form of equation (4.1.3) is not significant but only presence of a trajectory space $\mathcal{H}^+$ is important and everything will depend only on the properties of this set. Generally speaking, the nature of $\mathcal{H}^+$ may be different from the one described above.

Definition 4.2.1. A set $P \subset C([0,+\infty); E_0) \cap L_\infty(0,+\infty; E)$ is called *attracting* (for the trajectory space $\mathcal{H}^+$) if for any set $B \subset \mathcal{H}^+$ which is bounded in $L_\infty(0,+\infty; E)$, one has

$$\sup_{u\in B} \inf_{v\in P} \|T(h)u - v\|_{C([0,+\infty);E_0)} \underset{h\to\infty}{\rightarrow} 0.$$

Definition 4.2.2. A set $P \subset C([0,+\infty); E_0) \cap L_\infty(0,+\infty; E)$ is called *absorbing* (for the trajectory space $\mathcal{H}^+$) if for any set $B \subset \mathcal{H}^+$ which is bounded in $L_\infty(0,+\infty; E)$, there is an $h \geq 0$ such that for all $t \geq h$:

$$T(t)B \subset P.$$

It is easy to see that any absorbing set is attracting.

Definition 4.2.3. A set $P \subset C([0,+\infty); E_0) \cap L_\infty(0,+\infty; E)$ is called a *trajectory semiattractor* (for the trajectory space $\mathcal{H}^+$) if

i) P is compact in $C([0,+\infty); E_0)$ and bounded in $L_\infty(0,+\infty; E)$;

ii) $T(t)P \subset P$ for any $t \geq 0$;

iii) P is attracting in the sense of Definition 4.2.1.

Definition 4.2.4. A set $P \subset C([0,+\infty); E_0) \cap L_\infty(0,+\infty; E)$ is called a *trajectory quasiattractor* (for the trajectory space $\mathcal{H}^+$) if it satisfies conditions i), iii) of Definition 4.2.3 and ii') $T(t)P \supset P$ for any $t \geq 0$.

Definition 4.2.5. A set $P \subset C([0, +\infty); E_0) \cap L_\infty(0, +\infty; E)$ is called a *trajectory attractor* (for the trajectory space $\mathcal{H}^+$) if it is a trajectory semiattractor and a trajectory quasiattractor (for the trajectory space $\mathcal{H}^+$). A trajectory attractor is called *minimal* if it is contained in any other trajectory attractor.

Definition 4.2.6. A set $\mathcal{A} \subset E$ is called a *global attractor* (in E_0) for the trajectory space $\mathcal{H}^+$ of equation (4.1.3) if

i) $\mathcal{A}$ is compact in E_0 and bounded in E;

ii) for any set $B \subset \mathcal{H}^+$ which is bounded in $L_\infty(0, +\infty; E)$, the attraction property is fulfilled:

$$\sup_{u \in B} \inf_{v \in \mathcal{A}} \|u(t) - v\|_{E_0} \underset{t \to \infty}{\to} 0.$$

iii) $\mathcal{A}$ is the minimal set satisfying conditions i) and ii) (that is $\mathcal{A}$ is contained in every set satisfying conditions i) and ii)).

Remark 4.2.4. It is obvious that if there exists a minimal trajectory attractor or a global attractor, then it is unique.

Remark 4.2.5. If a trajectory attractor for the trajectory space $\mathcal{H}^+$ is contained in $\mathcal{H}^+$, then it is minimal. It follows from Lemma 4.2.10 (see below). In [66] (at the additional restriction that $T(h)\mathcal{H}^+ \subset \mathcal{H}^+$ for all $h \geq 0$) there were considered only trajectory attractors (in the sense of Definition 4.2.5) contained in $\mathcal{H}^+$. Nevertheless, the (more general) concept of minimal trajectory attractor used by us has many usual properties of trajectory attractors. In particular, a minimal trajectory attractor always generates a global attractor (see below, Theorem 4.2.2). Furthermore, under some conditions on the trajectory space $\mathcal{H}^+$ a minimal trajectory attractor (provided it exists) is always contained in $\mathcal{H}^+$ (see below, Remarks 4.2.11, 4.2.12).

Remark 4.2.6. A natural way to define a trajectory space is $\mathcal{H}^+$ is to take all the solutions (in a certain sense: strong, weak, etc.) for equation (4.1.3), which belong to some class $\mathcal{F} \subset C([0, +\infty); E_0) \cap L_\infty(0, +\infty; E)$. Usually, if u is a solution, then $T(h)u$, $h \geq 0$, is also a solution, so the condition $T(h)\mathcal{H}^+ \subset \mathcal{H}^+$ for all $h \geq 0$ may be violated only if (4.1.6) is violated. This may appear, for example, in the following situations:

a) $\mathcal{F}$ consists of the solutions satisfying some inequality which is not invariant with respect to the shifts $T(h)$. In particular, this may be an inequality containing integrals or other non-local functions, e.g. an inequality of the form

$$e^{Ct} \|u(t)\| \leq \max_{s \geq 0} \|u(s)\|, \quad t \geq 0.$$

Such issues appear at the study of the equations of motion for viscoelastic medium with Jeffreys' constitutive law (see Chapter 6).

b) $\mathcal{F}$ consists of the "surviving" solutions, i.e. the solutions u satisfying the property

$$u(t) \in M(u(0)), \quad t \in [0, T],$$

where $M : E \multimap E$ is a fixed multi-valued map and $T > 0$ is a fixed number.

c) $\mathcal{F}$ consists of the solutions which have some property at least at one point t or on some fixed subset of the positive axis, e.g. the solutions u for which there exists $t > 0$ such that $u(t) = u(0)$.

d) $\mathcal{F}$ consists of the solutions which do not have some property globally, e.g. all solutions except the ones satisfying the property

$$u(t) \in A, \quad t \geq 0,$$

where $A \subset E$ is a fixed set.

Let us introduce one more useful notion.

Definition 4.2.7. The *kernel* $\mathcal{K}(P)$ of a set $P \subset C([0, +\infty); E_0) \cap L_\infty(0, +\infty; E)$ is the set

$$\{u \in L_\infty(-\infty, +\infty; E) | \forall t \in \mathbb{R} : \Pi_+ T(t)u \in P\}.$$

Here Π_+ is the operator of restriction on the semi-axis $[0, +\infty)$.

Obviously, $\Pi_+ \mathcal{K}(P) \subset P$.

4.2.2 Simple properties of attracting sets and auxiliary statements

Lemma 4.2.1. *Let a set $P \subset C([0, +\infty); E_0) \cap L_\infty(0, +\infty; E)$ satisfy conditions i), iii) of Definition 4.2.3. Then $\Pi_+ \mathcal{K}(\mathcal{H}^+) \subset P$.*

Proof. Let $u \in \Pi_+ \mathcal{K}(\mathcal{H}^+)$, that is $u = \Pi_+ u_1$, where $u_1 \in \mathcal{K}(\mathcal{H}^+)$. Consider the set

$$B_u = \{\Pi_+ T(t) u_1 | t \in \mathbb{R}\}.$$

It suffices to show that $B_u \subset P$. The set B_u is contained in $\mathcal{H}^+$ and is bounded in $L_\infty(0, +\infty; E)$. Besides, $T(h) B_u \supset B_u$ for all $h \geq 0$. Really, if $v \in B_u$, then for some $t \in \mathbb{R}$ one has $v = \Pi_+ T(t) u_1 = T(h) \Pi_+ T(t - h) u_1 \in T(h) B_u$. Since P is an attracting set, for any $\varepsilon > 0$ there is an $h \geq 0$ such that

$$\sup_{w \in B_u} \inf_{v \in P} \|T(h)w - v\|_{C([0,+\infty); E_0)} < \varepsilon.$$

Since $T(h)B_u \supset B_u$,

$$\sup_{w \in B_u} \inf_{v \in P} \|w - v\|_{C([0,+\infty);E_0)} < \varepsilon.$$

The number $\varepsilon > 0$ was arbitrary, so

$$\inf_{v \in P} \|w - v\|_{C([0,+\infty);E_0)} = 0$$

for all $w \in B_u$. Since P is compact in $C([0, +\infty); E_0)$, $w \in P$. □

Roughly speaking, the kernel $\mathcal{K}(\mathcal{H}^+)$ is the set of solutions for equation (4.1.3) defined on the whole real axis, which are uniformly bounded in E and continuous with values in E_0. The following statement on properties of this set takes place.

Lemma 4.2.2. *Under the conditions of Lemma 4.2.1 the kernel $\mathcal{K}(\mathcal{H}^+)$ is relatively compact in $C((-\infty, +\infty); E_0)$ and bounded in $L_\infty(-\infty, +\infty; E)$.*

Proof. During the proof of the previous lemma it has been shown that the set

$$U = \{\Pi_+ T(t)u_1 | t \in \mathbb{R}, u_1 \in \mathcal{K}(\mathcal{H}^+)\}$$

is contained in P. Hence, U is contained in $C([0, +\infty); E_0)$ and is bounded in $L_\infty(0, +\infty; E)$. Therefore $\mathcal{K}(\mathcal{H}^+)$ is contained in $C((-\infty, +\infty); E_0)$ and is bounded in $L_\infty(-\infty, +\infty; E)$.

For $v \in \mathcal{K}(\mathcal{H}^+)$ and a natural number m we put

$$v_m(t) = v(t), \quad t \geq -m;$$
$$v_m(t) = v(-m), \quad t < -m.$$

Denote by $\mathcal{H}_m$ the set $\{v_m | v \in \mathcal{K}(\mathcal{H}^+)\}$. Since $\Pi_+\mathcal{K}(\mathcal{H}^+) \subset P$ is relatively compact in $C([0, +\infty); E_0)$, $\mathcal{H}_0$ is relatively compact in $C((-\infty, +\infty); E_0)$. It is easy to see that $T(m)\mathcal{K}(\mathcal{H}^+) = \mathcal{K}(\mathcal{H}^+)$. Therefore $\mathcal{H}_m = T(m)\mathcal{H}_0$ and it is relatively compact in $C((-\infty, +\infty); E_0)$. But for any $\varepsilon > 0$

$$\sup_{v \in \mathcal{K}(\mathcal{H}^+)} \inf_{w \in \mathcal{H}_m} \|w - v\|_{C((-\infty,+\infty);E_0)} \leq \sup_{v \in \mathcal{K}(\mathcal{H}^+)} \|v_m - v\|_{C((-\infty,+\infty);E_0)} \leq \frac{1}{2^m} < \varepsilon$$

for m large enough. Therefore $\mathcal{K}(\mathcal{H}^+)$ is relatively compact in $C((-\infty, +\infty); E_0)$ as a uniform limit of relatively compact sets $\mathcal{H}_m$ as $m \to \infty$ (Theorem 2.2.1 and Corollary 2.2.2 remain valid in Fréchet spaces: just replace norms with prenorms). □

Trajectory attractors possess the following interesting property.

Lemma 4.2.3. *If there exists a trajectory attractor P for the trajectory space $\mathcal{H}^+$, then $P = \Pi_+\mathcal{K}(P)$.*

Proof. As it was mentioned above, $P \supset \Pi_+\mathcal{K}(P)$. We shall show now the inverse inclusion. Let $u_0 \in P$. Since $T(1)P = P$, there exists $u_{-1} \in P$ such that $T(1)u_{-1} = u_0$. Similarly, there is $u_{-2} \in P$ such that $T(1)u_{-2} = u_{-1}$; $u_{-3} \in P$ such that $T(1)u_{-3} = u_{-2}$ and so on. Define $u \in L_\infty(-\infty, +\infty; E)$ as follows:

$$u(t) = u_0(t),\ t \geq 0; \quad u(t) = T(\{t\})u_{[t]}(0),\ t < 0.$$

Here the brackets $[\cdot]$ and $\{\cdot\}$ denote the integer part and the fractional part of a number, respectively. We shall show now that $u \in \mathcal{K}(P)$. Really, for $t \geq 0$, $\Pi_+T(t)u = T(t)u_0 \in T(t)P = P$, and, for $t < 0$, $\Pi_+T(t)u = T(\{t\})u_{[t]} \subset T(\{t\})P = P$. It remains to observe that $u_0 = \Pi_+u \in \Pi_+\mathcal{K}(P)$. □

Lemma 4.2.4. *a) Let sets $P_1, P_2 \subset C([0, +\infty); E_0) \cap L_\infty(0, +\infty; E)$ satisfy conditions i) or ii) of Definition 4.2.3. Then $P_1 \cap P_2$ also satisfies a corresponding condition. b) If P_1, P_2 are compact in $C([0, +\infty); E_0)$ and satisfy condition iii) of Definition 4.2.3, then $P_1 \cap P_2$ also satisfies condition iii).*

Proof. Statement a) is clear. Let us show b). Let P_1, P_2 be compact in $C([0, +\infty); E_0)$ and satisfy condition iii). We have to show that $P_1 \cap P_2$ is an attracting set. If it is not so, then for some $\delta > 0$ and some set $B \subset \mathcal{H}^+$ which is bounded in $L_\infty(0, +\infty; E)$, there is a sequence $h_m \to \infty$ such that

$$\sup_{u \in B} \inf_{v \in P_1 \cap P_2} \|T(h_m)u - v\|_{C([0,+\infty);E_0)} > \delta.$$

Then there are elements $u_m \in B$ such that

$$\inf_{v \in P_1 \cap P_2} \|T(h_m)u_m - v\|_{C([0,+\infty);E_0)} > \delta. \tag{4.2.1}$$

On the other hand, since P_1 and P_2 are attracting sets, for any natural number k there exist a number m_k and elements $v_k^1 \in P_1$, $v_k^2 \in P_2$ such that

$$\|T(h_{m_k})u_{m_k} - v_k^1\|_{C([0,+\infty);E_0)} < \frac{1}{k},$$
$$\|T(h_{m_k})u_{m_k} - v_k^2\|_{C([0,+\infty);E_0)} < \frac{1}{k}.$$

Since P_1 is compact in $C([0, +\infty); E_0)$, without loss of generality we may assume that the sequence v_k^1 converges to an element v_0 as $k \to \infty$. Then sequences $T(h_{m_k})u_{m_k}$ and v_k^2 also converge to v_0. Thus, $v_0 \in P_1 \cap P_2$ and $\|T(h_{m_k})u_{m_k} - v_0\|_{C([0,+\infty);E_0)} \underset{k\to\infty}{\to} 0$, which contradicts (4.2.1). □

Lemma 4.2.5. *Let a set* $P \subset C([0,+\infty); E_0) \cap L_\infty(0,+\infty; E)$ *satisfy one of conditions i), ii), ii') or iii) of Definitions 4.2.3, 4.2.4. Then* $T(h)P$ *also satisfies a corresponding condition for all* $h \geq 0$.

Proof. The statement of the lemma concerning conditions i), ii), ii') is obvious. Let P satisfy condition iii), that is it is attracting. Since the map $T(h)$ is bounded in $C([0,+\infty); E_0)$, one has

$$\|T(h)u\|_{C([0,+\infty);E_0)} \leq C\|u\|_{C([0,+\infty);E_0)}$$

for some constant C and all $u \in C([0,+\infty); E_0)$. Then for any set $B \subset \mathcal{H}^+$ which is bounded in $L_\infty(0,+\infty; E)$, one has

$$\begin{aligned}\sup_{u\in B}\ \inf_{v\in T(h)P} &\|T(t)u - v\|_{C([0,+\infty);E_0)} \\ &= \sup_{u\in B}\inf_{v\in P}\|T(h)(T(t-h)u - v)\|_{C([0,+\infty);E_0)} \\ &\leq C\sup_{u\in B}\inf_{v\in P}\|T(t-h)u - v\|_{C([0,+\infty);E_0)} \underset{t\to\infty}{\to} 0,\end{aligned}$$

that is $T(h)P$ is also attracting. □

We will need also the following statement.

Lemma 4.2.6. *Let* (X,ρ) *be a metric space and* $\{K_\alpha\}_{\alpha\in\Xi}$ *be a system of non-empty compact sets in* X. *Assume that for any* $\alpha_1,\alpha_2 \in \Xi$ *there is* $\alpha_3 \in \Xi$ *such that* $K_{\alpha_1} \cap K_{\alpha_2} = K_{\alpha_3}$. *Then* $K_0 = \bigcap_{\alpha\in\Xi} K_\alpha \neq \varnothing$ *and for any* $\varepsilon > 0$ *there is* $\alpha_\varepsilon \in \Xi$ *such that for any* $y \in K_{\alpha_\varepsilon}$:

$$\inf_{x\in K_0}\rho(x,y) < \varepsilon.$$

Proof. By induction one easily proves that an intersection of any finite number of sets from the system $\{K_\alpha\}$ belongs to this system.

Consider the set

$$K_\varepsilon = \{y \in X \mid \inf_{x\in K_0}\rho(x,y) < \varepsilon\} \quad \text{if} \quad K_0 \neq \varnothing;$$

$$K_\varepsilon = \varnothing \quad \text{if} \quad K_0 = \varnothing.$$

It is clear that K_ε is open and $K_0 \subset K_\varepsilon$. Note that if $K_\varepsilon \neq \varnothing$, then $K_0 \neq \varnothing$. Thus, to prove the lemma, it suffices to find a set $K_{\alpha_\varepsilon} \subset K_\varepsilon$ in the system $\{K_\alpha\}$.

Let $\alpha_0 \in \Xi$. Then $K_{\alpha_0}\backslash K_\varepsilon$ is compact and $\{X\backslash K_\alpha\}_{\alpha\in\Xi}$ is an open cover for $K_{\alpha_0}\backslash K_\varepsilon$. One can choose its finite subcover $X\backslash K_{\alpha_1}, X\backslash K_{\alpha_2}, \dots, X\backslash K_{\alpha_m}$. Thus, $K_{\alpha_0}\backslash K_\varepsilon \subset X\backslash \bigcap_{i=1}^m K_{\alpha_i}$. Therefore $\bigcap_{i=1}^m K_{\alpha_i} \subset X\backslash(K_{\alpha_0}\backslash K_\varepsilon)$. It implies that the set $\bigcap_{i=0}^m K_{\alpha_i}$ is contained in K_ε, so it can be taken as the required set K_{α_ε}. □

By analogy to the concept of minimal trajectory attractor for the trajectory space $\mathcal{H}^+$ it is possible to introduce the concept of *minimal trajectory semiattractor* as a trajectory semiattractor contained in any other trajectory semiattractor.

Lemma 4.2.7. *A minimal trajectory semiattractor is always a minimal trajectory attractor.*

Remark 4.2.7. We will prove below the inverse statement (Lemma 4.2.8).

Proof. Let $\mathfrak{U}$ be a minimal trajectory semiattractor, that is it is a trajectory semiattractor and it is contained in any other trajectory semiattractor. By Lemma 4.2.5 $T(h)\mathfrak{U}$ is a trajectory semiattractor for all $h \geq 0$, therefore $\mathfrak{U} \subset T(h)\mathfrak{U}$. Thus, $\mathfrak{U}$ is a trajectory semiattractor and a trajectory quasiattractor, that is it is a (minimal) trajectory attractor. □

4.2.3 Existence of a minimal trajectory attractor

Theorem 4.2.1 (see [75, 76]). *Assume that there exists a trajectory semiattractor P for the trajectory space $\mathcal{H}^+$. Then there exists a minimal trajectory attractor $\mathfrak{U}$ for the trajectory space $\mathcal{H}^+$. Here one has*

$$\Pi_+\mathcal{K}(\mathcal{H}^+) \subset \mathfrak{U} = \Pi_+\mathcal{K}(\mathfrak{U}) \subset \Pi_+\mathcal{K}(P) \subset P$$

and the kernel $\mathcal{K}(\mathcal{H}^+)$ is relatively compact in $C((-\infty,+\infty);E_0)$ and bounded in $L_\infty(-\infty,+\infty;E)$.

Proof. Take in Lemma 4.2.6 $X = C([0,+\infty);E_0)$ and let $\{K_\alpha\}_{\alpha\in\Xi}$ be the system of all trajectory semiattractors for the trajectory space $\mathcal{H}^+$. Denote by $\mathfrak{U}$ the intersection of all trajectory semiattractors. By Lemma 4.2.4 the intersection of any two sets from the system $\{K_\alpha\}$ belongs to this system. Thus, for the system of semiattractors $\{K_\alpha\}$ the conditions of Lemma 4.2.6 hold.

Let us show that $\mathfrak{U}$ is a trajectory semiattractor. Clearly, $\mathfrak{U}$ satisfies conditions i) and ii) of Definition 4.2.3. We shall show now that $\mathfrak{U}$ satisfies condition iii), that is it is attracting. Fix $\varepsilon > 0$ and a set $B \subset \mathcal{H}^+$ which is bounded in $L_\infty(0,+\infty;E)$. By Lemma 4.2.6 there is a semiattractor P_ε such that for any $v \in P_\varepsilon$:

$$\inf_{w\in\mathfrak{U}} \|w - v\|_{C([0,+\infty);E_0)} < \frac{\varepsilon}{2}.$$

Since P_ε is an attracting set, there exists $h \geq 0$ such that for $t \geq h$:

$$\sup_{u\in B}\inf_{v\in P_\varepsilon} \|T(t)u - v\|_{C([0,+\infty);E_0)} < \frac{\varepsilon}{2}.$$

Therefore for every $u \in B$ there exists $v(u) \in P_\varepsilon$ such that

$$\|T(t)u - v(u)\|_{C([0,+\infty);E_0)} < \frac{\varepsilon}{2}.$$

We have:

$$\sup_{u\in B}\inf_{w\in\mathfrak{U}}\|T(t)u-w\|_{C([0,+\infty);E_0)}$$
$$\le \sup_{u\in B}\big(\|T(t)u-v(u)\|_{C([0,+\infty);E_0)}+\inf_{w\in\mathfrak{U}}\|v(u)-w\|_{C([0,+\infty);E_0)}\big)$$
$$\le \frac{\varepsilon}{2}+\frac{\varepsilon}{2}=\varepsilon.$$

Thus, $\mathfrak{U}$ is a trajectory semiattractor. It is clear that it is minimal. By Lemma 4.2.7 $\mathfrak{U}$ is a minimal trajectory attractor. It remains to use Lemmas 4.2.1 – 4.2.3. □

Corollary 4.2.1. *Assume that there exists an absorbing set P for the trajectory space $\mathcal{H}^+$ which is compact in $C([0,+\infty);E_0)$ and bounded in $L_\infty(0,+\infty;E)$. Then there exists a minimal trajectory attractor $\mathfrak{U}$ for the trajectory space $\mathcal{H}^+$. Here one has $\Pi_+\mathcal{K}(\mathcal{H}^+)\subset\mathfrak{U}=\Pi_+\mathcal{K}(\mathfrak{U})\subset\Pi_+\mathcal{K}(P)$ and the kernel $\mathcal{K}(\mathcal{H}^+)$ is relatively compact in $C((-\infty,+\infty);E_0)$ and bounded in $L_\infty(-\infty,+\infty;E)$.*

Proof. Due to Theorem 4.2.1, it suffices to find a trajectory semiattractor $P_1\subset P$.

By Definition 4.2.2, for any set $B\subset\mathcal{H}^+$ which is bounded in $L_\infty(0,+\infty;E)$, there is a number $h(B)\ge 0$ such that for all $t\ge h(B)$ one has $T(t)B\subset P$. Denote by P_1 the closure in $C([0,+\infty);E_0)$ of the set

$$P_0=\bigcup_{B\in\mathcal{B}}\bigcup_{t\ge h(B)}T(t)B,$$

where $\mathcal{B}$ is the set of all sets contained in $\mathcal{H}^+$ which are bounded in $L_\infty(0,+\infty;E)$. The set P_1 is contained in P, so it is compact in $C([0,+\infty);E_0)$ and bounded in $L_\infty(0,+\infty;E)$. It is clear that it is absorbing. Moreover, $T(h)P_0\subset P_0$ for $h\ge 0$. This implies (cf. below, Lemma 4.2.9) that $T(h)P_1\subset P_1$. Thus, P_1 is a semiattractor. □

Lemma 4.2.8. *A minimal trajectory attractor is always a minimal trajectory semiattractor.*

Proof. Let $\mathfrak{U}$ be a minimal trajectory attractor and let P be a trajectory semiattractor. Since the minimal trajectory attractor is unique, by Theorem 4.2.1 $\mathfrak{U}\subset P$. Thus, $\mathfrak{U}$ is contained in any trajectory semiattractor, so it is a minimal trajectory semiattractor. □

4.2.4 Existence of a global attractor

Consider the sections of a trajectory attractor and the kernel at fixed $t\ge 0$:

$$\mathfrak{U}(t)=\{v(t)|v\in\mathfrak{U}\};$$

$$\mathcal{K}(P)(t) = \{v(t) | v \in \mathcal{K}(P)\}.$$

It is easy to see that these sets are contained in E (see Section 4.1.2).

Theorem 4.2.2 (see [75, 76]). *If there exists a minimal trajectory attractor $\mathcal{U}$ for the trajectory space $\mathcal{H}^+$, then there is a global attractor $\mathcal{A}$ for the trajectory space $\mathcal{H}^+$ of equation (4.1.3) and for all $t \geq 0$ one has*

$$\mathcal{K}(\mathcal{H}^+)(t) \subset \mathcal{A} = \mathcal{U}(t) = \mathcal{K}(\mathcal{U})(t).$$

Proof. Observe first that since $T(t)\mathcal{U} = \mathcal{U}$, $t \geq 0$, the set $\mathcal{A} = \mathcal{U}(t)$ does not depend on t. From Lemmas 4.2.1 and 4.2.3 it follows that $\mathcal{K}(\mathcal{H}^+)(t) \subset \mathcal{U}(t) = \mathcal{K}(\mathcal{U})(t)$. The set $\mathcal{U}$ is compact in $C([0, +\infty); E_0)$, so its section $\mathcal{A} = \mathcal{U}(0)$ is compact in E_0. Moreover, $\mathcal{U}$ is bounded in $L_\infty(0, +\infty; E)$, so for $u \in \mathcal{U}$ and almost all $t \geq 0$ (except those t which belong to some set $\mathcal{O}_u$ of zero measure)

$$\|u(t)\|_E \leq \sup_{u \in \mathcal{U}} \|u\|_{L_\infty(0,+\infty;E)}.$$

Since $u \in C_w([0, T]; E)$, one has $u(t) \to u(0)$ weakly in E as $t \to 0$. Thus,

$$\|u(0)\|_E \leq \liminf_{t \to 0, t \notin \mathcal{O}_u} \|u(t)\|_E \leq \sup_{u \in \mathcal{U}} \|u\|_{L_\infty(0,+\infty;E)},$$

so $\mathcal{A} = \mathcal{U}(0)$ is bounded in E.

Let a set $B \subset \mathcal{H}^+$ be bounded in $L_\infty(0, +\infty; E)$. Since $\mathcal{U}$ is an attracting set,

$$\sup_{u \in B} \inf_{v \in \mathcal{U}} \|T(h)u - v\|_{C([0,+\infty);E_0)} \underset{h \to \infty}{\to} 0.$$

It yields the pointwise convergence:

$$\sup_{u \in B} \inf_{v \in \mathcal{U}} \|(T(h)u - v)(t)\|_{E_0} \underset{h \to \infty}{\to} 0, \ t \geq 0.$$

At $t = 0$ we get

$$\sup_{u \in B} \inf_{v \in \mathcal{A} = \mathcal{U}(0)} \|u(h) - v\|_{E_0} \underset{h \to \infty}{\to} 0.$$

It remains to show that $\mathcal{A}$ is contained in every set $\mathcal{A}_0$ which is compact in E_0, bounded in E and possesses the attraction property

$$\sup_{u \in B} \inf_{v \in \mathcal{A}_0} \|u(t) - v\|_{E_0} \underset{t \to \infty}{\to} 0 \tag{4.2.2}$$

for every set $B \subset \mathcal{H}^+$ which is bounded in $L_\infty(0, +\infty; E)$. Let $\mathcal{U}_0 = \{u \in \mathcal{U} | u(t) \in \mathcal{A}_0 \ \forall t \geq 0\}$. It suffices to show that $\mathcal{U} \subset \mathcal{U}_0$. By Lemma 4.2.8 $\mathcal{U}$ is contained in every trajectory semiattractor. Hence, it is enough to show that $\mathcal{U}_0$ is a semiattractor. Since $\mathcal{U}_0 \subset \mathcal{U}$, $\mathcal{U}_0$ is relatively compact in $C([0, +\infty); E_0)$

and is bounded in $L_\infty(0,+\infty;E)$. For any sequence $\{u_m\} \subset \mathfrak{U}_0$ converging in $C([0,+\infty);E_0)$ the limit u_0 belongs to the (closed in $C([0,+\infty);E_0)$) set $\mathfrak{U}$. The convergence in $C([0,+\infty);E_0)$ yields the pointwise convergence: $u_m(t) \to u_0(t)$ in E_0, $t \geq 0$. Since $\mathcal{A}_0$ is compact in E_0, $u_0(t) \in \mathcal{A}_0$, $t \geq 0$. Thus, $\mathfrak{U}_0$ is closed and compact in $C([0,+\infty);E_0)$. Since $T(t)\mathfrak{U} \subset \mathfrak{U}$, one has $T(t)\mathfrak{U}_0 \subset \mathfrak{U}_0$, $t \geq 0$. It remains to show that $\mathfrak{U}_0$ is an attracting set. If not, for some $\delta > 0$ and some set $B \subset \mathcal{H}^+$ which is bounded in $L_\infty(0,+\infty;E)$, there exists a sequence $h_m \to \infty$ such that

$$\sup_{u\in B} \inf_{v\in\mathfrak{U}_0} \|T(h_m)u - v\|_{C([0,+\infty);E_0)} > \delta.$$

Then there are elements $u_m \in B$ such that

$$\inf_{v\in\mathfrak{U}_0} \|T(h_m)u_m - v\|_{C([0,+\infty);E_0)} > \delta. \tag{4.2.3}$$

On the other hand, since $\mathfrak{U}$ is an attracting set, for every natural k there exist a number m_k and an element $v_k \in \mathfrak{U}$ such that

$$\|T(h_{m_k})u_{m_k} - v_k\|_{C([0,+\infty);E_0)} < \frac{1}{k}.$$

Since $\mathfrak{U}$ is compact in $C([0,+\infty);E_0)$, without loss of generality the sequence v_k converges to an element $v_0 \in \mathfrak{U}$ as $k \to \infty$. Then

$$\|T(h_{m_k})u_{m_k} - v_0\|_{C([0,+\infty);E_0)} \underset{k\to\infty}{\to} 0. \tag{4.2.4}$$

Now (4.2.4) together with (4.2.3) yield $v_0 \notin \mathfrak{U}_0$, that is $v_0(t_0) \notin \mathcal{A}_0$ for some $t_0 \geq 0$. But from (4.2.2) it follows that

$$\inf_{v\in\mathcal{A}_0} \|T(h_{m_k})u_{m_k}(t_0) - v\|_{E_0} \underset{k\to\infty}{\to} 0.$$

Then there is a sequence $\{v_k^*\} \subset \mathcal{A}_0$ such that

$$\|T(h_{m_k})u_{m_k}(t_0) - v_k^*\|_{E_0} \underset{k\to\infty}{\to} 0.$$

Since $\mathcal{A}_0$ is compact, without loss of generality v_k^* converges to some element v^*. But (4.2.4) gives:

$$\|T(h_{m_k})u_{m_k}(t_0) - v_0(t_0)\|_{E_0} \underset{k\to\infty}{\to} 0.$$

Therefore $v_0(t_0) = v^* \in \mathcal{A}_0$, and we have a contradiction. □

Theorems 4.2.1 and 4.2.2 give existence of a global attractor under the conditions of Theorem 4.2.1. It appears that existence of a global attractor may be proved also under weaker assumptions.

Definition 4.2.8. A trajectory quasiattractor $\mathfrak{U}$ for the trajectory space $\mathcal{H}^+$ is called *homogeneous* if the set $\mathfrak{U}(t)$ does not depend on $t \geq 0$.

Theorem 4.2.3. *Assume that there exists a attracting set P for the trajectory space $\mathcal{H}^+$ which is compact in $C([0,+\infty); E_0)$ and bounded in $L_\infty(0,+\infty; E)$. Then there exists a homogeneous trajectory quasiattractor $\mathfrak{U} \subset P$ for the trajectory space $\mathcal{H}^+$ such that the set $\mathcal{A} = \mathfrak{U}(t)$ is a global attractor for the trajectory space $\mathcal{H}^+$ of equation (4.1.3).*

Proof. Take in Lemma 4.2.6 $X = C([0,+\infty); E_0)$ and let $\{K_\alpha\}_{\alpha\in\Xi}$ be the system of all attracting sets for the trajectory space $\mathcal{H}^+$ which are compact in $C([0,+\infty); E_0)$ and bounded in $L_\infty(0,+\infty; E)$. Denote by $\mathfrak{U}$ the intersection of all such sets. By Lemma 4.2.4 the intersection of any two sets from the system $\{K_\alpha\}$ belongs to this system. Thus, for system $\{K_\alpha\}$ the conditions of Lemma 4.2.6 hold.

Clearly, $\mathfrak{U}$ satisfies condition i) of Definition 4.2.3. As in the proof of Theorem 4.2.1 one shows that $\mathfrak{U}$ is an attracting set. By Lemma 4.2.5 for all $h \geq 0$, the set $T(h)\mathfrak{U}$ is attracting, compact in $C([0,+\infty); E_0)$ and bounded in $L_\infty(0,+\infty; E)$, therefore $\mathfrak{U} \subset T(h)\mathfrak{U}$. Thus, $\mathfrak{U}$ is a trajectory quasiattractor.

As in the proof of Theorem 4.2.2 one shows that $\mathfrak{U}(0)$ satisfies conditions i), ii) of Definition 4.2.6. Since $\mathfrak{U} \subset T(t)\mathfrak{U}$, $\mathfrak{U}(0) \subset \mathfrak{U}(t)$ for all $t \geq 0$. It remains to show that $\mathfrak{U}(t)$ is contained in any set $\mathcal{A}_0$ which satisfies conditions i), ii) of Definition 4.2.6 (in particular, in $\mathfrak{U}(0)$). Denote $\mathfrak{U}_0 = \{u \in \mathfrak{U} | u(t) \in \mathcal{A}_0 \ \forall t \geq 0\}$. It suffices to show that $\mathfrak{U} \subset \mathfrak{U}_0$. Since $\mathfrak{U}$ is contained in every set from the system $\{K_\alpha\}$, it is enough to show that $\mathfrak{U}_0$ belongs to this system, that is it is an attracting set for the trajectory space $\mathcal{H}^+$ which is compact in $C([0,+\infty); E_0)$ and bounded in $L_\infty(0,+\infty; E)$. It may be realized in the same way as the check of the corresponding statements in the proof of Theorem 4.2.2. □

Remark 4.2.8. For facilitation of check of conditions of Theorems 4.2.1 and 4.2.3, and of Corollary 4.2.1, the following simple statement may be used.

Lemma 4.2.9. *Let P be an attracting (or absorbing) set for the trajectory space $\mathcal{H}^+$ which is relatively compact in $C([0,+\infty); E_0)$ and bounded in $L_\infty(0,+\infty; E)$. Then its closure $\overline{P}$ in $C([0,+\infty); E_0)$ is an attracting (resp. absorbing) set for the trajectory space $\mathcal{H}^+$ which is compact in $C([0,+\infty); E_0)$ and bounded in $L_\infty(0,+\infty; E)$. If in addition $T(t)P \subset P$ for any $t \geq 0$, then $\overline{P}$ is a semiattractor.*

Proof. Since P is bounded in $L_\infty(0,+\infty; E)$ and its elements are weakly continuous functions, there is a constant C such that for any function $u \in P$ one has $\|u(t)\|_E \leq C$ for $t \geq 0$. Let $u_0 \in \overline{P}$. Then in P there is a sequence $u_m \to u_0$ in $C([0,+\infty); E_0)$. Hence, $u_m(t) \to u_0(t)$ in E_0 for all $t \geq 0$. Since $\|u_m(t)\|_E \leq C$ and the space E is reflexive, $u_m(t) \rightharpoonup u_0(t)$ weakly in E and $\|u_0(t)\|_E \leq C$. Thus, $\overline{P}$ is bounded in $L_\infty(0,+\infty; E)$. It is easy to see that $\overline{P}$ is a compact attracting (absorbing) set in

$C([0,+\infty);E_0)$. Let $T(h)P \subset P$ for any $t \geq 0$. We shall show now that $T(h)\overline{P} \subset \overline{P}$. Let $u \in T(h)\overline{P}$. Then there are $u_0 \in \overline{P}$ and a sequence $u_m \in P$ such that $u_m \to u_0$, $u = T(h)u_0$. Then $T(h)u_m \to T(h)u_0 = u$, that is $u \in \overline{T(h)P} \subset \overline{P}$. Thus, $\overline{P}$ is a semiattractor. □

4.2.5 The case when a trajectory attractor is contained in the trajectory space

Trajectory attractors which are contained in $\mathcal{H}^+$ have additional properties.

Lemma 4.2.10. *Let $\mathfrak{U} \subset \mathcal{H}^+$ be a trajectory quasiattractor for the trajectory space $\mathcal{H}^+$. Then*

a) *$\mathfrak{U}$ is contained in any compact attracting set $P \subset C([0,+\infty);E_0)$;*

b) *in $\mathcal{H}^+$ there are no trajectory quasiattractors different from $\mathfrak{U}$;*

c) *if in addition it is known that $T(h)\mathfrak{U} \subset \mathcal{H}^+$ for all $h \geq 0$, then $\mathfrak{U}$ is a minimal trajectory attractor.*

Proof. a) Since $\mathfrak{U} \subset \mathcal{H}^+$ is bounded in $L_\infty(0,+\infty;E)$, for any neighborhood

$$P_\varepsilon = \{y \in C([0,+\infty);E_0) | \inf_{x\in P} \|x-y\|_{C([0,+\infty);E_0)} < \varepsilon\}$$

of the attracting set P in $C([0,+\infty);E_0)$ one has $\mathfrak{U} \subset T(h)\mathfrak{U} \subset P_\varepsilon$ at $h \geq 0$ large enough. Since P is compact in $C([0,+\infty);E_0)$, it yields $\mathfrak{U} \subset P$.

b) If $\mathfrak{U}_1 \subset \mathcal{H}^+$ is another trajectory quasiattractor, then by point a) one has $\mathfrak{U}_1 \subset \mathfrak{U}$, $\mathfrak{U} \subset \mathfrak{U}_1$.

c) By Lemma 4.2.5 $T(h)\mathfrak{U}$ is a trajectory quasiattractor for all $h \geq 0$. If $T(h)\mathfrak{U} \subset \mathcal{H}^+$, then by point b) $T(h)\mathfrak{U} = \mathfrak{U}$ and $\mathfrak{U}$ is a trajectory attractor. By point a) it is minimal. □

Theorems 4.2.1 – 4.2.3 and Lemma 4.2.10 imply

Theorem 4.2.4. *Assume that there exists an attracting set P for the trajectory space $\mathcal{H}^+$ which is compact in $C([0,+\infty);E_0)$ and bounded in $L_\infty(0,+\infty;E)$. Let $T(h)P \subset \mathcal{H}^+$ for all $h \geq 0$. Then there exist a minimal trajectory attractor $\mathfrak{U} = \Pi_+\mathcal{K}(\mathcal{H}^+)$ for the trajectory space $\mathcal{H}^+$ and a global attractor $\mathcal{A} = \mathfrak{U}(t) = \mathcal{K}(\mathcal{H}^+)(t)$, $t \geq 0$ for the trajectory space $\mathcal{H}^+$ of equation (4.1.3).*

Remark 4.2.9. Under conditions of Theorem 4.2.4 $\mathfrak{U} = \Pi_+\mathcal{K}(\mathcal{H}^+)$, so the kernel $\mathcal{K}(\mathcal{H}^+)$ is non-empty.

Remark 4.2.10. In [66] (Theorem 1.1) this theorem was proved under the additional assumption $T(h)\mathcal{H}^+ \subset \mathcal{H}^+$ for all $h \geq 0$. In this case the trajectory attractor can be constructed using an expression resembling formula (4.1.2):

$$\mathcal{U} = \bigcap_{s>0} [\bigcup_{t\geq s} T(t)P], \tag{4.2.5}$$

where $[\cdot]$ stands for the closure in $C([0, +\infty); E_0)$.

4.2.6 Structure of the minimal trajectory attractor and of the homogeneous trajectory quasiattractor

The following theorem gives some characterization for the structure of a minimal trajectory attractor and helps to specify the connection of this concept with the trajectory attractor from [66].

Let us define a topology on the set $C([0, +\infty); E_0) \cap L_\infty(0, +\infty; E)$ as follows: a set $V \subset C([0, +\infty); E_0) \cap L_\infty(0, +\infty; E)$ is closed if any sequence of elements from V which is bounded in $L_\infty(0, +\infty; E)$ and converges in $C([0, +\infty); E_0)$ has a limit that belongs to V. Hereafter the square brackets $[\cdot]$ denote the closure of a set in this topology.

Lemma 4.2.11. *For every fixed $h \geq 0$, the map*

$$T(h) : C([0, +\infty); E_0) \cap L_\infty(0, +\infty; E) \to C([0, +\infty); E_0) \cap L_\infty(0, +\infty; E)$$

is continuous. As a corollary, for any $V \subset C([0, +\infty); E_0) \cap L_\infty(0, +\infty; E)$ one has

$$T(h)[V] \subset [T(h)V].$$

Proof. Let $W \subset C([0, +\infty); E_0) \cap L_\infty(0, +\infty; E)$, $[W] = W$. Take a sequence $\{u_m\} \subset T(h)^{-1}(W)$ which is bounded in $L_\infty(0, +\infty; E)$ and converges to a limit u_0 in $C([0, +\infty); E_0)$. Then the sequence $\{T(h)u_m\} \subset W$ is bounded in $L_\infty(0, +\infty; E)$ and converges to $T(h)u_0 \in W$ in $C([0, +\infty); E_0)$. Therefore $u_0 \in T(h)^{-1}(W)$. Thus, the set $T(h)^{-1}(W)$ is closed, so $T(h)$ is continuous.

Let $V \subset C([0, +\infty); E_0) \cap L_\infty(0, +\infty; E)$. Then $V \subset T(h)^{-1}([T(h)V])$. Since the set in the right-hand side is closed, $[V] \subset T(h)^{-1}([T(h)V])$. Hence, $T(h)[V] \subset [T(h)V]$. □

Theorem 4.2.5 (see [76]). *Assume that there exists a minimal trajectory attractor $\mathcal{U}$ for the trajectory space $\mathcal{H}^+$. Then*

$$\Pi_+\mathcal{K}(\mathcal{H}^+) \subset \mathcal{U} \subset \Pi_+\mathcal{K}([\bigcup_{t\geq 0} T(t)\mathcal{H}^+]). \tag{4.2.6}$$

Proof. The first inclusion is already proved (see Lemma 4.2.1). By Lemma 4.2.3 $\mathcal{U} = \Pi_+\mathcal{K}(\mathcal{U})$. Therefore it suffices to prove that $\mathcal{U} \subset [\bigcup_{t\geq 0} T(t)\mathcal{H}^+]$. By Lemma 4.2.8 $\mathcal{U}$ is contained in every trajectory semiattractor, so it suffices to show that $\mathcal{U} \cap [\bigcup_{t\geq 0} T(t)\mathcal{H}^+]$ is a semiattractor.

Observe first that $\mathcal{U} \cap [\bigcup_{t\geq 0} T(t)\mathcal{H}^+]$ is closed in $C([0,+\infty); E_0)$. But $\mathcal{U}$ is compact in $C([0,+\infty); E_0)$ and bounded in $L_\infty(0,+\infty; E)$, so $\mathcal{U} \cap [\bigcup_{t\geq 0} T(t)\mathcal{H}^+]$ satisfies condition i) of Definition 4.2.3.

We have for any $h \geq 0$:

$$T(h)[\bigcup_{t\geq 0} T(t)\mathcal{H}^+] \subset [T(h)\bigcup_{t\geq 0} T(t)\mathcal{H}^+] = [\bigcup_{t\geq h} T(t)\mathcal{H}^+] \subset [\bigcup_{t\geq 0} T(t)\mathcal{H}^+].$$

By Lemma 4.2.4 $\mathcal{U} \cap [\bigcup_{t\geq 0} T(t)\mathcal{H}^+]$ satisfies condition ii) of Definition 4.2.3.

It remains to show that $\mathcal{U} \cap [\bigcup_{t\geq 0} T(t)\mathcal{H}^+]$ satisfies condition iii) of Definition 4.2.3, that is it is attracting. If not, for some $\delta > 0$ and some set $B \subset \mathcal{H}^+$ which is bounded in $L_\infty(0,+\infty; E)$, there exists a sequence $h_m \to \infty$ such that

$$\sup_{u\in B} \inf_{v\in \mathcal{U}\cap[\bigcup_{t\geq 0} T(t)\mathcal{H}^+]} \|T(h_m)u - v\|_{C([0,+\infty);E_0)} > \delta.$$

Then there are elements $u_m \in B$ such that

$$\inf_{v\in \mathcal{U}\cap[\bigcup_{t\geq 0} T(t)\mathcal{H}^+]} \|T(h_m)u_m - v\|_{C([0,+\infty);E_0)} > \delta. \tag{4.2.7}$$

On the other hand, since $\mathcal{U}$ is an attracting set, for any natural k there are a number m_k and elements $v_k \in \mathcal{U}$ such that

$$\|T(h_{m_k})u_{m_k} - v_k\|_{C([0,+\infty);E_0)} < \frac{1}{k}.$$

Since $\mathcal{U}$ is compact in $C([0,+\infty); E_0)$, without loss of generality the sequence v_k converges to an element v_0 as $k \to \infty$. Then the sequence $T(h_{m_k})u_{m_k}$ also converges to v_0. This sequence is bounded in $L_\infty(0,+\infty; E)$ (since B is bounded in $L_\infty(0,+\infty; E)$) and it is contained in $\bigcup_{t\geq 0} T(t)\mathcal{H}^+$. Thus, $v_0 \in \mathcal{U} \cap [\bigcup_{t\geq 0} T(t)\mathcal{H}^+]$ and $\|T(h_{m_k})u_{m_k} - v_0\|_{C([0,+\infty);E_0)} \underset{k\to\infty}{\to} 0$, which contradicts (4.2.7). □

Remark 4.2.11. Let $T(h)\mathcal{H}^+ \subset \mathcal{H}^+$ at all $h \geq 0$ and $[\mathcal{H}^+] = \mathcal{H}^+$. If there exists a minimal trajectory attractor $\mathcal{U}$ for the trajectory space $\mathcal{H}^+$, then by Theorem 4.2.5 $\mathcal{U} = \Pi_+\mathcal{K}(\mathcal{H}^+) \subset \mathcal{H}^+$, that is $\mathcal{U}$ is a trajectory attractor in the sense of [66]. Using Theorem 4.2.2 or Theorem 4.2.4 we conclude that $\mathcal{A} = \mathcal{U}(t) = \mathcal{K}(\mathcal{H}^+)(t)\ \ (t \geq 0)$ is a global attractor for the trajectory space $\mathcal{H}^+$ of equation (4.1.3).

Remark 4.2.12. In [66] there was investigated existence of attractors for the Navier–Stokes problem. The set of weak solutions of the Navier–Stokes problem satisfying an energy estimate of differential type was taken as a trajectory space $\mathcal{H}^+$. In this situation it appears that $T(h)\mathcal{H}^+ \subset \mathcal{H}^+$ for all $h \geq 0$ and $[\mathcal{H}^+] = \mathcal{H}^+$ ([66], Proposition 3.3).

Some similar information can be obtained under the conditions of Theorem 4.2.3 for the homogeneous trajectory quasiattractor.

Theorem 4.2.6. *Under the conditions of Theorem 4.2.3, the homogeneous trajectory quasiattractor $\mathcal{U}$ is contained in the set* $[\bigcup_{t\geq 0} T(t)\mathcal{H}^+]$.

Proof. Since $\mathcal{U}$ is contained in every attracting set for the trajectory space $\mathcal{H}^+$ which is compact in $C([0,+\infty); E_0)$ and bounded in $L_\infty(0,+\infty; E)$, it suffices to show that $\mathcal{U} \cap [\bigcup_{t\geq 0} T(t)\mathcal{H}^+]$ is an attracting set which is compact in $C([0,+\infty); E_0)$ and bounded in $L_\infty(0,+\infty; E)$. It may be realized in the same way as the check of the corresponding statements in the proof of Theorem 4.2.5. □

Corollary 4.2.2. *Under the conditions of Theorem 4.2.3, let $T(h)\mathcal{H}^+ \subset \mathcal{H}^+$ at all $h \geq 0$ and $[\mathcal{H}^+] = \mathcal{H}^+$. Then the homogeneous trajectory quasiattractor $\mathcal{U}$ is a minimal trajectory attractor for the trajectory space $\mathcal{H}^+$, and*

$$\mathcal{U} = \Pi_+\mathcal{K}(\mathcal{H}^+), \tag{4.2.8}$$

whereas for the global attractor one has the expression

$$\mathcal{A} = \mathcal{K}(\mathcal{H}^+)(t) \tag{4.2.9}$$

for any $t \geq 0$.

Proof. By Theorem 4.2.6, $\mathcal{U} \subset \mathcal{H}^+$. Moreover, $T(h)\mathcal{U} \subset T(h)\mathcal{H}^+ \subset \mathcal{H}^+$ for all $h \geq 0$. By Lemma 4.2.10, c), $\mathcal{U}$ is a minimal trajectory attractor, so (4.2.6) implies (4.2.8). Since $\mathcal{A} = \mathcal{U}(t)$, $t \geq 0$, one has (4.2.9). □

4.2.7 Correspondence between two concepts of global attractor

The results of the current subsection give some description of the connection between Definitions 4.1.7 and 4.2.6.

Let the assumptions of Section 4.1.2 hold true. Then we have the semigroup S_t. In this situation we can define the trajectory space as

$$\mathcal{H}_0^+ = \{u_b \in C([0,+\infty); E_0) \cap L_\infty(0,+\infty; E) | u_b(t) = S_t b, b \in E\}. \quad (4.2.10)$$

Then (4.1.7) implies that

$$T(h)\mathcal{H}_0^+ \subset \mathcal{H}_0^+ \quad (4.2.11)$$

for all $h \geq 0$.

Lemma 4.2.12. *Any (E, E_0)-attracting set for the semigroup S_t has the attraction property from Definition 4.2.6, ii). If the semigroup S_t is bounded, then any set $P \subset E$ which is compact in E_0 and possesses the attraction property from Definition 4.2.6, ii), is (E, E_0)-attracting for the semigroup S_t.*

Proof. Let P be an (E, E_0)-attracting set for the semigroup S_t. Take a set $B \subset \mathcal{H}_0^+$ which is bounded in $L_\infty(0,+\infty; E)$. Due to (4.2.10), there is a bounded set $V \subset E$ such that

$$B = \{u_b \in C([0,+\infty); E_0) \cap L_\infty(0,+\infty; E) | u_b(t) = S_t b, b \in V\}. \quad (4.2.12)$$

We have

$$\sup_{u \in B} \inf_{v \in P} \|u(t) - v\|_{E_0} = \sup_{b \in V} \inf_{v \in P} \|S_t b - v\|_{E_0}. \quad (4.2.13)$$

But P is (E, E_0)-attracting, so, for any $\varepsilon > 0$, $S_t V$ is contained in the open neighborhood of P in E_0 of the form

$$\{y \in E_0 | \inf_{v \in P} \|y - v\|_{E_0} < \varepsilon\} \quad (4.2.14)$$

for sufficiently large t. Therefore the right-hand side of (4.2.13) tends to zero as $t \to \infty$. Hence, P has the attraction property from Definition 4.2.6, ii).

Conversely, let $P \subset E$ possess the attraction property from Definition 4.2.6, ii). Take any bounded set $V \subset E$. If the semigroup S_t is bounded, the set B given by formula (4.2.12) is bounded in $L_\infty(0,+\infty; E)$, and (4.2.10) yields $B \subset \mathcal{H}_0^+$. Now the attraction property from Definition 4.2.6, ii), implies that the left-hand side of (4.2.13) tends to zero as $t \to \infty$. Take an open neighborhood W of P in E_0. Since P is compact in E_0, W contains a neighborhood of P of form (4.2.14) for some $\varepsilon > 0$. But the right-hand side of (4.2.13) tends to zero as $t \to \infty$, so the set $S_t V$ is contained in this neighborhood for sufficiently large t. Hence, P is (E, E_0)-attracting for the semigroup S_t. □

Lemmas 4.2.12 and 4.1.1 yield

Corollary 4.2.3. *If the semigroup S_t is bounded, then any set $P \subset E$ which is compact in E_0 and possesses the attraction property from Definition 4.2.6, ii), contains all invariant sets of the semigroup S_t that are bounded in E. In particular, a global attractor for the trajectory space $\mathcal{H}_0^+$ of equation (4.1.3) (in the sense of Definition 4.2.6) contains all invariant sets of the semigroup S_t that are bounded in E.*

Corollary 4.2.4. *Let the semigroup S_t be bounded. If there exists a global (E, E_0)-attractor $\mathcal{A}$ of equation (4.1.3), then it is a global attractor for the trajectory space $\mathcal{H}_0^+$ of equation (4.1.3).*

Proof. Property i) of Definition 4.2.6 is fulfilled for $\mathcal{A}$, and property ii) follows from Lemma 4.2.12. Observe that property iii) also holds. Really, for any set $P \subset E$ satisfying properties i) and ii) of Definition 4.2.6, one has $\mathcal{A} \subset P$ by Corollary 4.2.3. □

The reciprocal connection between Definitions 4.1.7 and 4.2.6 is more delicate.

Corollary 4.2.5 (cf. [66], Corollary 2.1). *Let the conditions of Theorem 4.2.2 hold for the trajectory space $\mathcal{H}_0^+$. Let the semigroup S_t be bounded. If the minimal trajectory attractor $\mathcal{U}$ is contained in $\mathcal{H}_0^+$, then the global attractor $\mathcal{A} = \mathcal{U}(t)$, $t \geq 0$, is a global (E, E_0)-attractor of equation (4.1.3).*

Proof. In view of Lemma 4.2.12 it suffices to prove that $\mathcal{A}$ is invariant for the semigroup S_t.

The inclusion $\mathcal{U} \subset \mathcal{H}_0^+$ and formula (4.2.10) involve the representation:

$$\mathcal{U} = \{u_b \in C([0, +\infty); E_0) \cap L_\infty(0, +\infty; E) | u_b(t) = S_t b, b \in \mathcal{U}(0)\}. \quad (4.2.15)$$

Hence, for all $h \geq 0$, $S_h \mathcal{U}(0) = \mathcal{U}(h)$, i.e. $S_h \mathcal{A} = \mathcal{A}$. □

Corollaries 4.2.5 and 4.2.2 imply

Corollary 4.2.6. *Let the conditions of Theorem 4.2.3 hold for the trajectory space $\mathcal{H}_0^+$. Let the semigroup S_t be bounded. If $\mathcal{H}_0^+ = [\mathcal{H}_0^+]$, then the global attractor $\mathcal{A}$ (existing by Theorem 4.2.3) is a global (E, E_0)-attractor of equation (4.1.3).*

Definition 4.2.9. We say that the semigroup S_t is *closed* if for any $t > 0$ and any sequence $\{b_m\}$ bounded in E such that

i) $b_m \to b_0 \in E$ in the topology of E_0,

ii) $S_t b_m$ converges in E_0, one has $\lim_{m\to\infty} S_t b_m = S_t b_0$.

Corollary 4.2.7. *Let the conditions of Theorem 4.2.3 hold for the trajectory space* $\mathcal{H}_0^+$. *Let the semigroup* S_t *be bounded and closed. Then the global attractor* $\mathcal{A}$ *is a global* (E, E_0)*-attractor of equation (4.1.3).*

Proof. In view of Corollary 4.2.6 it is enough to show that $[\mathcal{H}_0^+] = \mathcal{H}_0^+$. Take any sequence of trajectories $v_m \in \mathcal{H}_0^+$, $v_m \to v_0$ in $C([0, +\infty); E_0)$ which is bounded in $L_\infty(0, +\infty; E)$. Then $v_m(t) = S_t b_m$, $b_m \in E$, $m \in \mathbb{N}$, $t \geq 0$. But $v_m(t) \to v_0(t)$ in E_0. Let $b_0 = v_0(0)$. Then $b_m = v_m(0) \to b_0$, $S_t b_m \to v_0(t)$ in E_0, so $v_0(t) = S_t b_0$, i.e. $v_0 \in \mathcal{H}_0^+$. □

Remark 4.2.13. If the semigroup S_t is not closed (but still bounded), the global (E, E_0)-attractor may not exist even if there exist a minimal trajectory attractor and a global attractor for the trajectory space $\mathcal{H}_0^+$. Such a situation appears in the following example. In the space $E = E_0 = \mathbb{R}$, consider the following combined differential equation:

$$
\begin{aligned}
&u'(t) = -(u(t) - 1)^2, \quad u(t) > 1,\\
&u'(t) = -u^2(t), \quad 0 \leq u(t) \leq 1, \qquad\qquad (4.2.16)\\
&u'(t) = u^2(t), \quad u(t) < 0.
\end{aligned}
$$

The semigroup S_t here acts as follows:

$$
\begin{aligned}
&S_t(0) \equiv 0,\\
&S_t b = \Big(t + \frac{1}{b-1}\Big)^{-1} + 1, \quad b > 1,\\
&S_t b = \Big(t + \frac{1}{b}\Big)^{-1}, \quad 0 < b \leq 1, \qquad\qquad (4.2.17)\\
&S_t b = -\Big(t - \frac{1}{b}\Big)^{-1}, \quad b < 0.
\end{aligned}
$$

The minimal trajectory attractor is the set $\{u_0(t) \equiv 0, u_1(t) \equiv 1\}$. Hence, the global attractor is the set $\{0, 1\}$. However, there is no global $(\mathbb{R}, \mathbb{R})$-attractor, since the only bounded invariant set is $\{0\}$, but it is not $(\mathbb{R}, \mathbb{R})$-attracting for the semigroup S_t.

4.3 Attractors of non-autonomous equations

Let E and E_0 be Banach spaces, $E \subset E_0$, E is reflexive. Following [14, 15], we write an abstract non-autonomous evolution differential equation in the form

$$
u'(t) = A_{\sigma(t)}(u(t)), \quad u(t) \in E. \qquad (4.3.1)
$$

Here σ is a functional parameter, which is called the time symbol of equation (4.3.1). Assume that σ belongs to some fixed parameter set Σ, which is called the *symbol space* and is usually a subset of some space of time-dependent functions. In applications, a function $\sigma(t)$ consists of all time-dependent coefficients, terms and right-hand sides of a considered equation.

We investigate attractors of solutions of equation (4.3.1) which belong to the space $C([0,+\infty); E_0) \cap L_\infty(0,+\infty; E)$.

Assume that for every $\sigma \in \Sigma$ we have a fixed set

$$\mathcal{H}_\sigma^+ \subset C([0,+\infty); E_0) \cap L_\infty(0,+\infty; E)$$

of solutions (strong, weak, etc.) for equation (4.3.1), defined on the positive axis $t \geq 0$. The sets $\mathcal{H}_\sigma^+$ are called *trajectory spaces* and their elements are called *trajectories*.

Consider the united trajectory space $\mathcal{H}_\Sigma^+ = \bigcup\limits_{\sigma\in\Sigma} \mathcal{H}_\sigma^+$.

Definition 4.3.1. A set $P \subset C([0,+\infty); E_0) \cap L_\infty(0,+\infty; E)$ is called *uniformly* (with respect to $\sigma \in \Sigma$) *attracting* (for equation (4.3.1)) if for any set $B \subset \mathcal{H}_\Sigma^+$ which is bounded in $L_\infty(0,+\infty; E)$, one has

$$\sup_{u\in B} \inf_{v\in P} \|T(h)u - v\|_{C([0,+\infty);E_0)} \underset{h\to\infty}{\to} 0.$$

Definition 4.3.2. A set $P \subset C([0,+\infty); E_0) \cap L_\infty(0,+\infty; E)$ is called *uniformly absorbing* if for any set $B \subset \mathcal{H}_\Sigma^+$, which is bounded in $L_\infty(0,+\infty; E)$, there is $h \geq 0$ such that for all $t \geq h$:

$$T(t)B \subset P.$$

It is easy to see that any uniformly absorbing set is uniformly attracting.

Definition 4.3.3. A set $P \subset C([0,+\infty); E_0) \cap L_\infty(0,+\infty; E)$ is called a *uniform trajectory semiattractor* (for equation (4.3.1)) if

i) P is compact in $C([0,+\infty); E_0)$ and bounded in $L_\infty(0,+\infty; E)$;

ii) $T(t)P \subset P$ for any $t \geq 0$;

iii) P is uniformly attracting in the sense of Definition 4.3.1.

Definition 4.3.4. A set $P \subset C([0,+\infty); E_0) \cap L_\infty(0,+\infty; E)$ is called a *uniform trajectory quasiattractor* (for equation (4.3.1)) if it satisfies conditions i), iii) of Definition 4.3.3 and

ii') $T(t)P \supset P$ for any $t \geq 0$.

Definition 4.3.5. A uniform trajectory quasiattractor $\mathcal{U}$ is called *homogeneous* if the set $\mathcal{U}(t)$ does not depend on $t \geq 0$.

Definition 4.3.6. A set $P \subset C([0, +\infty); E_0) \cap L_\infty(0, +\infty; E)$ is called a *uniform* (with respect to $\sigma \in \Sigma$) *trajectory attractor* (for equation (4.3.1)) if it is a uniform trajectory semiattractor and a uniform trajectory quasiattractor (i.e. $T(t)P = P$).

Definition 4.3.7. A uniform trajectory attractor $\mathcal{U}$ is called *minimal* if it is contained in any other uniform trajectory attractor.

Definition 4.3.8. A set $\mathcal{A} \subset E$ is called a *uniform* (with respect to $\sigma \in \Sigma$) *global attractor* (in E_0) for equation (4.3.1) if

i) $\mathcal{A}$ is compact in E_0 and bounded in E;

ii) for any set $B \subset \mathcal{H}^+_\Sigma$ which is bounded in $L_\infty(0, +\infty; E)$, the attraction property is fulfilled:

$$\sup_{u \in B} \inf_{v \in \mathcal{A}} \|u(t) - v\|_{E_0} \underset{t \to \infty}{\to} 0.$$

iii) $\mathcal{A}$ is the minimal set satisfying conditions i) and ii) (that is $\mathcal{A}$ is contained in every set satisfying conditions i) and ii)).

Remark 4.3.1. The minimal uniform trajectory attractor and the uniform global attractor depend on the symbol space Σ. It is easy to observe that if we have two symbol spaces $\Sigma_1 \subset \Sigma_2$, then for corresponding minimal uniform trajectory attractors and uniform global attractors one has $\mathcal{U}_{\Sigma_1} \subset \mathcal{U}_{\Sigma_2}$, $\mathcal{A}_{\Sigma_1} \subset \mathcal{A}_{\Sigma_2}$. Moreover, existence of uniform attractors may fail after an extension of the symbol space (see Remark 4.3.3).

Remark 4.3.2. We do not assume that the semigroup $T(t)$ *acts* on Σ, i.e. that for any $t \geq 0$ the map $T(t) : \sigma(s) \to \sigma(s + t)$ (remember that the time symbols σ are time-dependent functions) transforms Σ into itself, and the family $T(t)$ is a semigroup on Σ in the sense of Definition 4.1.1. We do not assume also that the family of trajectory spaces $\mathcal{H}^+_\sigma$, $\sigma \in \Sigma$, is *translation-coordinated*, i.e. $T(h)\mathcal{H}^+_\sigma \subset \mathcal{H}^+_{T(h)\sigma}$ for $h \geq 0$ (see [15]).

Remark 4.3.3. Since the action of the semigroup $T(t)$ is not required of the symbol space Σ, one may assume that the symbol space consists of only one element σ_* corresponding to the particular equation under consideration. However, in applications, in order to justify the word *uniform* in Definitions 4.3.1 – 4.3.8 it seems to be preferable to assume that $\Sigma \supset \Sigma_0$ where

$$\Sigma_0 = \{T(t)\sigma_* | t \geq 0\}.$$

But the uniform attractors in the case $\Sigma = \{\sigma_*\}$ (as well as in the case of any other possible Σ) are of independent interest. Let us illustrate this with the following example. In the space $E = E_0 = \mathbb{R}$ we consider the differential equation

$$u'(t) = \frac{u(t)}{t} \chi(u^2(t)t^2). \tag{4.3.2}$$

Here $\chi(s) = 1$ for $s > 2$, $\chi(s) = 0$ for $s < 1$ and $\chi(s) = -1$ elsewhere. The solutions of this equation are

$$u(t) = Ct, \quad u^2(t)t^2 > 2, \tag{4.3.3}$$

$$u(t) = \frac{C}{t}, \quad 1 \le u^2(t)t^2 \le 2, \tag{4.3.4}$$

$$u(t) = C, \quad u^2(t)t^2 < 1. \tag{4.3.5}$$

Here and below in this remark, C is a real number.

Let us rewrite (4.3.2) in the form (4.3.1):

$$u'(t) = u(t)\sigma(t) \tag{4.3.6}$$

where $\sigma(t)$ is a scalar function of $t > 0$. Equation (4.3.2) corresponds to the case

$$\sigma(t) = \sigma_*(t) = \frac{\chi(u^2(t)t^2)}{t}.$$

For each function σ, define the trajectory space $\mathcal{H}^+_\sigma$ as the subset of $C([0, +\infty); \mathbb{R})$ consisting of the functions which at almost all $t > 0$ are differentiable and satisfy (4.3.6). Note that the trajectories from $\mathcal{H}^+_{\sigma_*}$ are of the form

$$u(t) = C, \quad t|C| < 1, \tag{4.3.7}$$

$$u(t) = \frac{\operatorname{sign} C}{t}, \quad t|C| \ge 1. \tag{4.3.8}$$

Therefore, if $\Sigma = \{\sigma_*\}$, then the minimal uniform trajectory attractor is the set $\{u_0(t) \equiv 0\}$, and the uniform global attractor is the set $\{0\}$. But if $\Sigma \supset \Sigma_0$, $\mathcal{H}^+_\Sigma$ contains solutions of form (4.3.3), so there are no compact uniformly attracting sets.

Due to Remark 4.2.3, the analogues of all results of Section 4.2 (except the ones from Section 4.2.7) are true in the non-autonomous case. In particular, we have the following statements.

Theorem 4.3.1 (see [77]). *Assume that there exists a uniform trajectory semiattractor P for equation (4.3.1). Then there exists a minimal uniform trajectory attractor $\mathfrak{U}$ for equation (4.3.1). Here one has $\Pi_+\mathcal{K}(\mathcal{H}^+_\Sigma) \subset \mathfrak{U} = \Pi_+\mathcal{K}(\mathfrak{U}) \subset \Pi_+\mathcal{K}(P)$ and the kernel $\mathcal{K}(\mathcal{H}^+_\Sigma)$ is relatively compact in $C((-\infty, +\infty); E_0)$ and bounded in $L_\infty(-\infty, +\infty; E)$.*

Theorem 4.3.2. *Assume that there exists a uniformly absorbing set P for equation (4.3.1) which is compact in $C([0, +\infty); E_0)$ and bounded in $L_\infty(0, +\infty; E)$. Then there exists a minimal uniform trajectory attractor $\mathfrak{U}$ for equation (4.3.1). Here one has $\Pi_+\mathcal{K}(\mathcal{H}^+_\Sigma) \subset \mathfrak{U} = \Pi_+\mathcal{K}(\mathfrak{U}) \subset \Pi_+\mathcal{K}(P)$ and the kernel $\mathcal{K}(\mathcal{H}^+_\Sigma)$ is relatively compact in $C((-\infty, +\infty); E_0)$ and bounded in $L_\infty(-\infty, +\infty; E)$.*

Theorem 4.3.3 (see [77]). *If there exists a minimal uniform trajectory attractor $\mathfrak{U}$ for equation (4.3.1), then there is a uniform global attractor $\mathcal{A}$ for equation (4.3.1) and for all $t \geq 0$ one has*

$$\mathcal{K}(\mathcal{H}_\Sigma^+)(t) \subset \mathcal{A} = \mathfrak{U}(t) = \mathcal{K}(\mathfrak{U})(t).$$

Theorem 4.3.4. *Assume that there exists a uniformly attracting set P for equation (4.3.1) which is compact in $C([0, +\infty); E_0)$ and bounded in $L_\infty(0, +\infty; E)$. Then there exists a homogeneous uniform trajectory quasiattractor $\mathfrak{U} \subset P$ for equation (4.3.1) such that the set $\mathcal{A} = \mathfrak{U}(t)$, $t \geq 0$, is a uniform global attractor for equation (4.3.1).*

Lemma 4.3.1. *Let P be a uniformly attracting (or absorbing) set for equation (4.3.1) which is relatively compact in $C([0, +\infty); E_0)$ and bounded in $L_\infty(0, +\infty; E)$. Then its closure $\overline{P}$ in $C([0, +\infty); E_0)$ is a uniformly attracting (resp. absorbing) set for equation (4.3.1) which is compact in $C([0, +\infty); E_0)$ and bounded in $L_\infty(0, +\infty; E)$. If in addition $T(t)P \subset P$ for any $t \geq 0$, then $\overline{P}$ is a uniform trajectory semiattractor.*

Lemma 4.3.2. *Let $\mathfrak{U} \subset \mathcal{H}_\Sigma^+$ be a uniform trajectory quasiattractor. Then*

a) *$\mathfrak{U}$ is contained in any compact uniformly attracting set $P \subset C([0, +\infty); E_0)$;*
b) *in $\mathcal{H}_\Sigma^+$ there is no uniform trajectory quasiattractors different from $\mathfrak{U}$;*
c) *if, in addition, it is known that $T(h)\mathfrak{U} \subset \mathcal{H}_\Sigma^+$ for all $h \geq 0$, then $\mathfrak{U}$ is a minimal uniform trajectory attractor.*

Theorem 4.3.5. *Assume that there exists a uniformly attracting set P for equation (4.3.1) which is compact in $C([0, +\infty); E_0)$ and bounded in $L_\infty(0, +\infty; E)$. Let $T(h)P \subset \mathcal{H}_\Sigma^+$ for all $h \geq 0$. Then there exist a minimal uniform trajectory attractor $\mathfrak{U} = \Pi_+\mathcal{K}(\mathcal{H}_\Sigma^+)$ for equation (4.3.1) and a uniform global attractor $\mathcal{A} = \mathfrak{U}(t) = \mathcal{K}(\mathcal{H}_\Sigma^+)(t)$, $t \geq 0$ for equation (4.3.1).*

Remark 4.3.4. In [66] (Theorem 4.1, p. 210) this theorem was proved under additional assumptions that the family of trajectory spaces is translation-coordinated (see Remark 4.3.2) and that the symbol space Σ is a compact metric space.

Theorem 4.3.6. *Assume that there exists a minimal uniform trajectory attractor $\mathfrak{U}$ for equation (4.3.1). Then*

$$\Pi_+\mathcal{K}(\mathcal{H}_\Sigma^+) \subset \mathfrak{U} \subset \Pi_+\mathcal{K}\Big(\Big[\bigcup_{t\geq 0} T(t)\mathcal{H}_\Sigma^+\Big]\Big). \tag{4.3.9}$$

Assume now that Σ is a Hausdorff topological space, and the semigroup $T(t)$ acts on Σ.

Definition 4.3.9 (cf. [15]). The family of trajectory spaces $\{\mathcal{H}_\sigma^+\}$ is said to be *closed* if for any sequence of symbols $\sigma_m \to \sigma_0$ and for any sequence of trajectories $u_m \in \mathcal{H}_{\sigma_m}^+$ converging in $C([0,+\infty); E_0)$ and being bounded in $L_\infty(0,+\infty; E)$, one has $\lim_{m\to+\infty} u_m \in \mathcal{H}_{\sigma_0}^+$.

Corollary 4.3.1. *Let the family of trajectory spaces $\{\mathcal{H}_\sigma^+\}$ be closed and translation-coordinated. Let Σ be compact. If there exists a minimal uniform trajectory attractor $\mathfrak{U}$ for equation (4.3.1), then*

$$\mathfrak{U} = \Pi_+\mathcal{K}(\mathcal{H}_\Sigma^+). \tag{4.3.10}$$

Proof. Since $\{\mathcal{H}_\sigma^+\}$ is translation-coordinated, we have $T(h)\mathcal{H}_\Sigma^+ \subset \mathcal{H}_\Sigma^+$ for all $h \geq 0$. Let us prove that $[\mathcal{H}_\Sigma^+] = \mathcal{H}_\Sigma^+$. Take any sequence of trajectories $u_m \in \mathcal{H}_{\sigma_m}^+$, $u_m \to u_0$ in $C([0,+\infty); E_0)$ that is bounded in $L_\infty(0,+\infty; E)$. Since Σ is compact, without loss of generality there is $\sigma_0 \in \Sigma$ such that $\sigma_m \to \sigma_0$. Since the family $\{\mathcal{H}_\sigma^+\}$ is closed, $u_0 \in \mathcal{H}_{\sigma_0}^+ \subset \mathcal{H}_\Sigma^+$. Now (4.3.9) yields (4.3.10). □

Thus an analogue of Corollary 4.2.2 may be written as

Corollary 4.3.2. *Under the conditions of Theorem 4.3.4, let the family of trajectory spaces $\{\mathcal{H}_\sigma^+\}$ be closed and translation-coordinated and let Σ be compact. Then the homogeneous uniform trajectory quasiattractor $\mathfrak{U}$ is a minimal uniform trajectory attractor, and*

$$\mathfrak{U} = \Pi_+\mathcal{K}(\mathcal{H}_\Sigma^+), \tag{4.3.11}$$

whereas for the uniform global attractor one has the expression

$$\mathcal{A} = \mathcal{K}(\mathcal{H}_\Sigma^+)(t) \tag{4.3.12}$$

for any $t \geq 0$.

Chapter 5

Strong solutions for equations of motion of viscoelastic medium

5.1 The Guillopé–Saut theorem

This section is concerned with the initial-boundary value problem describing the dynamics of a homogeneous incompressible viscoelastic medium in a bounded domain in $\mathbb{R}^n$, $n = 2, 3$. The constitutive relation which we consider here is Jeffreys' law (1.3.32) with Oldroyd's derivative (1.3.36). Combining it with the equation of motion (1.1.12), the incompressibility condition (1.1.10) and the no-slip condition (1.1.15), we get the following problem:

$$\frac{\partial u}{\partial t} + \sum_{i=1}^{n} u_i \frac{\partial u}{\partial x_i} - \operatorname{Div} \sigma + \operatorname{grad} p = f_0, \ (t,x) \in [0,T] \times \Omega, \tag{5.1.1}$$

$$\sigma + \lambda_1 \frac{D_{\mathbf{a}}\sigma}{Dt} = 2\eta\big(\mathcal{E}(u) + \lambda_2 \frac{D_{\mathbf{a}}\mathcal{E}(u)}{Dt}\big), \ (t,x) \in [0,T] \times \Omega, \tag{5.1.2}$$

$$\operatorname{div} u = 0, \ (t,x) \in [0,T] \times \Omega, \tag{5.1.3}$$

$$u(t,x) = 0, \ (t,x) \in [0,T] \times \partial\Omega. \tag{5.1.4}$$

Here u is the unknown velocity vector, σ is the unknown extra-stress tensor, p is the unknown scalar pressure function, f_0 is the given body force (all of them depend on a point x of a sufficiently regular bounded domain $\Omega \subset \mathbb{R}^n$, $n = 2, 3$ and on a moment of time t). The divergences div and Div and the gradient grad are taken with respect to the variable x. Besides, $\mathcal{E}(u) = (\mathcal{E}_{ij}(u))$, $\mathcal{E}_{ij}(u) = \frac{1}{2}(\frac{\partial u_i}{\partial x_j} + \frac{\partial u_j}{\partial x_i})$ is the strain velocity tensor, $\eta > 0$ is the viscosity of the medium, λ_1 is the relaxation time, λ_2 is the retardation time, $0 < \lambda_2 < \lambda_1$, $-1 \leq \mathbf{a} \leq 1$.

Denote $\mu_1 = \eta \frac{\lambda_2}{\lambda_1}$, $\eta_1 = \eta - \mu_1$, $\tau = \sigma - 2\mu_1 \mathcal{E}(u)$. Proceeding as in Section 1.5.1, and remembering (1.3.36), (1.3.7) and (1.3.34), we can rewrite (5.1.1) and (5.1.2) as follows:

$$\frac{\partial u}{\partial t} + \sum_{i=1}^{n} u_i \frac{\partial u}{\partial x_i} - \mu_1 \Delta u - \operatorname{Div} \tau + \operatorname{grad} p = f_0, \tag{5.1.5}$$

$$\tau + \lambda_1\big(\frac{\partial \tau}{\partial t} + \sum_{i=1}^{n} u_i \frac{\partial \tau}{\partial x_i} + \tau W - W\tau - \mathbf{a}(\tau\mathcal{E} + \mathcal{E}\tau)\big) = 2\eta_1 \mathcal{E}(u). \tag{5.1.6}$$

Remember that here $W = (W_{ij}(u))$, $W_{ij}(u) = \frac{1}{2}(\frac{\partial u_i}{\partial x_j} - \frac{\partial u_j}{\partial x_i})$.

The main result of this section is

Theorem 5.1.1. *Given*

$$f_0 \in L_2(0,T; H^1(\Omega)^n) \cap W_2^1(0,T; H^{-1}(\Omega)^n),$$

and

$$a \in H^2(\Omega)^n \cap V, \ \tau_0 \in H_M^2(\Omega),$$

there exist $t_0 > 0$ *and a triple* (u, p, τ) *from the class*

$$u \in L_2(0,t_0; H^3(\Omega)^n) \cap C([0,t_0]; H^2(\Omega)^n \cap V) \\ \cap W_2^1(0,t_0; V) \cap C^1([0,t_0]; L_2(\Omega)^n), \tag{5.1.7}$$

$$p \in L_2(0,t_0; H^2(\Omega)), \tag{5.1.8}$$

$$\tau \in C([0,t_0]; H_M^2(\Omega)) \cap C^1([0,t_0]; H_M^1(\Omega)) \tag{5.1.9}$$

which is a solution to problem (5.1.3) – (5.1.6) and satisfies the initial condition

$$u(0) = a, \quad \tau(0) = \tau_0. \tag{5.1.10}$$

Remark 5.1.1. The result is due to C. Guillopé and J.C. Saut [21], Theorem 2.4. R. Talhouk [60] generalized it onto the case of unbounded domains. Some other local existence and uniqueness results for problem (5.1.3) – (5.1.6) may be found in [13, 24, 25].

Remark 5.1.2. The uniqueness of u and τ in Theorem 5.1.1 can be proved by standard methods ([21], Theorem 2.5, [25], Theorem 3.1, see also the proof of uniqueness in Lemma 5.4.4 of the current chapter).

Remark 5.1.3. As in Theorem 3.2.1, an a priori estimate for $\|u(t)\|_2$ and $\|\tau(t)\|_2$ in Theorem 5.1.1 yields that $t_0 = T$. Such an a priori estimate can be proved provided the numbers η_1, $\|a\|_2$, $\|\tau_0\|_2$, $\|f_0\|_{L_2(0,T;H^1(\Omega)^n)}$ and $\|f_0'\|_{L_2(0,T;H^{-1}(\Omega)^n)}$ are small enough, see [21], Theorem 3.3, (or [25], for more general results of this type). In the case $\Omega = \mathbb{R}^n$ the global in time existence may be proved for arbitrary $\eta_1 > 0$ even for more general models (see below, Section 5.2). Let us mention also paper [13] with global existence results for small a, τ_0, f_0 in Besov spaces (Ω is $\mathbb{R}^n$ or torus).

Remark 5.1.4. If f is small and independent of time, one can prove also the existence of time-independent (*stationary*) solutions for problem (5.1.3) – (5.1.6) (see [50, 51, 21, 22, 25]).

Remark 5.1.5. Conditions (5.1.3), (5.1.4) are understood in the sense

$$u(t) \in V,\ t \in [0, T]$$

(cf. Section 2.1.2).

The proof of Theorem 5.1.1 requires some lemmas on solvability of linear Stokes' and transport problems.

Consider the Stokes problem in the form

$$\frac{\partial u}{\partial t} - \mu_1 \Delta u + \operatorname{grad} p = f, \quad (t, x) \in [0, T] \times \Omega, \tag{5.1.11}$$

$$\operatorname{div} u = 0, \quad (t, x) \in [0, T] \times \Omega, \tag{5.1.12}$$

$$u(t, x) = 0, \quad (t, x) \in [0, T] \times \partial\Omega, \tag{5.1.13}$$

$$u(0, x) = a(x), \quad x \in \Omega. \tag{5.1.14}$$

Lemma 5.1.1 (see [61], p. 181). *Given $f \in L_2(0, T; L_2(\Omega)^n)$ and $a \in V$, there exists a solution (u, p) to problem (5.1.11) – (5.1.14) from the class*

$$u \in L_2(0, T; H^2(\Omega)^n) \cap C([0, T]; V) \cap W_2^1(0, T; H), \tag{5.1.15}$$

$$p \in L_2(0, T; H^1(\Omega)). \tag{5.1.16}$$

The pair $(u, \operatorname{grad} p)$ is unique. The solution (u, p) is arbitrary regular with respect to t and x is provided Ω, f and a are regular enough.

Denote by Ω^k, $k = 1, 2, \ldots$, the connected components of Ω. Fix a function $\vartheta \in L_2(\Omega)$ with $\int_{\Omega^k} \vartheta(x)\, dx \neq 0$ for each k. Denote by $E^m_{\vartheta,\Omega}$ the subspace of $H^m(\Omega)$, $m = 1, 2$, which is the intersection of the kernels of the functionals $(\cdot, \vartheta)_{L_2(\Omega^k)}$. Lemma 5.1.1 and Corollary 3.1.3 imply

Corollary 5.1.1. *The operator*

$$S : L_2(0, T; H^2(\Omega)^n) \cap C([0, T]; V) \cap W_2^1(0, T; H) \times L_2(0, T; E^1_{\vartheta,\Omega})$$
$$\to L_2(0, T; L_2(\Omega)^n) \times V,$$
$$S(u, p) = (\frac{\partial u}{\partial t} - \mu_1 \Delta u + \operatorname{grad} p, u(0)),$$

is an isomorphism.

Corollary 5.1.2. *S^{-1} is a bounded operator from*

$$L_2(0, T; H^1(\Omega)^n) \cap C([0, T]; L_2(\Omega)^n) \cap W_2^1(0, T; H^{-1}(\Omega)^n) \times H^2(\Omega)^n \cap V$$

into

$$L_2(0,T;H^3(\Omega)^n)\cap C([0,T];H^2(\Omega)^n\cap V)\cap W_2^1(0,T;V)\cap C^1([0,T];H)$$
$$\times L_2(0,T;E^2_{\vartheta,\Omega}).$$

Proof. If f and a are more regular with respect to t and x, $S^{-1}(f,a)$ is also more regular. Since the smooth functions are dense in $L_2(0,T;H^1(\Omega)^n)\cap C([0,T];L_2(\Omega)^n)\cap W_2^1(0,T;H^{-1}(\Omega)^n)\times H^2(\Omega)^n\cap V$, it suffices to show that S^{-1} is bounded on the sufficiently regular functions.

Denote $S^{-1}(f,a)$ by (u,p), where (f,a) are regular enough. We have:

$$\frac{\partial u}{\partial t}-\mu_1\Delta u+\operatorname{grad}p=f. \tag{5.1.17}$$

This implies

$$u'(0)-\mu_1\Delta a+\operatorname{grad}p(0)=f(0),$$

so

$$\|u'(0)+\operatorname{grad}p(0)\|\le\mu_1\|\Delta a\|+\|f(0)\|.$$

But u' and $\operatorname{grad}p$ are orthogonal in $L_2(\Omega)^n$ (this follows from Lemma 3.1.1 or direct integration by parts using the condition $\operatorname{div}u'=0$). Thus,

$$\|u'(0)\|\le\mu_1\|a\|_2+\|f\|_{C([0,T];L_2(\Omega)^n)}. \tag{5.1.18}$$

Differentiate (5.1.17) with respect to t:

$$\frac{\partial u'}{\partial t}-\mu_1\Delta u'+\frac{\partial}{\partial t}\operatorname{grad}p=f'.$$

Taking the scalar product of each term with u' in $L_2(\Omega)^n$ for each $t\in[0,T]$, we get:

$$\frac{d}{dt}(u'(t),u'(t))-\mu_1\int_\Omega\Delta u'(t)(x)\,u'(t)(x)\,dx+\int_\Omega\frac{\partial\operatorname{grad}p(t)(x)}{\partial t}u'(t)(x)\,dx$$
$$=(f'(t),u'(t)).$$

Integrating by parts (cf. Section 6.1.2) in the second and the third terms and taking into account that $\operatorname{div}u'=0$, we obtain:

$$\frac{d}{dt}(u',u')+\mu_1(\nabla u',\nabla u')=(f',u').$$

Integration from 0 to t yields

$$\|u'(t)\|^2+\mu_1\int_0^t\|\nabla u'(s)\|^2ds\le\|u'(0)\|^2+\int_0^t\|f'(s)\|_{-1}\|u'(s)\|_1\,ds,$$

so by inequality (2.1.28)

$$\|u'(t)\|^2 + \mu_1 \int_0^t \|\nabla u'(s)\|^2 ds \\ \leq \|u'(0)\|^2 + (1 + K_0^2(\Omega))^{1/2} \int_0^t \|f'(s)\|_{-1} \|\nabla u'(s)\| \, ds.$$

Application of Cauchy's inequality $ab \leq \frac{a^2}{2c} + \frac{b^2 c}{2}$ to the last integral implies

$$\|u'(t)\|^2 + \mu_1 \int_0^t \|\nabla u'(s)\|^2 ds \\ \leq \|u'(0)\|^2 + (1 + K_0^2(\Omega))^{1/2} \int_0^t \Big(\frac{(1 + K_0^2(\Omega))^{1/2}}{2\mu_1} \|f'(s)\|_{-1}^2 \\ + \frac{\mu_1}{2(1 + K_0^2(\Omega))^{1/2}} \|\nabla u'(s)\|^2\Big) \, ds,$$

so

$$\|u'(t)\|^2 + \frac{\mu_1}{2} \int_0^t \|\nabla u'(s)\|^2 ds \leq \|u'(0)\|^2 + \frac{(1 + K_0^2(\Omega))}{2\mu_1} \int_0^t \|f'(s)\|_{-1}^2 \, ds. \tag{5.1.19}$$

Now, if we take a pair (f, a) from a fixed bounded set in $L_2(0, T; H^1(\Omega)^n) \cap C([0, T]; L_2(\Omega)^n) \cap W_2^1(0, T; H^{-1}(\Omega)^n) \times H^2(\Omega)^n \cap V$, then, by estimates (5.1.18) and (5.1.19), u' is bounded in $L_2(0, T; H^1(\Omega)^n) \cap C([0, T]; L_2(\Omega)^n)$ by a constant independent of (f, a). But $-\mu_1 \Delta u + \operatorname{grad} p = f - u'$, so, by a classical regularity result for the stationary Stokes problem ([61], Proposition 1.2.2), u is bounded in $L_2(0, T; H^3(\Omega)^n) \cap C([0, T]; H^2(\Omega)^n)$ and p is bounded in $L_2(0, T; H^2(\Omega))$. The proof is complete. □

Now, let us give an existence result for the following equation of transport type:

$$\tau + \lambda_1\Big(\frac{\partial \tau}{\partial t} + \sum_{i=1}^n v_i \frac{\partial \tau}{\partial x_i} + \tau W(v) - W(v)\tau - \mathbf{a}(\tau \mathcal{E}(v) + \mathcal{E}(v)\tau)\Big) \\ = 2\eta_1 \mathcal{E}(v), \ (t, x) \in [0, T] \times \Omega. \tag{5.1.20}$$

Here $v(t, x)$ is a given velocity field, and τ is unknown.

Lemma 5.1.2 (see [21], Lemma 2.3). *Given $v \in L_1(0, T; H^3(\Omega)^n) \cap L_\infty(0, T; H^2(\Omega)^n \cap V)$, $\tau_0 \in H_M^2(\Omega)$, there exists a unique (e.g. in the class $C([0, T]; H_M^1(\Omega))$ solution $\tau \in C([0, T]; H_M^2(\Omega)) \cap W_\infty^1(0, T; H_M^1(\Omega))$ to equation (5.1.20), satisfying the initial condition*

$$\tau(0) = \tau_0. \tag{5.1.21}$$

One has the following estimate for this solution:

$$\|\tau\|_{C([0,T];H^2_M(\Omega))} \le (\|\tau_0\|_2 + K_1)\exp(K_2\|v\|_{L_1(0,T;H^3(\Omega)^n)}) \tag{5.1.22}$$

$$\begin{aligned} &\|\tau'\|_{L_\infty(0,T;H^1_M(\Omega))} \\ &\quad \le K_3(\|v\|_{L_\infty(0,T;H^2(\Omega)^n)} + K_4)(\|\tau_0\|_2 + K_1) \\ &\qquad \cdot \exp(K_2\|v\|_{L_1(0,T;H^3(\Omega)^n)}). \end{aligned} \tag{5.1.23}$$

Here $K_1, \dots, K_4$ depend only on Ω, λ_1, η_1.

If, in addition, $v \in C([0,T]; H^2(\Omega)^n \cap V)$, then $\tau \in C^1([0,T]; H^1_M(\Omega))$.

Now we are ready to turn to

Proof of Theorem 5.1.1. Consider the set $Q = Q(t_0, R)$ which consists of the pairs

$$\begin{aligned} (v,\varsigma);\quad & v \in L_2(0,t_0; H^3(\Omega)^n) \cap L_\infty(0,t_0; H^2(\Omega)^n \cap V) \\ &\qquad \cap C_w([0,t_0]; H^2(\Omega)^n) \cap W^1_2(0,t_0; V) \cap W^1_\infty(0,t_0; L_2(\Omega)^n); \\ &\varsigma \in L_\infty(0,t_0; H^2_M(\Omega)) \cap C_w([0,t_0]; H^2_M(\Omega)) \cap W^1_\infty(0,t_0; H^1_M(\Omega)); \end{aligned}$$

$$\begin{aligned} \|v\|^2_{L_2(0,t_0;H^3(\Omega)^n)} + \|v\|^2_{L_\infty(0,t_0;H^2(\Omega)^n)} + \|v\|^2_{W^1_2(0,t_0;V)} + \|v\|^2_{W^1_\infty(0,t_0;L_2(\Omega)^n)} \\ + \|\varsigma\|_{L_\infty(0,t_0;H^2_M(\Omega))} + \|\varsigma\|_{W^1_\infty(0,t_0;H^1_M(\Omega))} \le R; \\ v(0) = a;\ \varsigma(0) = \tau_0 \end{aligned}$$

(the parameters $t_0, R > 0$ will be defined below). This set is convex and non-empty for R large enough (it contains, for example, the pair (u_*, τ_*) where $(u_*, p_*) = S^{-1}(0,a)$ and $\tau_*(t) \equiv \tau_0$). By Theorem 2.2.6 it is relatively compact in the space $X = C([0,t_0]; V) \times C([0,t_0]; H^1_M(\Omega))$. Moreover, it is closed in X. Really, for any sequence $(v_m, \varsigma_m) \in Q$ converging to a pair (v_0, ς_0) in the topology of X, one has $v_0(0) = a$, $\varsigma_0(0) = \tau_0$; $v_m \rightharpoonup v_0$ weakly in $L_2(0,t_0; H^3(\Omega)^n)$ and in $W^1_2(0,t_0; V)$, $*$-weakly in $L_\infty(0,t_0; H^2(\Omega)^n)$ and in $W^1_\infty(0,t_0; L_2(\Omega)^n)$; $\varsigma_m \rightharpoonup \varsigma_0$ $*$-weakly in $L_\infty(0,t_0; H^2_M(\Omega))$ and in $W^1_\infty(0,t_0; H^1_M(\Omega))$. Hence, v_0 belongs to $L_\infty(0,t_0; H^2(\Omega)^n)$ and to $W^1_2(0,t_0; V) \subset C_w([0,t_0]; V)$, so, by Lemma 2.2.6, $v_0 \in C_w([0,t_0]; H^2(\Omega)^n)$. Analogously, $\varsigma_0 \in C_w([0,t_0]; H^2_M(\Omega))$. Thus, $(v_0, \varsigma_0) \in Q$.

Consider the mapping:

$$\phi : Q \subset X \to X,$$

$$\phi(v,\varsigma) = (u,\tau),$$

u is the first component of the solution of Stokes' problem $(u,p) = [S^{-1}(f,a)]\big|_{[0,t_0]}$ where $f = f_0 + \operatorname{Div}\varsigma - \sum_{i=1}^n v_i \frac{\partial v}{\partial x_i}$; τ is the solution of the transport problem (5.1.20) – (5.1.21), restricted to the segment $[0,t_0]$. Observe that if (v,ς) is a fixed point of ϕ, then the corresponding triple (u, p, τ) is a solution to (5.1.3) – (5.1.6) in class (5.1.7) – (5.1.9).

Let us show that $\phi(Q) \subset Q$ for appropriate t_0 and R. Below in the proof C stands for various constants which do not depend on t_0 and R.

Take a pair $(v, \varsigma) \in Q$ and let $(u, \tau) = \phi(v, \varsigma)$.

Estimate (2.1.21) gives $\| \sum_{i=1}^{n} v_i \frac{\partial v}{\partial x_i} \|_1 \le C \|v\|_2^2$.

Using Lemma 2.2.1, a), we obtain:

$$
\begin{aligned}
&\|f\|_{L_2(0,t_0;H^1(\Omega)^n)} \\
&\quad\le \|f_0\|_{L_2(0,t_0;H^1(\Omega)^n)} + \|\operatorname{Div} \varsigma\|_{L_2(0,t_0;H^1(\Omega)^n)} \\
&\qquad + \|\sum_{i=1}^{n} v_i \frac{\partial v}{\partial x_i}\|_{L_2(0,t_0;H^1(\Omega)^n)} \\
&\quad\le \|f_0\|_{L_2(0,t_0;H^1(\Omega)^n)} + t_0^{1/2} \|\operatorname{Div} \varsigma\|_{L_\infty(0,t_0;H^1(\Omega)^n)} \\
&\qquad + t_0^{1/2} \|\sum_{i=1}^{n} v_i \frac{\partial v}{\partial x_i}\|_{L_\infty(0,t_0;H^1(\Omega)^n)} \\
&\quad\le \|f_0\|_{L_2(0,t_0;H^1(\Omega)^n)} + C t_0^{1/2} \|\varsigma\|_{L_\infty(0,t_0;H^2_M(\Omega))} \\
&\qquad + C t_0^{1/2} \|v\|^2_{L_\infty(0,t_0;H^2(\Omega)^n)} \\
&\quad\le C + C R t_0^{1/2}.
\end{aligned}
\tag{5.1.24}
$$

Take a function $\varphi \in H^1(\Omega)^n$. Applying integration by parts and estimate (2.1.19), and remembering that $\operatorname{div} v = 0$, we get

$$
\begin{aligned}
\Big|\Big(\sum_{i=1}^{n} \frac{\partial}{\partial t}\Big(v_i(t) \frac{\partial v(t)}{\partial x_i}\Big), \varphi\Big)\Big| &= \Big|\Big(\sum_{i=1}^{n} v_i'(t) \frac{\partial v(t)}{\partial x_i}, \varphi\Big) + \Big(\sum_{i=1}^{n} v_i(t) \frac{\partial v'(t)}{\partial x_i}, \varphi\Big)\Big| \\
&= \Big|\Big(\sum_{i=1}^{n} v_i'(t) v(t), \frac{\partial \varphi}{\partial x_i}\Big) + \Big(\sum_{i=1}^{n} v_i(t) v'(t), \frac{\partial \varphi}{\partial x_i}\Big)\Big| \\
&\le C \|v'\|_0 \|v\|_2 \|\varphi\|_1,
\end{aligned}
$$

so

$$
\|\sum_{i=1}^{n} \frac{\partial}{\partial t}\Big(v_i \frac{\partial v}{\partial x_i}\Big)\|_{-1}(t) \le C \|v'(t)\|_0 \|v(t)\|_2.
$$

Thus we have

$$
\begin{aligned}
&\|f'\|_{L_2(0,t_0;H^{-1}(\Omega)^n)} \\
&\quad \le \|f_0'\|_{L_2(0,t_0;H^{-1}(\Omega)^n)} + \|\operatorname{Div}\varsigma'\|_{L_2(0,t_0;H^{-1}(\Omega)^n)} \\
&\qquad + \Big\|\sum_{i=1}^n \frac{\partial}{\partial t}\Big(v_i\frac{\partial v}{\partial x_i}\Big)\Big\|_{L_2(0,t_0;H^{-1}(\Omega)^n)} \\
&\quad \le \|f_0'\|_{L_2(0,t_0;H^{-1}(\Omega)^n)} + t_0^{1/2}\|\operatorname{Div}\varsigma'\|_{L_\infty(0,t_0;H^{-1}(\Omega)^n)} \\
&\qquad + t_0^{1/2}\Big\|\sum_{i=1}^n \frac{\partial}{\partial t}\Big(v_i\frac{\partial v}{\partial x_i}\Big)\Big\|_{L_\infty(0,t_0;H^{-1}(\Omega)^n)} \\
&\quad \le \|f_0\|_{W_2^1(0,t_0;H^{-1}(\Omega)^n)} + Ct_0^{1/2}\|\varsigma\|_{W_\infty^1(0,t_0;H_M^1(\Omega))} \\
&\qquad + Ct_0^{1/2}\|v\|_{L_\infty(0,t_0;H^2(\Omega)^n)}\|v\|_{W_\infty^1(0,t_0;H)} \\
&\quad \le C + CRt_0^{1/2}.
\end{aligned} \tag{5.1.25}
$$

By Lemma 2.2.7, $f_0, f \in C([0,T]; L_2(\Omega)^n)$. Then, Lemma 2.2.7 and the Newton–Leibnitz formula give:

$$
\begin{aligned}
\max_{t\in[0,t_0]}\|f(t)\|_0^2 &\le \|f(0)\|_0^2 + \Big|2\int_0^{t_0}\langle f'(s), f(s)\rangle\,ds\Big| \\
&\le C\Big(\|f_0(0)\|_0^2 + \|\operatorname{Div}\tau_0\|_0^2 + \Big\|\sum_{i=1}^n a_i\frac{\partial a}{\partial x_i}\Big\|_0^2 \\
&\qquad + \|f'\|_{L_2(0,t_0;H^{-1}(\Omega)^n)}\|f\|_{L_2(0,t_0;H^1(\Omega)^n)}\Big) \\
&\le C\Big(\|f_0\|^2_{C([0,T];L_2(\Omega)^n)} + \|\tau_0\|_1^2 + \|a\|_2^4 \\
&\qquad + \|f'\|_{L_2(0,t_0;H^{-1}(\Omega)^n)}\|f\|_{L_2(0,t_0;H^1(\Omega)^n)}\Big) \\
&\le C(C + (C+CRt_0^{1/2})^2) \le (C+CRt_0^{1/2})^2.
\end{aligned} \tag{5.1.26}
$$

Estimates (5.1.24) – (5.1.26) and Corollary 5.1.2 imply that

$$
\|u\|^2_{L_2(0,t_0;H^3(\Omega)^n)} + \|u\|^2_{L_\infty(0,t_0;H^2(\Omega)^n)} + \|u\|^2_{W_2^1(0,t_0;V)} + \|u\|^2_{W_\infty^1(0,t_0;L_2(\Omega)^n)} \le (C+CRt_0^{1/2})^2. \tag{5.1.27}
$$

By Lemma 2.2.1, a), $\|v\|_{L_1(0,t_0;H^3(\Omega)^n)} \le t_0^{1/2}\|v\|_{L_2(0,t_0;H^3(\Omega)^n)} \le (Rt_0)^{1/2}$. Applying estimates (5.1.22) and (5.1.23), we get:

$$
\|\tau\|_{C([0,t_0];H_M^2(\Omega))} \le C\exp(CR^{1/2}t_0^{1/2}) \tag{5.1.28}
$$

$$
\|\tau'\|_{L_\infty(0,t_0;H_M^1(\Omega))} \le C(R^{1/2}+C)\exp(CR^{1/2}t_0^{1/2}). \tag{5.1.29}
$$

Estimates (5.1.27) – (5.1.29) give that, for sufficiently large R and sufficiently small t_0 (depending on R), one has $(u, \tau) \in Q$.

Let us prove now that ϕ is continuous. Take a sequence $(v_m, \varsigma_m) \in Q$ converging to a pair (v_0, ς_0) in the topology of X. Let $(u_m, \tau_m) = \phi(v_m, \varsigma_m)$ and $f^m = f_0 + \operatorname{Div} \varsigma_m - \sum_{i=1}^{n} v_{mi} \frac{\partial v_m}{\partial x_i}$. Then

$$
\begin{aligned}
&\| f^m - f^0 \|_{L_2(0,t_0;L_2(\Omega)^n)} \\
&\quad \le \| \operatorname{Div}(\varsigma_m - \varsigma_0) \|_{L_2(0,t_0;L_2(\Omega)^n)} + \Big\| \sum_{i=1}^{n} (v_{mi} \frac{\partial v_m}{\partial x_i} - v_{0i} \frac{\partial v_0}{\partial x_i}) \Big\|_{L_2(0,t_0;L_2(\Omega)^n)} \\
&\quad \le \| \varsigma_m - \varsigma_0 \|_{L_2(0,t_0;H^1_M(\Omega))} + \Big\| \sum_{i=1}^{n} v_{mi} \frac{\partial (v_m - v_0)}{\partial x_i} \Big\|_{L_2(0,t_0;L_2(\Omega)^n)} \\
&\qquad + \Big\| \sum_{i=1}^{n} (v_{mi} - v_{0i}) \frac{\partial v_0}{\partial x_i} \Big\|_{L_2(0,t_0;L_2(\Omega)^n)} \\
&\quad \le t_0^{1/2} \Big(\| \varsigma_m - \varsigma_0 \|_{C([0,t_0];H^1_M(\Omega))} + \Big\| \sum_{i=1}^{n} v_{mi} \frac{\partial (v_m - v_0)}{\partial x_i} \Big\|_{L_\infty(0,t_0;L_2(\Omega)^n)} \\
&\qquad + \Big\| \sum_{i=1}^{n} (v_{mi} - v_{0i}) \frac{\partial v_0}{\partial x_i} \Big\|_{L_\infty(0,t_0;L_2(\Omega)^n)} \Big) \\
&\quad \le t_0^{1/2} \Big(\| \varsigma_m - \varsigma_0 \|_{C([0,t_0];H^1_M(\Omega))} \\
&\qquad + \| v_m \|_{L_\infty(0,t_0;H^2(\Omega)^n)} \| v_m - v_0 \|_{C([0,t_0];H^1(\Omega)^n)} \\
&\qquad + \| v_m - v_0 \|_{C([0,t_0];H^1(\Omega)^n)} \| v_0 \|_{L_\infty(0,t_0;H^2(\Omega)^n)} \Big) \\
&\quad \le t_0^{1/2} \Big(\| \varsigma_m - \varsigma_0 \|_{C([0,t_0];H^1_M(\Omega))} + R^{1/2} \| v_m - v_0 \|_{C([0,t_0];H^1(\Omega)^n)} \\
&\qquad + R^{1/2} \| v_m - v_0 \|_{C([0,t_0];H^1(\Omega)^n)} \Big) \xrightarrow[m \to \infty]{} 0.
\end{aligned}
$$

Thus, the map $(v, \varsigma) \mapsto f$ is continuous from X to $L_2(0, t_0; L_2(\Omega)^n)$. By Corollary 5.1.1, the map $f \mapsto u$ is continuous from $L_2(0, t_0; L_2(\Omega)^n)$ to $C([0, t_0]; V)$. Hence, $u_m \to u_0$ in $C([0, t_0]; V)$.

Further, the inclusion $\phi(Q) \subset Q$ implies that the sequence $\{\tau_m\}$ is relatively compact in $C([0, t_0]; H^1_M(\Omega))$. Any accumulation point of this sequence is a solution to (5.1.20) – (5.1.21) with $v = v_0$. In fact, if $\tau_{m_k} \xrightarrow[k \to \infty]{} \tau_*$ in $C([0, t_0]; H^1_M(\Omega))$, then we can pass to the limit in the equality

$$
\begin{aligned}
&\tau_{m_k} + \lambda_1 \Big(\frac{\partial \tau_{m_k}}{\partial t} + \sum_{i=1}^{n} v_{m_k i} \frac{\partial \tau_{m_k}}{\partial x_i} + \tau_{m_k} W(v_{m_k}) - W(v_{m_k}) \tau_{m_k} - \\
&\qquad \mathbf{a}(\tau_{m_k} \mathcal{E}(v_{m_k}) + \mathcal{E}(v_{m_k}) \tau_{m_k}) \Big) = 2\eta_1 \mathcal{E}(v_{m_k})
\end{aligned}
\tag{5.1.30}
$$

at least in the sense of distributions.

But, since τ_0 is the only solution of (5.1.20) – (5.1.21) with $v = v_0$, we conclude that $\tau_m \to \tau_0$ in $C([0, t_0]; H^1_M(\Omega))$.

Thus, the conditions of Theorem 3.2.4 are fulfilled for ϕ, so ϕ has a required fixed point. □

5.2 Initial-value problem for the combined model of nonlinear viscoelastic medium

5.2.1 Formulation of the initial-value problem

In the remainder of the chapter we investigate the initial-value problem describing the motion of homogeneous incompressible nonlinear viscoelastic medium with the combined constitutive law (1.5.15) – (1.5.19) described in Section 1.5.2. We shall consider the case of motion in the whole space $\mathbb{R}^n$. Combining the constitutive relations with the equation of motion (1.1.9) and the incompressibility condition (1.1.10), we get the following problem:

$$\frac{\partial u}{\partial t} + \sum_{i=1}^{n} u_i \frac{\partial u}{\partial x_i} = \text{Div } \mathrm{T} + f_0, \quad (t, x) \in [0, T] \times \mathbb{R}^n, \tag{5.2.1}$$

$$\operatorname{div} u = 0, \quad (t, x) \in [0, T] \times \mathbb{R}^n, \tag{5.2.2}$$

$$\mathrm{T} = -pI + \sum_{k=0}^{r} \tau^k, \tag{5.2.3}$$

$$\tau^k + \lambda_k \frac{D_0 \tau^k}{Dt} + \beta_k(\tau^k, \mathcal{E}) = 2\eta_k \mathcal{E}(u), \quad k = 1, \dots, r, \tag{5.2.4}$$

$$\tau^0 = \Psi(\mathcal{E}) = \varphi_1 \mathcal{E} + \varphi_2 \mathcal{E}^2, \tag{5.2.5}$$

$$\begin{aligned} \beta_k(\tau, \mathcal{E}) = {} & \alpha_0^k I + \alpha_1^k \mathcal{E} + \alpha_2^k \mathcal{E}^2 + \alpha_3^k \tau + \alpha_4^k \tau^2 + \alpha_5^k (\mathcal{E}\tau + \tau\mathcal{E}) + \\ & + \alpha_6^k (\mathcal{E}^2 \tau + \tau \mathcal{E}^2) + \alpha_7^k (\mathcal{E}\tau^2 + \tau^2 \mathcal{E}) + \alpha_8^k (\mathcal{E}^2 \tau^2 + \tau^2 \mathcal{E}^2). \end{aligned} \tag{5.2.6}$$

Here u is the unknown velocity vector, T is the unknown stress tensor, f_0 is the given body force (all of them depend on a point x of the space $\mathbb{R}^n$, $n = 2, 3$, and on a moment of time t); $\lambda_k > 0$ are relaxation times, $\eta_k > 0$ are viscosities; φ_1, φ_2 and α_j^k are scalar functions:

$$\varphi_i = \varphi_i(\mathrm{Tr}(\mathcal{E}^2), \det \mathcal{E}), \quad i = 1, 2, \tag{5.2.7}$$

$$\begin{aligned} \alpha_j^k = \alpha_j^k \big(& \mathrm{Tr}\, \mathcal{E}^2, \mathrm{Tr}\, \mathcal{E}^3, \mathrm{Tr}(\tau), \mathrm{Tr}(\tau^2), \mathrm{Tr}(\tau^3), \mathrm{Tr}(\tau\mathcal{E}), \mathrm{Tr}(\tau^2\mathcal{E}), \\ & \mathrm{Tr}(\tau\mathcal{E}^2), \mathrm{Tr}(\tau^2\mathcal{E}^2)\big), \quad k = 1, \dots, r; \ j = 0, \dots, 8. \end{aligned} \tag{5.2.8}$$

Remark 5.2.1. For convenience, we omitted tildes over p and φ_1 (cf. Remark 1.4.4).

Generally, the pressure $p(t,x)$ may be determined up to an arbitrary scalar function of time (cf. Lemma 5.1.1). To provide uniqueness, fix a function $\vartheta \in L_2(\mathbb{R}^n)$ with a compact support and satisfying $\int_{\mathbb{R}^n} \vartheta(x)\, dx \neq 0$, and assume that

$$\int_{\mathbb{R}^n} p(t,x)\vartheta(x)dx \equiv 0. \tag{5.2.9}$$

The initial conditions have the following form:

$$u(0,x) = a(x),\ \tau^k(0,x) = \tau_0^k(x), \quad x \in \mathbb{R}^n, \quad k = 1,\dots,r, \tag{5.2.10}$$

where a and τ_0^k are prescribed functions.

5.2.2 The Leray projection in $\mathbb{R}^n$ and some additional notations

Below in this chapter, in addition to the notations already introduced in the book, we use also the following notations.

We shall use the function spaces of Sobolev–Slobodetskii type

$$H_V^s = \{u \in H^s(\mathbb{R}^n, \mathbb{R}^n),\ \operatorname{div} u = 0\}$$

and

$$H_M^s = H_M^s(\mathbb{R}^n) = H^s(\mathbb{R}^n, \mathbb{R}_S^{n\times n}),$$

where $s \in \mathbb{R}, s \geq 0$.

A scalar product in these spaces can be given by the equality

$$(u,v)_s = (B_0^{\frac{s}{2}} u, B_0^{\frac{s}{2}} v)_0 \tag{5.2.11}$$

where B_0 is formally defined as $I - \Delta$, identity operator minus Laplacian. The operators $B_0 : H_V^2 \subset H_V^0 \to H_V^0$ and $B_0 : H_M^2 \subset H_M^0 \to H_M^0$ are self-adjoint and strongly positive. The powers of such operators were defined in Section 3.1.4.

Remark 5.2.2. In the case of natural s the notation $(u,v)_s$ was already introduced in Section 2.1.1:

$$(u,v)_s = \sum_{|\alpha| \leq s} (D^\alpha u, D^\alpha v)_0.$$

However, it is easy to observe that this expression coincides with (5.2.11) for natural s.

The Euclidean norm in both spaces will be denoted as $\|\cdot\|_s$. This norm is equivalent to the Sobolev–Slobodetskii norm introduced in Section 2.1.1 (see e.g. [34]).

Now we are going to give a characterization of the space H^s_V. For this purpose, consider the *Leray projection* P, which is formally defined as:

$$(Pu)_i = \sum_{j=1}^{n} \mathcal{F}^{-1}(\delta_{ij} - \frac{\xi_i \xi_j}{|\xi|^2})\mathcal{F}(u_j), \qquad (5.2.12)$$

where $\mathcal{F}_{x\to\xi}$ is the Fourier transform of $\mathbb{R}^n$, and δ_{ij} is the Kronecker delta, $i, j = 1, \dots, n$.

Sometimes (5.2.12) is written as

$$(Pu)_i = \sum_{j=1}^{n} (\delta_{ij} - \frac{\partial}{\partial x_i}\frac{\partial}{\partial x_j}\Delta^{-1})u_j$$

or

$$Pu = u - \operatorname{grad} \operatorname{div} \Delta^{-1} u.$$

It is easy to see that $P^2 = P$ and P is a continuous automorphism of $H^s(\mathbb{R}^n, \mathbb{R}^n)$ (we may assume that the scalar product in this space is given by (5.2.11)). Furthermore,

$$P(H^s) = H^s_V.$$

Really, (5.2.12) implies that for $u \in H^s$ one has

$$\operatorname{div} Pu \equiv 0. \qquad (5.2.13)$$

This implies the inclusion $P(H^s) \subset H^s_V$. But $P|_{H^s_V} = I$, so $H^s_V \subset P(H^s)$.

The operators P and B_0 commute. Therefore [84] P and the fractional powers of B_0 also commute.

Let us show that the norm of the operator P in H^s does not exceed 1. It suffices to show that

$$(Pu, Pu)_s \le (u, u)_s. \qquad (5.2.14)$$

Since P is continuous, it is enough to prove (5.2.14) for smooth u.

We have:

$$(u, u)_s = (Pu, Pu)_s + 2((I - P)u, Pu)_s + ((I - P)u, (I - P)u)_s. \qquad (5.2.15)$$

But

$$\begin{aligned}((I - P)u, Pu)_s &= \int_{\mathbb{R}^n} B_0^{\frac{s}{2}}(I - P)u(x) B_0^{\frac{s}{2}} Pu(x)dx \\ &= \int_{\mathbb{R}^n} (I - P) B_0^{\frac{s}{2}} u(x) P B_0^{\frac{s}{2}} u(x)dx \\ &= \int_{\mathbb{R}^n} \operatorname{grad} \operatorname{div} \Delta^{-1} B_0^{\frac{s}{2}} u(x) P B_0^{\frac{s}{2}} u(x)dx \\ &= -\int_{\mathbb{R}^n} \operatorname{div} \Delta^{-1} B_0^{\frac{s}{2}} u(x) \cdot \operatorname{div} P B_0^{\frac{s}{2}} u(x)dx = 0\end{aligned}$$

(we have integrated by parts and used (5.2.13)). So (5.2.15) yields

$$(u,u)_s = (Pu,Pu)_s + ((I-P)u,(I-P)u)_s \geq (Pu,Pu)_s.$$

Thus, $\|P\| \leq 1$. But $P^2 = P$. So [84] P is an orthogonal projector in H^s.

In addition to the introduced notations, we shall use the following ones. The symbol ∇_ξ stands for the Fréchet derivative of functions or matrix functions of one or two matrix arguments $\varphi : \mathbb{R}_S^{n\times n} \to \mathbb{R}$, $\beta : \mathbb{R}_S^{n\times n} \times \mathbb{R}_S^{n\times n} \to \mathbb{R}_S^{n\times n}$ etc. The partial derivative of a function $\varphi(\xi)$ of matrix argument $\xi = (\xi_{ij})$ with respect to an element ξ_{ij} will be denoted as $\frac{\partial\varphi}{\partial\xi_{ij}}$.

5.2.3 The main existence and uniqueness theorem

Let us turn now to the formulation of the existence and uniqueness theorem for a solution of problem (5.2.1) – (5.2.10).

Let $\eta_0 = \frac{\varphi_1(0,0)}{2}$. Assume that $\eta_0 > 0$. This is a natural condition since the physical meaning of η_0 is a viscosity parameter.

Let us also assume that φ_i and α_j^k are C^4- and C^3-smooth functions, respectively, and

$$\alpha_0^k(\theta) = \alpha_1^k(\theta) = \alpha_3^k(\theta) = 0, \qquad \frac{\partial\alpha_0^k(\theta)}{\partial \mathrm{Tr}(\tau)} = 0$$

(θ stands for the point $(0,0,0,0,0,0,0,0,0)$). This assumption is also natural in the considered model since the functions β_k in it correspond to "nonlinear" effects, i.e. effects of the second order and higher. Therefore the coefficients α_1^k and α_3^k at the first order terms $\mathcal{E}$ and τ should be "of the first order", i.e. they should vanish in the point θ, and the coefficient α_0^k at the zero order term I should be of the second order, i.e. it should vanish in θ and its partial derivative with respect to the "linear" argument $\mathrm{Tr}(\tau)$ should vanish in θ.

Remark 5.2.3. Instead of the C^4- smoothness of φ_1 and φ_2 it is sufficient for these functions to be differentiable in zero provided the function $\Psi : \mathbb{R}_S^{n\times n} \to \mathbb{R}_S^{n\times n}$ is C^4 - smooth or has locally Lipschitz third derivatives (see Section 5.4.2, proof of Lemma 5.4.2).

Remark 5.2.4. All these assumptions hold for the Oldroyd "8 constants" model, models of Larson, Giesekus, Phan-Thien and Tanner, the Jeffreys model with Oldroyd's derivative (the model from Section 5.1) as well as with Jaumann's or Spriggs' derivative, models of Prandtl and Eyring and any combinations of all these models (see Sections 1.4.4, 1.5.1, 1.5.2).

Theorem 5.2.1 (see [73]). *Let the described assumptions hold true. Let $T > 0$ be an* *arbitrary fixed moment of time. Then, for any*

$$a \in H^3_V,$$
$$\tau_0^k \in H^3_M, \quad k = 1, \dots, r,$$
$$f_0 \in L_1(0, T; H^3(\mathbb{R}^n, \mathbb{R}^n)) \cap L_2(0, T; H^2(\mathbb{R}^n, \mathbb{R}^n)),$$

there exists a constant $K_5 > 0$*, independent on* T*, such that provided*

$$\|a\|_3 + \sum_{k=1}^{r} \|\tau_0^k\|_3 + \|f_0\|_{L_1(0,T;H^3)} < K_5, \tag{5.2.16}$$

problem (5.2.1) – (5.2.10) has a solution in the class

$$u \in L_2(0, T; H^4_V) \cap C([0, T]; H^3_V) \cap W_2^1(0, T; H^2_V), \tag{5.2.17}$$
$$\mathrm{T} \in L_2(0, T; H^3_{M,\mathrm{loc}}), \tag{5.2.18}$$
$$p \in L_2(0, T; H^3_{\mathrm{loc}}(\mathbb{R}^n)). \tag{5.2.19}$$

Furthermore, we have the following information about the constituents of the stress tensor:

$$\tau^0 \in L_2(0, T; H^3_M) \cap C([0, T]; H^2_M) \cap W_2^1(0, T; H^1_M), \tag{5.2.20}$$
$$\tau^k \in L_\infty(0, T; H^3_M) \cap C([0, T]; H^2_M) \cap C^1([0, T]; H^1_M),\ k = 1, \dots, r. \tag{5.2.21}$$

This solution is unique in class (5.2.17) – (5.2.21).

Corollary 5.2.1. *If* $f_0 \in L_1(0, +\infty; H^3(\mathbb{R}^n, \mathbb{R}^n)) \cap L_2(0, +\infty; H^2(\mathbb{R}^n, \mathbb{R}^n))$*, and*

$$\|a\|_3 + \sum_{k=1}^{r} \|\tau_0^k\|_3 + \|f_0\|_{L_1(0,+\infty;H^3)} < K_5, \tag{5.2.22}$$

then problem (5.2.1) – (5.2.10) has a unique solution in class (5.2.17) – (5.2.21) for every $T > 0$.

The following three sections are devoted to the proof of the theorem.

5.3 Operator treatment of the problem

In this section problem (5.2.1) – (5.2.10), describing the motion of a nonlinear viscoelastic medium, will be rewritten as a Cauchy problem in a Banach space.

Below for simplicity we shall consider the case $r = 1$ (concerning $r > 1$, see Section 5.5.3). Let for brevity $\tau = \tau^1$, $\lambda = \lambda_1$, $\eta = \frac{\eta_1}{\lambda_1}$. Let also

$$\Phi(\mathcal{E}) = \Psi(\mathcal{E}) - 2\eta_0\mathcal{E}. \tag{5.3.1}$$

Introduce the function $g : \mathbb{R}_S^{n\times n} \times \mathbb{R}^{n\times n} \to \mathbb{R}_S^{n\times n}$ by the formula

$$g(\xi_1, \xi_2) = \xi_1\Big(\frac{\xi_2 - \xi_2^\top}{2}\Big) - \Big(\frac{\xi_2 - \xi_2^\top}{2}\Big)\xi_1 + \frac{\beta_1(\xi_1, \frac{\xi_2+\xi_2^\top}{2})}{\lambda}.$$

Then

$$g(\tau, \nabla u) = \tau W - W\tau + \frac{\beta_1(\tau, \mathcal{E}(u))}{\lambda}. \tag{5.3.2}$$

Note that Div $pI = \operatorname{grad} p$ and, due to condition (5.2.2), $2\operatorname{Div}\mathcal{E}(u) = \Delta u$. We have from (5.2.1) – (5.2.5), (5.3.1) – (5.3.2):

$$\frac{\partial u}{\partial t} + \sum_{i=1}^{n} u_i \frac{\partial u}{\partial x_i} + \operatorname{grad} p - \eta_0 \Delta u - \operatorname{Div}(\Phi(\mathcal{E}) + \tau) = f_0 \tag{5.3.3}$$

$$\frac{\tau}{\lambda} + \frac{\partial \tau}{\partial t} + \sum_{i=1}^{n} u_i \frac{\partial \tau}{\partial x_i} + g(\tau, \nabla u) = 2\eta\mathcal{E}. \tag{5.3.4}$$

Let $\Phi = (\Phi_{ij})$. Let us introduce the following notations

$$(\widetilde{A}(u)v)_j = \sum_{i,k,l=1}^{n} \frac{\partial \Phi_{ij}}{\partial \xi_{kl}}(\mathcal{E}(u))\frac{\partial \mathcal{E}_{kl}(v)}{\partial x_i}, \quad j = 1, \dots, n, \tag{5.3.5}$$

$$A(u)v = -P\widetilde{A}(u)v + v - \eta_0\Delta v, \tag{5.3.6}$$

$$F_1(u, v) = -P\sum_{i=1}^{n} u_i \frac{\partial v}{\partial x_i}, \tag{5.3.7}$$

$$F(u, \tau) = -\sum_{i=1}^{n} u_i \frac{\partial \tau}{\partial x_i}, \tag{5.3.8}$$

$$\widetilde{F}(u) = F_1(u, u) + u, \tag{5.3.9}$$

$$G(u, \tau) = -g(\tau, \nabla u), \tag{5.3.10}$$

$$N_1(\tau) = P(\operatorname{Div} \tau), \tag{5.3.11}$$

$$N_2(u) = 2\eta\mathcal{E}(u), \tag{5.3.12}$$

$$B_0 = I - \Delta, \tag{5.3.13}$$

$$A_0 = A(a), \tag{5.3.14}$$

$$f = Pf_0. \tag{5.3.15}$$

Consider the problem

$$\frac{du}{dt} + A(u)u = \widetilde{F}(u) + N_1(\tau) + f, \tag{5.3.16}$$

$$\frac{d\tau}{dt} + \frac{\tau}{\lambda} = F(u,\tau) + N_2(u) + G(u,\tau), \tag{5.3.17}$$

$$u(0) = a, \qquad \tau(0) = \tau_0. \tag{5.3.18}$$

Remark 5.3.1. Here and below $\frac{d}{dt}$ is not a substantial derivative but the time derivative of a time-dependent function u with values in a Banach space.

Remark 5.3.2. Equation (5.3.16) is a formal consequence of (5.3.3) after application of the Leray projection. Equation (5.3.17) is formally equivalent to (5.3.4).

The statement of Theorem 5.2.1 will be deduced from the following result.

Theorem 5.3.1. *Given $a \in H^3_V$, $\tau_0 \in H^3_M$, $f \in L_1(0,T;H^3_V) \cap L_2(0,T;H^2_V)$, there exists a constant $K_6 > 0$, independent on T, such that provided*

$$\|a\|_3 + \|\tau_0\|_3 + \|f\|_{L_1(0,T;H^3_V)} < K_6 \tag{5.3.19}$$

problem (5.3.16) – (5.3.18) has a unique solution in the class

$$u \in L_2(0,T;H^4_V) \cap C([0,T];H^3_V) \cap W^1_2(0,T;H^2_V), \tag{5.3.20}$$

$$\tau \in L_\infty(0,T;H^3_M) \cap C([0,T];H^2_M) \cap C^1([0,T];H^1_M). \tag{5.3.21}$$

5.4 Auxiliary problem

5.4.1 Solvability of the auxiliary problem

Before proving Theorems 5.2.1 and 5.3.1, we investigate the solvability of an auxiliary problem.

Introduce a family of operators

$$A_\varepsilon(\tau) = \frac{\tau}{\lambda} - \varepsilon\Delta\tau, \ \varepsilon > 0, \tag{5.4.1}$$

and consider the following equation

$$\frac{d\tau}{dt} + A_\varepsilon\tau = F(u,\tau) + N_2(u) + G(u,\tau). \tag{5.4.2}$$

Theorem 5.4.1. *Given $a \in H_V^4$, $\tau_0 \in H_M^4$, $f \in L_1(0,T;H_V^3) \cap C^1([0,T];H_V^2)$, there exists a constant $K_7 > 0$, independent on T and ε, such that provided*

$$\|a\|_3 + \|\tau_0\|_3 + \|f\|_{L_1(0,T;H_V^3)} < K_7$$

there is a unique solution of problem (5.3.16), (5.4.2), (5.3.18) in the class

$$\begin{aligned} &u \in C^1([0,T];H_V^2) \cap C([0,T];H_V^4),\\ &\tau \in C^1([0,T];H_M^2) \cap C([0,T];H_M^4). \end{aligned} \tag{5.4.3}$$

In order to get this theorem we have to prove some auxiliary facts.

5.4.2 Operator estimates

We need some estimates for the operators introduced in Section 5.3.

Lemma 5.4.1. *The following estimates take place.*
For $l = 1,2$, $u \in H_V^2$, $v \in H_V^{l+1}$, $\tau \in H_M^{l+1}$

$$\|F_1(u,v)\|_l \le K_8\|u\|_2\|v\|_{l+1}, \tag{5.4.4}$$

$$\|F(u,\tau)\|_l \le K_8\|u\|_2\|\tau\|_{l+1}. \tag{5.4.5}$$

For $l = 2,3$, $u \in H_V^3$, $v \in H_V^{l+1}$, $\tau \in H_M^{l+1}$

$$\begin{aligned} |(F_1(u,v),v)_l| &\le K_9\|\nabla u\|_2\|\nabla v\|_{l-1}^2,\\ |(F(u,\tau),\tau)_l| &\le K_9\|\nabla u\|_2\|\nabla \tau\|_{l-1}^2. \end{aligned} \tag{5.4.6}$$

For $l = 0,1$ and $\gamma = \frac{7}{4}$ or $l = \gamma = 2$

$$\begin{aligned} \|G(u_1,\tau_1) - G(u_2,\tau_2)\|_l \le K_{10} &\cdot (\|\nabla u_1 - \nabla u_2\|_l + \|\tau_1 - \tau_2\|_l)\\ &\cdot (\|\nabla u_1\|_\gamma + \|\nabla u_2\|_\gamma + \|\tau_1\|_\gamma + \|\tau_2\|_\gamma) \end{aligned} \tag{5.4.7}$$

where $u_1, u_2 \in H_V^{\gamma+1}$, $\tau_1, \tau_2 \in H_M^\gamma$.
For $u \in H_V^4$, $\tau \in H_M^3$

$$\|G(u,\tau)\|_3 \le K'_{10} \cdot (\|u\|_3 + \|\tau\|_2)(\|\nabla u\|_3 + \|\tau\|_3). \tag{5.4.8}$$

Here K_{10} and K'_{10} depend continuously on $(\|\nabla u_1\|_\gamma + \|\nabla u_2\|_\gamma + \|\tau_1\|_\gamma + \|\tau_2\|_\gamma)$ and on $(\|u\|_3 + \|\tau\|_2)$, respectively.

Lemma 5.4.2. *Let $l = 0,1,2$ and $\alpha_0(0) = \frac{11}{8}$, $\alpha_0(1) = \frac{7}{8}$, $\alpha_0(2) = \frac{1}{2}$.*

1) *If*

$$\|B_0^{\alpha_0(l)}v\|_l,\ \|B_0^{\alpha_0(l)}w\|_l < 1$$

then the following estimates take place

$$\begin{aligned}\|(A(v)-A(w))h\|_l &\le K_{11}\|\nabla B_0^{\alpha_0(l)}(v-w)\|_{l-1}\|\nabla h\|_{l+1},\\ \|(A(v)-A(w))h\|_l &\le K_{11}\|B_0^{\alpha_0(l)}(v-w)\|_l\|B_0 h\|_l.\end{aligned} \tag{5.4.9}$$

Furthermore, for $l=0$ *there is another variant of the second estimate:*

$$\|(A(v)-A(w))h\|_0 \le K_{11}\|v-w\|_{\frac{3}{2}}\|h\|_3. \tag{5.4.10}$$

2) *There exist* $K_{12}, K_{13}, K_{14}, K_{15} > 0$ *such that provided*

$$\|B_0^{\alpha_0(l)}a\|_l < K_{12}$$

one has

$$(A_0 v, B_0 v)_l \ge K_{13}\|B_0 v\|_l^2, \tag{5.4.11}$$

and provided

$$\|u\|_3,\ \|v\|_3 < K_{14}$$

one has

$$(A(u)u - A(v)v, B_0(u-v))_1 \ge \frac{1}{2}K_{13}\|B_0(u-v)\|_1^2, \tag{5.4.12}$$

$$((A(u)-I)u, B_0 u)_2 \ge K_{15}\|\nabla u\|_3^2. \tag{5.4.13}$$

Proof of Lemma 5.4.1. Using inequality (2.1.21), we have:

$$\|F_1(u,v)\|_1 \le \sum_{i=1}^n \Big\|u_i\frac{\partial v}{\partial x_i}\Big\|_1 \le C\sum_{i=1}^n \|u_i\|_2\Big\|\frac{\partial v}{\partial x_i}\Big\|_1 \le C\|u\|_2\|v\|_2.$$

Using (2.1.22), we have:

$$\|F_1(u,v)\|_2 \le \sum_{i=1}^n \Big\|u_i\frac{\partial v}{\partial x_i}\Big\|_2 \le C\sum_{i=1}^n \|u_i\|_2\Big\|\frac{\partial v}{\partial x_i}\Big\|_2 \le C\|u\|_2\|v\|_3.$$

Estimate (5.4.4) is proved. Similarly one proves (5.4.5).

Let us prove now estimate (5.4.6). We have, using the Leibnitz rule:

$$\begin{aligned}|(F_1(u,v),v)_2| &\le \sum_{|\alpha|\le 2}\sum_{i=1}^n \Big|\Big(D^\alpha\Big(u_i\frac{\partial v}{\partial x_i}\Big), D^\alpha v\Big)_0\Big|\\ &\le \sum_{|\alpha|\le 2}\sum_{i=1}^n \Big|\Big(u_i D^\alpha\frac{\partial v}{\partial x_i}, D^\alpha v\Big)_0\Big| + \sum_{0<|\alpha|\le 2}\sum_{i=1}^n \Big|\Big(D^\alpha u_i\frac{\partial v}{\partial x_i}, D^\alpha v\Big)_0\Big|\\ &\quad + \sum_{|\alpha|=2,|\beta|=1,\beta\le\alpha}\sum_{i=1}^n K_{16}(\alpha,\beta)\Big|\Big(D^\beta u_i D^{\alpha-\beta}\frac{\partial v}{\partial x_i}, D^\alpha v\Big)_0\Big|.\end{aligned}$$

Since $\operatorname{div} u = 0$, integration by parts easily shows that the first sum is equal to zero. With the help of (2.1.20) and Cauchy–Buniakowski inequality we observe that the second sum does not exceed the expression

$$C\sum_{i=1}^{n} \|\nabla u_i\|_2 \|\frac{\partial v}{\partial x_i}\|_1 \|\nabla v\|_1. \tag{5.4.14}$$

Using inequality (2.1.19), we conclude that the third sum does not exceed an expression of form (5.4.14). Estimate (5.4.6) for the case $l = 2$ is proved. The case $l = 3$ is examined analogously.

Let us prove now estimate (5.4.7).

The conditions on coefficients α_j^k from Section 5.2.3 imply that the function

$$\beta_1 : \mathbb{R}_S^{n\times n} \times \mathbb{R}_S^{n\times n} \to \mathbb{R}_S^{n\times n}$$

is C^3-smooth, and

$$\beta_1(0,0) = 0, \quad \nabla_\xi \beta_1(0,0) = 0.$$

Then the function

$$g : \mathbb{R}_S^{n\times n} \times \mathbb{R}^{n\times n} \to \mathbb{R}_S^{n\times n}$$

introduced in Section 5.3 is also C^3 - smooth, and

$$g(0,0) = 0, \quad \nabla_\xi g(0,0) = 0.$$

We have:

$$\|G(u_1,\tau_1) - G(u_2,\tau_2)\|_l \le \sum_{|\alpha|\le l} \|D^\alpha(G(u_1,\tau_1) - G(u_2,\tau_2))\|_0.$$

First we estimate the term with $|\alpha| = 0$. Taking into account embedding (2.1.18) and the Lagrange theorem we have:

$$\begin{aligned}
&\|G(u_1,\tau_1) - G(u_2,\tau_2)\|_0 = \|g(\nabla u_1,\tau_1) - g(\nabla u_2,\tau_2)\|_0 \\
&\le \sup_{\substack{\|\xi^1\|\le\|\nabla u_1\|_{L_\infty}+\|\nabla u_2\|_{L_\infty} \\ \|\xi^2\|\le\|\tau_1\|_{L_\infty}+\|\tau_2\|_{L_\infty}}} |\nabla_\xi g(\xi^1,\xi^2)|(\|\nabla(u_1-u_2)\|_0 + \|\tau_1-\tau_2\|_0) \\
&= \sup_{\substack{\|\xi^1\|\le\|\nabla u_1\|_{L_\infty}+\|\nabla u_2\|_{L_\infty} \\ \|\xi^2\|\le\|\tau_1\|_{L_\infty}+\|\tau_2\|_{L_\infty}}} |\nabla_\xi g(\xi^1,\xi^2) - \nabla_\xi g(0,0)|(\|\nabla(u_1-u_2)\|_0 + \|\tau_1-\tau_2\|_0) \\
&\le C(\|\nabla u_1\|_{L_\infty} + \|\nabla u_2\|_{L_\infty} + \|\tau_1\|_{L_\infty} + \|\tau_2\|_{L_\infty})(\|\nabla(u_1-u_2)\|_0 + \|\tau_1-\tau_2\|_0) \\
&\le C\big(\|\nabla u_1\|_{\frac{7}{4}} + \|\nabla u_2\|_{\frac{7}{4}} + \|\tau_1\|_{\frac{7}{4}} + \|\tau_2\|_{\frac{7}{4}}\big)(\|\nabla(u_1-u_2)\|_0 + \|\tau_1-\tau_2\|_0).
\end{aligned}$$

Note that for $u \in H_V^2, \tau \in H_V^1$ one has

$$\nabla g(\nabla u,\tau) = \frac{\partial g}{\partial \xi_{\nabla u}}(\nabla u,\tau)\nabla\nabla u + \frac{\partial g}{\partial \xi_\tau}(\nabla u,\tau)\nabla\tau,$$

and let us denote it by

$$\nabla_\xi g(\nabla u, \tau)(\nabla^2 u, \nabla \tau).$$

Using Hölder's inequality (2.1.1) in the form

$$\|\psi_1 \psi_2\|_0 \le \|\psi_1\|_{L_6} \|\psi_2\|_{L_3}$$

and the embeddings

$$H^{1/2} \subset L_3, \quad H^1 \subset L_6, \quad H^{7/4} \subset L_\infty,$$

we estimate the first derivatives:

$$\begin{aligned}
&\|\nabla(G(u_1, \tau_1) - G(u_2, \tau_2))\|_0 \\
&\quad = \|\nabla_\xi g(\nabla u_1, \tau_1)(\nabla^2 u_1, \nabla \tau_1) - \nabla_\xi g(\nabla u_2, \tau_2)(\nabla^2 u_2, \nabla \tau_2)\|_0 \\
&\quad \le \|\nabla_\xi g(\nabla u_1, \tau_1)(\nabla^2 (u_1 - u_2), \nabla(\tau_1 - \tau_2))\|_0 \\
&\qquad + \|(\nabla_\xi g(\nabla u_1, \tau_1) - \nabla_\xi g(\nabla u_2, \tau_2))(\nabla^2 u_2, \nabla \tau_2)\|_0 \\
&\quad \le \|\nabla_\xi g(\nabla u_1, \tau_1)\|_{L_\infty} (\|\nabla(u_1 - u_2)\|_1 + \|\tau_1 - \tau_2\|_1) \\
&\qquad + \|\nabla_\xi g(\nabla u_1, \tau_1) - \nabla_\xi g(\nabla u_2, \tau_2)\|_{L_6} (\|\nabla^2 u_2\|_{L_3} + \|\nabla \tau_2\|_{L_3}) \\
&\quad \le \|\nabla_\xi g(\nabla u_1, \tau_1) - \nabla_\xi g(0, 0)\|_{L_\infty} (\|\nabla(u_1 - u_2)\|_1 + \|\tau_1 - \tau_2\|_1) \\
&\qquad + \|\nabla_\xi g(\nabla u_1, \tau_1) - \nabla_\xi g(\nabla u_2, \tau_2)\|_{L_6} (\|\nabla^2 u_2\|_{L_3} + \|\nabla \tau_2\|_{L_3}) \\
&\quad \le C(\|\nabla u_1\|_{L_\infty} + \|\tau_1\|_{L_\infty})(\|\nabla(u_1 - u_2)\|_1 + \|\tau_1 - \tau_2\|_1) \\
&\qquad + C(\|\nabla(u_1 - u_2)\|_{L_6} + \|\tau_1 - \tau_2\|_{L_6})(\|\nabla^2 u_2\|_{\frac{1}{2}} + \|\nabla \tau_2\|_{\frac{1}{2}}) \\
&\quad \le C\big(\|\nabla(u_1 - u_2)\|_1 + \|\tau_1 - \tau_2\|_1\big)\big(\|\nabla u_1\|_{\frac{7}{4}} + \|\tau_1\|_{\frac{7}{4}} + \|\nabla u_2\|_{\frac{7}{4}} + \|\tau_2\|_{\frac{7}{4}}\big).
\end{aligned}$$

In the same way the terms $\|D^\alpha(G(u_1, \tau_1) - G(u_2, \tau_2))\|_0$ with $|\alpha| = 2$ are estimated and we obtain estimate (5.4.7).

Let us turn to estimate (5.4.8). We have:

$$\begin{aligned}
\|G(u, \tau)\|_3 &\le \|g(\nabla u, \tau)\|_2 + \|\nabla^3 g(\nabla u, \tau)\|_0 \\
&\le \|g(\nabla u, \tau)\|_2 + \|\nabla_\xi g(\nabla u, \tau)(\nabla^4 u, \nabla^3 \tau)\|_0 \\
&\quad + \|\nabla_\xi^2 g(\nabla u, \tau)(\nabla^3 u, \nabla^2 \tau)(\nabla^2 u, \nabla \tau)\|_0 \\
&\quad + \|\nabla_\xi^2 g(\nabla u, \tau)(\nabla^2 u, \nabla \tau)(\nabla^3 u, \nabla^2 \tau)\|_0 \\
&\quad + \|\nabla_\xi^3 g(\nabla u, \tau)(\nabla^2 u, \nabla \tau)(\nabla^2 u, \nabla \tau)(\nabla^2 u, \nabla \tau)\|_0.
\end{aligned}$$

All these items except the last one are bounded at least by

$$C \cdot (\|u\|_3 + \|\tau\|_2)(\|\nabla u\|_3 + \|\tau\|_3)$$

where C depends on $\|u\|_3 + \|\tau\|_2$ (it may be checked in a similar way as above). Using Hölder's inequality in the form

$$\|\psi_1\psi_2\psi_3\|_0 \le \|\psi_1\|_{L_6}\|\psi_2\|_{L_6}\|\psi_3\|_{L_6},$$

we conclude that the last item does not exceed

$$\begin{aligned}
&\sup_{\substack{\|\xi^1\|\le\|\nabla u\|_{L_\infty}\\ \|\xi^2\|\le\|\tau\|_{L_\infty}}} |\nabla_\xi^3 g(\xi^1,\xi^2)|(\|\nabla^2 u\|_{L_6} + \|\nabla\tau\|_{L_6})^3\\
&\qquad \le C\cdot(\|\nabla^2 u\|_1 + \|\nabla\tau\|_1)^3\\
&\qquad \le C\cdot(\|u\|_3 + \|\tau\|_2)(\|\nabla u\|_3 + \|\tau\|_3)
\end{aligned}$$

where both C depend on $\|u\|_3 + \|\tau\|_2$. □

Proof of Lemma 5.4.2. The C^4-smoothness of coefficients φ_i (see Section 5.2.3) implies that the function

$$\Phi : \mathbb{R}_S^{n\times n} \to \mathbb{R}_S^{n\times n}$$

introduced by formula (5.3.1) is C^4-smooth. Moreover, it is easy to check that

$$\Phi(0) = 0, \nabla_\xi\Phi(0) = 0.$$

We have:

$$\|(A(v) - A(w))h\|_l = \|(\widetilde{A}(v) - \widetilde{A}(w))h\|_l \le \sum_{|\alpha|\le l} \|D^\alpha(\widetilde{A}(v) - \widetilde{A}(w))h\|_0.$$

First we estimate the term with $|\alpha| = 0$:

$$\begin{aligned}
&\|(\widetilde{A}(v) - \widetilde{A}(w))h\|_0\\
&\quad\le \sum_{i,j,k,l=1}^n \left\| \left(\frac{\partial\Phi_{ij}}{\partial\xi_{kl}}(\mathcal{E}(v)) - \frac{\partial\Phi_{ij}}{\partial\xi_{kl}}(\mathcal{E}(w))\right)\frac{\partial\mathcal{E}_{kl}(h)}{\partial x_i}\right\|_0\\
&\quad\le \sum_{i,j,k,l=1}^n \left\|\frac{\partial\Phi_{ij}}{\partial\xi_{kl}}(\mathcal{E}(v)) - \frac{\partial\Phi_{ij}}{\partial\xi_{kl}}(\mathcal{E}(w))\right\|_{L_\infty}\left\|\frac{\partial\mathcal{E}_{kl}(h)}{\partial x_i}\right\|_0 \qquad (5.4.15)\\
&\quad\le C\|\mathcal{E}(v) - \mathcal{E}(w)\|_{L_\infty}\|\nabla h\|_1\\
&\quad\le C\|\mathcal{E}(v) - \mathcal{E}(w)\|_{\frac{7}{4}}\|\nabla h\|_1 \le C\|\nabla B_0^{\frac{11}{8}}(v-w)\|_{-1}\|\nabla h\|_1.
\end{aligned}$$

Furthermore,

$$\begin{aligned}
&\|(\widetilde{A}(v)-\widetilde{A}(w))h\|_0 \\
&\quad\le \sum_{i,j,k,l=1}^{n} \Big\| \Big(\frac{\partial \Phi_{ij}}{\partial \xi_{kl}}(\mathcal{E}(v)) - \frac{\partial \Phi_{ij}}{\partial \xi_{kl}}(\mathcal{E}(w))\Big) \frac{\partial \mathcal{E}_{kl}(h)}{\partial x_i} \Big\|_0 \\
&\quad\le \sum_{i,j,k,l=1}^{n} \Big\| \frac{\partial \Phi_{ij}}{\partial \xi_{kl}}(\mathcal{E}(v)) - \frac{\partial \Phi_{ij}}{\partial \xi_{kl}}(\mathcal{E}(w)) \Big\|_{L_3} \Big\| \frac{\partial \mathcal{E}_{kl}(h)}{\partial x_i} \Big\|_{L_6} \qquad (5.4.16)\\
&\quad\le C\|\mathcal{E}(v)-\mathcal{E}(w)\|_{L_3} \Big\| \frac{\partial \mathcal{E}_{kl}(h)}{\partial x_i} \Big\|_1 \\
&\quad\le C\|v-w\|_{\frac{3}{2}} \|h\|_3,
\end{aligned}$$

and this implies (5.4.10).

For $|\alpha| = 1$ we obtain

$$\begin{aligned}
&\|D^\alpha((\widetilde{A}(v)-\widetilde{A}(w))h)\|_0 \\
&\quad\le \sum_{i,j,k,l=1}^{n} \Big(\Big\| D^\alpha \Big(\frac{\partial \Phi_{ij}}{\partial \xi_{kl}}(\mathcal{E}(v)) - \frac{\partial \Phi_{ij}}{\partial \xi_{kl}}(\mathcal{E}(w))\Big) \frac{\partial \mathcal{E}_{kl}(h)}{\partial x_i} \Big\|_0 \\
&\qquad + \Big\| \Big(\frac{\partial \Phi_{ij}}{\partial \xi_{kl}}(\mathcal{E}(v)) - \frac{\partial \Phi_{ij}}{\partial \xi_{kl}}(\mathcal{E}(w))\Big) D^\alpha \frac{\partial \mathcal{E}_{kl}(h)}{\partial x_i} \Big\|_0 \Big) \\
&\quad\le \sum_{i,j,k,l=1}^{n} \Big(\Big\| D^\alpha \Big(\frac{\partial \Phi_{ij}}{\partial \xi_{kl}}(\mathcal{E}(v)) - \frac{\partial \Phi_{ij}}{\partial \xi_{kl}}(\mathcal{E}(w))\Big) \Big\|_{L_3} \Big\| \frac{\partial \mathcal{E}_{kl}(h)}{\partial x_i} \Big\|_{L_6} \\
&\qquad + \Big\| \frac{\partial \Phi_{ij}}{\partial \xi_{kl}}(\mathcal{E}(v)) - \frac{\partial \Phi_{ij}}{\partial \xi_{kl}}(\mathcal{E}(w)) \Big\|_{L_\infty} \Big\| D^\alpha \frac{\partial \mathcal{E}_{kl}(h)}{\partial x_i} \Big\|_0 \Big) \\
&\quad\le \sum_{i,j,k,l=1}^{n} C\Big(\Big\| \sum_{k_1,l_1=1}^{n} \frac{\partial^2 \Phi_{ij}}{\partial \xi_{kl} \partial \xi_{k_1 l_1}}(\mathcal{E}(v)) D^\alpha \mathcal{E}_{k_1 l_1}(v) \\
&\qquad - \frac{\partial^2 \Phi_{ij}}{\partial \xi_{kl} \partial \xi_{k_1 l_1}}(\mathcal{E}(w)) D^\alpha \mathcal{E}_{k_1 l_1}(w) \Big\|_{L_3} \|\nabla h\|_2 + \|\mathcal{E}(v)-\mathcal{E}(w)\|_{L_\infty} \|\nabla h\|_2 \Big) \\
&\quad\le \sum_{i,j,k,l=1}^{n} C\Big(\sum_{k_1,l_1=1}^{n} \Big[\Big\| \frac{\partial^2 \Phi_{ij}}{\partial \xi_{kl} \partial \xi_{k_1 l_1}}(\mathcal{E}(v)) \Big\|_{L_\infty} \|D^\alpha(\mathcal{E}_{k_1 l_1}(v) - \mathcal{E}_{k_1 l_1}(w))\|_{L_3} \\
&\qquad + \Big\| \frac{\partial^2 \Phi_{ij}}{\partial \xi_{kl} \partial \xi_{k_1 l_1}}(\mathcal{E}(v)) - \frac{\partial^2 \Phi_{ij}}{\partial \xi_{kl} \partial \xi_{k_1 l_1}}(\mathcal{E}(w)) \Big\|_{L_\infty} \|D^\alpha \mathcal{E}_{k_1 l_1}(w)\|_{L_3} \Big] \\
&\qquad + \|\mathcal{E}(v)-\mathcal{E}(w)\|_{L_\infty} \Big) \|\nabla h\|_2
\end{aligned}$$

$$
\begin{aligned}
&\le C\Big[\Big(\Big\|\frac{\partial^2\Phi_{ij}}{\partial\xi_{kl}\partial\xi_{k_1l_1}}(\mathcal{E}(v)) - \frac{\partial^2\Phi_{ij}}{\partial\xi_{kl}\partial\xi_{k_1l_1}}(0)\Big\|_{L_\infty} \\
&\quad + \Big\|\frac{\partial^2\Phi_{ij}}{\partial\xi_{kl}\partial\xi_{k_1l_1}}(0)\Big\|_{L_\infty}\Big)\|\nabla(v-w)\|_{\frac{7}{4}} + \|\mathcal{E}(v)-\mathcal{E}(w)\|_{L_\infty}\|\nabla w\|_{\frac{7}{4}} \\
&\quad + \|\mathcal{E}(v)-\mathcal{E}(w)\|_{L_\infty}\Big]\|\nabla h\|_2 \\
&\le C\Big[(\|\mathcal{E}(v)\|_{L_\infty} + C)\|\nabla(v-w)\|_{\frac{7}{4}} + \|\mathcal{E}(v)-\mathcal{E}(w)\|_{L_\infty}\|w\|_{\frac{11}{4}} \\
&\quad + \|\mathcal{E}(v)-\mathcal{E}(w)\|_{\frac{7}{4}}\Big]\|\nabla h\|_2 \\
&\le C\Big(\|v\|_{\frac{11}{4}}\|\nabla(v-w)\|_{\frac{7}{4}} + \|\nabla(v-w)\|_{\frac{7}{4}} + \|\nabla(v-w)\|_{\frac{7}{4}}\|w\|_{\frac{11}{4}}\Big)\|\nabla h\|_2 \\
&\le C\|\nabla B_0^{\frac{7}{8}}(v-w)\|_0\|\nabla h\|_2.
\end{aligned} \tag{5.4.17}
$$

For $|\alpha| = 2$ in a similar way one obtains

$$
\big\|D^\alpha\big((\widetilde{A}(v)-\widetilde{A}(w))h\big)\big\|_0 \le C\|\nabla B_0^{1/2}(v-w)\|_1\|\nabla h\|_3. \tag{5.4.18}
$$

Estimates (5.4.15), (5.4.17), (5.4.18) yield estimate (5.4.9) for all l.

Notations (5.3.5), (5.3.6) imply $A(0) = I - \eta_0\Delta$. Therefore

$$
(A(0)v, B_0v)_l \ge 2K_{13}\|B_0v\|_l^2
$$

for some $K_{13} > 0$.

On the other hand, for K_{12} small enough, estimate (5.4.9) yields

$$
\begin{aligned}
\|((A_0 - A(0))v, B_0v)_l\| &\le \|(A(a)-A(0))v\|_l\|B_0v\|_l \\
&\le K_{11}\|B_0^{\alpha_0(l)}a\|_l\|B_0v\|_l^2 \le K_{13}\|B_0v\|_l^2
\end{aligned}
$$

what implies (5.4.11).

Let $\|u\|_3$, $\|v\|_3 < K_{14}$ and K_{14} be small enough. Then estimate (5.4.11) yields

$$
(A(u)(u-v), B_0(u-v))_1 \ge K_{13}\|B_0(u-v)\|_1^2.
$$

But (5.4.9) implies

$$
\begin{aligned}
|((A(u)-A(v))v, B_0(u-v))_1| &\le K_{11}\|B_0^{\frac{7}{8}}(u-v)\|_1\|B_0(u-v)\|_1^2 \\
&\le \frac{K_{13}}{2}\|B_0(u-v)\|_1^2
\end{aligned}
$$

for K_{14} small enough, and it yields (5.4.12).

Now, note that

$$
((A(0)-I)u, B_0u)_2 = -\eta_0(\Delta u, B_0u)_2 \ge 2K_{15}\|\nabla u\|_3^2
$$

for some $K_{15} > 0$.

But estimate (5.4.9) and triangle inequality yield

$$\begin{aligned}|((A(u) - A(0))u, B_0 u)|_2 &\le K_{11}\|\nabla B_0^{\frac{1}{2}} u\|_1 \|\nabla u\|_3 \|(I - \Delta)u\|_2 \\ &\le C(\|\nabla u\|_2\|\nabla u\|_3\|u\|_2 + \|u\|_3\|\nabla u\|_3\|\Delta u\|_2) \\ &\le K_{15}\|\nabla u\|_3^2\end{aligned}$$

for K_{14} small enough. This implies (5.4.13). □

5.4.3 Properties of the operator A_0

Lemma 5.4.3. *Under the conditions and notations of Lemma 5.4.2 there exist constants K_{17} and K_{18} such that provided*

$$\|B_0^{\alpha_0(l)} a\|_l < K_{12}, \quad \lambda \in \mathbb{C}, \quad \operatorname{Re}\lambda \ge 0,$$

the following estimates are valid:

$$\|(A_0 + \lambda I)v\|_l \le K_{17}(\lambda)\|B_0 v\|_l, \tag{5.4.19}$$

$$\|(A_0 + \lambda I)v\|_l \ge K_{18}\|B_0 v\|_l. \tag{5.4.20}$$

Under the same conditions and $l = 1, 2$ the operator

$$A_0 : H_V^{l+2} \simeq D(A_0) \subset H_V^l \to H_V^l$$

is strongly positive, and the constant K_1 in (3.1.11) does not depend on a (but may depend on K_{12}).

Remark 5.4.1. Here we have to use complexifications of Sobolev spaces and differential operators (cf. Remark 3.1.2).

Proof of Lemma 5.4.3. Let $\|B_0^{\alpha_0(l)} a\|_l < K_{12}$. Then estimate (5.4.9) implies

$$\|(A(a) - A(0))v\|_l \le C\|B_0 v\|_l.$$

But for all $\lambda \in \mathbb{C}, \operatorname{Re}\lambda \ge 0$:

$$\|(A(0) + \lambda I)v\|_l = \|(-\eta_0\Delta + (\lambda + 1)I)v\|_l \le K_{18}(\lambda)\|B_0 v\|_l,$$

and these two estimates yield (5.4.19).

From (5.4.11) we have:

$$|((A_0 + \lambda I)v, B_0 v)_l| \ge \operatorname{Re}((A_0 + \lambda I)v, B_0 v)_l \ge K_{13}\|B_0 v\|_l^2. \tag{5.4.21}$$

This estimate and the Cauchy–Buniakowski inequality imply (5.4.20).

From (5.4.20) it follows that $\mathrm{Ker}(A_0+\lambda I) = \{0\}$. Let us show that $\mathrm{Im}(A_0+\lambda I)$ is dense in H_V^l. If not, then in H_V^l there is a nonzero vector h orthogonal to $\mathrm{Im}(A_0+\lambda I)$. Let $\zeta = B_0^{-1}h \in D(A_0)$. Then (5.4.21) implies

$$\mathrm{Re}((A_0 + \lambda I)\zeta, B_0\zeta)_l \geq K_{13}\|B_0\zeta\|_l^2 > 0. \tag{5.4.22}$$

But since $(A_0 + \lambda I)\zeta \in \mathrm{Im}(A_0 + \lambda I)$ and $B_0\zeta = h$, the left-hand side of (5.4.22) is equal to zero. Hence, $\mathrm{Im}(A_0 + \lambda I)$ is dense in H_V^l.

Let us show now that $\mathrm{Im}(A_0 + \lambda I)$ is closed. Let $h_i \underset{i\to\infty}{\to} h_0$, $h_i \in \mathrm{Im}(A_0 + \lambda I)$. We have to show that $h_0 \in \mathrm{Im}(A_0 + \lambda I)$. Note that there exist ζ_i such that $(A_0 + \lambda I)\zeta_i = h_i$. Then the sequence $(A_0 + \lambda I)\zeta_i$ converges in H_V^l, and from (5.4.20) it follows that $B_0\zeta_i$ also converges in H_V^l: $B_0\zeta_i \to \xi_0$. Let $\zeta_0 = B_0^{-1}\xi_0 \in D(A_0)$. We have: $B_0(\zeta_i - \zeta_0) \to 0$. And estimate (5.4.19) yields $(A_0 + \lambda I)(\zeta_i - \zeta_0) \to 0$, i.e. $h_0 = (A_0 + \lambda I)\zeta_0$.

Thus, $A_0 + \lambda I$ is surjective.

Now, for $|\lambda| \geq 1, \mathrm{Re}\,\lambda \geq 0$, $u \in D(A_0)$ we obtain

$$\begin{aligned}
(u,u)_l &= \mathrm{Re}\,\frac{1}{|\lambda|^2}(\lambda u, \lambda u)_l \\
&\leq \mathrm{Re}\,\frac{1}{|\lambda|^2}(\lambda u, \lambda u)_l + \mathrm{Re}\,\frac{1}{|\lambda|^2}(A_0 u, \lambda B_0 u)_{l-1} \\
&\quad + \mathrm{Re}\,\frac{1}{|\lambda|^2}(\lambda B_0 u, A_0 u)_{l-1} + \mathrm{Re}\,\frac{1}{|\lambda|^2}(A_0 u, A_0 u)_l \\
&= \mathrm{Re}\,\frac{1}{|\lambda|^2}((A_0 + \lambda I)u, (A_0 + \lambda I)u)_l
\end{aligned}$$

whence

$$\|u\|_l \leq \frac{1}{|\lambda|}\|(A_0 + \lambda I)u\|_l \leq \frac{2}{1 + |\lambda|}\|(A_0 + \lambda I)u\|_l. \tag{5.4.23}$$

Similarly, for $|\lambda| \leq 1, \mathrm{Re}\,\lambda \geq 0$, by (5.4.20)

$$\|u\|_l \leq \|B_0 u\|_l \leq K_{18}^{-1}\|(A_0 + \lambda I)u\|_l \leq \frac{2}{1 + |\lambda|}K_{18}^{-1}\|(A_0 + \lambda I)u\|_l.$$

This estimate and (5.4.23) imply

$$\|(A_0 + \lambda I)^{-1}\| \leq \frac{C}{1 + |\lambda|}, \quad \mathrm{Re}\,\lambda \geq 0,$$

with C independent of a. □

5.4.4 Uniqueness lemma

Lemma 5.4.4. *There exists a constant $K_{19} > 0$, independent of T and ε, such that if a solution (u_1, τ_1) of problem (5.3.16), (5.4.2), (5.3.18) exists in the class*

$$u_1 \in C^1([0,T]; H_V^1) \cap C([0,T]; H_V^3),$$
$$\tau_1 \in C^1([0,T]; H_M^1) \cap C([0,T]; H_M^3), \tag{5.4.24}$$

and

$$\|u_1(t)\|_3 < K_{19}, \quad t \in [0,T], \tag{5.4.25}$$

then it is unique in class (5.4.24).

Proof. Assume that in addition to (u_1, τ_1) there is another solution (u_2, τ_2) of problem (5.3.16), (5.4.2), (5.3.18).

Suppose first that

$$\|u_2(t)\|_3 \le K_{19}, \quad t \in [0,T] \tag{5.4.26}$$

(K_{19} will be defined below).

The following identity takes place:

$$(N_1(\tau), u)_m + \frac{1}{2\eta}(N_2(u), \tau)_m = 0, \quad u \in H_V^{m+1}, \quad \tau \in H_M^{m+1}, \quad m = 0, 1, \dots \tag{5.4.27}$$

Indeed,

$$\begin{aligned}
&(u, N_1(\tau))_m + \frac{1}{2\eta}(\tau, N_2(u))_m \\
&= \sum_{i,j=1}^{n} \Big[\Big(u_i, \frac{\partial \tau_{ij}}{\partial x_j}\Big)_m + \Big(\tau_{ij}, \frac{1}{2}\Big(\frac{\partial u_i}{\partial x_j} + \frac{\partial u_j}{\partial x_i}\Big)\Big)_m\Big] \\
&= \sum_{i,j=1}^{n} \Big[\Big(u_i, \frac{\partial \tau_{ij}}{\partial x_j}\Big)_m + \Big(\tau_{ij}, \frac{\partial u_i}{\partial x_j}\Big)_m\Big] \\
&= \sum_{i,j=1, |\alpha| \le m}^{n} \int_{\mathbb{R}^n} \Big(D^\alpha u_i D^\alpha \frac{\partial \tau_{ij}}{\partial x_j} + D^\alpha \tau_{ij} D^\alpha \frac{\partial u_i}{\partial x_j}\Big)\, dx = 0
\end{aligned}$$

by Green's formula (cf. Lemma 6.1.1).

Let $w = u_1 - u_2$, $\sigma_* = \tau_1 - \tau_2$. Substitute (u_1, τ_1) and (u_2, τ_2) into (5.3.16) and take the difference of the obtained equalities. Treat (5.4.2) in the same way. We have:

$$\begin{aligned}
\frac{dw(t)}{dt} &= A(u_2)u_2 - A(u_1)u_1 + w + F_1(w, u_1) + F_1(u_2, w) + N_1(\sigma_*) \\
\frac{d\sigma_*(t)}{dt} &= -A_\varepsilon(\sigma_*) + F(w, \tau_1) + F(u_2, \sigma_*) + N_2(w) + G(u_1, \tau_1) - G(u_2, \tau_2).
\end{aligned}$$

Taking the scalar product of the first equation with $B_0 w(t)$ in H_V^1 and of the second one with $\frac{1}{2\eta} B_0 \sigma_*(t)$ in H_M^1 at every $t \in [0, T]$, adding the obtained equalities and taking into account (5.4.27) we get

$$\begin{aligned}
&\Big(\frac{dw}{dt}, B_0 w\Big)_1 + \frac{1}{2\eta}\Big(\frac{d\sigma_*}{dt}, B_0\sigma_*\Big)_1 \\
&\quad = -(A(u_1)u_1 - A(u_2)u_2, B_0 w)_1 + (w, B_0 w)_1 + (F_1(w, u_1), w)_2 \\
&\qquad + (F_1(u_2, w), w)_2 + \frac{1}{2\eta}[-(A_\varepsilon(\sigma_*), B_0\sigma_*)_1 + (F(w, \tau_1), \sigma_*)_2 \\
&\qquad + (F(u_2, \sigma_*), \sigma_*)_2 + (G(u_1, \tau_1) - G(u_2, \tau_2), \sigma_*)_2].
\end{aligned}$$

Obviously,

$$(A_\varepsilon(\sigma_*), B_0\sigma_*)_1 = \Big(\frac{\sigma_*}{\lambda} - \varepsilon\Delta\sigma_*, \sigma_*\Big)_2 \geq \frac{1}{\lambda}\|\sigma_*\|_2.$$

Now, using estimates (5.4.4) – (5.4.7), (5.4.12) (we can assume K_{19} to be small enough so that estimate (5.4.12) is also valid) we obtain:

$$\begin{aligned}
&\frac{1}{2}\frac{d}{dt}\|w\|_2^2 + \frac{1}{4\eta}\frac{d}{dt}\|\sigma_*\|_2^2 + \frac{1}{2}K_{13}\|w\|_3^2 + \frac{1}{2\eta\lambda}\|\sigma_*\|_2^2 \\
&\quad \leq K_{20}[\|w\|_2^2(1 + \|u_1\|_3 + \|u_2\|_3) + \|w\|_2\|\tau_1\|_3\|\sigma_*\|_2 + \|u_2\|_3\|\sigma_*\|_2^2 \\
&\qquad + \|\sigma_*\|_2(\|\sigma_*\|_2 + \|w\|_3)(\|\tau_1\|_2 + \|\tau_2\|_2 + \|u_1\|_3 + \|u_2\|_3)].
\end{aligned}$$

It is obvious that there exists K_{21} such that

$$\begin{aligned}
&\frac{1}{2}K_{13}\|w\|_3^2 + K_{21}\|\sigma_*\|_2^2 \\
&\qquad - K_{20}\|\sigma_*\|_2\|w\|_3(\|\tau_1\|_2 + \|\tau_2\|_2 + \|u_1\|_3 + \|u_2\|_3) \geq 0.
\end{aligned}$$

Therefore

$$\begin{aligned}
&\frac{1}{2}\Big(\frac{d}{dt}\|w\|_2^2 + \frac{1}{2\eta}\frac{d}{dt}\|\sigma_*\|_2^2\Big) \\
&\quad \leq \Big(K_{21} - \frac{1}{2\eta\lambda}\Big)\|\sigma_*\|_2^2 + K_{20}[\|w\|_2^2(1 + \|u_1\|_3 + \|u_2\|_3) + \|w\|_2\|\tau_1\|_3\|\sigma_*\|_2 \\
&\qquad + \|\sigma_*\|_2^2\|u_2\|_3 + \|\sigma_*\|_2^2(\|\tau_1\|_2 + \|\tau_2\|_2 + \|u_1\|_3 + \|u_2\|_3)] \\
&\quad \leq C\big(\|w\|_2^2 + \frac{1}{2\eta}\|\sigma_*\|_2^2\big).
\end{aligned}$$

Since $w(0) = 0,\ \sigma_*(0) = 0$, by the Gronwall lemma $w \equiv 0,\ \tau \equiv 0$.

If (5.4.26) is not valid for some t, let t_1 be the infimum of such t. It is evident that $\|u_2(t_1)\|_3 = K_{19}$. But on the other hand, we have just proved that on $[0, t_1]$ the solution is unique. Hence $u_1(t_1) = u_2(t_1),\ \tau_1(t_1) = \tau_2(t_1)$ what contradicts (5.4.25). □

5.4.5 A priori estimate

Lemma 5.4.5. *There exists a constant $K_{22} > 0$, independent of T and ε, such that for any positive number $\kappa < K_{22}$, and*

$$\|a\|_3 + \frac{1}{\sqrt{2\eta}}\|\tau_0\|_3 + \|f\|_{L_1(0,T;H^3_V)} < \kappa$$

one has the following bounds for every solution u of (5.3.16), (5.4.2), (5.3.18) in class (5.4.3):

$$\|u(t)\|_3^2 + \frac{1}{2\eta}\|\tau(t)\|_3^2 < \kappa^2 < K_{22}^2, \tag{5.4.28}$$

$$\int_0^T \|\nabla u\|_3^2\,ds < \frac{3\kappa^2}{K_{15}}. \tag{5.4.29}$$

Proof. Denote by $t_1 = t_1(\kappa)$ the infimum of those t at which (5.4.28) is not valid (K_{22} will be defined later, κ is an arbitrary positive number which is less than K_{22}).

Taking the scalar product of (5.3.16) with $B_0u(t)$ in H^2_V and of (5.4.2) with $\frac{1}{2\eta}B_0\tau(t)$ in H^2_M at every $t \in [0, t_1]$, adding the obtained equalities and taking into account (5.4.27) we get

$$\begin{aligned}\Big(\frac{du}{dt}, B_0u\Big)_2 &+ \frac{1}{2\eta}\Big(\frac{d\tau}{dt}, B_0\tau\Big)_2 \\ &= -(A(u)u, B_0u)_2 + (u, B_0u)_2 + (F_1(u,u), B_0u)_2 \\ &\quad + \frac{1}{2\eta}[-(A_\varepsilon(\tau), B_0\tau)_2 + (F(u,\tau), B_0\tau)_2 + (G(u,\tau), B_0\tau)_2] + (f, B_0u)_2.\end{aligned}$$

Then we have

$$\begin{aligned}\frac{1}{2}\frac{d}{dt}\|u\|_3^2 + \frac{1}{4\eta}\frac{d}{dt}\|\tau\|_3^2 \le{}& -((A(u)-I)u, B_0u)_2 + (F_1(u,u), u)_3 - \frac{1}{2\eta\lambda}\|\tau\|_3^2 \\ &+ \frac{1}{2\eta}(F(u,\tau),\tau)_3 + \frac{1}{2\eta}(G(u,\tau),\tau)_3 + (f,u)_3.\end{aligned}$$

Using (5.4.6), (5.4.8), (5.4.13) (we can assume K_{22} to be small enough so that estimate (5.4.13) is also valid) we obtain:

$$\begin{aligned}\frac{1}{2}\frac{d}{dt}\|u\|_3^2 &+ \frac{1}{4\eta}\frac{d}{dt}\|\tau\|_3^2 \\ &\le -K_{15}\|\nabla u\|_3^2 - \frac{1}{2\eta\lambda}\|\tau\|_3^2 + K_9\|\nabla u\|_2^3 + \frac{K_9}{2\eta}\|\nabla u\|_2\|\tau\|_3^2 \\ &\quad + \frac{1}{2\eta}K'_{10}\cdot(\|\nabla u\|_3 + \|\tau\|_3)(\|u\|_3 + \|\tau\|_2)\|\tau\|_3 + \|f\|_3\|u\|_3.\end{aligned}$$

Note that K'_{10} is uniformly bounded for $\|u(t)\|_3 + \|\tau(t)\|_2$ is bounded.

For K_{22} small enough this yields

$$\frac{1}{2}\frac{d}{dt}\|u\|_3^2 + \frac{1}{4\eta}\frac{d}{dt}\|\tau\|_3^2 \le \|f\|_3\|u\|_3 - \frac{1}{2}K_{15}\|\nabla u\|_3^2. \tag{5.4.30}$$

Let $\psi(t) = \sqrt{\|u\|_3^2(t) + \frac{1}{2\eta}\|\tau\|_3^2(t)}$. Then from (5.4.30) it follows that

$$\frac{1}{2}\frac{d}{dt}\psi^2(t) \le \|f\|_3\psi(t), \qquad t \in [0, t_1],$$
$$\frac{d}{dt}\psi(t) \le \|f\|_3.$$

Integrate from 0 to t_1:

$$\psi(t_1) \le \sqrt{\|a\|_3^2 + \frac{1}{2\eta}\|\tau_0\|_3^2} + \int_0^{t_1} \|f\|_3(s)ds$$
$$\le \|a\|_3 + \frac{1}{\sqrt{2\eta}}\|\tau_0\|_3 + \int_0^T \|f\|_3(s)ds < \kappa,$$

which contradicts the definition of t_1. Thus, (5.4.28) is valid for all $t \in [0, T]$.

Integrating (5.4.30) from 0 to T we obtain

$$\frac{1}{2}\|u\|_3^2(T) + \frac{1}{4\eta}\|\tau\|_3^2(T) - \frac{1}{2}\|a\|_3^2 - \frac{1}{4\eta}\|\tau_0\|_3^2 \le \int_0^T (\|f\|_3\|u\|_3 - \frac{1}{2}K_{15}\|\nabla u\|_3^2)\, ds.$$

This inequality and (5.4.28) imply

$$\frac{1}{2}K_{15}\int_0^T \|\nabla u\|_3^2\, ds \le \frac{1}{2}\|a\|_3^2 + \frac{1}{4\eta}\|\tau_0\|_3^2 + \int_0^T \|f\|_3\|u\|_3\, ds$$
$$\le \frac{1}{2}\kappa^2 + \kappa\int_0^T \|f\|_3\, ds \le \frac{3}{2}\kappa^2,$$

and we get estimate (5.4.29). □

5.4.6 Proof of Theorem 5.4.1

Since the operators F_1 and F are bilinear, estimates (5.4.4) and (5.4.5) imply

$$\|\widetilde{F}(u) - \widetilde{F}(u)\|_l \le C\|u - v\|_{l+1}, \tag{5.4.31}$$
$$\|F(u,\tau) - F(v,\tau_*)\|_l \le C(\|u - v\|_{l+1} + \|\tau - \tau_*\|_{l+1}) \tag{5.4.32}$$

for $l = 1, 2,\ u, v \in H_V^{l+1},\ \tau, \tau_* \in H_M^{l+1},\ \|u\|_{l+1} + \|v\|_{l+1} < 1,\ \|\tau\|_{l+1} + \|\tau_*\|_{l+1} < 1$.

Now we are going to use Theorem 3.2.1 twice. First, take

$$
\begin{aligned}
E &= H_V^1 \times H_M^1,\ v_0 = (a, \tau_0),\\
f_1(t,(u,\tau)) &= (f(t) + \widetilde{F}(u) + N_1(\tau), F(u,\tau) + N_2(u) + G(u,\tau)),\\
\mathcal{B}(u,\tau) &= (B_0 u, B_0 \tau),\ D(\mathcal{B}) = H_V^3 \times H_M^3, \qquad (5.4.33)\\
\mathcal{A}(u,\tau)(v_1,v_2) &= (A(u)v_1, A_\varepsilon v_2), \mathcal{A}_0(v_1,v_2) = (A_0 v_1, A_\varepsilon v_2),\\
\alpha &= \frac{7}{8},\quad \beta = 1,\quad R \text{ small enough.}
\end{aligned}
$$

Then estimate (3.2.1) follows from (5.4.9) with $l = 1$, estimate (3.2.3) follows from (5.4.19) and (5.4.20) with $l = 1$, and estimate (3.2.4) follows from C^1-smoothness of f and estimates (5.4.31), (5.4.32), (5.4.7) with $l = 1$.

Thus, the conditions of Theorem 3.2.1, a) hold. Hence, for K_7 small enough, system (5.3.16), (5.4.2), (5.3.18) has a solution (u,τ) in class (5.4.24) on some interval $[0,t_0]$. Moreover, by Theorem 3.1.5, formula (3.1.29), the function

$$\mathcal{A}_0^{\delta_1} \mathcal{A}(u,\tau)(u,\tau) : [0,t_0] \to E$$

is continuous for some $\delta_1 > 0$. But by Theorem 3.1.2 the operator

$$\mathcal{B}^{1-\delta_1+\delta}[\mathcal{A}(u,\tau)]^{-1+\delta_1}\mathcal{A}_0^{-\delta_1}$$

is bounded uniformly with respect to t for $0 < \delta < \delta_1$. This yields the estimate

$$\|\mathcal{B}^{\frac{1}{2}-\delta_1+\delta}[\mathcal{A}(u,\tau)]^{\delta_1}(u,\tau)\|_2 < C, \quad t \in [0,t_0] \qquad (5.4.34)$$

where C does not depend on t.

Now, let us apply Theorem 3.2.1 with

$$
\begin{aligned}
E &= H_V^2 \times H_M^2,\ D(\mathcal{B}) = H_V^4 \times H_M^4,\\
\alpha &= \frac{1}{2},\ \beta = \frac{1}{2} + \delta_2 \quad (0 < \delta_2 < \delta),
\end{aligned}
$$

and with v_0, f_1, $\mathcal{A}$, $\mathcal{B}$, R as in (5.4.33). Estimates (5.4.9), (5.4.19), (5.4.20), (5.4.31), (5.4.32), (5.4.7) with $l = 2$ imply again that the conditions of Theorem 3.2.1, a) hold. Hence, problem (5.3.16), (5.4.2), (5.3.18) has a solution in class (5.4.3) on some interval $[0,t_*]$. Furthermore, estimate (5.4.28) with κ small enough enables to apply Lemma 5.4.4. Therefore this solution is unique in class (5.4.24) on $[0,t_*]$, i.e. it coincides with (u,τ). But by Theorem 3.1.2 the operator

$$\mathcal{B}^{\beta}[\mathcal{A}(u,\tau)]^{-\delta_1}\mathcal{B}^{-\frac{1}{2}+\delta_1-\delta}$$

is uniformly (with respect to t) bounded in $E = H_V^2 \times H_M^2$. Then estimate (5.4.34) implies

$$\|\mathcal{B}^{\beta}(u,\tau)(t)\|_{H_V^2 \times H_M^2} < K_{23},\ t \in [0,t_*]$$

where K_{23} does not depend on t_* and t. Furthermore, by estimate (5.4.28) this solution is a priori bounded by a constant independent on t, t_*, t_0:

$$\|\mathcal{B}^{1/2}(u,\tau)(t)\|_{H^2_V \times H^2_M} \leq K_{24}, \tag{5.4.35}$$

and K_{24} is arbitrary small provided K_7 is small. By Theorem 3.2.1, b)

$$t_* = t_0,$$

so the solution (u,τ) belongs to class (5.4.3) on the segment $[0,t_0]$. Rewrite (5.4.35) as

$$\|\mathcal{B}(u,\tau)(t)\|_{H^1_V \times H^1_M} \leq K_{24}. \tag{5.4.36}$$

If K_{24} is sufficiently small, (5.4.36) yields

$$\|\mathcal{B}^{\frac{7}{8}}(u,\tau)(t)\|_{H^1_V \times H^1_M} < K_{25} < R. \tag{5.4.37}$$

Thus, by Theorem 3.2.1, b) applied to (5.4.33), t_0 is equal to T.

5.5 Proof of the main theorems

5.5.1 Proof of Theorem 5.3.1

Evidently, there exist sequences

$$a_m \underset{m\to\infty}{\to} a \text{ in } H^3_V,\ a_m \in H^4_V,$$

$$\tau_{0m} \underset{m\to\infty}{\to} \tau_0 \text{ in } H^3_M,\ \tau_{0m} \in H^4_M,$$

$$f_m \underset{m\to\infty}{\to} f \text{ in } L_1(0,T;H^3_V),\ f_m \in L_1(0,T;H^3_V) \cap C^1([0,T];H^2_V).$$

Without loss of generality all triplets (a_m, τ_{0m}, f_m) satisfy estimate (5.3.19). Consider problems (5.3.16), (5.4.2), (5.3.18) with the data a_m, τ_{0m}, f_m and $\varepsilon = \frac{1}{m}$ for all natural m. If K_6 is sufficiently small, by Theorem 5.4.1 each of these problems has a unique solution (u_m, τ_m) in class (5.4.3) and all these solutions are bounded uniformly with respect to m by estimates (5.4.28), (5.4.29). Therefore without loss of generality (see Remark 2.1.1) we may assume that there exists a pair (u_*, τ_*) such that

$$u_m \to u_* \text{ in } L_\infty(0,T;H^3_V) \ * - \text{weakly},$$

$$\tau_m \to \tau_* \text{ in } L_\infty(0,T;H^3_M) \ * - \text{weakly},$$

$$u_m \to u_* \text{ in } L_2(0,T;H^4_V) \text{ weakly}.$$

Let us show that the sequence (u_m, τ_m) is fundamental in $C([0,T];H^2_V \times H^2_M)$.

Let $w_{ij} = u_i - u_j$, $\sigma_{ij} = \tau_i - \tau_j$. Apply the same procedure as in the proof of Lemma 5.4.4: substitute (u_i, τ_i) into (5.3.16), (5.4.2) with $\varepsilon = \frac{1}{i}$ and (u_j, τ_j) into (5.3.16), (5.4.2) with $\varepsilon = \frac{1}{j}$, take the differences of the corresponding equations, take the scalar product of the first of the obtained equations with $w_{ij}(t)$ in H_V^2 and of the second one with $\frac{\sigma_{ij}}{2\eta}$ in H_M^2 at every $t \in [0, T]$, and taking into account (5.4.27) add the obtained equalities :

$$\begin{aligned}
&\Big(\frac{dw_{ij}}{dt}, w_{ij}\Big)_2 + \frac{1}{2\eta}\Big(\frac{d\sigma_{ij}}{dt}, \sigma_{ij}\Big)_2 \\
&\quad = -(A(u_i)u_i - A(u_j)u_j, B_0 w_{ij})_1 + (w_{ij}, w_{ij})_2 + (F_1(w_{ij}, u_i), w_{ij})_2 \\
&\qquad + (F_1(u_j, w_{ij}), w_{ij})_2 - \frac{1}{2\eta\lambda}(\sigma_{ij}, \sigma_{ij})_2 + \frac{1}{2\eta}\Big(\frac{\Delta\tau_i}{i} - \frac{\Delta\tau_j}{j}, \sigma_{ij}\Big)_2 \\
&\qquad + (F(w_{ij}, \tau_i), \sigma_{ij})_2 + (F(u_j, \sigma_{ij}), \sigma_{ij})_2 + (G(u_i, \tau_i) \\
&\qquad - G(u_j, \tau_j), \sigma_{ij})_2 + (f_i - f_j, w_{ij})_2.
\end{aligned}$$

Let K_6 be small enough. Using the estimates and arguing as in the proof of Lemma 5.4.4 we conclude:

$$\begin{aligned}
&\frac{d}{dt}\|w_{ij}\|_2^2 + \frac{1}{2\eta}\frac{d}{dt}\|\sigma_{ij}\|_2^2 \\
&\quad \le C\big(\|w_{ij}\|_2^2 + \frac{1}{2\eta}\|\sigma_{ij}\|_2^2 + \|f_i - f_j\|_2\big) + \frac{1}{i\eta}\|\tau_i\|_3\|\sigma_{ij}\|_3 + \frac{1}{j\eta}\|\tau_j\|_3\|\sigma_{ij}\|_3 \\
&\quad \le C\big(\|w_{ij}\|_2^2 + \frac{1}{2\eta}\|\sigma_{ij}\|_2^2 + \|f_i - f_j\|_2 + \frac{1}{i} + \frac{1}{j}\big).
\end{aligned}$$

Integrate from 0 to t:

$$\begin{aligned}
&\|w_{ij}\|_2^2 + \frac{1}{2\eta}\|\sigma_{ij}\|_2^2 \\
&\quad \le \|a_i - a_j\|_2^2 + \frac{1}{2\eta}\|\tau_{0i} - \tau_{0j}\|_2^2 \\
&\qquad + C\int_0^t \big(\|w_{ij}\|_2^2(s) + \frac{1}{2\eta}\|\sigma_{ij}\|_2^2(s) + \|f_i - f_j\|_2(s) + \frac{1}{i} + \frac{1}{j}\big)ds
\end{aligned}$$

and by Gronwall's lemma $\|w_{ij}\|_2^2(t) + \frac{1}{2\eta}\|\sigma_{ij}\|_2^2(t) \underset{\min(i,j)\to\infty}{\longrightarrow} 0$ uniformly with respect to $t \in [0, T]$.

Thus,

$$\begin{aligned}
u_m &\to u_* \text{ in } C([0, T]; H_V^2), \\
\tau_m &\to \tau_* \text{ in } C([0, T]; H_M^2).
\end{aligned}$$

This implies

$$\frac{1}{m}\Delta\tau_m \to 0 \text{ in } C([0,T];H^0_M),$$
$$A_{\frac{1}{m}}\tau_m \to \frac{\tau_*}{\lambda} \text{ in } C([0,T];H^0_M),$$
$$N_1(\tau_m) \to N_1(\tau_*) \text{ in } C([0,T];H^1_V),$$

$$N_2(u_m) \to N_2(u_*) \text{ in } C([0,T];H^1_M),$$
$$\frac{du_m}{dt} \to \frac{du_*}{dt} \text{ in the sense of distributions,}$$
$$\frac{d\tau_m}{dt} \to \frac{d\tau_*}{dt} \text{ in the sense of distributions.}$$

From estimates (5.4.4), (5.4.5), (5.4.7) we have also

$$\widetilde{F}(u_m) \to \widetilde{F}(u_*) \text{ in } C([0,T];H^1_V),$$
$$F(u_m,\tau_m) \to F(u_*,\tau_*) \text{ in } C([0,T];H^1_M),$$
$$G(u_m,\tau_m) \to G(u_*,\tau_*) \text{ in } C([0,T];H^1_M).$$

Furthermore, using (5.4.9), (5.4.10) we obtain

$$\begin{aligned}
&\operatorname{vrai}\max_{t\in[0,T]} \|A(u_m)u_m - A(u_*)u_*\|_0(t)\\
&\quad\le \operatorname{vrai}\max_{t\in[0,T]} \big(\|A(0)(u_m-u_*)(t)\|_0 + \|(A(u_*)-A(0))(u_m-u_*)\|_0(t)\\
&\qquad + \|(A(u_m)-A(u_*))u_m\|_0(t)\big)\\
&\quad\le C\operatorname{vrai}\max_{t\in[0,T]} \big(\|u_m-u_*\|_2 + \|u_*\|_{\frac{11}{4}}\|u_m-u_*\|_2 + \|u_m-u_*\|_{\frac{3}{2}}\|u_m\|_3\big).
\end{aligned}$$

Therefore

$$A(u_m)u_m \to A(u_*)u_* \text{ in } L_\infty(0,T;H^0_V).$$

Passing to the limit we conclude that the pair (u_*,τ_*) is a solution of problem (5.3.16) – (5.3.18).

Remark 5.5.1. Substituting u_m, f_m, τ_m into (5.3.16) and (5.3.17) we observe that the sequences $\frac{du_m}{dt}$, $\frac{d\tau_m}{dt}$ are bounded (and fundamental) in $L_1(0,T;H^0_V)$ and in $C([0,T];H^1_M)$, respectively. But $u_m \to u_*$ in $L_\infty(0,T;H^3_V)$ $*-$ weakly, and $\{u_m\}$ is a bounded sequence in $L_2(0,T;H^4_V)$. Since the embedding $H^4(B) \subset H^3(B)$ is compact for any ball B in $\mathbb{R}^n$, due to Theorem 2.2.7 without loss of generality we may assume that $u_m\big|_B \to u_*\big|_B$ in $L_p(0,T;H^3(B,\mathbb{R}^n))$ strongly for all $1 \le p < \infty$ and every ball B. Furthermore, the embedding $H^3(B) \subset H^{3-\delta}(B)$ is compact for any $\delta > 0$, and $\tau_m \to \tau_*$ in $L_\infty(0, T; H^3_M)$ $*-$ weakly, so by Theorem 2.2.6 without loss of generality we may assume that $\tau_m\big|_B \to \tau_*\big|_B$ in $C([0, T]; H^{3-\delta}(B, \mathbb{R}^{n\times n}_S))$ strongly. These facts will be used in Section 5.6.

Let us return to the proof of the theorem. Estimate (5.4.9) yields that

$$\int_0^T \|A(u_*)u_*\|_2^2 \le 2\int_0^T \|(A(u_*) - A(0))u_*\|_2^2 + 2\int_0^T \|A(0)u_*\|_2^2$$
$$\le C\int_0^T (\|u_*\|_3^2\|u_*\|_4^2 + \|u_*\|_4^2).$$

Therefore $A(u_*)u_*$ belongs to $L_2(0, T; H^2_V)$. From estimate (5.4.4) we have $\widetilde{F}(u_*) \in L_\infty(0, T; H^2_V)$. Note that $N_1(\tau_*) \in L_\infty(0, T; H^2_V)$ and $f \in L_2(0, T; H^2_V)$. Substituting (u_*, τ_*) into (5.3.16), we conclude that $\frac{du_*}{dt} \in L_2(0, T; H^2_V)$. But $u_* \in L_2(0, T; H^4_V)$ and by Lemma 2.2.7 we have $u_* \in C([0, T]; H^3_V)$.

Substituting u_*, τ_* into (5.3.17) and taking into account estimates (5.4.5) and (5.4.7) we conclude that all terms in (5.3.17) except $\frac{d\tau_*}{dt}$ belong to $C([0, T]; H^1_M)$. Hence, $\tau_* \in C^1([0, T]; H^1_M)$.

The uniqueness of the solution may be proved in exactly the same way as Lemma 5.4.4.

5.5.2 Proof of Theorem 5.2.1

Estimates (5.4.7), (5.4.8) in the particular case $G(u, \tau) = \Phi(\mathcal{E}(u))$ have the form

$$\|\Phi(\mathcal{E}(u_1)) - \Phi(\mathcal{E}(u_2))\|_l \le K_{26}\cdot(\|u_1\|_{\gamma+1} + \|u_2\|_{\gamma+1})\|u_1 - u_2\|_{l+1}, \quad (5.5.1)$$
$$\|\Phi(\mathcal{E}(u))\|_3 \le K_{27}\cdot\|u\|_3\|u\|_4 \quad (5.5.2)$$

where K_{26}, K_{27} depend continuously on $\|\nabla u_1\|_\gamma + \|\nabla u_2\|_\gamma$ and on $\|u\|_3$, respectively, and l,γ are as in estimate (5.4.7).

Let $\tau_0 = \tau_0^1$, $K_5 = K_6$. Then (5.2.16) implies estimate (5.3.19). Hence, by Theorem 5.3.1 there exists a unique solution (u, τ) of problem (5.3.16) – (5.3.18) in class (5.3.20), (5.3.21). Denote

$$g = \frac{\partial u}{\partial t} + \sum_{i=1}^n u_i\frac{\partial u}{\partial x_i} - \eta_0\Delta u - \mathrm{Div}\,(\Phi(\mathcal{E}) + \tau) - f_0.$$

Taking into account notations (5.3.5) – (5.3.15) we conclude (cf. Remark 5.3.2) that (5.3.4) is valid and

$$Pg = 0.$$

Since P is an orthogonal projector in $H^0(\mathbb{R}^n, \mathbb{R}^n)$, g satisfies condition (3.1.3) with $\Omega = \mathbb{R}^n$.

Since u belongs to class (5.2.17), estimates (5.5.1), (5.5.2), and Lemma 2.2.9 yield that $\Phi(\mathcal{E}(u))$ belongs to class (5.2.20). Thus we see that $g \in L_2(0, T; H^2)$.

By Corollary 3.1.4 there exists unique p from class (5.2.19) satisfying (5.2.9) such that

$$\operatorname{grad} p(t) = -g(t) \text{ almost everywhere in } (0, T).$$

Therefore we have (5.3.3).

Let

$$\tau^1 = \tau, \quad \tau^0 = 2\eta_0 \mathcal{E}(u) + \Phi(\mathcal{E}(u)), \quad \mathrm{T} = -pI + \tau^0 + \tau^1.$$

Obviously, the triplet (u, T, p) is a solution of problem (5.2.1) – (5.2.10); τ^0 belongs to class (5.2.20), and T belongs to (5.2.18).

5.5.3 The case $r > 1$

The case $r > 1$ is investigated exactly in the same way as the case $r = 1$. During the operator treatment, problem (5.3.16) – (5.3.18) is replaced by the problem

$$\frac{du}{dt} + A(u)u = \widetilde{F}(u) + N_1\Big(\sum_{i=1}^{r} \tau^i\Big) + f, \tag{5.5.3}$$

$$\frac{d\tau^i}{dt} + \frac{\tau^i}{\lambda} = F(u, \tau^i) + N_2(u) + G_i(u, \tau^i), \tag{5.5.4}$$

$$u(0) = a, \quad \tau^i(0) = \tau_0^i, \quad i = 1, \dots, r, \tag{5.5.5}$$

where

$$G_i(u, \tau) = -\Big(\tau W - W\tau + \frac{\beta_i(\tau, \mathcal{E})}{\lambda}\Big).$$

Then one considers the following system

$$\frac{d\tau^i}{dt} + A_\varepsilon \tau^i = F(u, \tau^i) + N_2(u) + G_i(u, \tau^i) \tag{5.5.6}$$

and the auxiliary problem (5.5.3), (5.5.6), (5.5.5). Just as in the case $r = 1$, the analogues of Theorem 5.4.1 and Lemmas 5.4.4, 5.4.5 are proved and the passage to the limit is carried out.

5.6 Continuous dependence of solutions on data

In this section, we study continuous dependence of solutions to the initial problem for the equations of motion of nonlinear viscoelastic medium on the initial data and the body force. For simplicity, we consider the case $r = 1$. Thus, let us deal with the operator problem (5.3.16) – (5.3.18).

Theorem 5.6.1. *Let triples* $(a_k, \tau_{0k}, f_k), k = 0, 1, 2, \ldots$, *satisfy the conditions of Theorem 5.3.1, including estimate (5.3.19). Let* (u_k, τ_k) *be the corresponding solutions to problem (5.3.16) – (5.3.18). Let*

$$a_k \to a_0 \text{ in } H_V^3,$$
$$\tau_{0k} \to \tau_{00} \text{ in } H_M^3,$$
$$f_k \to f_0 \text{ in } L_1(0, T; H_V^3),$$

as $k \to \infty$. *Then*

$$u_k \to u_0 \text{ in } C([0, T]; H_V^2) \text{ and in } L_p(0, T; H_{V,\mathrm{loc}}^3),$$
$$\tau_k \to \tau_0 \text{ in } C([0, T]; H_M^2) \text{ and in } C([0, T]; H_{M,\mathrm{loc}}^{3-\delta}),$$

for all $1 \leq p < \infty, \delta > 0$.

Proof. Take some sequences

$$a_{m,k} \underset{m\to\infty}{\to} a_k \text{ in } H_V^3,\ a_{m,k} \in H_V^4,$$
$$\tau_{0m,k} \underset{m\to\infty}{\to} \tau_{0k} \text{ in } H_M^3,\ \tau_{0m,k} \in H_M^4,$$
$$f_{m,k} \underset{m\to\infty}{\to} f_k \text{ in } L_1(0, T; H_V^3),\ f_{m,k} \in L_1(0, T; H_V^3) \cap C^1([0, T]; H_V^2).$$

Without loss of generality the triples $(a_{m,k}, \tau_{0m,k}, f_{m,k})$ satisfy estimate (5.3.19). Consider problems (5.3.16), (5.4.2), (5.3.18) with data $a_{m,k}$, $\tau_{0m,k}$, $f_{m,k}$ and $\varepsilon = \frac{1}{m}$ for every natural m and for all k. If K_6 is sufficiently small, by Theorem 5.4.1 each of these problems possesses a unique solution $(u_{m,k}, \tau_{m,k})$. As is shown in the proof of Theorem 5.3.1,

$$u_{m,k} \to u_k \text{ in } C([0, T]; H_V^2), u_{m,k}\big|_B \to u_k\big|_B \text{ in } L_p(0, T; H^3(B)),$$
$$\tau_{m,k} \to \tau_k \text{ in } C([0, T]; H_M^2), \tau_{m,k}\big|_B \to \tau_k\big|_B \text{ in } C([0, T]; H^{3-\delta}(B)),$$

as $m \to \infty$ for all $1 \leq p < \infty, \delta > 0$ and for any ball B in $\mathbb{R}^n$.

Fix an arbitrary ball B in $\mathbb{R}^n$. Denote by

$$\rho(u^1, \tau^1; u^2, \tau^2)$$

the expression

$$\|u^1 - u^2\|_{C([0,T];H_V^2)} + \|(u^1 - u^2)\big|_B\|_{L_p(0,T;H^3(B))}$$
$$+ \|\tau^1 - \tau^2\|_{C([0,T];H_M^2)} + \|(\tau^1 - \tau^2)\big|_B\|_{C([0,T];H^{3-\delta}(B))},$$

and denote by

$$\rho(a^1, \tau_0^1, f^1; a^2, \tau_0^2, f^2)$$

the expression

$$\|a^1 - a^2\|_{H_V^3} + \|\tau_0^1 - \tau_0^2\|_{H_M^3} + \|f^1 - f^2\|_{L_1(0,T;H_V^3)}$$

for any $u^1, \tau^1, u^2, \tau^2, a^1, a^2, \tau_0^1, \tau_0^2, f^1, f^2$ for which these expressions make sense.

Thus it suffices to prove that

$$\rho(u_k, \tau_k; u_0, \tau_0) \underset{k\to\infty}{\to} 0. \tag{5.6.1}$$

Fix $\varepsilon > 0$. For every natural k there is a number $L(k)$ such that if $m \geq L(k)$, then

$$\rho(u_{m,k}, \tau_{m,k}; u_k, \tau_k) \leq \frac{\varepsilon}{2}. \tag{5.6.2}$$

Besides, for any k there is $M(k)$ such that if $m \geq M(k)$, then

$$\rho(a_{m,k}, \tau_{0m,k}, f_{m,k}; a_k, \tau_{0k}, f_k) \leq \frac{1}{k}.$$

Let $P(k) = \max(L(k), M(k))$. Then

$$\begin{aligned}
&\rho(a_{P(k),k}, \tau_{0P(k),k}, f_{P(k),k}; a_0, \tau_{00}, f_0) \\
&\quad \leq \rho(a_{P(k),k}, \tau_{0P(k),k}, f_{P(k),k}; a_k, \tau_{0k}, f_k) + \rho(a_k, \tau_{0k}, f_k; a_0, \tau_{00}, f_0) \\
&\quad \leq \frac{1}{k} + \rho(a_k, \tau_{0k}, f_k; a_0, \tau_{00}, f_0) \underset{k\to\infty}{\to} 0.
\end{aligned}$$

As is shown in the proof of Theorem 5.3.1, this implies

$$\rho(u_{P(k),k}, \tau_{P(k),k}; u_0, \tau_0) \underset{k\to\infty}{\to} 0.$$

Then there is N such that if $k > N$, then

$$\rho(u_{P(k),k}, \tau_{P(k),k}; u_0, \tau_0) \leq \frac{\varepsilon}{2}.$$

Taking into account (5.6.2) and triangle inequality, we conclude

$$\rho(u_k, \tau_k; u_0, \tau_0) \leq \varepsilon,$$

so we have (5.6.1). □

Chapter 6

Weak solutions for equations of motion of viscoelastic medium

6.1 Preliminaries

6.1.1 Weak solutions for equations of fluid dynamics: general scheme

The feature of the most part of equations in fluid dynamics is that in the general case (i.e. without restrictions on the domain Ω and its dimension, the body force, and the initial data) the problem of existence of strong solutions to the initial-boundary value problems for these equations is open. A possible way to break this deadlock is to investigate generalized, *weak* solutions to these problems. There is a great variety of approaches to weak formulation of the problems of fluid mechanics (mostly of the Navier–Stokes problem; for instance, weak and generalized solutions in many senses, including the *Leray–Hopf* solutions, very weak and mild solutions, variational inequalities, see e.g. [6, 61, 37, 36]). Therefore, let us first describe a general scheme of weak problem formulation (not only for the problems of fluid dynamics), which includes most of the approaches.

Let $\mathcal{U}, \mathcal{P}, \mathcal{Z}$ be some sets, let $\mathcal{X} = \mathcal{U} \times \mathcal{P}$, and let $\mathcal{G} : \mathcal{X} \to \mathcal{Z}$ be a map. Denote by $\mathcal{J}$ the projection

$$\mathcal{J} : \mathcal{X} \to \mathcal{U},\ \mathcal{J}(U, P) = U.$$

Consider an abstract equation

$$\mathcal{G}(U, P) = Z. \tag{6.1.1}$$

Here $U \in \mathcal{U}$ and $P \in \mathcal{P}$ are the unknowns, and $Z \in \mathcal{Z}_0 \subset \mathcal{Z}$ is prescribed ($\mathcal{Z}_0$ is some subset of $\mathcal{Z}$ containing the possible "right-hand sides" of the equation under consideration). Denote by $\mathcal{O}_s(Z) \subset \mathcal{X}$ the set of solutions to (6.1.1).

Remark 6.1.1. The set $\mathcal{P}$ can be one-point. Then (6.1.1) can be abbreviated:

$$\mathcal{G}_*(U) = Z \tag{6.1.2}$$

where $\mathcal{G}_*(U) = \mathcal{G}(U, P),\ P \in \mathcal{P}$.

Remark 6.1.2. It is clear that Cauchy problems, boundary value problems, and initial-boundary value problems for partial differential equations can be expressed in form (6.1.1).

Let $\mathfrak{W}$ be some set containing $\mathfrak{U}$, let $\mathfrak{L}$ be some set containing $\mathfrak{Z}_0$, let $\mathfrak{M}$ be a fixed subset of some set $\mathfrak{Q}$, and let $\mathfrak{N} : \mathfrak{W} \times \mathfrak{L} \multimap \mathfrak{Q}$ be a multi-valued map. Consider the inclusion

$$\mathfrak{N}(W, L) \subset \mathfrak{M}. \tag{6.1.3}$$

Here $W \in \mathfrak{W}$ is unknown, and $L \in \mathfrak{L}$ is given. Denote by $\mathfrak{O}_w(L) \subset \mathfrak{W}$ the set of solutions to (6.1.3).

Assume that it can be *a priori* checked that for all $Z \in \mathfrak{Z}_0$ one has

$$\mathcal{J}(\mathfrak{O}_s(Z)) = \mathfrak{O}_w(Z) \bigcap \mathfrak{U}. \tag{6.1.4}$$

Then (6.1.3) is called the *weak statement* of (6.1.1), and the solutions to (6.1.3) are called *weak solutions* of (6.1.1).

Remark 6.1.3. Note that (6.1.3) may be studied not only for $L \in \mathfrak{Z}_0$ but for arbitrary L from $\mathfrak{L}$.

Let us illustrate this scheme with the weak formulation of the initial-boundary value problem for the Navier–Stokes system (1.1.14), (1.1.10), (1.1.15):

$$\frac{\partial u}{\partial t} + \sum_{i=1}^{n} u_i \frac{\partial u}{\partial x_i} - \eta \Delta u + \operatorname{grad} p = f, \tag{6.1.5}$$

$$\operatorname{div} u = 0, \tag{6.1.6}$$

$$u\big|_{\partial\Omega} = 0, \tag{6.1.7}$$

$$u|_{t=0} = a. \tag{6.1.8}$$

Here u is an unknown velocity vector, p is an unknown pressure function, f is the given body force (all of them depend on a point x in a domain $\Omega \subset \mathbb{R}^n$, $n = 2, 3$, and on a moment of time t), $\eta > 0$ is the viscosity, a is a given function of $x \in \Omega$.

Let $T > 0$ be a fixed moment of time. Denote $H_V^2 = H^2(\Omega)^n \cap V$.

Let

$$\mathfrak{U} = C^1([0, T]; H_V^2),$$
$$\mathcal{P} = C([0, T]; H^1_{\mathrm{loc}}(\Omega)),$$
$$\mathfrak{Z} = \mathfrak{Z}_0 = C([0, T]; L_2(\Omega)^n) \times H_V^2,$$
$$Z = (f, a),$$
$$\mathcal{G}(U, P) = \Big(\frac{\partial U}{\partial t} + \sum_{i=1}^{n} U_i \frac{\partial U}{\partial x_i} - \eta \Delta U + \operatorname{grad} P, U|_{t=0}\Big)$$

Then we can rewrite the Navier–Stokes problem in form (6.1.1). For the weak statement, take

$$\begin{aligned}
\mathfrak{W} &= L_2(0,T;V)\bigcap C_w([0,T];H)\bigcap W_1^1(0,T;V^*),\\
\mathfrak{L} &= L_2(0,T;V^*)\times H,\\
\mathfrak{Q} &= L_1(0,T)\times H,\\
\mathfrak{M} &= \{(0 \text{ of } L_1(0,T), 0 \text{ of } H)\},
\end{aligned}$$

and define the multi-valued map $\mathfrak{N}$ by the formula

$$\begin{aligned}
\mathfrak{N}(W,L) = \Big\{\Big(\frac{d}{dt}(W,\varphi) + \eta(\nabla W,\nabla\varphi) - \sum_{i=1}^{n}\big(W_i W, \frac{\partial\varphi}{\partial x_i}\big)\\
- \langle f,\varphi\rangle_{V^*\times V}, W|_{t=0} - a\Big)\Big|\varphi\in V\Big\}
\end{aligned}$$

for $W \in \mathfrak{W}$ and $L = (f,a) \in \mathfrak{L}$.

Then we arrive at the classical

Definition 6.1.1 (see e.g. [61]). A function

$$u \in L_2(0,T;V)\bigcap C_w([0,T];H)\bigcap W_1^1(0,T;V^*),$$

is a *weak solution* of problem (6.1.5) – (6.1.8) if it satisfies condition (6.1.8), and if the equality

$$\frac{d}{dt}(u,\varphi) + \eta(\nabla u,\nabla\varphi) - \sum_{i=1}^{n}\big(u_i u, \frac{\partial\varphi}{\partial x_i}\big) = \langle f,\varphi\rangle \tag{6.1.9}$$

is valid for all $\varphi \in V$ almost everywhere on $(0,T)$.

It remains to check (6.1.4). Let $Z = (f,a) \in \mathfrak{Z}_0 = C([0,T];L_2(\Omega)^n)\times H_V^2$, and let $(u,p) \in \mathfrak{X} = \mathfrak{U}\times\mathfrak{P}$ be a solution to (6.1.5) – (6.1.8). Taking the scalar product of (6.1.5) with an arbitrary function $\varphi \in V$ in $L_2(\Omega)^n$, we obtain

$$\frac{d}{dt}(u,\varphi) - \eta(\Delta u,\varphi) + \sum_{i=1}^{n}\big(u_i\frac{\partial u}{\partial x_i},\varphi\big) + (\operatorname{grad} p,\varphi) = (f,\varphi) = \langle f,\varphi\rangle. \tag{6.1.10}$$

The second equality follows from (2.2.27). Integrating by parts in the second, third, and forth terms (cf. Section 6.1.2), we arrive at

$$\frac{d}{dt}(u,\varphi) + \eta(\nabla u,\nabla\varphi) - \sum_{i=1}^{n}\big(u_i u, \frac{\partial\varphi}{\partial x_i}\big) - (p,\operatorname{div}\varphi) = \langle f,\varphi\rangle. \tag{6.1.11}$$

But div $\varphi = 0$, so we obtain (6.1.9), and u is a weak solution to (6.1.5) – (6.1.8). Thus, $\mathcal{J}(\mathcal{O}_s(Z)) \subset \mathcal{O}_w(Z) \bigcap \mathcal{U}$.

Conversely, if $u \in \mathcal{U} = C^1([0,T]; H^2_V)$ is a weak solution, then integrating by parts in the second and the third terms of (6.1.9) we get

$$(u', \varphi) - \eta(\Delta u, \varphi) + \sum_{i=1}^{n} (u_i \frac{\partial u}{\partial x_i}, \varphi) - (f, \varphi) = 0 \tag{6.1.12}$$

for all $\varphi \in V$. Since $u' - \eta\Delta u + \sum_{i=1}^{n} u_i \frac{\partial u}{\partial x_i} - f$ belongs to $C([0,T]; L_2)$, by Corollary 3.1.5 there is $q \in C([0,T]; H^1_{\text{loc}}(\Omega))$ such that

$$\frac{\partial u}{\partial t} + \sum_{i=1}^{n} u_i \frac{\partial u}{\partial x_i} - \eta\Delta u - f = \operatorname{grad} q(t). \tag{6.1.13}$$

Then the pair $(u, -q)$ belongs to $\mathcal{X}$ and is a solution to (6.1.5) – (6.1.8). Hence, $\mathcal{O}_w(Z) \bigcap \mathcal{U} \subset \mathcal{J}(\mathcal{O}_s(Z))$.

6.1.2 Integration by parts

As we see, integration by parts plays an important role in the analysis of weak solutions to the problems of fluid dynamics. Let us state a simple lemma about it.

Lemma 6.1.1. *Let Ω be an open domain in $\mathbb{R}^n$, $n \in \mathbb{N}$. Let $u_j \in Z_j(\Omega)$ (where Z_j stands for some function space, e.g. W^r_p) for any $j = 1, \dots, k$. Assume that the set of smooth functions on Ω is dense in each $Z_j(\Omega)$, and, moreover, $C_0^\infty(\Omega)$ is dense in $Z_m(\Omega)$ for some m, $1 \le m \le k$.*

Then, for $l = 1, \dots, n$,

$$\sum_{j=1}^{k} (\frac{\partial u_j}{\partial x_l}, u_1 u_2 \cdot \ldots \cdot u_{j-1} u_{j+1} \cdot \ldots \cdot u_k) = 0 \tag{6.1.14}$$

provided the k-linear form in the left-hand side of (6.1.14) is continuous on $Z_1(\Omega) \times \ldots \times Z_k(\Omega)$.

Proof. Since the smooth functions are dense in $Z_j(\Omega)$, and $C_0^\infty(\Omega)$ is dense in $Z_m(\Omega)$, it remains to check (6.1.14) for smooth u_j, $j = 1, \dots, k$, and $u_m \in C_0^\infty(\Omega)$.

But in this case the function $u_1 \cdot \ldots \cdot u_k$ has compact support in Ω, so

$$\begin{aligned}\sum_{j=1}^{k} &(\frac{\partial u_j}{\partial x_l}, u_1 u_2 \cdot \ldots \cdot u_{j-1} u_{j+1} \cdot \ldots \cdot u_k) \\ &= \sum_{j=1}^{k} \int_{\Omega} u_1 u_2 \cdot \ldots \cdot u_{j-1} \frac{\partial u_j}{\partial x_l} u_{j+1} \cdot \ldots \cdot u_k \, dx \\ &= \int_{\Omega} \frac{\partial}{\partial x_l} (u_1 u_2 \cdot \ldots \cdot u_k) \, dx = 0\end{aligned}$$

by Green's formula. □

Corollary 6.1.1. *Formula (6.1.14) is valid in the following situations:*

a) *Ω is an arbitrary domain, $Z_j = W^1_{p_j}(\Omega)$ for $j = 1, \ldots, k$, $j \neq m$; $Z_m = \overset{\circ}{W}{}^1_{p_m}(\Omega)$ for some m; and*

$$\sum_{j=1}^{k} \frac{1}{p_j} = 1, \quad 1 \leq p_j \leq \infty \quad (\frac{1}{\infty} = 0). \tag{6.1.15}$$

b) *Ω is an arbitrary domain, $Z_j = \overset{\circ}{W}{}^1_{p_j}(\Omega)$ for every $j = 1, \ldots, k$, and there are $q_j \geq 0$ such that*

$$\begin{aligned}&\frac{n}{p_j} - 1 \leq \frac{n}{q_j}, \quad 1 \leq p_j < \infty, \\ &\frac{1}{p_j} + \frac{1}{q_1} + \frac{1}{q_2} + \cdots + \frac{1}{q_{j-1}} + \frac{1}{q_{j+1}} + \cdots + \frac{1}{q_k} \leq 1.\end{aligned}$$

c) *Ω is a sufficiently regular domain, $Z_j = W^1_{p_j}(\Omega)$ for $j = 1, \ldots, k$, $j \neq m$; $Z_m = \overset{\circ}{W}{}^1_{p_m}(\Omega)$ for some m; and for every $j = 1, \ldots, k$ there are $q_j \geq 0$ such that*

$$\begin{aligned}&\frac{n}{p_j} - 1 \leq \frac{n}{q_j}, \quad 1 \leq p_j < \infty, \\ &\frac{1}{p_j} + \frac{1}{q_1} + \frac{1}{q_2} + \cdots + \frac{1}{q_{j-1}} + \frac{1}{q_{j+1}} + \cdots + \frac{1}{q_k} \leq 1.\end{aligned}$$

Proof. We have to show that the k-linear form in the left-hand side of (6.1.14) is continuous on $Z_1(\Omega) \times \ldots \times Z_k(\Omega)$. It suffices to check that the forms $(\frac{\partial u_j}{\partial x_l}, u_1 u_2 \cdot$

$\dots \cdot u_{j-1}u_{j+1} \cdot \dots \cdot u_k)$ are bounded. In the case a), it's a direct consequence of Hölder's inequality (2.1.1):

$$
\begin{aligned}
&\left(\frac{\partial u_j}{\partial x_l}, u_1 u_2 \cdot \dots \cdot u_{j-1} u_{j+1} \cdot \dots \cdot u_k\right) \\
&\quad = \int_\Omega \frac{\partial u_j}{\partial x_l} u_1 u_2 \cdot \dots \cdot u_{j-1} u_{j+1} \cdot \dots \cdot u_k \, dx \\
&\quad \leq \left\| \frac{\partial u_j}{\partial x_l} u_1 u_2 \cdot \dots \cdot u_{j-1} u_{j+1} \cdot \dots \cdot u_k \right\|_{L_1(\Omega)} \\
&\quad \leq \left\| \frac{\partial u_j}{\partial x_l} \right\|_{L_{p_j}(\Omega)} \|u_1\|_{L_{p_1}(\Omega)} \|u_2\|_{L_{p_2}(\Omega)} \cdot \dots \\
&\qquad \cdot \|u_{j-1}\|_{L_{p_{j-1}}(\Omega)} \|u_{j+1}\|_{L_{p_{j+1}}(\Omega)} \cdot \dots \cdot \|u_k\|_{L_{p_k}(\Omega)} \\
&\quad \leq \|u_j\|_{W^1_{p_j}(\Omega)} \|u_1\|_{W^1_{p_1}(\Omega)} \|u_2\|_{W^1_{p_2}(\Omega)} \cdot \dots \\
&\qquad \cdot \|u_{j-1}\|_{W^1_{p_{j-1}}(\Omega)} \|u_{j+1}\|_{W^1_{p_{j+1}}(\Omega)} \cdot \dots \cdot \|u_k\|_{W^1_{p_k}(\Omega)}.
\end{aligned}
$$

In the cases b) and c), let us assume without loss of generality that $j \neq 1$. Let s be such that

$$
\frac{1}{p_j} + \frac{1}{s} + \frac{1}{q_2} + \dots + \frac{1}{q_{j-1}} + \frac{1}{q_{j+1}} + \dots + \frac{1}{q_k} = 1.
$$

It is easy to see that $1 \leq s \leq q_1$ and $\frac{n}{p_1} - 1 \leq \frac{n}{s}$. By Theorem 2.1.1 a), we have continuous embeddings $Z_j(\Omega) \subset L_{q_j}$ and $Z_1(\Omega) \subset L_s$. Applying Hölder's inequality (2.1.1), we conclude:

$$
\begin{aligned}
&\left(\frac{\partial u_j}{\partial x_l}, u_1 u_2 \cdot \dots \cdot u_{j-1} u_{j+1} \cdot \dots \cdot u_k\right) \\
&\quad \leq \left\| \frac{\partial u_j}{\partial x_l} u_1 u_2 \cdot \dots \cdot u_{j-1} u_{j+1} \cdot \dots \cdot u_k \right\|_{L_1(\Omega)} \\
&\quad \leq \left\| \frac{\partial u_j}{\partial x_l} \right\|_{L_{p_j}(\Omega)} \|u_1\|_{L_s(\Omega)} \|u_2\|_{L_{q_2}(\Omega)} \cdot \dots \\
&\qquad \cdot \|u_{j-1}\|_{L_{q_{j-1}}(\Omega)} \|u_{j+1}\|_{L_{q_{j+1}}(\Omega)} \cdot \dots \cdot \|u_k\|_{L_{q_k}(\Omega)} \\
&\quad \leq C \|u_j\|_{W^1_{p_j}(\Omega)} \|u_1\|_{Z_1(\Omega)} \|u_2\|_{Z_2(\Omega)} \cdot \dots \\
&\qquad \cdot \|u_{j-1}\|_{Z_{j-1}(\Omega)} \|u_{j+1}\|_{Z_{j+1}(\Omega)} \cdot \dots \cdot \|u_k\|_{Z_k(\Omega)} \\
&\quad = C \|u_1\|_{Z_1(\Omega)} \cdot \dots \cdot \|u_k\|_{Z_k(\Omega)}.
\end{aligned}
$$

□

Remark 6.1.4. Formula (6.1.14) may be rewritten as a formula of integration by parts:

$$\begin{aligned}(\frac{\partial u_r}{\partial x_l}, u_1 u_2 \cdot \ldots \cdot u_{r-1} u_{r+1} \cdot \ldots \cdot u_k) \\ = -\sum_{j=1}^{r-1} (\frac{\partial u_j}{\partial x_l}, u_1 u_2 \cdot \ldots \cdot u_{j-1} u_{j+1} \cdot \ldots \cdot u_k) \\ - \sum_{j=r+1}^{k} (\frac{\partial u_j}{\partial x_l}, u_1 u_2 \cdot \ldots \cdot u_{j-1} u_{j+1} \cdot \ldots \cdot u_k)\end{aligned} \tag{6.1.16}$$

for any $r = 1, \ldots, k$.

Corollary 6.1.2. *Let Ω be an arbitrary domain in $\mathbb{R}^n$. The following identities hold:*

$$(u, \operatorname{grad} q) = -(\operatorname{div} u, q), \tag{6.1.17}$$

$$\sum_{i=1}^{n} (u_i u, \frac{\partial u}{\partial x_i}) = 0, \tag{6.1.18}$$

$$\sum_{i=1}^{n} (u_i \tau, \frac{\partial \tau}{\partial x_i}) = 0, \tag{6.1.19}$$

$$(\tau, \nabla u) + (u, \operatorname{Div} \tau) = 0, \tag{6.1.20}$$

$$\sum_{i=1}^{n} \Big(\frac{u_i u}{1 + \delta\big(\frac{|\tau|^2}{2\mu_2} + |u|^2\big)}, \frac{\partial u}{\partial x_i} \Big) + \frac{1}{2\mu_2} \Big(\frac{u_i \tau}{1 + \delta\big(\frac{|\tau|^2}{2\mu_2} + |u|^2\big)}, \frac{\partial \tau}{\partial x_i} \Big) = 0. \tag{6.1.21}$$

In identities (6.1.17), (6.1.20) $u \in H_0^1(\Omega)^n, q \in H^1(\Omega)$, $\tau \in W_2^1(\Omega, \mathbb{R}_S^{n\times n})$. In identities (6.1.18), (6.1.19), (6.1.21) $u \in V$. In (6.1.19) $\tau \in W_4^1(\Omega, \mathbb{R}_S^{n\times n})$ or $H_0^1(\Omega, \mathbb{R}_S^{n\times n})$. In (6.1.21) $\tau \in W_2^1(\Omega, \mathbb{R}_S^{n\times n})$, $\mu_2, \delta > 0$. In identities (6.1.18), (6.1.19) $n \le 4$.

Proof. Equality (6.1.17) is shown by a direct application of integration by parts (the case a) of Corollary 6.1.1):

$$\begin{aligned}(u, \operatorname{grad} q) &= \sum_{i=1}^{n} (u_i, \frac{\partial q}{\partial x_i}) \\ &= -\sum_{i=1}^{n} (\frac{\partial u_i}{\partial x_i}, q) = -(\operatorname{div} u, q).\end{aligned}$$

Similarly, we have

$$\begin{aligned}(\tau, \nabla u) &= \sum_{i,j=1}^{n} \big(\tau_{ij}, \frac{\partial u_i}{\partial x_j}\big) \\ &= -\sum_{i,j=1}^{n} \big(u_i, \frac{\partial \tau_{ij}}{\partial x_j}\big) = -(u, \operatorname{Div} \tau),\end{aligned}$$

and this yields (6.1.20).

To show (6.1.21), we transform its left-hand side as follows

$$\begin{aligned}&\sum_{i=1}^{n}\Big(\frac{u_i u}{1+\delta\big(\frac{|\tau|^2}{2\mu_2}+|u|^2\big)}, \frac{\partial u}{\partial x_i}\Big) + \frac{1}{2\mu_2}\Big(\frac{u_i \tau}{1+\delta\big(\frac{|\tau|^2}{2\mu_2}+|u|^2\big)}, \frac{\partial \tau}{\partial x_i}\Big) \\ &= \sum_{i=1}^{n}\int_\Omega \frac{u_i}{1+\delta\big(\frac{|\tau|^2}{2\mu_2}+|u|^2\big)}\Big[\big(u, \frac{\partial u}{\partial x_i}\big)_{\mathbb{R}^n} + \frac{1}{2\mu_2}\big(\tau, \frac{\partial \tau}{\partial x_i}\big)_{\mathbb{R}^{n\times n}_S}\Big]\,dx \\ &= \frac{1}{2}\sum_{i=1}^{n}\int_\Omega \frac{u_i}{1+\delta\big(\frac{|\tau|^2}{2\mu_2}+|u|^2\big)} \frac{\partial}{\partial x_i}\big(|u|^2 + \frac{1}{2\mu_2}|\tau|^2\big)\,dx \\ &= \frac{1}{2\delta}\sum_{i=1}^{n}\int_\Omega u_i \frac{\partial}{\partial x_i} \ln\big(1+\delta\big(\frac{|\tau|^2}{2\mu_2}+|u|^2\big)\big)\,dx \\ &= -\frac{1}{2\delta}\int_\Omega \sum_{i=1}^{n} \frac{\partial u_i}{\partial x_i} \ln\big(1+\delta\big(\frac{|\tau|^2}{2\mu_2}+|u|^2\big)\big)\,dx \\ &= -\frac{1}{2\delta}\int_\Omega \operatorname{div} u \cdot \ln\big(1+\delta\big(\frac{|\tau|^2}{2\mu_2}+|u|^2\big)\big)\,dx = 0.\end{aligned}$$

We have applied integration by parts in the case a) of Corollary 6.1.1.

If $n \le 4$, by integration by parts, using case a) of Corollary 6.1.1 if $\tau \in W_4^1(\Omega, \mathbb{R}^{n\times n}_S)$ and case b) if $\tau \in H_0^1(\Omega, \mathbb{R}^{n\times n}_S)$, one has

$$\begin{aligned}\sum_{i=1}^{n}\big(u_i\tau, \frac{\partial \tau}{\partial x_i}\big) &= -\sum_{i=1}^{n}\big(\frac{\partial u_i}{\partial x_i}, |\tau|^2\big) - \sum_{i=1}^{n}\big(\frac{\partial \tau}{\partial x_i}, u_i\tau\big) \\ &= -(\operatorname{div} u, |\tau|^2) - \sum_{i=1}^{n}\big(u_i\tau, \frac{\partial \tau}{\partial x_i}\big) = -\sum_{i=1}^{n}\big(u_i\tau, \frac{\partial \tau}{\partial x_i}\big),\end{aligned}$$

and we obtain (6.1.19). The proof of (6.1.18) is similar. □

6.2 Initial-boundary value problem for equations of motion of a viscoelastic medium with Jeffreys' constitutive law and its weak formulation

6.2.1 Statement of the problem

Let Ω be an arbitrary domain in the space $\mathbb{R}^n$, $n = 2, 3$, which, in particular, may be unbounded.

We consider the initial-boundary value problem which describes the motion of an incompressible viscoelastic medium with Jeffreys' constitutive law (1.3.12) which corresponds to the substantial derivative (1.3.7):

$$\frac{\partial u}{\partial t} + \sum_{i=1}^{n} u_i \frac{\partial u}{\partial x_i} + \operatorname{grad} p = \operatorname{Div} \sigma + f, \tag{6.2.1}$$

$$\sigma + \lambda_1\Big(\frac{\partial \sigma}{\partial t} + \sum_{i=1}^{n} u_i \frac{\partial \sigma}{\partial x_i}\Big) = 2\eta\Big(\mathcal{E} + \lambda_2\Big(\frac{\partial \mathcal{E}}{\partial t} + \sum_{i=1}^{n} u_i \frac{\partial \mathcal{E}}{\partial x_i}\Big)\Big), \tag{6.2.2}$$

$$\operatorname{div} u = 0, \tag{6.2.3}$$

$$u\big|_{\partial\Omega} = 0, \tag{6.2.4}$$

$$u|_{t=0} = a, \ \sigma|_{t=0} = \sigma_0. \tag{6.2.5}$$

Here, as usual, u is an unknown velocity vector, p is an unknown pressure function, σ is an unknown deviatoric stress tensor, f is the given body force (all of them depend on a point x in the domain Ω, and on a moment of time t); $\mathcal{E}(u)$, $\mathcal{E}(u) = (\mathcal{E}_{ij}(u))$, $\mathcal{E}_{ij}(u) = \frac{1}{2}(\frac{\partial u_i}{\partial x_j} + \frac{\partial u_j}{\partial x_i})$, is the strain velocity tensor, $\eta > 0$ is the viscosity of the medium, λ_1 is the relaxation time, λ_2 is the retardation time, $0 < \lambda_2 < \lambda_1$, a and σ_0 are given functions.

Equation (6.2.1) is the equation of motion (1.1.12). Equation (6.2.2) is the Jeffreys constitutive law (1.3.12). Equation (6.2.3) is the equation of continuity (1.1.10). Equation (6.2.4) is the no-slip condition (1.1.15). Equation (6.2.5) is simply an initial condition.

6.2.2 Weak formulation of the problem

First let us point out that C_0^∞ will stand only for $C_0^\infty(\Omega, \mathbb{R}_S^{n\times n})$ in Chapter 6.

Let, as in Section 6.1.1, $T > 0$ be a fixed moment of time. Following the general scheme of weak setting, let

$$
\begin{aligned}
\mathfrak{U} &= C^1([0,T]; H_V^2) \times C^1([0,T]; H^1(\Omega, \mathbb{R}_S^{n\times n})),\\
\mathcal{P} &= C([0,T]; H^1_{\mathrm{loc}}(\Omega)),\\
\mathfrak{Z} &= C([0,T]; L_2(\Omega)^n) \times C([0,T]; L_2(\Omega, \mathbb{R}_S^{n\times n})) \times H_V^2 \times H^1(\Omega, \mathbb{R}_S^{n\times n}),\\
\mathfrak{Z}_0 &= C([0,T]; L_2(\Omega)^n) \times \{0\} \times H_V^2 \times H^1(\Omega, \mathbb{R}_S^{n\times n}),\\
Z &= (f, 0, a, \sigma_0),\\
\mathcal{G}(U,P) &= \mathcal{G}((u,\sigma),P)\\
&= \Big(\frac{\partial u}{\partial t} + \sum_{i=1}^n u_i \frac{\partial u}{\partial x_i} + \operatorname{grad} P - \operatorname{Div}\sigma,\\
&\qquad \sigma + \lambda_1\Big(\frac{\partial \sigma}{\partial t} + \sum_{i=1}^n u_i \frac{\partial \sigma}{\partial x_i}\Big) - 2\eta\Big(\mathcal{E}(u) + \lambda_2\Big(\frac{\partial \mathcal{E}(u)}{\partial t} + \sum_{i=1}^n u_i \frac{\partial \mathcal{E}(u)}{\partial x_i}\Big)\Big),\\
&\qquad u|_{t=0}, \sigma|_{t=0}\Big).
\end{aligned}
$$

Then we can rewrite problem (6.2.1) – (6.2.5) in form (6.1.1). In order to define weak solutions, take

$$
\begin{aligned}
\mathfrak{W} &= L_2(0,T;V) \bigcap C_w([0,T];H) \bigcap W_1^1(0,T;V^*)\\
&\qquad \times L_2(0,T;L_2(\Omega,\mathbb{R}_S^{n\times n})) \bigcap C_w([0,T]; H^{-1}(\Omega,\mathbb{R}_S^{n\times n})),\\
\mathfrak{L} &= L_2(0,T;V^*) \times \{0\} \times H \times W_2^{-1}(\Omega,\mathbb{R}_S^{n\times n}),\\
\mathfrak{Q} &= \mathcal{D}'(0,T) \times \mathcal{D}'(0,T) \times H \times W_2^{-1}(\Omega,\mathbb{R}_S^{n\times n}),\\
\mathfrak{M} &= \{(0,0,0,0)\},\\
\mathfrak{N}(W,L) &= \Big\{\Big(\frac{d}{dt}(u,\varphi) + (\sigma,\nabla\varphi) - \sum_{i=1}^n \Big(u_i u, \frac{\partial \varphi}{\partial x_i}\Big) - \langle f,\varphi\rangle,\\
&\qquad (\sigma,\Phi) + \lambda_1 \frac{d}{dt}(\sigma,\Phi) - \lambda_1 \sum_{i=1}^n \Big(u_i\sigma, \frac{\partial \Phi}{\partial x_i}\Big)\\
&\qquad\qquad + 2\eta(u, \operatorname{Div}\Phi) + 2\eta\lambda_2\Big(\frac{d}{dt}(u,\operatorname{Div}\Phi) + \sum_{i=1}^n \Big(u_i\mathcal{E}(u), \frac{\partial \Phi}{\partial x_i}\Big)\Big),\\
&\qquad u|_{t=0} - a, \sigma|_{t=0} - \sigma_0\Big) \Big| \varphi \in \mathcal{V}, \Phi \in C_0^\infty\Big\}
\end{aligned}
$$

for $W = (u,\sigma) \in \mathfrak{W}$ and $L = (f, 0, a, \sigma_0) \in \mathfrak{L}$.

Then we arrive at

Definition 6.2.1. A pair of functions (u, σ),

$$
\begin{aligned}
&u \in L_2(0,T;V)\bigcap C_w([0,T];H),\ \frac{du}{dt} \in L_1(0,T;V^*),\\
&\sigma \in L_2(0,T;L_2(\Omega,\mathbb{R}_S^{n\times n}))\bigcap C_w([0,T];H^{-1}(\Omega,\mathbb{R}_S^{n\times n}))
\end{aligned}
\tag{6.2.6}
$$

is a *weak solution* of problem (6.2.1) – (6.2.5) if it satisfies condition (6.2.5), and if the equalities

$$
\frac{d}{dt}(u,\varphi) + (\sigma,\nabla\varphi) - \sum_{i=1}^{n}(u_i u, \frac{\partial\varphi}{\partial x_i}) = \langle f,\varphi\rangle, \tag{6.2.7}
$$

$$
\begin{aligned}
&(\sigma,\Phi) + \lambda_1\frac{d}{dt}(\sigma,\Phi) - \lambda_1\sum_{i=1}^{n}(u_i\sigma, \frac{\partial\Phi}{\partial x_i})\\
&\quad = -2\eta(u,\operatorname{Div}\Phi) - 2\eta\lambda_2(\frac{d}{dt}(u,\operatorname{Div}\Phi) + \sum_{i=1}^{n}(u_i\mathcal{E}(u), \frac{\partial\Phi}{\partial x_i}))
\end{aligned}
\tag{6.2.8}
$$

are true for all $\varphi \in \mathcal{V}$ and $\Phi \in C_0^\infty$ in the sense of distributions on $(0,T)$ (i.e. in $\mathcal{D}'(0,T)$; see, however, Remark 6.5.3).

Condition (6.1.4) is checked just as for the Navier–Stokes problem (using integration by parts again).

6.2.3 An existence result

Now we are ready to formulate one of the main results of this chapter.

Theorem 6.2.1 (see [72]). *Given* $f \in L_2(0,T;V^*)$, $a \in H$, $\sigma_0 \in W_2^{-1}(\Omega,\mathbb{R}_S^{n\times n})$, $\sigma_0 - 2\eta\frac{\lambda_2}{\lambda_1}\mathcal{E}(a) \in L_2(\Omega,\mathbb{R}_S^{n\times n})$, *there exists a weak solution of problem (6.2.1) – (6.2.5) in class (6.2.6) .*

We shall prove this theorem in Section 6.5.

Remark 6.2.1. As a matter of fact one can show a bit more regularity than in (6.2.6) for the solution of problem (6.2.1) – (6.2.5) even under the conditions of Theorem 6.2.1. See Theorem 6.5.1 and Remark 6.8.1.

Remark 6.2.2. Existence of a weak solution for problem (6.2.1) – (6.2.5) (in a slightly different weak setting) was proved by Turganbaev [65] in the case when Ω is a bounded sufficiently regular domain and $f \in L_2(0,T;L_2)$.

Remark 6.2.3. In Chapters 6 and 7 it is possible to replace the space $\mathbb{R}_S^{n\times n}$ with $\mathbb{R}_D^{n\times n} = \{\xi \in \mathbb{R}_S^{n\times n} | \operatorname{Tr}\xi = 0\}$.

6.3 Auxiliary problem

Before proving Theorem 6.2.1, we study an auxiliary problem. Let us begin with an equivalent transformation of system (6.2.7), (6.2.8). Denote $\mu_1 = \eta\frac{\lambda_2}{\lambda_1}$, $\mu_2 = \frac{\eta-\mu_1}{\lambda_1}$, $\tau = \sigma - 2\mu_1\mathcal{E}(u)$. Then we can rewrite (6.2.8) and (6.2.7) as follows (cf. Section 1.5.1):

$$\frac{d}{dt}(\tau,\Phi) + \frac{1}{\lambda_1}(\tau,\Phi) - \sum_{i=1}^{n}(u_i\tau, \frac{\partial\Phi}{\partial x_i}) + 2\mu_2(u, \operatorname{Div}\Phi) = 0 \tag{6.3.1}$$

$$\frac{d}{dt}(u,\varphi) - \sum_{i=1}^{n}(u_i u, \frac{\partial\varphi}{\partial x_i}) + \mu_1(\nabla u, \nabla\varphi) + (\tau, \nabla\varphi) = \langle f, \varphi\rangle \tag{6.3.2}$$

for all $\varphi \in \mathcal{V}$ and $\Phi \in C_0^\infty$ (cf. Remark 6.5.1).

Remark 6.3.1. During the study of many issues concerning the weak solutions of problem (6.2.1) – (6.2.4) (see Sections 6.5 – 6.8) it appears to be convenient to pass to the variables (u, τ) and to investigate these issues for problem (6.3.1) – (6.3.2).

Consider the following auxiliary problem

$$\begin{aligned}\frac{d}{dt}(\tau,\Phi) + \frac{1}{\lambda_1}(\tau,\Phi) - \xi\sum_{i=1}^{n}\Big(\frac{u_i\tau}{1+\delta\big(\frac{|\tau|^2}{2\mu_2}+|u|^2\big)}, \frac{\partial\Phi}{\partial x_i}\Big)&\\ + 2\mu_2\xi(u, \operatorname{Div}\Phi) + \frac{\varepsilon}{\lambda_1}(\nabla\tau, \nabla\Phi) = 0,&\end{aligned} \tag{6.3.3}$$

$$\begin{aligned}\frac{d}{dt}(u,\varphi) - \xi\sum_{i=0}^{n}\Big(\frac{u_i u}{1+\delta\big(\frac{|\tau|^2}{2\mu_2}+|u|^2\big)}, \frac{\partial\varphi}{\partial x_i}\Big)&\\ + \mu_1(\nabla u, \nabla\varphi) + \xi(\tau, \nabla\varphi) = \langle f, \varphi\rangle&\end{aligned} \tag{6.3.4}$$

for all $\varphi \in V$, $\Phi \in H_0^1$ almost everywhere in $(0, T)$;

$$u|_{t=0} = a, \quad \tau|_{t=0} = \tau_0. \tag{6.3.5}$$

The numbers $\delta > 0$, $0 \le \xi \le 1, 0 < \varepsilon \le 1$ are parameters.

Let us introduce the following spaces,

$$W = \{u \in L_2(0,T;V), \ \frac{du}{dt} \in L_2(0,T;V^*)\},$$
$$\|u\|_W = \|u\|_{L_2(0,T;V)} + \|u'\|_{L_2(0,T;V^*)},$$

$$W_M = \{\tau \in L_2(0,T; H_0^1(\Omega, \mathbb{R}_S^{n\times n})),\ \frac{d\tau}{dt} \in L_2(0,T; H^{-1}(\Omega, \mathbb{R}_S^{n\times n}))\},$$
$$\|\tau\|_{W_M} = \|\tau\|_{L_2(0,T;H_0^1)} + \|\tau'\|_{L_2(0,T;H^{-1})}.$$

By Corollary 2.2.3, W and W_M are continuously embedded into $C([0,T]; H)$ and $C([0,T]; L_2)$, respectively. If Ω is bounded, the embeddings $V \subset H$ and $H_0^1 \subset L_2$ are compact, so the embeddings $W \subset L_2(0,T;H)$ and $W_M \subset L_2(0,T;L_2)$ are also compact by Theorem 2.2.6.

Lemma 6.3.1. *Let $a \in H,\ \tau_0 \in L_2,\ f \in L_2(0,T;V^*)$ and let a pair $(u \in W, \tau \in W_M)$ be a solution of problem (6.3.3) – (6.3.5). Then the following estimate holds,*

$$\max_{t\in[0,T]} \|u\|(t) + \max_{t\in[0,T]} \|\tau\|(t) + \int_0^T \|u\|_1^2(t)dt + \varepsilon \int_0^T \|\tau\|_1^2(t)dt \le K_0(\|a\|, \|\tau_0\|, \|f\|_{L_2(0,T;V^*)}) \tag{6.3.6}$$

where the constant K_0 does not depend on $\Omega, \varepsilon, \delta, \xi$.

Proof. It follows from Lemma 2.2.8 that

$$\langle \frac{du}{dt}, \varphi\rangle = \frac{d}{dt}(u,\varphi), \langle \frac{d\tau}{dt}, \Phi\rangle = \frac{d}{dt}(\tau, \Phi). \tag{6.3.7}$$

So, by (2.2.28),

$$\frac{d}{dt}(u,\varphi)\Big|_{\varphi=u(t)} = \langle \frac{du(t)}{dt}, u(t)\rangle = \frac{1}{2}\frac{d}{dt}(u(t),u(t)).$$

Analogously

$$\frac{d}{dt}(\tau,\Phi)\Big|_{\Phi=\tau(t)} = \frac{1}{2}\frac{d}{dt}(\tau,\tau).$$

Put $\Phi = \dfrac{\tau(t)}{2\mu_2}$ in (6.3.3) and $\varphi = u(t)$ in (6.3.4) for almost all $t \in [0,T]$, and add the results:

$$\frac{1}{2}\frac{d}{dt}(u,u) + \frac{1}{4\mu_2}\frac{d}{dt}(\tau,\tau) + \mu_1(\nabla u, \nabla u) + \frac{1}{2\lambda_1\mu_2}(\tau,\tau)$$
$$-\xi\sum_{i=1}^n \Big(\frac{u_i u}{1+\delta\Big(\frac{|\tau|^2}{2\mu_2} + |u|^2\Big)}, \frac{\partial u}{\partial x_i}\Big)$$
$$-\xi\frac{1}{2\mu_2}\Big(\frac{u_i \tau}{1+\delta\Big(\frac{|\tau|^2}{2\mu_2} + |u|^2\Big)}, \frac{\partial \tau}{\partial x_i}\Big) + \frac{\varepsilon}{2\lambda_1\mu_2}(\nabla\tau, \nabla\tau)$$
$$+\xi(\tau, \nabla u) + \xi(u, \operatorname{Div}\tau) = \langle f, u\rangle.$$

Taking into account (6.1.20) and (6.1.21), we obtain

$$\begin{aligned}&\frac{1}{2}\frac{d}{dt}(u,u)+\frac{1}{4\mu_2}\frac{d}{dt}(\tau,\tau)+\mu_1(\nabla u,\nabla u)+\frac{1}{2\lambda_1\mu_2}(\tau,\tau)\\&\quad+\frac{\varepsilon}{2\lambda_1\mu_2}(\nabla\tau,\nabla\tau)=\langle f,u\rangle.\end{aligned}\tag{6.3.8}$$

We are going to use the following simple inequality:

$$\begin{aligned}&\frac{1}{k}\big(\max_{s\in J}\phi_1(s)+\max_{s\in J}\phi_2(s)+\cdots+\max_{s\in J}\phi_k(s)\big)\\&\quad\le\max_{s\in J}(\phi_1(s)+\phi_2(s)+\cdots+\phi_k(s))\end{aligned}\tag{6.3.9}$$

for scalar functions $\phi_1,\phi_2,\ldots,\phi_k : J\subset\mathbb{R}\to[0,+\infty)$, $k\in\mathbb{N}$.

Integration of the terms in (6.3.8) from 0 to t yields:

$$\begin{aligned}&\frac{1}{2}\|u\|^2(t)+\frac{1}{4\mu_2}\|\tau\|^2(t)+\int_0^t\frac{1-\varepsilon}{2\lambda_1\mu_2}\|\tau\|^2\,ds+\int_0^t\frac{\varepsilon}{2\lambda_1\mu_2}\|\tau\|_1^2\,ds\\&\quad+\int_0^t\mu_1(\|u\|_1^2-\|u\|^2)\,ds\le\frac{1}{2}\|a\|^2+\frac{1}{4\mu_2}\|\tau_0\|^2+\int_0^t\|f\|_{V^*}\|u\|_1\,ds.\end{aligned}$$

Then we get:

$$\begin{aligned}&\frac{1}{10}\max_{t\in[0,T]}\|u\|^2(t)+\frac{1}{20\mu_2}\max_{t\in[0,T]}\|\tau\|^2(t)+\int_0^T\frac{1-\varepsilon}{10\lambda_1\mu_2}\|\tau\|^2\,dt\\&\quad+\int_0^T\frac{\varepsilon}{10\lambda_1\mu_2}\|\tau\|_1^2\,dt+\frac{1}{5}\int_0^T\mu_1(\|u\|_1^2-\|u\|^2)\,dt\\&\le\max_{t\in[0,T]}\Big[\frac{1}{2}\|u\|^2(t)+\frac{1}{4\mu_2}\|\tau\|^2(t)+\int_0^t\frac{1-\varepsilon}{2\lambda_1\mu_2}\|\tau\|^2\,ds\\&\quad+\int_0^t\frac{\varepsilon}{2\lambda_1\mu_2}\|\tau\|_1^2\,ds+\int_0^t\mu_1(\|u\|_1^2-\|u\|^2)\,ds\Big]\\&\le\frac{1}{2}\|a\|^2+\frac{1}{4\mu_2}\|\tau_0\|^2+\int_0^T\|f\|_{V^*}\|u\|_1\,dt\\&\le\frac{1}{2}\|a\|^2+\frac{1}{4\mu_2}\|\tau_0\|^2+\|f\|_{L_2(0,T;V^*)}\Big(\int_0^T\|u\|_1^2\,dt\Big)^{1/2}.\end{aligned}\tag{6.3.10}$$

Note that the following inequality is valid:

$$\max_{t\in[0,T]}\frac{1}{20}\|u\|^2(t)+\frac{1}{5}\int_0^T\mu_1(\|u\|_1^2-\|u\|^2)\,dt\ge\gamma\int_0^T\|u\|_1^2\,dt\tag{6.3.11}$$

where $\gamma = \min\left(\frac{1}{20T}, \frac{\mu_1}{5}\right)$.

For its proof it is enough to add inequalities:

$$\max_{t\in[0,T]} \frac{1}{20}\|u\|^2(t) \ge \frac{1}{20T}\int_0^T \|u\|^2\,dt \ge \gamma \int_0^T \|u\|^2\,dt$$

and

$$\frac{1}{5}\int_0^T \mu_1(\|u\|_1^2 - \|u\|^2)\,dt \ge \gamma \int_0^T (\|u\|_1^2 - \|u\|^2)\,dt.$$

Now, we have from (6.3.10) and (6.3.11):

$$\gamma \int_0^T \|u\|_1^2\,dt \le \frac{1}{2}\|a\|^2 + \frac{1}{4\mu_2}\|\tau_0\|^2 + \|f\|_{L_2(0,T;V^*)}\Big(\int_0^T \|u\|_1^2\,dt\Big)^{1/2}.$$

This yields $\left(\int_0^T \|u\|_1^2\,dt\right)^{1/2} \le y_2$, where y_2 is the greater root of the quadratic equation

$$\gamma y^2 = \frac{1}{2}\|a\|^2 + \frac{1}{4\mu_2}\|\tau_0\|^2 + \|f\|_{L_2(0,T;V^*)}y.$$

Then from (6.3.10) and (6.3.11) it follows that

$$\frac{1}{20}\max_{t\in[0,T]}\|u\|^2(t) + \frac{1}{20\mu_2}\max_{t\in[0,T]}\|\tau\|^2(t) + \gamma\int_0^T \|u\|_1^2\,dt + \frac{\varepsilon}{10\lambda_1\mu_2}\int_0^T \|\tau\|_1^2\,dt$$
$$\le \frac{1}{2}\|a\|^2 + \frac{1}{4\mu_2}\|\tau_0\|^2 + \|f\|_{L_2(0,T;V^*)}y_2,$$

what yields the statement of the lemma. □

Theorem 6.3.1. *Let Ω be bounded and a, τ_0, f satisfy the conditions of Lemma 6.3.1. Then problem (6.3.3)–(6.3.5) possesses a solution $u \in W$, $\tau \in W_M$.*

Proof. Let us introduce auxiliary operators by the following formulas (in these formulas φ and Φ are arbitrary elements of V and $H_0^1(\Omega, \mathbb{R}_S^{n\times n})$ respectively):

$$N_1 : W_M \to L_2(0,T;V^*),\ \langle N_1(\tau), \varphi\rangle = (\tau, \nabla\varphi)$$
$$N_2 : W \to L_2(0,T;H^{-1}),\ \langle N_2(u), \Phi\rangle = 2\mu_2(u, \operatorname{Div}\Phi)$$
$$K_\delta : W \times W_M \to L_2(0,T;V^*),$$

$$\langle K_\delta(u,\tau), \varphi\rangle = -\sum_{i=1}^n \Bigg(\frac{u_i u}{1 + \delta\Big(|u|^2 + \dfrac{|\tau|^2}{2\mu_2}\Big)}, \frac{\partial\varphi}{\partial x_i}\Bigg),$$

$$\tilde{K_\delta} : W \times W_M \to L_2(0,T;H^{-1}),$$

$$\langle \tilde{K_\delta}(u,\tau),\Phi\rangle = -\sum_{i=1}^{n}\Big(\frac{u_i\tau}{1+\delta\Big(|u|^2+\frac{|\tau|^2}{2\mu_2}\Big)},\frac{\partial\Phi}{\partial x_i}\Big),$$

$$A : V \to V^*,\ \langle A(u),\varphi\rangle = \mu_1(\nabla u,\nabla\varphi),$$

$$A_\varepsilon : H_0^1 \to H^{-1},\ \langle A_\varepsilon(\tau),\Phi\rangle = \varepsilon(\nabla\tau,\nabla\Phi)+\frac{1}{\lambda_1}(\tau,\Phi),$$

$$\tilde{A} : W \times W_M \to L_2(0,T;V^*) \times L_2(0,T;H^{-1}) \times H \times L_2,$$

$$\tilde{A}(u,\tau) = \Big(\frac{du(t)}{dt}+A(u(t)),\frac{d\tau(t)}{dt}+A_\varepsilon(\tau(t)),u|_{t=0},\tau|_{t=0}\Big),$$

$$Q : W \times W_M \to L_2(0,T;V^*) \times L_2(0,T;H^{-1}) \times H \times L_2,$$

$$Q(u,\tau) = (K_\delta(u,\tau)+N_1(\tau),\tilde{K_\delta}(u,\tau)+N_2(u),0,0).$$

Then problem (6.3.3) – (6.3.5) is equivalent to the operator equation

$$\tilde{A}(u,\tau)+\xi Q(u,\tau) = (f,0,a,\tau_0). \tag{6.3.12}$$

Since the embeddings $W \subset L_2(0,T;H)$ and $W_M \subset L_2(0,T;L_2)$ are compact, the operators N_1, N_2 are compact. Let us show that the operator $\tilde{K_\delta}$ is also compact. This operator may be considered as a superposition of the embedding operator

$$j : W \times W_M \to L_2(0,T;H) \times L_2(0,T;L_2)$$

and of the operator

$$\tilde{K_\delta} : L_2(0,T;H) \times L_2(0,T;L_2) \to L_2(0,T;H^{-1}).$$

The first operator is compact, so it suffices to show that the second one is continuous. Observe that for this purpose it is enough to know that the following Nemytskii operators are continuous:

$$\phi_{ijk} : L_2((0,T)\times\Omega,\mathbb{R}^n)\times L_2((0,T)\times\Omega,\mathbb{R}_S^{n\times n}) \to L_2((0,T)\times\Omega),$$

$$\phi_{ijk}(v,\varsigma)(t,x) = \frac{v_i(t,x)\varsigma_{jk}(t,x)}{1+\delta\Big(|v(t,x)|^2+\frac{|\varsigma(t,x)|^2}{2\mu_2}\Big)},\ i,j,k=1,\dots,n.$$

But by Cauchy's inequality for all $v \in \mathbb{R}^n$ and $\varsigma \in \mathbb{R}_S^{n\times n}$ one has

$$\left|\frac{v_i\varsigma_{jk}}{1+\delta\Big(|v|^2+\frac{|\varsigma|^2}{2\mu_2}\Big)}\right| \le \left|\frac{\frac{1}{2}(2\mu_2)^{1/2}v_i^2+\frac{1}{2(2\mu_2)^{1/2}}\varsigma_{jk}^2}{1+\delta\Big(|v|^2+\frac{|\varsigma|^2}{2\mu_2}\Big)}\right| \le \frac{(2\mu_2)^{1/2}}{2\delta}.$$

Therefore, by Krasnoselskii's theorem [31], the Nemytskii operators ϕ_{ijk} are continuous.

Similarly one checks that the operator K_δ is compact.

Hence, the operator Q is also compact. But the operator $\tilde{A}$ is continuously invertible by Lemma 3.1.3. Rewrite equation (6.3.12) as

$$(u,\tau) + \xi \tilde{A}^{-1} Q(u,\tau) = \tilde{A}^{-1}(f, 0, a, \tau_0). \tag{6.3.13}$$

By Lemma 6.3.1 equation (6.3.13) has no solutions on the boundary of a sufficiently large ball B in $W \times W_M$, independent on ξ. Without loss of generality $a_0 = \tilde{A}^{-1}(f, 0, a, \tau_0)$ belongs to this ball. Then we can consider the Leray–Schauder degree (see Section 3.2.2) of the map $I + \xi \tilde{A}^{-1} Q$ on the ball B with respect to the point a_0,

$$\deg_{LS}(I + \xi \tilde{A}^{-1} Q, B, a_0),$$

where I is the identity operator. By the homotopic invariance property of the degree we have

$$\deg_{LS}(I + \xi \tilde{A}^{-1} Q, B, a_0) = \deg_{LS}(I, B, a_0) = 1.$$

By Theorem 3.2.3, equation (6.3.13) (and therefore, problem (6.3.3) – (6.3.5)) has a solution in the ball B for every ξ. □

We need the following estimates on the time derivatives of the solutions of problem (6.3.3) – (6.3.5).

Lemma 6.3.2. *Under the conditions of the previous theorem the following estimates of the solutions are valid:*

$$\left\|\frac{du}{dt}\right\|_{L_1(0,T;V^*)} \le K_1\big(\|a\|, \|\tau_0\|, \|f\|_{L_2(0,T;V^*)}\big), \tag{6.3.14}$$

$$\left\|\frac{d\tau}{dt}\right\|_{L_1(0,T;H^{-1})} \le K_2\big(\|a\|, \|\tau_0\|, \|f\|_{L_2(0,T;V^*)}, \varepsilon\big) \tag{6.3.15}$$

where the constants K_1, K_2 do not depend on Ω, δ, ξ, and K_1 is independent of ε.

Proof. We have from (6.3.4):

$$\begin{aligned}
\|\langle u', \varphi\rangle\|_{L_1(0,T)} &= \int_0^T \Big|\frac{d}{dt}(u,\varphi)\Big| dt \\
&\le \int_0^T \Big|\xi \sum_{i=1}^n \Big(\frac{u_i u}{1 + \delta\Big(\frac{|\tau|^2}{2\mu_2} + |u|^2\Big)}, \frac{\partial \varphi}{\partial x_i}\Big)\Big| + |\mu_1(\nabla u, \nabla \varphi)| \\
&\quad + |\xi(\tau, \nabla\varphi)| + |\langle f, \varphi\rangle| dt.
\end{aligned}$$

Note that $V \subset L_4$ and $\frac{\xi}{1+\delta(\frac{|\tau|^2}{2\mu_2}+|u|^2)} \leq 1$. Using Hölder's inequality (2.1.1), Cauchy's inequality and estimate (6.3.6), we conclude that the right-hand side does not exceed

$$\begin{aligned}
&\|\varphi\|_1 \int_0^T (\|(u(t)(\cdot), u(t)(\cdot))_{\mathbb{R}^n}\| + \mu_1\|u(t)\|_1 + \|\tau(t)\| + \|f(t)\|_{V^*})dt \\
&\quad \leq \|\varphi\|_1 \int_0^T (\|u(t)\|_{L_4}^2 + \frac{1}{2} + \frac{\mu_1}{2}\|u(t)\|_1^2 + \|\tau(t)\| + \|f(t)\|_{V^*})dt \\
&\quad \leq C\|\varphi\|_1 \big(1 + \|u\|_{L_2(0,T;V)}^2 + \|\tau\|_{L_\infty(0,T;L_2)} + \|f\|_{L_2(0,T;V^*)}\big) \\
&\quad \leq K_1(\|a\|, \|\tau_0\|, \|f\|_{L_2(0,T;V^*)})\|\varphi\|_1,
\end{aligned}$$

so we get (6.3.14).

Using embedding $H_0^1 \subset L_4$, (6.3.3) and (6.3.6), we have:

$$\begin{aligned}
&\|\langle \tau', \Phi\rangle\|_{L_1(0,T)} \\
&\quad = \Big\|\frac{d}{dt}(\tau, \Phi)\Big\|_{L_1(0,T)} \\
&\quad \leq \Big\|\frac{1}{\lambda_1}(\tau, \Phi)\Big\|_{L_1(0,T)} + \frac{\varepsilon}{\lambda_1}\|(\nabla\tau, \nabla\Phi)\|_{L_1(0,T)} \\
&\qquad + \Big\|\xi \sum_{i=1}^n \Big(\frac{u_i\tau}{1+\delta\big(\frac{|\tau|^2}{2\mu_2}+|u|^2\big)}, \frac{\partial\Phi}{\partial x_i}\Big)\Big\|_{L_1(0,T)} \\
&\qquad + \|2\xi\mu_2(u, \operatorname{Div}\Phi)\|_{L_1(0,T)} \\
&\quad \leq \frac{1}{\lambda_1}\|\tau\|_{L_1(0,T;L_2)}\|\Phi\|_{L_2} + \frac{\varepsilon}{\lambda_1}\|\Phi\|_1\|\tau\|_{L_1(0,T;H^1)} \\
&\qquad + \|u\|_{L_2(0,T;L_4)}\|\tau\|_{L_2(0,T;L_4)}\Big\|\frac{\partial\Phi}{\partial x_i}\Big\|_{L_2} + 2\mu_2\|u\|_{L_1(0,T;L_2)}\|\operatorname{Div}\Phi\|_{L_2} \\
&\quad \leq C\|\Phi\|_1\big(\|\tau\|_{L_\infty(0,T;L_2)} + \varepsilon\|\tau\|_{L_2(0,T;H^1)} \\
&\qquad + \|u\|_{L_2(0,T;V)}\|\tau\|_{L_2(0,T;V)} + \|u\|_{L_\infty(0,T;L_2)}\big) \\
&\quad \leq \|\Phi\|_1 K_2\big(\|a\|, \|\tau_0\|, \|f\|_{L_2(0,T;V^*)}, \varepsilon\big),
\end{aligned}$$

and so we arrive at (6.3.15). □

6.4 Passage to the limit.

Consider one more auxiliary system

$$\frac{d}{dt}(\tau, \Phi) + \frac{1}{\lambda_1}(\tau, \Phi) - \sum_{i=1}^n \Big(u_i\tau, \frac{\partial\Phi}{\partial x_i}\Big) + 2\mu_2(u, \operatorname{Div}\Phi) + \frac{\varepsilon}{\lambda_1}(\nabla\tau, \nabla\Phi) = 0, \quad (6.4.1)$$

$$\frac{d}{dt}(u,\varphi) - \sum_{i=1}^{n} \left(u_i u, \frac{\partial \varphi}{\partial x_i}\right) + \mu_1(\nabla u, \nabla \varphi) + (\tau, \nabla \varphi) = \langle f, \varphi \rangle \tag{6.4.2}$$

for all $\varphi \in \mathcal{V}$ and $\Phi \in C_0^\infty$ almost everywhere in $(0,T)$; $0 < \varepsilon \le 1$.

Theorem 6.4.1. *Let Ω, a, τ_0, f satisfy the conditions of Theorem 6.3.1. Then the problem (6.4.1), (6.4.2), (6.3.5) possesses a solution in the class*

$$u \in L_2(0,T;V), \quad \tau \in L_2(0,T;H_0^1),$$
$$\frac{du}{dt} \in L_1(0,T;V^*), \quad \frac{d\tau}{dt} \in L_1(0,T;H^{-1}) \tag{6.4.3}$$

which satisfies estimates (6.3.14), (6.3.15) and

$$\operatorname{vrai}\max_{t\in[0,T]} \|u\|(t) + \operatorname{vrai}\max_{t\in[0,T]} \|\tau\|(t) + \int_0^T \|u\|_1^2(t)dt + \varepsilon \int_0^T \|\tau\|_1^2(t)dt$$
$$\le K_0(\|a\|, \|\tau_0\|, \|f\|_{L_2(0,T;V^*)}). \tag{6.4.4}$$

Proof. Consider problems (6.3.3) – (6.3.5) with $\xi = 1$ and $\delta = \frac{1}{m}$, $m = 1, 2, \dots$. By Theorem 6.3.1 there exist solutions (u_m, τ_m) of these problems. Taking into account estimate (6.3.6), without loss of generality (see Remark 2.1.1) we may assume that there exists a pair (u_*, τ_*) such that

$$u_m \to u_* \text{ weakly in } L_2(0,T;V),$$
$$u_m \to u_* \text{ *-weakly in } L_\infty(0,T;H),$$
$$\tau_m \to \tau_* \text{ weakly in } L_2(0,T;H_0^1),$$
$$\tau_m \to \tau_* \text{ *-weakly in } L_\infty(0,T;L_2),$$

as $m \to \infty$.

By Lemma 6.3.2 the sequence $\left\{\frac{du_m}{dt}\right\}$ is bounded in $L_1(0,T;V^*)$, and the sequence $\left\{\frac{d\tau_m}{dt}\right\}$ is bounded in $L_1(0,T;H^{-1})$. Then by Theorem 2.2.6

$$u_m \to u_* \text{ strongly in } L_2(0,T;H),$$
$$\tau_m \to \tau_* \text{ strongly in } L_2(0,T;L_2),$$

so without loss of generality we may assume that

$$u_m(t)(x) \to u_*(t)(x) \text{ almost everywhere in } (0,T) \times \Omega,$$
$$\tau_m(t)(x) \to \tau_*(t)(x) \text{ almost everywhere in } (0,T) \times \Omega.$$

It is obvious that estimate (6.4.4) is valid for (u_*, τ_*).

Substitute (u_m, τ_m) in equalities (6.3.3) and (6.3.4) with $\delta = \frac{1}{m}$, $\xi = 1$. Taking the scalar product of these equalities in $L_2(0,T)$ with a smooth scalar function $\psi(t)$, $\psi(T) = 0$ and integrating by parts the first terms, we obtain

$$-\int_0^T (\tau_m, \Phi\psi'(t))dt + \int_0^T \Big(\frac{1}{\lambda_1}(\tau_m, \psi\Phi) - \sum_{i=1}^n \Big(\frac{(u_m)_i\ \tau_m}{1+\frac{1}{m}(\frac{|\tau_m|^2}{2\mu_2}+|u_m|^2)}, \psi\frac{\partial\Phi}{\partial x_i}\Big)$$

$$+ 2\mu_2(u_m, \psi \operatorname{Div} \Phi) + \frac{\varepsilon}{\lambda_1}(\nabla\tau_m, \psi\nabla\Phi)\Big)dt = (\tau_0, \Phi)\psi(0), \tag{6.4.5}$$

$$-\int_0^T (u_m, \varphi\psi'(t))dt + \int_0^T \Big(\mu_1(\nabla u_m,\ \psi\nabla\varphi)$$

$$-\sum_{i=1}^n \Big(\frac{(u_m)_i\ u_m}{1+\frac{1}{m}(\frac{|\tau_m|^2}{2\mu_2}+|u_m|^2)}, \psi\frac{\partial\varphi}{\partial x_i}\Big) + (\tau_m, \psi\nabla\varphi)\Big)\, dt \tag{6.4.6}$$

$$= \int_0^T \langle f, \varphi\psi\rangle dt + (a, \varphi)\psi(0).$$

Let us check that

$$\int_0^T \sum_{i=1}^n \Big(\frac{(u_m)_i\ \tau_m}{1+\frac{1}{m}(\frac{|\tau_m|^2}{2\mu_2}+|u_m|^2)}, \psi\frac{\partial\Phi}{\partial x_i}\Big)dt \to \int_0^T \sum_{i=1}^n \Big((u_*)_i\tau_*,\ \psi\frac{\partial\Phi}{\partial x_i}\Big)dt, \tag{6.4.7}$$

$$\int_0^T \sum_{i=1}^n \Big(\frac{(u_m)_i u_m}{1+\frac{1}{m}(\frac{|\tau_m|^2}{2\mu_2}+|u_m|^2)}, \psi\frac{\partial\varphi}{\partial x_i}\Big)dt \to \int_0^T \sum_{i=1}^n ((u_*)_i u_*, \psi\frac{\partial\varphi}{\partial x_i})dt \tag{6.4.8}$$

as $m \to \infty$.

We have:

$$\Big|\int_0^T \sum_{i=1}^n \Big(\frac{(u_m)_i\ \tau_m}{1+\frac{1}{m}(\frac{|\tau_m|^2}{2\mu_2}+|u_m|^2)} - (u_*)_i\tau_*, \psi\frac{\partial\Phi}{\partial x_i}\Big)dt\Big|$$

$$= \Big|\int_0^T \sum_{i,j,k=1}^n \Big(\frac{(u_m)_i\ (\tau_m)_{jk}}{1+\frac{1}{m}(\frac{|\tau_m|^2}{2\mu_2}+|u_m|^2)} - (u_*)_i(\tau_*)_{jk}, \psi\frac{\partial\Phi_{jk}}{\partial x_i}\Big)dt\Big|$$

$$\leq C \max_{i,j,k} \left\| \frac{(u_m)_i\,(\tau_m)_{jk}}{1+\frac{1}{m}(\frac{|\tau_m|^2}{2\mu_2}+|u_m|^2)} - (u_*)_i(\tau_*)_{jk} \right\|_{L_1(0,T;L_1)}$$
$$\cdot \|\psi\|_{L_\infty(0,T;\mathbb{R})}\|\Phi\|_{C^1(\Omega,\mathbb{R}_S^{n\times n})}.$$

Then,

$$\left\| \frac{(u_m)_i\,(\tau_m)_{jk}}{1+\frac{1}{m}(\frac{|\tau_m|^2}{2\mu_2}+|u_m|^2)} - (u_*)_i(\tau_*)_{jk} \right\|_{L_1(0,T;L_1)}$$
$$\leq \left\| \frac{(u_m)_i\,(\tau_m)_{jk} - (u_*)_i(\tau_*)_{jk}}{1+\frac{1}{m}(\frac{|\tau_m|^2}{2\mu_2}+|u_m|^2)} \right\|_{L_1(0,T;L_1)}$$
$$+ \left\| \frac{(u_*)_i(\tau_*)_{jk}}{1+\frac{1}{m}(\frac{|\tau_m|^2}{2\mu_2}+|u_m|^2)} - (u_*)_i(\tau_*)_{jk} \right\|_{L_1(0,T;L_1)}$$
$$\leq \left\|(u_m)_i\,(\tau_m)_{jk} - (u_*)_i(\tau_*)_{jk}\right\|_{L_1(0,T;L_1)}$$
$$+ \left\| \frac{(\frac{|\tau_m|^2}{2\mu_2}+|u_m|^2)(u_*)_i(\tau_*)_{jk}}{m+\frac{|\tau_m|^2}{2\mu_2}+|u_m|^2} \right\|_{L_1(0,T;L_1)}$$

$$\leq \|(u_m)_i\|_{L_2(0,T;L_2)}\|(\tau_m)_{jk}-(\tau_*)_{jk}\|_{L_2(0,T;L_2)}$$
$$+ \|(u_m)_i-(u_*)_i\|_{L_2(0,T;L_2)}\|(\tau_*)_{jk}\|_{L_2(0,T;L_2)}$$
$$+ \left\| \frac{(\frac{|\tau_m|^2}{2\mu_2}+|u_m|^2)(u_*)_i(\tau_*)_{jk}}{m+\frac{|\tau_m|^2}{2\mu_2}+|u_m|^2} \right\|_{L_1(0,T;L_1)}.$$

The first and the second terms tend to zero as $m\to\infty$ for $u_m \to u_*$ in $L_2(0,T;H)$, and $\tau_m \to \tau_*$ in $L_2(0,T;L_2)$. It remains to prove that the third term also tends to zero. In fact, the convergences $u_m(t)(x) \to u_*(t)(x)$ almost everywhere in $(0,T)\times\Omega$, $\tau_m(t)(x) \to \tau_*(t)(x)$ almost everywhere in $(0,T)\times\Omega$ imply that

$$\frac{(\frac{|\tau_m(t)(x)|^2}{2\mu_2}+|u_m(t)(x)|^2)(u_*)_i(t)(x)(\tau_*)_{jk}(t)(x)}{m+\frac{|\tau_m(t)(x)|^2}{2\mu_2}+|u_m(t)(x)|^2}$$
$$\to 0 \text{ almost everywhere in } (0,T)\times\Omega.$$

Furthermore,

$$\left|\frac{(\frac{|\tau_m(t)(x)|^2}{2\mu_2}+|u_m(t)(x)|^2)(u_*)_i(t)(x)(\tau_*)_{jk}(t)(x)}{m+\frac{|\tau_m(t)(x)|^2}{2\mu_2}+|u_m(t)(x)|^2}\right| \leq \left|(u_*)_i(t)(x)(\tau_*)_{jk}(t)(x)\right|,$$

and $(u_*)_i(\tau_*)_{jk}(t)(x) \in L_1((0,T)\times\Omega)$. Thus, by the Lebesgue theorem,

$$\frac{\left(\frac{|\tau_m(t)(x)|^2}{2\mu_2} + |u_m(t)(x)|^2\right)(u_*)_i(t)(x)(\tau_*)_{jk}(t)(x)}{m + \frac{|\tau_m(t)(x)|^2}{2\mu_2} + |u_m(t)(x)|^2} \to 0$$

in $L_1((0,T)\times\Omega)$ as $m \to \infty$.

We have proved (6.4.7). Observe that (6.4.8) can be shown just in the same way.

Now, passing to the limit in (6.4.5), (6.4.6) as $m \to \infty$, we conclude

$$\begin{aligned}
&-\int_0^T (\tau_*, \Phi\psi'(t))dt + \int_0^T \Big(\frac{1}{\lambda_1}(\tau_*, \psi\Phi) - \sum_{i=1}^n \Big((u_*)_i\tau_*, \psi\frac{\partial\Phi}{\partial x_i}\Big)\\
&\quad + 2\mu_2(u_*, \psi \operatorname{Div} \Phi) + +\frac{\varepsilon}{\lambda_1}(\nabla\tau_*, \psi\nabla\Phi)\Big)dt = (\tau_0, \Phi)\psi(0),
\end{aligned} \tag{6.4.9}$$

$$\begin{aligned}
&-\int_0^T (u_*, \varphi\psi'(t))dt + \int_0^T \Big(\mu_1(\nabla u_*, \psi\nabla\varphi) - \sum_{i=1}^n \Big((u_*)_i u_*, \psi\frac{\partial\varphi}{\partial x_i}\Big)\\
&\quad + (\tau_*, \psi\nabla\varphi)\Big)dt = \int_0^T \langle f, \varphi\psi\rangle dt + (a, \varphi)\psi(0).
\end{aligned} \tag{6.4.10}$$

Since it has been carried out, in particular, for every $\psi \in C_0^\infty(0,T)$, the function (u_*, τ_*) satisfies (6.4.1), (6.4.2) in the sense of distributions on $(0,T)$.

Substitute (u_*, τ_*) into equalities (6.4.1), (6.4.2). Since all terms in the obtained equalities are integrable on $(0,T)$, these equalities are valid almost everywhere on $(0,T)$. Taking the scalar product of these equalities in $L_2(0,T)$ with a smooth scalar function $\psi(t)$, $\psi(T) = 0$, $\psi(0) \neq 0$ and comparing the result with (6.4.9), (6.4.10), we see that

$$\begin{aligned}
(u_*|_{t=0}, \varphi)\psi(0) &= (a, \varphi)\psi(0),\\
(\tau_*|_{t=0}, \Phi)\psi(0) &= (\tau_0, \Phi)\psi(0).
\end{aligned}$$

Since Φ and φ are arbitrary, u_* and τ_* satisfy (6.3.5). Repeating the proof of Lemma 6.3.2 with $\delta = 0$, $\xi = 1$, we see that the solutions of problem (6.4.1), (6.4.2), (6.3.5) satisfy estimates (6.3.14), (6.3.15). Thus, (u_*, τ_*) is the desirable solution. □

Remark 6.4.1. There is another way to check that the limit (u_*, τ_*) satisfies the initial condition (6.3.5). One has to observe that the the sequence $\{\frac{du_m}{dt}\}$ is in fact bounded in $L_{4/3}(0,T;V^*)$, and the sequence $\{\frac{d\tau_m}{dt}\}$ is bounded in $L_2(0,T;H^{-2})$ (see Remark 6.8.1 below). But since $u_m \to u_*$ *-weakly in $L_\infty(0,T;H)$, $\tau_m \to \tau_*$ *-weakly in $L_\infty(0,T;L_2)$, Theorem 2.2.6 gives that $u_m \to u_*$ strongly in $C(0,T;V^*)$,

$\tau_m \to \tau_*$ strongly in $C(0, T; H^{-1})$. In particular, there is the pointwise convergence. Hence, $a = u_m(0) \to u_*(0)$ strongly in V^*, i.e. $a = u_*(0)$, and $\tau_0 = \tau_m(0) \to \tau_*(0)$ strongly in H^{-1}, i.e. $\tau_0 = \tau_*(0)$. Such an approach may be used also in the proof of Theorem 6.5.1, where we, however, shall apply the first way of proof again.

6.5 Existence of a weak solution for the Jeffreys model

6.5.1 Existence of velocity and stress

Let us prove a statement which yields Theorem 6.2.1 immediately.

Theorem 6.5.1 (see [72]). *Given $f \in L_2(0, T; V^*)$, $a \in H$, $\tau_0 \in L_2$, there exists a pair of functions (u, τ),*

$$\begin{aligned} &u \in L_2(0, T; V) \bigcap L_\infty(0, T; H) \bigcap C_w([0, T], H), \ \frac{du}{dt} \in L_1(0, T; V^*), \\ &\tau \in L_\infty(0, T; L_2) \bigcap C_w([0, T], L_2), \ \frac{d\tau}{dt} \in L_2(0, T; H^{-2}) \end{aligned} \tag{6.5.1}$$

satisfying (6.3.1), (6.3.2) almost everywhere in $(0, T)$, the initial condition (6.3.5) and the estimate

$$\begin{aligned} &\|u\|_{L_2(0,T;V)} + \|u\|_{L_\infty(0,T;H)} + \left\|\frac{du}{dt}\right\|_{L_1(0,T;V^*)} + \|\tau\|_{L_\infty(0,T;L_2)} \\ &\quad + \left\|\frac{d\tau}{dt}\right\|_{L_2(0,T;H^{-2})} \le K_3(\|a\|, \|\tau_0\|, \|f\|_{L_2(0,T;V^*)}), \end{aligned} \tag{6.5.2}$$

where K_3 does not depend on Ω.

Remark 6.5.1. As a matter of fact, the solution (u, τ) satisfies (6.3.1) for all $\Phi \in \overset{\circ}{W}{}^1_4(\Omega, \mathbb{R}^{n\times n}_S)$ (not only from C_0^∞), and (6.3.2) holds for all $\varphi \in V$ (not only from $\mathcal{V}$). Really, as $u \in L_2(0, T; V) \subset L_2(0, T; H_0^1) \subset L_2(0, T; L_4)$, one has $u_i u \in L_1(0, T; L_2)$ by inequality (2.2.1). But $\nabla u \in L_2(0, T; L_2)$, $\tau \in L_\infty(0, T; L_2)$, $f \in L_2(0, T; V^*)$. Applying Lemma 2.2.8 (implication i) $\to$ ii)) with $X = V^*$, $X^* = V$ (V is reflexive), $Z = \mathcal{V}$, and with (6.3.2) instead of (2.2.29), we get the second claim. Furthermore, since $u \in L_2(0, T; L_4)$, $\tau \in L_\infty(0, T; L_2)$, one has $u_i \tau \in L_2(0, T; L_{4/3})$; and applying Lemma 2.2.8 with $X = W^{-1}_{4/3}(\Omega, \mathbb{R}^{n\times n}_S)$, $X^* = \overset{\circ}{W}{}^1_4(\Omega, \mathbb{R}^{n\times n}_S)$, $Z = C_0^\infty$, and with (6.3.1) instead of (2.2.29), we get the first claim.

Proof of Theorem 6.5.1. Denote by Ω_m the intersection of Ω with the ball B_m centered at the origin in the space $\mathbb{R}^n$ of radius $m = 1, 2, \dots$. From the definition of the space H (Section 2.1.2) it easily follows that there exists a sequence $\{a_m\}$,

$$a_m \in C_0^\infty(\Omega)^n, \ \operatorname{div} a_m = 0, \ \operatorname{supp} a_m \subset \Omega_m, \ a_m \to a \text{ in } L_2(\Omega)^n, \ \|a_m\| \le \|a\|.$$

Consider on Ω_m, for every m, problem (6.4.1), (6.4.2) with $\varepsilon = \frac{1}{m}$ and the initial condition

$$u|_{t=0} = a_m, \ \tau|_{t=0} = \tau_0|_{\Omega_m}. \tag{6.5.3}$$

By Theorem 6.4.1 there exists a solution (u_m, τ_m) of this problem. All these solutions are bounded by estimates (6.3.14) and (6.4.4) with $\varepsilon = \frac{1}{m}$. Denote by $\tilde{u}_m$ and $\tilde{\tau}_m$ the functions which coincide with u_m and τ_m, respectively, in Ω_m, and are identically zero in $\Omega \backslash \Omega_m$. Without loss of generality (see Remark 2.1.1) we may assume that there exists a pair (u_*, τ_*) such that

$$\begin{aligned} &\tilde{u}_m \to u_* \text{ weakly in } L_2(0, T; V), \\ &\tilde{u}_m \to u_* \text{ *-weakly in } L_\infty(0, T; H), \\ &\tilde{\tau}_m \to \tau_* \text{ *-weakly in } L_\infty(0, T; L_2). \end{aligned}$$

Furthermore, by Theorem 2.2.6:

$$\tilde{u}_m|_{\Omega_k} \to u_*|_{\Omega_k} \text{ strongly in } L_2(0, T; L_2(\Omega_k))$$

for every k. Obviously, (u_*, τ_*) satisfies estimate (6.4.4).

Take arbitrary $\varphi \in \mathcal{V}$, $\Phi \in C_0^\infty$. Fix k large enough, such that the supports of φ and Φ are contained in Ω_k.

Substitute (u_m, τ_m) in equalities (6.4.1), (6.4.2) with $m \geq k$, $\varepsilon = \frac{1}{m}$. Take the scalar product of these equalities in $L_2(0, T)$ with a smooth scalar function $\psi(t)$, $\psi(T) = 0$ and integrate by parts the first terms. Due to the choice of k we can replace u_m and τ_m in these equalities by $\tilde{u}_m$ and $\tilde{\tau}_m$. We have:

$$\begin{aligned} -\int_0^T (\tilde{\tau}_m, \Phi\varphi'(t))dt + \int_0^T \Big(\frac{1}{\lambda_1}(\tilde{\tau}_m, \psi\Phi) - \sum_{i=1}^n \Big((\tilde{u}_m)_i \, \tilde{\tau}_m, \psi \frac{\partial \Phi}{\partial x_i}\Big) \\ + 2\mu_2(\tilde{u}_m, \psi \operatorname{Div} \Phi) + \frac{1}{\lambda_1 m}(\nabla \tilde{\tau}_m, \psi \nabla \Phi)\Big)dt = (\tau_0, \Phi)\psi(0), \end{aligned} \tag{6.5.4}$$

$$\begin{aligned} -\int_0^T \Big(\tilde{u}_m, \varphi\psi'(t)\Big)dt + \int_0^T \Big(\mu_1(\nabla \tilde{u}_m, \psi \nabla \varphi) - \sum_{i=1}^n \Big((\tilde{u}_m)_i \tilde{u}_m, \psi \frac{\partial \varphi}{\partial x_i}\Big) \\ + (\tilde{\tau}_m, \psi \nabla \varphi)\Big) dt = \int_0^T \langle f, \varphi\psi \rangle dt + (a_m, \varphi)\psi(0). \end{aligned} \tag{6.5.5}$$

Observe that

$$\begin{aligned} \Big|\frac{1}{m}\int_0^T (\nabla \tilde{\tau}_m, \psi \nabla \Phi)\Big| &= \Big|\frac{1}{m}\int_0^T (\tilde{\tau}_m, \psi \Delta \Phi)\Big| \\ &\leq \frac{1}{m}\|\tau_m\|_{L_\infty(0,T;L_2)} \int_0^T \|\psi \Delta \Phi\| \underset{m\to\infty}{\longrightarrow} 0. \end{aligned}$$

Now, let us show that

$$\int_0^T \sum_{i=1}^n \Big((\tilde{u}_m)_i \tilde{\tau}_m, \psi \frac{\partial \Phi}{\partial x_i}\Big) dt \to \int_0^T \sum_{i=1}^n \Big((\tilde{u}_*)_i \tilde{\tau}_*, \psi \frac{\partial \Phi}{\partial x_i}\Big) dt, \tag{6.5.6}$$

$$\int_0^T \sum_{i=1}^n \Big((\tilde{u}_m)_i \tilde{u}_m, \psi \frac{\partial \varphi}{\partial x_i}\Big) dt \to \int_0^T \sum_{i=1}^n \Big((\tilde{u}_*)_i \tilde{u}_*, \psi \frac{\partial \varphi}{\partial x_i}\Big) dt. \tag{6.5.7}$$

Really, using inequality (2.2.1), we get:

$$\begin{aligned}
&\Big|\int_0^T \sum_{i=1}^n \Big((\tilde{u}_m)_i \tilde{\tau}_m - (\tilde{u}_*)_i \tilde{\tau}_*, \psi \frac{\partial \Phi}{\partial x_i}\Big) dt\Big| \\
&\le \Big|\int_0^T \sum_{i=1}^n \Big(((\tilde{u}_m)_i - (\tilde{u}_*)_i)\tilde{\tau}_m, \psi \frac{\partial \Phi}{\partial x_i}\Big) dt\Big| \\
&\quad + \Big|\int_0^T \sum_{i=1}^n \Big((\tilde{u}_*)_i (\tilde{\tau}_m - \tilde{\tau}_*), \psi \frac{\partial \Phi}{\partial x_i}\Big) dt\Big| \\
&\le \|(u_m - u_*)\big|_{\Omega_k}\|_{L_2(0,T;L_2)} \|\tau_m\|_{L_2(0,T;L_2)} \\
&\quad \times \|\psi \nabla \Phi\|_{L_\infty(0,T;L_\infty)} + \Big|\Big\langle \tilde{\tau}_m - \tilde{\tau}_*, \sum_{i=1}^n (\tilde{u_*})_i \psi \frac{\partial \Phi}{\partial x_i}\Big\rangle_{L_\infty(0,T;L_2)\times L_1(0,T;L_2)}\Big|.
\end{aligned}$$

Both terms tend to zero, and (6.5.6) is proved. Similarly one shows (6.5.7).

Now, let m tend to infinity in (6.5.4) and (6.5.5). We obtain:

$$\begin{aligned}
&-\int_0^T (\tau_*, \Phi\psi'(t)) dt + \int_0^T \Big(\frac{1}{\lambda_1}(\tau_*, \psi\Phi) - \sum_{i=1}^n \Big((u_*)_i \tau_*, \psi \frac{\partial \Phi}{\partial x_i}\Big) \\
&\quad + 2\mu_2 (u_*, \psi \operatorname{Div} \Phi)\Big) dt = (\tau_0, \Phi)\psi(0),
\end{aligned} \tag{6.5.8}$$

$$\begin{aligned}
&-\int_0^T \Big(u_*, \varphi\psi'(t)\Big) dt + \int_0^T \Big(\mu_1 (\nabla u_*, \psi \nabla \varphi) - \sum_{i=1}^n \Big((u_*)_i u_*, \psi \frac{\partial \varphi}{\partial x_i}\Big) \\
&\quad + (\tau_*, \psi \nabla \varphi)\Big) dt = \int_0^T \langle f, \varphi\psi\rangle dt + (a, \varphi)\psi(0).
\end{aligned} \tag{6.5.9}$$

As in the proof of Theorem 6.4.1, this implies that (u_*, τ_*) is a solution of (6.3.1), (6.3.2), (6.3.5) and u_* satisfies estimate (6.3.14).

To prove (6.5.2), it remains to estimate the fifth term in its left-hand side. Substitute (u_*, τ_*) in (6.3.1). Using inequality (2.2.1), the embedding $V \subset L_4$, and estimate

(6.4.4) for (u_*, τ_*) , we have:

$$
\begin{aligned}
\|\langle\tau'_*, \Phi\rangle\|_{L_2(0,T)} &= \Big\|\frac{d}{dt}(\tau_*, \Phi)\Big\|_{L_2(0,T)} \\
&\le \Big\|\frac{1}{\lambda_1}(\tau_*, \Phi)\Big\|_{L_2(0,T)} + \Big\|\sum_{i=1}^{n}((u_*)_i\tau_*, \frac{\partial\Phi}{\partial x_i})\Big\|_{L_2(0,T)} \\
&\quad + \|2\mu_2(u_*, \operatorname{Div}\Phi)\|_{L_2(0,T)} \\
&\le \frac{1}{\lambda_1}\|\tau_*\|_{L_2(0,T.L_2)}\|\Phi\|_{L_2} + \|u_*\|_{L_2(0,T;L_4)}\|\tau_*\|_{L_\infty(0,T;L_2)}\big\|\nabla\Phi\big\|_{L_4} \\
&\quad + 2\mu_2\|u_*\|_{L_2(0,T;L_2)}\|\operatorname{Div}\Phi\|_{L_2} \\
&\le C\|\Phi\|_{H_0^2}(\|\tau_*\|_{L_\infty(0,T;L_2)} + \|u_*\|_{L_2(0,T;V)}\|\tau_*\|_{L_\infty(0,T;L_2)} \\
&\quad + \|u_*\|_{L_2(0,T;V)}) \\
&\le K_4(\|a\|, \|\tau_0\|, \|f\|_{L_2(0,T;V^*)})\|\Phi\|_{H_0^2},
\end{aligned}
$$

and we obtain the desirable estimate.

By Lemma 2.2.8, almost everywhere on $(0, T)$ u_* is equal to a continuous function with values in V^*. Thus, without loss of generality we may assume that u_* itself is continuous on $[0, T]$ with values in this space. And since it belongs to $L_\infty(0, T; H)$, by Lemma 2.2.6 it is weakly continuous on $[0, T]$ as a function with values in H. Similarly, $\tau \in C_w([0, T]; L_2)$. The proof of the theorem is complete. □

Remark 6.5.2. By the same scheme it is easy to show that in Theorems 6.3.1 and 6.4.1 the condition of boundedness of Ω is not necessary.

Proof of Theorem 6.2.1. Take $\tau_0 = \sigma_0 - 2\mu_1\mathcal{E}(a)$. By the condition of the theorem $\tau_0 \in L_2$. By Theorem 6.5.1 there exists a solution (u, τ) of problem (6.3.1), (6.3.2), (6.3.5) in class (6.5.1).

It is easy to see that $\mathcal{E}(u) \in L_2(0, T; L_2) \bigcap C_w([0, T]; H^{-1})$.

Take $\sigma = \tau + 2\mu_1\mathcal{E}(u)$. Then the pair (u, σ) belongs to class (6.2.6) and satisfies (6.2.5), (6.2.7), (6.2.8), i.e. it is a weak solution of problem (6.2.1) – (6.2.5). □

Remark 6.5.3. Since (6.3.1) and (6.3.2) are satisfied almost everywhere in $(0, T)$, (6.2.7) is also satisfied almost everywhere in $(0, T)$. But we cannot assert that (6.2.8) is also satisfied almost everywhere in $(0, T)$, since (6.2.8) contains two time derivatives, and Lemma 2.2.8 is not applicable here.

6.5.2 Existence of pressure

The pressure p was removed from problem (6.2.1) – (6.2.5) by the weak setting procedure. Condition (6.1.4) gives an opportunity to restore it when a weak solution is sufficiently regular. Is it possible to restore the pressure (maybe in some generalized sense) in general? In this subsection we examine this question.

We begin with a "positive" result.

Theorem 6.5.2. *Let Ω be as in Corollary 3.1.1, $n = 2, 3$, and let $f \in L_2(0, T; H^{-1}(\Omega)^n)$. Then for any weak solution (u, σ) to problem (6.2.1) – (6.2.5) there exists a function $p \in W_\infty^{-1}(0, T; L_2(\Omega))$ such that equality (6.2.1) holds (e.g. in the space $W_\infty^{-1}(0, T; H^{-1}(\Omega)^n)$).*

Remark 6.5.4. Under the conditions of the theorem, use of Definition 6.2.1 is possible since one can always assume that $f \in L_2(0, T; V^*)$. Really, for almost all $t \in (0, T)$,

$$f(t) = (f_1(t), \dots, f_n(t)) \in H^{-1}(\Omega)^n = ((H_0^1(\Omega))^*)^n.$$

Then $f(t)$ may be considered as a linear continuous functional on V (i.e. as an element of V^*) according to the formula

$$\langle f(t), \varphi \rangle_{V^* \times V} = \sum_{i=1}^n \langle f_i(t), \varphi_i \rangle_{H^{-1} \times H_0^1}, \quad \varphi \in V.$$

Proof of Theorem 6.5.2. Let (u, σ) be a weak solution to problem (6.2.1) – (6.2.5). Then, in particular,

$$u \in L_2(0, T; V) \bigcap L_\infty(0, T; H),$$
$$\sigma \in L_2(0, T; L_2(\Omega, \mathbb{R}_S^{n \times n})),$$

and

$$\frac{d}{dt}(u, \varphi) + (\sigma, \nabla \varphi) - \sum_{i=1}^n (u_i u, \frac{\partial \varphi}{\partial x_i}) = \langle f, \varphi \rangle \tag{6.5.10}$$

for all $\varphi \in \mathcal{V}$.

Thus,

$$u \in L_2(0, T; L_4), \quad \nabla u \in L_2(0, T; L_2).$$

Then by (2.2.1),

$$u_i \frac{\partial u}{\partial x_i} \in L_1(0, T; L_{4/3}) \subset L_1(0, T; H^{-1}) \subset W_\infty^{-1}(0, T; H^{-1}), \quad i = 1, \dots, n.$$

The first embedding follows from Theorem 2.1.1 a). Let us check the second one. By Corollary 2.2.1,

$$W_1^1(0, T; H_0^1) \subset C([0, T]; H_0^1).$$

The embedding is dense and continuous. Hence,

$$L_1(0,T;H^{-1}) \subset (C([0,T];H_0^1))^* \subset (W_1^1(0,T;H_0^1))^* = W_\infty^{-1}(0,T;H^{-1}).$$

Observe that

$$u' \in W_\infty^{-1}(0,T;L_2),$$
$$\text{Div } \sigma \in L_2(0,T;H^{-1}).$$

Integrating by parts (see Section 6.1.2) in the second and the third terms of (6.5.10) we conclude that

$$(u',\varphi) - (\text{Div } \sigma,\varphi) + \Big(\sum_{i=1}^n u_i \frac{\partial u}{\partial x_i},\varphi\Big) = \langle f,\varphi\rangle \tag{6.5.11}$$

for all $\varphi \in \mathcal{V}$. Let $g = f - u' + \text{Div } \sigma - \sum_{i=1}^n u_i \frac{\partial u}{\partial x_i}$. Then $g \in W_\infty^{-1}(0,T;H^{-1}(\Omega)^n)$, and

$$\langle g(t),\varphi\rangle = 0 \tag{6.5.12}$$

for all $\varphi \in \mathcal{V}$. By Corollary 3.1.3, there exists $p \in W_\infty^{-1}(0,T;L_2)$ such that

$$\text{grad } p = g,$$

which is equivalent to (6.2.1). The proof is complete. □

However the answer to the question which was put at the beginning of this subsection is not positive in general. Even in the case of bounded sufficiently regular domain Ω with connected boundary, (6.2.1) cannot hold a priori for body forces from $L_2(0,T;V^*(\Omega))$. Let us illustrate this using the ideas from [55].

Lemma 6.5.1 (see [55], p. 228). *If Ω is bounded sufficiently regular and its boundary is connected, there exists a sequence of functions $\{v_m\} \subset \mathcal{V}$ such that*

$$v_m \to 0 \ \text{ in } \ V^*, \quad v_m \to v \neq 0 \ \text{ in } \ H^{-1}(\Omega)^n. \tag{6.5.13}$$

Corollary 6.5.1. *Let Ω be bounded sufficiently regular and let its boundary be connected.*

a) *There is no Hausdorff topological vector space $\mathcal{Y}$ such that*

$$V^*(\Omega) \subset \mathcal{Y}, \quad H^{-1}(\Omega)^n \subset \mathcal{Y}$$

with continuous embeddings;

b) *there is no Hausdorff topological vector space* $\mathcal{Y}$ *such that*

$$L_2(0,T;V^*(\Omega)) \subset \mathcal{Y}, \quad L_2(0,T;H^{-1}(\Omega)^n) \subset \mathcal{Y}$$

with continuous embeddings.

Since different members of (6.2.1) belong to different spaces (in particular, $u' \in W_\infty^{-1}(0,T;L_2)$, $\operatorname{Div}\sigma \in L_2(0,T;H^{-1}(\Omega)^n)$, $f \in L_2(0,T;V^*)$), one needs some "large" uniting space which contains them all (in order to interpret equality (6.2.1) in this space). Due to Corollary 6.5.1 b), it turns out to be impossible to find such a space even with minimal restrictions of its properties.

However, a "uniting" space exists for $\Omega = \mathbb{R}^n$: for example, it is the space $L_1(0,T;H^{-1}(\mathbb{R}^n)^n)$. Observe first that for $u \in H(\mathbb{R}^n)$, $\varphi \in H^1(\mathbb{R}^n,\mathbb{R}^n)$ one has

$$\langle u,\varphi\rangle_{H^{-1}\times H^1} = (u,\varphi) = (u,P\varphi) = \langle u,P\varphi\rangle_{H^*\times H} = \langle u,P\varphi\rangle_{V^*\times V},$$

where P is the Leray projection (5.2.12). Note that Lemma 3.1.1 and Liouville's theorem imply

$$P(L_2(\mathbb{R}^n)) = H_V^0 = H(\mathbb{R}^n),\ P(H^1(\mathbb{R}^n)) = H_V^1 = V(\mathbb{R}^n).$$

Thus, we arrive at the embedding

$$V^*(\mathbb{R}^n) \subset H^{-1}(\mathbb{R}^n)^n, \tag{6.5.14}$$

i.e. any $u \in V^*$ determines a linear continuous functional on $H^1(\mathbb{R}^n)^n$ by the formula

$$\langle u,\varphi\rangle_{H^{-1}(\mathbb{R}^n)^n\times H^1(\mathbb{R}^n)^n} = \langle u,P\varphi\rangle_{V^*\times V} \tag{6.5.15}$$

for $\varphi \in H^1(\mathbb{R}^n)^n$.

Theorem 6.5.3. *Let* $f \in L_2(0,T;V^*(\mathbb{R}^n)^n)$. *Then for any weak solution* (u,σ) *to problem (6.2.1) – (6.2.5) there exists a function* $p \in L_1(0,T;L_{2,\mathrm{loc}}(\mathbb{R}^n))$ *such that equality (6.2.1) holds (e.g. in the space* $L_1(0,T;H^{-1}(\mathbb{R}^n)^n)$*).*

Proof. Let (u,σ) be a weak solution to problem (6.2.1) – (6.2.5). Then, due to (6.5.14),

$$u' \in L_1(0,T;V^*) \subset L_1(0,T;H^{-1}),$$
$$f \in L_2(0,T;V^*) \subset L_2(0,T;H^{-1}).$$

But as in the proof of Theorem 6.5.2,

$$u_i\frac{\partial u}{\partial x_i} \in L_1(0,T;H^{-1}),$$

$$\operatorname{Div}\sigma \in L_2(0,T;H^{-1}),$$

and one has (6.5.10) and (6.5.11).

Let $g = f - u' + \operatorname{Div} \sigma - \sum_{i=1}^{n} u_i \frac{\partial u}{\partial x_i}$. Then $g \in L_1(0, T; H^{-1}(\mathbb{R}^n)^n)$, and (6.5.12) holds for all $\varphi \in \mathcal{V}$. By Corollary 3.1.4, there exists $p \in L_1(0, T; L_{2,\text{loc}})$ such that

$$\operatorname{grad} p = g,$$

which is equivalent to (6.2.1). □

6.6 Differential energy estimate and uniqueness of the weak solution for the Jeffreys model

The problem of uniqueness of weak solutions for the majority of the equations of hydrodynamics generally remains open. For example, for the equations of Navier–Stokes (Newtonian fluid) in the two-dimensional case a weak solution is unique, and in the three dimensions there are only conditional results. For instance, the classical result by Sather and Serrin ([61], Theorem III.3.9) says that if a weak solution to the initial-boundary value for the equations of Navier–Stokes belongs, in addition, to $L_8(0, T; L_4)$, then it is unique in the class of weak solutions satisfying an energy inequality. In this section we shall prove a similar result for the Jeffreys model.

6.6.1 Differential energy inequality

First we derive an energy inequality for the constructed in Theorem 6.4.1 solution to problem (6.4.1), (6.4.2), (6.3.5).

Lemma 6.6.1. *The solution to problem (6.4.1), (6.4.2), (6.3.5) constructed in Theorem 6.4.1 satisfies the following inequality for almost all $t \in [0, T]$:*

$$\begin{aligned} &\frac{1}{2}\|u_*\|^2(t) + \frac{1}{4\mu_2}\|\tau_*\|^2(t) + \int_0^t \frac{1}{2\lambda_1\mu_2}\left((1-\varepsilon)\|\tau_*\|^2 + \varepsilon\|\tau_*\|_1^2\right) ds \\ &\quad + \int_0^t \mu_1\|u_*\|_Y^2\, ds \le \frac{1}{2}\|a\|^2 + \frac{1}{4\mu_2}\|\tau_0\|^2 + \int_0^t \langle f(s), u_*(s)\rangle\, ds. \end{aligned} \tag{6.6.1}$$

Remark 6.6.1. The space $Y = Y(\Omega)$ was introduced in Section 2.1.2 and will be often used from now on.

Proof. Consider the pairs (u_m, τ_m) which were used in the proof of Theorem 6.4.1 for the construction of the solution. Each of them satisfies equality (6.3.8). Having integrated this equality from 0 to t, we get:

$$\frac{1}{2}\|u_m\|^2(t)+\frac{1}{4\mu_2}\|\tau_m\|^2(t)+\int_0^t\frac{1}{2\lambda_1\mu_2}\left((1-\varepsilon)\|\tau_m\|^2+\varepsilon\|\tau_m\|_1^2\right)ds$$
$$+\int_0^t\mu_1\|u_m\|_Y^2\,ds=\frac{1}{2}\|a\|^2+\frac{1}{4\mu_2}\|\tau_0\|^2+\int_0^t\langle f(s),u_m(s)\rangle\,ds.$$

Take the scalar product in $L_2(0,T)$ of this equality with a smooth function $\psi\in\mathcal{D}(0,T)$ with non-negative values:

$$\int_0^T\Big\{\frac{1}{2}\|u_m\|^2(t)+\frac{1}{4\mu_2}\|\tau_m\|^2(t)+\int_0^t\frac{1}{2\lambda_1\mu_2}\left((1-\varepsilon)\|\tau_m\|^2+\varepsilon\|\tau_m\|_1^2\right)ds$$
$$+\int_0^t\mu_1\|u_m\|_Y^2\,ds\Big\}\psi(t)\,dt$$
$$=\int_0^T\Big(\frac{1}{2}\|a\|^2+\frac{1}{4\mu_2}\|\tau_0\|^2+\int_0^t\langle f(s),u_m(s)\rangle\,ds\Big)\psi(t)\,dt.$$

Passing in this relation to the inferior limit as $m\to\infty$, and using the fact that the norm of a weak limit of a sequence does not exceed the inferior limit of the norms, we arrive at:

$$\int_0^T\Big\{\frac{1}{2}\|u_*\|^2(t)+\frac{1}{4\mu_2}\|\tau_*\|^2(t)+\int_0^t\frac{1}{2\lambda_1\mu_2}\left((1-\varepsilon)\|\tau_*\|^2+\varepsilon\|\tau_*\|_1^2\right)ds$$
$$+\int_0^t\mu_1\|u_*\|_Y^2\,ds\Big\}\psi(t)\,dt$$
$$\le\int_0^T\Big(\frac{1}{2}\|a\|^2+\frac{1}{4\mu_2}\|\tau_0\|^2+\int_0^t\langle f(s),u_*(s)\rangle\,ds\Big)\psi(t)\,dt.$$

Since ψ was chosen arbitrarily, this yields (6.6.1). □

It is also possible to get an energy inequality for the solution of problem (6.3.1), (6.3.2), (6.3.5):

Lemma 6.6.2. *The solution to problem (6.3.1), (6.3.2), (6.3.5) constructed in Theorem 6.5.1 satisfies the following inequality at almost all $t\in[0,T]$:*

$$\frac{1}{2}\|u\|^2(t)+\frac{1}{4\mu_2}\|\tau\|^2(t)+\int_0^t\frac{1}{2\lambda_1\mu_2}\|\tau\|^2\,ds+\int_0^t\mu_1\|u\|_Y^2\,ds$$
$$\le\frac{1}{2}\|a\|^2+\frac{1}{4\mu_2}\|\tau_0\|^2+\int_0^t\langle f(s),u(s)\rangle\,ds. \tag{6.6.2}$$

Proof. Consider the pairs (u_m,τ_m) which were used in the proof of Theorem 6.5.1 for the construction of the solution. Each of them satisfies inequality (6.6.1), and, therefore, (6.6.2). The rest of the proof (the passage to the limit in this inequality) is similar to the arguments in the proof of Lemma 6.6.1. □

6.6.2 Uniqueness of the weak solution

Theorem 6.6.1 (see [69]). *Assume that $n = 3$ and, under the conditions of Theorem 6.2.1, there is a weak solution (u_1, σ_1) to problem (6.2.1) – (6.2.5). If*

$$u_1 \in L_8(0,T; L_4(\Omega)^n), \quad \tau_1 = [\sigma_1 - 2\mu_1 \mathcal{E}(u_1)] \in L_4(0,T; \overset{\circ}{W}{}^1_4(\Omega, \mathbb{R}^{n\times n}_S)), \tag{6.6.3}$$

then this solution is unique in the class of weak solutions (u, σ) (in the sense of Definition 6.2.1) which satisfy inequality (6.6.2) with $\tau = \sigma - 2\mu_1\mathcal{E}(u)$.

Proof. Let there be another weak solution (u_2, σ_2) to problem (6.2.1) – (6.2.5) satisfying inequality (6.6.2) with $\tau_2 = \sigma_2 - 2\mu_1\mathcal{E}(u_2)$. Let us show that it coincides with (u_1, σ_1), i.e. that (u_2, τ_2) coincides with (u_1, τ_1).

Obviously, both pairs (u_1, τ_1) and (u_2, τ_2) are solutions of problem (6.3.1), (6.3.2), (6.3.5).

As $u_1 \in L_8(0,T; L_4)$, one has $u_{1i}u_1 \in L_4(0,T; L_2)$ by inequality (2.2.1). Furthermore, since $u_1 \in L_\infty(0,T; L_2)$, $\tau_1 \in L_4(0,T; \overset{\circ}{W}{}^1_4(\Omega, \mathbb{R}^{n\times n}_S))$, $\overset{\circ}{W}{}^1_4(\Omega, \mathbb{R}^{n\times n}_S) \subset L_\infty(\Omega, \mathbb{R}^{n\times n}_S)$, one has $u_{1i}\tau_1 \in L_4(0,T; L_2)$. Using Lemma 2.2.8 (cf. Remark 6.5.1) we conclude that (u_1, τ_1) satisfies (6.3.1), (6.3.2) at all $\varphi \in V$, $\Phi \in H^1_0$, and $\frac{du_1}{dt} \in L_2(0,T; V^*)$, $\frac{d\tau_1}{dt} \in L_2(0,T; H^{-1})$.

Put $\tau = \tau_1$, $\Phi = \frac{\tau_1(t)}{2\mu_2}$ in (6.3.1) and $u(t) = \varphi = u_1(t)$ in (6.3.2) at almost all $t \in [0,T]$, and add the results. Taking into account (6.1.18) – (6.1.20) we get:

$$\frac{1}{2}\frac{d}{dt}(u_1, u_1) + \frac{1}{4\mu_2}\frac{d}{dt}(\tau_1, \tau_1) + \mu_1(\nabla u_1, \nabla u_1) + \frac{1}{2\lambda_1\mu_2}(\tau_1, \tau_1) = \langle f, u_1\rangle. \tag{6.6.4}$$

Integrating this equality from 0 to arbitrary $t \in [0,T]$, we obtain the energy equality

$$\begin{aligned} \frac{1}{2}\|u_1\|^2(t) + \frac{1}{4\mu_2}\|\tau_1\|^2(t) + \int_0^t \frac{1}{2\lambda_1\mu_2}\|\tau_1\|^2\,ds + \int_0^t \mu_1\|u_1\|_Y^2\,ds \\ = \frac{1}{2}\|a\|^2 + \frac{1}{4\mu_2}\|\tau_0\|^2 + \int_0^t \langle f(s), u_1(s)\rangle\,ds. \end{aligned} \tag{6.6.5}$$

Integration by parts in the third and fourth term of equality (6.3.1) with (u_1, τ_1) implies that (u_1, τ_1) satisfies the identity:

$$\frac{d}{dt}(\tau_1, \Phi) + \frac{1}{\lambda_1}(\tau_1, \Phi) + \sum_{i=1}^n \Big(u_{1i}\frac{\partial \tau_1}{\partial x_i}, \Phi\Big) - 2\mu_2(\mathcal{E}(u_1), \Phi) = 0. \tag{6.6.6}$$

By (2.2.1), $u_{1i}\frac{\partial\tau_1}{\partial x_i} \in L_2(0,T; L_2)$. Then Lemma 2.2.8 yields that (6.6.6) is fulfilled for all $\Phi \in L_2$.

Take $\Phi = \dfrac{\tau_2(t)}{2\mu_2}$ in (6.6.6) and $u = u_1, \varphi = u_2(t)$ in (6.3.2) at almost all $t \in [0, T]$, and add the results (taking into account (6.3.7)):

$$\begin{aligned}&\Big\langle\frac{d}{dt}u_1, u_2\Big\rangle + \frac{1}{2\mu_2}\Big\langle\frac{d}{dt}\tau_1, \tau_2\Big\rangle + \frac{1}{2\lambda_1\mu_2}(\tau_1, \tau_2) + \frac{1}{2\mu_2}\sum_{i=1}^n \Big(u_{1i}\frac{\partial \tau_1}{\partial x_i}, \tau_2\Big)\\ &\qquad - (\nabla u_1, \tau_2) - \sum_{i=1}^n \Big(u_{1i}u_1, \frac{\partial u_2}{\partial x_i}\Big) + \mu_1(\nabla u_1, \nabla u_2) + (\tau_1, \nabla u_2)\\ &= \langle f, u_2\rangle.\end{aligned} \tag{6.6.7}$$

Integrating by parts in the second term of equality (6.3.2) with (u_2, τ_2), we conclude that (u_2, τ_2) satisfies the identity:

$$\frac{d}{dt}(u_2, \varphi) + \sum_{i=1}^n \Big(u_{2i}\frac{\partial u_2}{\partial x_i}, \varphi\Big) + \mu_1(\nabla u_2, \nabla\varphi) + (\tau_2, \nabla\varphi) = \langle f, \varphi\rangle. \tag{6.6.8}$$

Note that (6.6.8) holds true for all $\varphi \in V$ (see Remark 6.5.1).

Take $\tau = \tau_2$, $\Phi = \frac{\tau_1(t)}{2\mu_2}$ in (6.3.1) (we can do it due to Remark 6.5.1) and $\varphi = u_1(t)$ in (6.6.8) at almost all $t \in [0, T]$, and add the results, taking into account (6.3.7):

$$\begin{aligned}&\Big\langle\frac{d}{dt}u_2, u_1\Big\rangle + \frac{1}{2\mu_2}\Big\langle\frac{d}{dt}\tau_2, \tau_1\Big\rangle + \frac{1}{2\lambda_1\mu_2}(\tau_2, \tau_1) - \frac{1}{2\mu_2}\sum_{i=1}^n \Big(u_{2i}\tau_2, \frac{\partial \tau_1}{\partial x_i}\Big)\\ &\qquad - (\nabla u_2, \tau_1) + \mu_1(\nabla u_2, \nabla u_1) + \sum_{i=1}^n \Big(u_{2i}\frac{\partial u_2}{\partial x_i}, u_1\Big) + (\tau_2, \nabla u_1)\\ &= \langle f, u_1\rangle.\end{aligned} \tag{6.6.9}$$

Inequality (2.1.25) implies

$$\|u_2\|_{L_{8/3}(0,T;L_4)} \le 2^{1/2}\|u_2\|^{\frac14}_{L_\infty(0,T;L_2)}\|u_2\|^{\frac34}_{L_2(0,T;V)} < +\infty.$$

But $\nabla u_2 \in L_2(0, T; L_2)$, so, by (2.2.1), $u_{2i}\frac{\partial u_2}{\partial x_i} \in L_{8/7}(0, T; L_{4/3})$. Then (6.6.8) yields $\frac{du_2}{dt} \in L_{8/7}(0, T; L_{4/3}) + L_2(0, T; V^*)$. Therefore

$$u_2 = u_{21} + u_{22}, \qquad \frac{du_{21}}{dt} \in L_{8/7}(0, T; L_{4/3}), \qquad \frac{du_{22}}{dt} \in L_2(0, T; V^*).$$

Similarly $u_{1i}\frac{\partial\tau_1}{\partial x_i} \in L_{8/3}(0, T; L_2)$ and $\frac{d\tau_1}{dt} \in L_2(0, T; L_2)$, $u_{2i}\tau_2 \in L_{8/3}(0, T; L_{4/3})$ and $\frac{d\tau_2}{dt} \in L_2(0, T; W^{-1}_{4/3})$. Hence, all terms of equalities (6.6.7) and (6.6.9) are integrable on $(0, T)$.

Let us check the following formula:

$$\Big\langle\frac{d}{dt}u_2,u_1\Big\rangle+\frac{1}{2\mu_2}\Big\langle\frac{d}{dt}\tau_2,\tau_1\Big\rangle+\Big\langle\frac{d}{dt}u_1,u_2\Big\rangle+\frac{1}{2\mu_2}\Big\langle\frac{d}{dt}\tau_1,\tau_2\Big\rangle$$
$$=\frac{d}{dt}\Big[(u_1,u_2)+\frac{1}{2\mu_2}(\tau_1,\tau_2)\Big]. \tag{6.6.10}$$

We are going to use the time averaging. Let

$$\psi_h(t)=\frac{1}{h}\int_{\frac{T-h}{T}t}^{\frac{T-h}{T}t+h}\psi(s)\,ds,$$

where ψ is a function of scalar argument with values in a Banach space X, and h is a small positive parameter (cf. (2.2.7)). It is easy to see that if $\psi\in L_p(0,T;X)$, then $\psi_h\in W_p^1(0,T;X)$, and $\psi_h\to\psi$ in $L_p(0,T;X)$ at $h\to 0$. Besides, this operation of averaging commutes with the derivation with respect to t. Therefore, if $\psi\in W_p^1(0,T;X)$, then $\psi_h\in W_p^2(0,T;X)$ and $\psi_h\to\psi$ in $W_p^1(0,T;X)$ at $h\to 0$.

Now, observe that

$$\Big\langle\frac{d}{dt}u_{2h},u_{1h}\Big\rangle+\frac{1}{2\mu_2}\Big\langle\frac{d}{dt}\tau_{2h},\tau_{1h}\Big\rangle+\Big\langle\frac{d}{dt}u_{1h},u_{2h}\Big\rangle+\frac{1}{2\mu_2}\Big\langle\frac{d}{dt}\tau_{1h},\tau_{2h}\Big\rangle$$
$$=\frac{d}{dt}\Big[(u_{1h},u_{2h})+\frac{1}{2\mu_2}(\tau_{1h},\tau_{2h})\Big]. \tag{6.6.11}$$

We have that $u_{1h}\to u_1$ in $L_8(0,T;L_4)$ and in $L_2(0,T;V)$. Furthermore, $u_2=u_{21}+u_{22}$, $\frac{du_{21h}}{dt}\to\frac{du_{21}}{dt}$ in $L_{8/7}(0,T;L_{4/3})$, $\frac{du_{22h}}{dt}\to\frac{du_{22}}{dt}$ in $L_2(0,T;V^*)$. Therefore, almost all on $(0,T)$,

$$\Big\langle\frac{d}{dt}u_{2h},u_{1h}\Big\rangle=\Big\langle\frac{d}{dt}u_{21h},u_{1h}\Big\rangle+\Big\langle\frac{d}{dt}u_{22h},u_{1h}\Big\rangle$$
$$\underset{h\to\infty}{\to}\Big\langle\frac{d}{dt}u_{21},u_1\Big\rangle+\Big\langle\frac{d}{dt}u_{22},u_1\Big\rangle=\Big\langle\frac{d}{dt}u_2,u_1\Big\rangle.$$

Similarly, the remaining terms of the left-hand side of (6.6.11) converge to the corresponding terms of the left-hand side of (6.6.10). Besides,

$$(u_{1h},u_{2h})+\frac{1}{2\mu_2}(\tau_{1h},\tau_{2h})\to(u_1,u_2)+\frac{1}{2\mu_2}(\tau_1,\tau_2).$$

Passing to the limit as $h\to\infty$ in (6.6.11) in the sense distributions on $(0,T)$, we obtain (6.6.10).

Let $w=u_1-u_2,\sigma_*=\tau_1-\tau_2$. Observe that

$$-\sum_{i=1}^n\Big(u_{1i}u_1,\frac{\partial u_2}{\partial x_i}\Big)+\sum_{i=1}^n\Big(u_{2i}\frac{\partial u_2}{\partial x_i},u_1\Big)=-\sum_{i=1}^n\Big(w_i\frac{\partial u_2}{\partial x_i},u_1\Big)$$
$$=\sum_{i=1}^n\Big(w_i\frac{\partial w}{\partial x_i},u_1\Big)-\sum_{i=1}^n\Big(w_i\frac{\partial u_1}{\partial x_i},u_1\Big).$$

Due to (6.1.18) the last term vanishes. Thus,

$$-\sum_{i=1}^{n}\left(u_{1i}u_1, \frac{\partial u_2}{\partial x_i}\right) + \sum_{i=1}^{n}\left(u_{2i}\frac{\partial u_2}{\partial x_i}, u_1\right) = \sum_{i=1}^{n}\left(w_i\frac{\partial w}{\partial x_i}, u_1\right). \tag{6.6.12}$$

Besides,

$$\begin{aligned}\sum_{i=1}^{n}\left(u_{1i}\frac{\partial \tau_1}{\partial x_i}, \tau_2\right) - \sum_{i=1}^{n}\left(u_{2i}\tau_2, \frac{\partial \tau_1}{\partial x_i}\right) &= \sum_{i=1}^{n}\left(w_i\tau_2, \frac{\partial \tau_1}{\partial x_i}\right)\\ &= -\sum_{i=1}^{n}\left(w_i\sigma_*, \frac{\partial \tau_1}{\partial x_i}\right) + \sum_{i=1}^{n}\left(w_i\tau_1, \frac{\partial \tau_1}{\partial x_i}\right).\end{aligned}$$

By (6.1.19) the last term is zero. Thus,

$$\sum_{i=1}^{n}\left(u_{1i}\frac{\partial \tau_1}{\partial x_i}, \tau_2\right) - \sum_{i=1}^{n}\left(u_{2i}\tau_2, \frac{\partial \tau_1}{\partial x_i}\right) = -\sum_{i=1}^{n}\left(w_i\sigma_*, \frac{\partial \tau_1}{\partial x_i}\right). \tag{6.6.13}$$

Now, adding (6.6.7) and (6.6.9), integrating from 0 to any $t \in [0, T]$, and using (6.6.10), (6.6.12), and (6.6.13), we obtain the equality:

$$\begin{aligned}&(u_1, u_2)(t) + \frac{1}{2\mu_2}(\tau_1, \tau_2)(t) + \sum_{i=1}^{n}\int_0^t \left(w_i\frac{\partial w}{\partial x_i}, u_1\right) ds\\ &\quad - \frac{1}{2\mu_2}\sum_{i=1}^{n}\int_0^t \left(w_i\sigma_*, \frac{\partial \tau_1}{\partial x_i}\right) ds + \int_0^t \frac{1}{\lambda_1\mu_2}(\tau_1, \tau_2)\, ds\\ &\quad + 2\int_0^t \mu_1(u_1, u_2)_Y\, ds\\ &= \|a\|^2 + \frac{1}{2\mu_2}\|\tau_0\|^2 + \int_0^t \langle f(s), u_1(s) + u_2(s)\rangle\, ds.\end{aligned} \tag{6.6.14}$$

But (u_2, τ_2) satisfies inequality (6.6.2):

$$\begin{aligned}&\frac{1}{2}\|u_2\|^2(t) + \frac{1}{4\mu_2}\|\tau_2\|^2(t) + \int_0^t \frac{1}{2\lambda_1\mu_2}\|\tau_2\|^2\, ds + \int_0^t \mu_1\|u_2\|_Y^2\, ds\\ &\quad \le \frac{1}{2}\|a\|^2 + \frac{1}{4\mu_2}\|\tau_0\|^2 + \int_0^t \langle f(s), u_2(s)\rangle\, ds.\end{aligned} \tag{6.6.15}$$

Adding it with (6.6.4), and subtracting (6.6.14), we get:

$$\begin{aligned}&\frac{1}{2}\|w\|^2(t) + \frac{1}{4\mu_2}\|\sigma_*\|^2(t) + \int_0^t \frac{1}{2\lambda_1\mu_2}\|\sigma_*\|^2\, ds + \int_0^t \mu_1\|w\|_Y^2\, ds\\ &\quad \le \sum_{i=1}^{n}\int_0^t \left(w_i\frac{\partial w}{\partial x_i}, u_1\right) ds - \frac{1}{2\mu_2}\sum_{i=1}^{n}\int_0^t \left(w_i\sigma_*, \frac{\partial \tau_1}{\partial x_i}\right) ds.\end{aligned} \tag{6.6.16}$$

Applying Hölder's inequality, we obtain:

$$\begin{aligned}&\frac{1}{2}\|w\|^2(t)+\frac{1}{4\mu_2}\|\sigma_*\|^2(t)+\int_0^t\frac{1}{2\lambda_1\mu_2}\|\sigma_*\|^2\,ds+\int_0^t\mu_1\|w\|_Y^2\,ds\\&\quad\le\int_0^t\Big(\|w\|_{L_4}\|w\|_Y\|u_1\|_{L_4}+\frac{1}{2\mu_2}\|w\|_{L_4}\|\sigma_*\|\|\tau_1\|_{W_4^1}\Big)\,ds.\end{aligned}\tag{6.6.17}$$

We recall Young's inequality for scalars: for any $\varepsilon>0$ and $p,q>1$ such that $\frac{1}{p}+\frac{1}{q}=1$, there is a constant C such that for all $a,b\ge 0$ one has

$$ab\le\varepsilon a^p+Cb^q.$$

But, by (2.1.24),

$$\|w\|_{L_4}\|w\|_Y\|u_1\|_{L_4}\le 2^{1/2}\|w\|^{1/4}\|w\|_Y^{7/4}\|u_1\|_{L_4}.$$

Due to Young's inequality there is a constant K_1 such that the last expression does not exceed $\frac{1}{2}\mu_1\|w\|_Y^2+K_1\|w\|^2\|u_1\|_{L_4}^8$.

Furthermore, (2.1.24) and Young's inequality give

$$\begin{aligned}&\|w\|_{L_4}\|\sigma_*\|\|\tau_1\|_{W_4^1}\\&\quad\le\|w\|^{1/4}\|\sigma_*\|^{1/4}\|w\|_Y^{3/4}\|\sigma_*\|^{3/4}\|\tau_1\|_{W_4^1}\\&\quad\le\sqrt{\frac{\mu_1}{\lambda_1\mu_2}}\|w\|_Y\|\sigma_*\|+K_2\|w\|\|\sigma_*\|\|\tau_1\|_{W_4^1}^4\\&\quad\le\frac{1}{2}\mu_1\|w\|_Y^2+\frac{1}{2\lambda_1\mu_2}\|\sigma_*\|^2+\frac{1}{2}K_2(\|w\|^2+\|\sigma_*\|^2)\|\tau_1\|_{W_4^1}^4.\end{aligned}$$

Then (6.6.17) implies:

$$\begin{aligned}&\frac{1}{2}\|w\|^2(t)+\frac{1}{4\mu_2}\|\sigma_*\|^2(t)+\int_0^t\frac{1}{2\lambda_1\mu_2}\|\sigma_*\|^2\,ds+\int_0^t\mu_1\|w\|_Y^2\,ds\\&\quad\le\int_0^t\Big(\mu_1\|w\|_Y^2+\frac{1}{2\lambda_1\mu_2}\|\sigma_*\|^2\\&\qquad+\frac{1}{2}K_2(\|w\|^2+\|\sigma_*\|^2)\|\tau_1\|_{W_4^1}^4+K_1\|w\|^2\|u_1\|_{L_4}^8\Big)\,ds.\end{aligned}$$

Therefore

$$\begin{aligned}&\frac{1}{2}\|w\|^2(t)+\frac{1}{4\mu_2}\|\sigma_*\|^2(t)\\&\quad\le\int_0^t\Big(\frac{1}{2}K_2(\|w\|^2+\|\sigma_*\|^2)\|\tau_1\|_{W_4^1}^4+K_1\|w\|^2\|u_1\|_{L_4}^8\Big)\,ds.\end{aligned}\tag{6.6.18}$$

But the functions $\|u_1\|_{L_4}^8(t)$ and $\|\tau_1\|_{W_4^1}^4(t)$ are integrable on $(0,T)$. Then, by the Gronwall lemma, $w\equiv 0$, $\sigma_*\equiv 0$, and the theorem is proved. □

In a similar way one can prove another uniqueness result:

Theorem 6.6.2. *Assume that $n = 3$ and, under the conditions of Theorem 6.2.1, there are two weak solutions (u_1, σ_1), (u_2, σ_2) to problem (6.2.1) – (6.2.5). If*

$$\begin{aligned} u_j &\in L_8(0, T; L_4(\Omega)^n), \\ \tau_j = [\sigma_j - 2\mu_1 \mathcal{E}(u_j)] &\in L_4(0, T; W_4^1(\Omega, \mathbb{R}_S^{n\times n})), \quad j = 1, 2, \end{aligned} \tag{6.6.19}$$

then $u_1 = u_2, \sigma_1 = \sigma_2$.

Proof. Let $w = u_1 - u_2, \sigma_* = \tau_1 - \tau_2$. Note that (6.6.19) implies

$$\frac{d}{dt}(\tau_j, \Phi) + \frac{1}{\lambda_1}(\tau_j, \Phi) + \sum_{i=1}^{n} \Big(u_{ji}\frac{\partial \tau_j}{\partial x_i}, \Phi\Big) - 2\mu_2(\mathcal{E}(u_j), \Phi) = 0 \tag{6.6.20}$$

for all $\Phi \in L_2$ (cf. (6.6.6)). Furthermore, by Lemma 2.2.8, the pairs (u_j, τ_j) satisfy (6.3.2) for all $\varphi \in V$ almost all on $(0, T)$.

Put $\Phi = \dfrac{\tau_j(t)}{2\mu_2}$ in (6.6.20) and $u(t) = \varphi = u_j(t)$ in (6.3.2) at almost all $t \in [0, T]$, and add the results:

$$\frac{1}{2}\frac{d}{dt}(u_j, u_j) + \frac{1}{4\mu_2}\frac{d}{dt}(\tau_j, \tau_j) + \mu_1(\nabla u_j, \nabla u_j) + \frac{1}{2\lambda_1\mu_2}(\tau_j, \tau_j) = \langle f, u_j\rangle. \tag{6.6.21}$$

Integrating this equality from 0 to arbitrary $t \in [0, T]$, we obtain the energy equality (cf. (6.6.5)):

$$\begin{aligned} &\frac{1}{2}\|u_j\|^2(t) + \frac{1}{4\mu_2}\|\tau_j\|^2(t) + \int_0^t \frac{1}{2\lambda_1\mu_2}\|\tau_j\|^2\,ds + \int_0^t \mu_1\|u_j\|_Y^2\,ds \\ &\quad = \frac{1}{2}\|a\|^2 + \frac{1}{4\mu_2}\|\tau_0\|^2 + \int_0^t \langle f(s), u_j(s)\rangle\,ds. \end{aligned} \tag{6.6.22}$$

Taking $j = 1, \Phi = \frac{\tau_2(t)}{2\mu_2}$ in (6.6.20) and $u = u_1, \varphi = u_2(t)$ in (6.3.2) at almost all $t \in [0, T]$, and adding the results, we get (6.6.7). Taking $j = 2, \Phi = \frac{\tau_1(t)}{2\mu_2}$ in (6.6.20) and $u = u_2, \varphi = u_1(t)$ in (6.3.2) at almost all $t \in [0, T]$, and adding the results, we get (6.6.9). From (6.6.7) and (6.6.9) we obtain (6.6.14).

Adding (6.6.22) for $j = 1, 2$, and subtracting (6.6.14), we get (6.6.16), and, hence, (6.6.18). □

6.7 Minimal trajectory and global attractors for the Jeffreys model

6.7.1 Integral energy estimate: autonomous case

In Sections 6.7 and 6.8 we study the attractors for weak solutions of the Jeffreys model. In these two sections we assume that Ω is an arbitrary bounded domain in $\mathbb{R}^n$, $n = 2, 3$ (in this situation the spaces V and Y coincide up to equivalent norm). According to Remark 6.3.1, we investigate here problem (6.3.1), (6.3.2). First we have to derive an integral energy estimate for this problem.

Theorem 6.7.1 (see [76]). *Let $f \in V^*$. Given $a \in H$, $\tau_0 \in L_2(\Omega, \mathbb{R}^{n\times n}_S)$, the solution (u, τ) to problem (6.3.1), (6.3.2), (6.3.5) constructed in Theorem 6.5.1 satisfies the energy inequality:*

$$\frac{1}{4}\|u\|^2_{L_\infty(t,t+1;H)} + \frac{1}{8\mu_2}\|\tau\|^2_{L_\infty(t,t+1;L_2)} + \frac{\mu_1}{2}\|u\|^2_{L_2(t,t+1;Y)} \leq e^{-2\gamma t}\left(\|a\|^2 + \frac{1}{2\mu_2}\|\tau_0\|^2\right) + \frac{\gamma+1}{2\mu_1\gamma}\|f\|^2_{Y^*} \tag{6.7.1}$$

for $t \in [0, T-1]$. Here $\gamma = \min(\frac{1}{\lambda_1}, \frac{\mu_1}{2K_0(\Omega)^2})$, where K_0 is the constant from Friedrichs' inequality (2.1.26).

Proof. Consider the pairs (u_k, τ_k), $k \in \mathbb{N}$ which were used in the proof of Theorem 6.5.1 for the construction of the solution. Each of them, in turn, was constructed in Theorem 6.4.1 using the sequences of solutions $(u_{m,k}, \tau_{m,k})$ to problem (6.3.3), (6.3.4) with $\delta = \frac{1}{m}$ and $\varepsilon = \frac{1}{k}$. It suffices to show that these solutions satisfy (6.7.1).

But these solutions $(u_{m,k}, \tau_{m,k})$ satisfy (6.3.8):

$$\frac{1}{2}\frac{d}{dt}(u_{m,k}, u_{m,k}) + \frac{1}{4\mu_2}\frac{d}{dt}(\tau_{m,k}, \tau_{m,k}) + \mu_1(\nabla u_{m,k}, \nabla u_{m,k}) + \frac{1}{2\lambda_1\mu_2}(\tau_{m,k}, \tau_{m,k}) + \frac{1}{2k\lambda_1\mu_2}(\nabla\tau_{m,k}, \nabla\tau_{m,k}) = \langle f, u_{m,k}\rangle. \tag{6.7.2}$$

Let $\overline{u}_{m,k}(t) = e^{\gamma t}u_{m,k}(t)$, $\overline{\tau}_{m,k}(t) = e^{\gamma t}\tau_{m,k}(t)$ where $t \in [0, T]$. Then we have:

$$\frac{1}{2}\frac{d}{dt}e^{-2\gamma t}\|\overline{u}_{m,k}\|^2 + \frac{1}{4\mu_2}\frac{d}{dt}e^{-2\gamma t}\|\overline{\tau}_{m,k}\|^2 + \mu_1 e^{-2\gamma t}\|\overline{u}_{m,k}\|^2_Y + \frac{1}{2\lambda_1\mu_2}e^{-2\gamma t}\|\overline{\tau}_{m,k}\|^2 + \frac{1}{2k\lambda_1\mu_2}e^{-2\gamma t}\|\overline{\tau}_{m,k}\|^2_Y = \langle f, e^{-\gamma t}\overline{u}_{m,k}\rangle.$$

Applying the formula of derivative of a product and multiplying by $e^{2\gamma t}$ we obtain

$$\begin{aligned}\frac{1}{2}\frac{d}{dt}\|\overline{u}_{m,k}\|^2 - \gamma\|\overline{u}_{m,k}\|^2 + \frac{1}{4\mu_2}\frac{d}{dt}\|\overline{\tau}_{m,k}\|^2 - \frac{\gamma}{2\mu_2}\|\overline{\tau}_{m,k}\|^2 \\ + \mu_1\|\overline{u}_{m,k}\|_Y^2 + \frac{1}{2\lambda_1\mu_2}\|\overline{\tau}_{m,k}\|^2 + \frac{1}{2k\lambda_1\mu_2}\|\overline{\tau}_{m,k}\|_Y^2 = \langle e^{\gamma t} f, \overline{u}_{m,k}\rangle.\end{aligned} \tag{6.7.3}$$

Since $\|\overline{u}_{m,k}\| \le K_0(\Omega)\|\overline{u}_{m,k}\|_Y$, (6.7.3) implies

$$\frac{1}{2}\frac{d}{dt}\|\overline{u}_{m,k}\|^2 + \frac{1}{4\mu_2}\frac{d}{dt}\|\overline{\tau}_{m,k}\|^2 + \frac{\mu_1}{2}\|\overline{u}_{m,k}\|_Y^2 \le e^{\gamma t}\|f\|_{Y^*}\|\overline{u}_{m,k}\|_Y.$$

Applying Cauchy's inequality we get

$$\frac{1}{2}\frac{d}{dt}\|\overline{u}_{m,k}\|^2 + \frac{1}{4\mu_2}\frac{d}{dt}\|\overline{\tau}_{m,k}\|^2 + \frac{\mu_1}{2}\|\overline{u}_{m,k}\|_Y^2 \le \frac{e^{2\gamma t}}{2\mu_1}\|f\|_{Y^*}^2 + \frac{\mu_1}{2}\|\overline{u}_{m,k}\|_Y^2,$$

that is

$$\frac{1}{2}\frac{d}{dt}\|\overline{u}_{m,k}\|^2 + \frac{1}{4\mu_2}\frac{d}{dt}\|\overline{\tau}_{m,k}\|^2 \le \frac{e^{2\gamma t}}{2\mu_1}\|f\|_{Y^*}^2.$$

Integrating from 0 to t we obtain

$$\frac{1}{2}\|\overline{u}_{m,k}\|^2 + \frac{1}{4\mu_2}\|\overline{\tau}_{m,k}\|^2 \le \frac{1}{2}\|a\|^2 + \frac{1}{4\mu_2}\|\tau_0\|^2 + \frac{1}{2\mu_1}\|f\|_{Y^*}^2\int_0^t e^{2\gamma s}\,ds.$$

This implies

$$\frac{1}{2}\|e^{\gamma t}u_{m,k}\|^2 + \frac{1}{4\mu_2}\|e^{\gamma t}\tau_{m,k}\|^2 \le \frac{1}{2}\|a\|^2 + \frac{1}{4\mu_2}\|\tau_0\|^2 + \frac{1}{4\gamma\mu_1}\|f\|_{Y^*}^2(e^{2\gamma t}-1).$$

Multiplying by $2e^{-2\gamma t}$ and taking the maximum along the interval $(t, t+1)$ we conclude

$$\begin{aligned}\max_{s\in(t,t+1)}\Big(\|u_{m,k}(s)\|^2 + \frac{1}{2\mu_2}\|\tau_{m,k}(s)\|^2\Big) \\ \le e^{-2\gamma t}\big(\|a\|^2 + \frac{1}{2\mu_2}\|\tau_0\|^2\big) + \frac{1}{2\gamma\mu_1}\|f\|_{Y^*}^2.\end{aligned} \tag{6.7.4}$$

Applying to the right-hand side of (6.7.2) Cauchy's inequality we get:

$$\begin{aligned}\frac{1}{2}\frac{d}{dt}\|u_{m,k}\|^2 + \frac{1}{4\mu_2}\frac{d}{dt}\|\tau_{m,k}\|^2 + \mu_1\|u_{m,k}\|_Y^2 + \frac{1}{2\lambda_1\mu_2}\|\tau_{m,k}\|^2 \\ + \frac{1}{2k\lambda_1\mu_2}\|\tau_{m,k}\|_Y^2 \le \frac{1}{2\mu_1}\|f\|_{Y^*}^2 + \frac{\mu_1}{2}\|u_{m,k}\|_Y^2.\end{aligned}$$

Hence,

$$\frac{1}{2}\frac{d}{dt}\|u_{m,k}\|^2 + \frac{1}{4\mu_2}\frac{d}{dt}\|\tau_{m,k}\|^2 + \frac{\mu_1}{2}\|u_{m,k}\|_Y^2 + \frac{1}{2k\lambda_1\mu_2}\|\tau_{m,k}\|_Y^2 \le \frac{1}{2\mu_1}\|f\|_{Y^*}^2.$$

Integrating from t to $t+1$ we obtain

$$\frac{1}{2}\|u_{m,k}(t+1)\|^2 - \frac{1}{2}\|u_{m,k}(t)\|^2 + \frac{1}{4\mu_2}\|\tau_{m,k}(t+1)\|^2 - \frac{1}{4\mu_2}\|\tau_{m,k}(t)\|^2 + \frac{\mu_1}{2}\int_t^{t+1}\|u_{m,k}(s)\|_Y^2\,ds + \frac{1}{2k\lambda_1\mu_2}\int_t^{t+1}\|\tau_{m,k}(s)\|_Y^2\,ds \le \frac{1}{2\mu_1}\|f\|_{Y^*}^2. \tag{6.7.5}$$

Adding (6.7.4) and (6.7.5) we get:

$$\begin{aligned}&\frac{1}{2}\|u_{m,k}(t+1)\|^2 + \frac{1}{4\mu_2}\|\tau_{m,k}(t+1)\|^2 \\ &\quad + \max_{s\in(t,t+1)}\Big(\frac{1}{2}\|u_{m,k}(s)\|^2 + \frac{1}{4\mu_2}\|\tau_{m,k}(s)\|^2\Big) \\ &\quad + \frac{\mu_1}{2}\int_t^{t+1}\|u_{m,k}(s)\|_Y^2\,ds + \frac{1}{2k\lambda_1\mu_2}\int_t^{t+1}\|\tau_{m,k}(s)\|_Y^2\,ds \\ &\le e^{-2\gamma t}\big(\|a\|^2 + \frac{1}{2\mu_2}\|\tau_0\|^2\big) + \frac{1+\gamma}{2\gamma\mu_1}\|f\|_{Y^*}^2, \quad 0 \le t \le T-1.\end{aligned} \tag{6.7.6}$$

Taking into account inequality (6.3.9) and ignoring the first two positive terms in the left-hand side we obtain

$$\begin{aligned}&\frac{1}{4}\max_{s\in(t,t+1)}\|u_{m,k}(s)\|^2 + \frac{1}{8\mu_2}\max_{s\in(t,t+1)}\|\tau_{m,k}(s)\|^2 \\ &\quad + \frac{\mu_1}{2}\int_t^{t+1}\|u_{m,k}(s)\|_Y^2\,ds + \frac{1}{2k\lambda_1\mu_2}\int_t^{t+1}\|\tau_{m,k}(s)\|_Y^2\,ds \\ &\le e^{-2\gamma t}\big(\|a\|^2 + \frac{1}{2\mu_2}\|\tau_0\|^2\big) + \frac{1+\gamma}{2\gamma\mu_1}\|f\|_{Y^*}^2, \ 0 \le t \le T-1,\end{aligned} \tag{6.7.7}$$

which is even stronger than (6.7.1). □

Let us also state the following

Lemma 6.7.1. *Let $f \in V^*$. Assume that a pair (u, τ), where*

$$u \in L_2(0,T;V)\bigcap L_\infty(0,T;H), \ \tau \in L_\infty(0,T;L_2), \tag{6.7.8}$$

satisfies either identities (6.3.1), (6.3.2) or identities (6.3.3), (6.3.4) almost everywhere on $(0, T)$ *for all* $\varphi \in \mathcal{V}$ *and* $\Phi \in C_0^\infty$. *Then for all* $t \in [0, T]$ *the following estimate is valid:*

$$\begin{aligned}&\|u'\|_{L_{4/3}(t,T;Y^*)} + \|\tau'\|_{L_2(t,T;H^{-2})}\\&\quad\le K_1(\|u\|_{L_\infty(t,T;H)}, \|u\|_{L_2(t,T;Y)}, \|\tau\|_{L_\infty(t,T;L_2)}, \|f\|_{Y^*}, T-t, \Omega).\end{aligned}$$

A more general variant of this lemma will be proved in Section 6.8.

6.7.2 Existence and structure of attractors

In this subsection we construct the minimal trajectory attractor and the global attractor for problem (6.3.1), (6.3.2) in the autonomous case ($f \in V^*$). We are going to apply the abstract results of Section 4.2. Let us choose $H \times L_2(\Omega, \mathbb{R}_S^{n\times n})$ as the space E and the space $V_\delta^* \times H^{-\delta}(\Omega, \mathbb{R}_S^{n\times n})$ as the space E_0, where $\delta \in (0, 1]$ is a fixed number. We have also to define the trajectory space $\mathcal{H}^+$ for the Jeffreys model. This will be the set of pairs of functions (u, τ) which

i) belong to the class

$$u \in L_{2,\mathrm{loc}}(0, +\infty; V) \bigcap L_\infty(0, +\infty; H), \ \ \tau \in L_\infty(0, +\infty; L_2); \tag{6.7.9}$$

ii) satisfy identities (6.3.1), (6.3.2) for almost everywhere $t \in (0, +\infty)$ and for all $\varphi \in \mathcal{V}$ and $\Phi \in C_0^\infty$;

iii) satisfy the energy inequality:

$$\begin{aligned}&\frac{1}{4}\|u\|^2_{L_\infty(t,t+1;H)} + \frac{1}{8\mu_2}\|\tau\|^2_{L_\infty(t,t+1;L_2)} + \frac{\mu_1}{2}\|u\|^2_{L_2(t,t+1;Y)}\\&\quad\le e^{-2\gamma t}\big(\|u\|^2_{L_\infty(0,+\infty;H)} + \frac{1}{2\mu_2}\|\tau\|^2_{L_\infty(0,+\infty;L_2)}\big) + \frac{\gamma+1}{2\mu_1\gamma}\|f\|^2_{Y^*}\end{aligned} \tag{6.7.10}$$

for all $t \geq 0$ where γ is as in Theorem 6.7.1.

Remark 6.7.1. In Section 4.2 it was supposed that $\mathcal{H}^+ \subset C([0, +\infty); E_0) \cap L_\infty(0, +\infty; E)$. Let us show that this condition holds for the Jeffreys model. In fact, every pair (u, τ) from $\mathcal{H}^+$ belongs to $L_\infty(0, +\infty; H \times L_2) = L_\infty(0, +\infty; E)$. For any $T \geq 0$ the function $u \in L_2(0, T; V)$. By Lemma 6.7.1 $u' \in L_{4/3}(0, T; V^*)$, $\tau' \in L_2(0, T; H^{-2})$. Since Ω is bounded, $H_0^\delta \subset L_2$ compactly. Therefore $V_\delta \subset H$ compactly, $L_2 \subset H^{-\delta}$ compactly, $H \subset V_\delta^*$ compactly. By Theorem 2.2.6 one has: $u \in C([0, T]; V_\delta^*)$, $\tau \in C([0, T]; H^{-\delta})$, that is $(u, \tau) \in C([0, T]; E_0)$ for any $T \geq 0$.

Remark 6.7.2. On account of inequality (6.7.10) the trajectory space $\mathcal{H}^+$ for the Jeffreys model is not invariant with respect to the translation operator $T(h)$.

Theorem 6.7.2 (see [75, 76]). *Let $f \in V^*$. Given $a \in H$, $\tau_0 \in L_2(\Omega, \mathbb{R}_S^{n\times n})$, there is a pair of functions (a trajectory) $(u, \tau) \in \mathcal{H}^+$ satisfying initial condition (6.3.5).*

Proof. Take an increasing sequence of positive numbers $T_m \to \infty$. By Theorem 6.5.1, for every natural m there is a pair (u_m, τ_m) in class (6.5.1) which satisfies (6.3.1), (6.3.2) almost everywhere in $(0, T_m)$ as well as the initial condition (6.3.5). Denote by $\widetilde{u}_m$ and $\widetilde{\tau}_m$ the functions which are equal to u_m and τ_m in $[0, T_m]$ and are equal to zero on $(T_m, +\infty)$. Since, by Theorem 6.7.1, all pairs (u_m, τ_m) satisfy inequality (6.7.1), the sequences $\widetilde{u}_m$ and $\widetilde{\tau}_m$ are bounded in $L_\infty(0, +\infty; H)$ and $L_\infty(0, +\infty; L_2)$, respectively. Thus, without loss of generality one may assume that there exist limits

$$u = \lim_{m\to\infty} \widetilde{u}_m, \quad \text{which is } *\text{-weak in} L_\infty(0, +\infty; H);$$

$$\tau = \lim_{m\to\infty} \widetilde{\tau}_m, \quad \text{which is } *\text{-weak in} L_\infty(0, +\infty; L_2).$$

Fix an arbitrary interval $[0, T]$. Then estimate (6.7.1) for all pairs (u_m, τ_m) implies the uniform boundedness of the sequence $\{\widetilde{u}_m\}$ in $L_2(0, T; V)$. Therefore without loss of generality we may consider that $\widetilde{u}_m \rightharpoonup u$ weakly in $L_2(0, T; V)$. It is easy to see that the pair (u, τ) satisfies inequality (6.7.1) (and, hence, (6.7.10)) and for $T_m \geq T$ the functions $\widetilde{u}_m$ and $\widetilde{\tau}_m$ coincide with u_m and τ_m on the segment $[0, T]$.

Just as in the proof of Theorems 6.4.1 and 6.5.1, estimate (6.7.1) for all pairs (u_m, τ_m) together with Lemma 6.7.1 ensure the convergence of all terms in identities (6.3.1), (6.3.2) with (u_m, τ_m) substituted there (m should be sufficiently large: so that $T_m \geq T$) to the corresponding terms in (6.3.1), (6.3.2) with (u, τ) in the sense of scalar distributions on $(0, T)$ as $m \to \infty$. Therefore the pair (u, τ) satisfies identities (6.3.1), (6.3.2) almost everywhere on $(0, T)$ for all $\varphi \in \mathcal{V}$ and $\Phi \in C_0^\infty$, belongs to class (6.7.8) and satisfies condition (6.3.5). Since T was arbitrary, the pair (u, τ) satisfies conditions i) – iii) of the definition of $\mathcal{H}^+$. □

The main result of this section is

Theorem 6.7.3 (see [75, 76]). *Let $f \in V^*$. There exists a minimal trajectory attractor $\mathcal{U}_J$ for the trajectory space $\mathcal{H}^+$ and (4.2.6) is fulfilled.*

Proof. It suffices to show existence of a trajectory semiattractor and to apply Theorems 4.2.1 and 4.2.5.

Consider the set P which consists of pairs

$$(u, \tau) \in C([0, +\infty); E_0) \cap L_\infty(0, +\infty; E)$$

satisfying inequalities

$$\frac{1}{4}\|u\|^2_{L_\infty(t,t+1;H)} + \frac{1}{8\mu_2}\|\tau\|^2_{L_\infty(t,t+1;L_2)} + \frac{\mu_1}{2}\|u\|^2_{L_2(t,t+1;Y)} \le \frac{\gamma+1}{\mu_1\gamma}\|f\|^2_{Y^*},$$

$$\|u'\|_{L_{4/3}(t,t+1;Y^*)} + \|\tau'\|_{L_2(t,t+1;H^{-2})}$$
$$\le K_1(\|u\|_{L_\infty(t,t+1;H)}, \|u\|_{L_2(t,t+1;Y)}, \|\tau\|_{L_\infty(t,t+1;L_2)}, \|f\|_{Y^*}, 1, \Omega)$$

for all $t \ge 0$, where K_1 is the constant from Lemma 6.7.1.

The set P is bounded in $L_\infty(0,+\infty;E)$. Let us show that P is relatively compact in $C([0,+\infty);E_0)$.

Fix $\varepsilon > 0$. Let a natural number M be such that $\frac{1}{2^{M-1}} < \varepsilon$. By Theorem 2.2.6 the set $P_M = \{v = u|_{[0,M]} : u \in P\}$ is relatively compact in $C([0,M];E_0)$. Then there is an $\frac{\varepsilon}{4}$-net $\{u_1,\dots,u_k\} \subset P_M$ in $C([0,M];E_0)$ for the set P_M. Let $\widetilde{u}_j(t) = u_j(t), 0 \le t \le M, \widetilde{u}_j(t) = u_j(M), t > M$, $j = 1,\dots,k$. Then for any $u \in P$:

$$\min_{j=1,\dots,k}\|u-\widetilde{u}_j\|_{C([0,+\infty);E_0)} = \min_{j=1,\dots,k}\sum_{i=0}^{+\infty}2^{-i}\frac{\|u-\widetilde{u}_j\|_{C([0,i];E_0)}}{1+\|u-\widetilde{u}_j\|_{C([0,i];E_0)}}$$

$$\le \min_{j=1,\dots,k}\sum_{i=0}^{M}2^{-i}\|u-\widetilde{u}_j\|_{C([0,i];E_0)} + \sum_{i=M+1}^{+\infty}2^{-i} \le 2\cdot\frac{\varepsilon}{4}+\frac{\varepsilon}{2} = \varepsilon.$$

Thus, $\{\widetilde{u}_1,\dots,\widetilde{u}_k\}$ is ε-net for P. Hence, P is relatively compact in $C([0,+\infty);E_0)$.

By (6.7.10) and Lemma 6.7.1 the set P is absorbing for the trajectory space $\mathcal{H}^+$. Furthermore, it is clear that $T(h)P \subset P$ for all $h \ge 0$. By Lemma 4.2.9, $\overline{P}$ (its closure in $C([0,+\infty);E_0)$) is a semiattractor for the trajectory space $\mathcal{H}^+$. □

Remark 6.7.3. Since the set P (and, hence, $\overline{P}$) is absorbing, we could apply Corollary 4.2.1 instead of Theorem 4.2.1.

Theorems 4.2.2 and 6.7.3 imply

Theorem 6.7.4. *Let $f \in V^*$. In the space $H \times L_2$ there is a global attractor $\mathcal{A}_J$ for problem (6.3.1) – (6.3.2), i.e. a minimal compact in $V_\delta^* \times H^{-\delta}$ and bounded in $H \times L_2$ set, which attracts all trajectories from $\mathcal{H}^+$ (see Definition 4.2.6). For all $t \ge 0$ we have*

$$\mathcal{K}(\mathcal{H}^+)(t) \subset \mathcal{A}_J = \mathcal{U}_J(t) = \mathcal{K}(\mathcal{U}_J)(t).$$

Remark 6.7.4. We have established existence of minimal trajectory and global attractors for the space $\mathcal{H}^+$ of solutions for problem (6.3.1) – (6.3.2) on the positive axis, which satisfy the integral energy inequality. At the same time it is not known whether there exist weak solutions of this problem which do not satisfy the energy inequality. Such a problem is open even for the Navier–Stokes system.

6.8 Uniform attractors for the Jeffreys model

6.8.1 Integral energy estimate: non-autonomous case

In this section we are going to investigate attractors of problem (6.3.1), (6.3.2) in the case when the body force depends on time. First let us introduce the space $\mathcal{X}$ which consists of all elements from $L_{2,\mathrm{loc}}(0,+\infty;V^*)$ for which the norm

$$\|\phi\|_{\mathcal{X}} = \sup_{t\geq 0} \|\phi\|_{L_2(t,t+1;Y^*)}$$

is finite.

We have to derive an integral energy estimate in the non-autonomous case.

Theorem 6.8.1 (see [77]). *Let $f \in \mathcal{X}$. Given $a \in H$, $\tau_0 \in L_2(\Omega, \mathbb{R}_S^{n\times n})$, the solution (u,τ) to problem (6.3.1), (6.3.2), (6.3.5) constructed in Theorem 6.5.1 satisfies the energy inequality:*

$$\begin{aligned}\frac{1}{4}\|u\|^2_{L_\infty(t,t+1;H)} &+ \frac{1}{8\mu_2}\|\tau\|^2_{L_\infty(t,t+1;L_2)} + \frac{\mu_1}{2}\|u\|^2_{L_2(t,t+1;Y)} \\ &\leq e^{-2\gamma t}\big(\|a\|^2 + \frac{1}{2\mu_2}\|\tau_0\|^2\big) + \frac{2e^{4\gamma}+e^{2\gamma}-1}{2\mu_1(e^{2\gamma}-1)}\|f\|^2_{\mathcal{X}}\end{aligned} \tag{6.8.1}$$

for $t \in [0, T-1]$ where γ is as in Theorem 6.7.1.

Proof. Consider the pairs (u_k,τ_k), $k \in \mathbb{N}$ which were used in the proof of Theorem 6.5.1 for the construction of the solution. Each of them, in turn, was constructed in Theorem 6.4.1 using the sequences of solutions $(u_{m,k},\tau_{m,k})$ to problem (6.3.3), (6.3.4) with $\delta = \frac{1}{m}$ and $\varepsilon = \frac{1}{k}$. It suffices to show that these solutions satisfy (6.8.1).

As in the proof of Theorem 6.7.1, these solutions $(u_{m,k},\tau_{m,k})$ satisfy (6.7.2). Let $\overline{u}_{m,k}(t) = e^{\gamma t}u_{m,k}(t)$, $\overline{\tau}_{m,k}(t) = e^{\gamma t}\tau_{m,k}(t)$ where $t \in [0,T]$. Then, as in Theorem 6.7.1, we have:

$$\frac{1}{2}\frac{d}{dt}\|\overline{u}_{m,k}\|^2 + \frac{1}{4\mu_2}\frac{d}{dt}\|\overline{\tau}_{m,k}\|^2 \leq \frac{e^{2\gamma t}}{2\mu_1}\|f(t)\|^2_{Y^*}.$$

Integrating from 0 to t we obtain

$$\frac{1}{2}\|\overline{u}_{m,k}\|^2 + \frac{1}{4\mu_2}\|\overline{\tau}_{m,k}\|^2 \leq \frac{1}{2}\|a\|^2 + \frac{1}{4\mu_2}\|\tau_0\|^2 + \frac{1}{2\mu_1}\int_0^t e^{2\gamma s}\|f(s)\|^2_{Y^*}\,ds. \tag{6.8.2}$$

Let us use the following simple inequality (see its proof after the proof of this theorem)

$$\int_0^t a^s\phi(s)\,ds \leq \frac{a^{t+2}}{a-1}\sup_{s\in[0,t-1]}\int_s^{s+1}|\phi(\zeta)|\,d\zeta, \tag{6.8.3}$$

where $a \geq 1$ and $\phi(s)$ is a scalar function. Then from (6.8.2) we get:

$$\begin{aligned}
&\frac{1}{2}\|e^{\gamma t}u_{m,k}\|^2 + \frac{1}{4\mu_2}\|e^{\gamma t}\tau_{m,k}\|^2 \\
&\qquad \leq \frac{1}{2}\|a\|^2 + \frac{1}{4\mu_2}\|\tau_0\|^2 + \frac{e^{2\gamma(t+2)}}{2\mu_1(e^{2\gamma}-1)}\sup_{s\in[0,t-1]}\int_s^{s+1}\|f(\zeta)\|^2_{Y^*}\,d\zeta.
\end{aligned}$$

Multiplying by $2e^{-2\gamma t}$ and taking the maximum along the interval $(t, t+1)$ we conclude

$$\begin{aligned}
&\max_{s\in(t,t+1)}\left(\|u_{m,k}(s)\|^2 + \frac{1}{2\mu_2}\|\tau_{m,k}(s)\|^2\right) \\
&\qquad \leq e^{-2\gamma t}\left(\|a\|^2 + \frac{1}{2\mu_2}\|\tau_0\|^2\right) + \frac{e^{4\gamma}}{\mu_1(e^{2\gamma}-1)}\|f\|^2_{\mathfrak{X}}.
\end{aligned} \tag{6.8.4}$$

As in the proof of Theorem 6.7.1, applying to the right-hand side of (6.7.2) Cauchy's inequality we get:

$$\begin{aligned}
&\frac{1}{2}\frac{d}{dt}\|u_{m,k}\|^2 + \frac{1}{4\mu_2}\frac{d}{dt}\|\tau_{m,k}\|^2 + \frac{\mu_1}{2}\|u_{m,k}\|^2_Y + \frac{1}{2k\lambda_1\mu_2}\|\tau_{m,k}\|^2_Y \\
&\qquad \leq \frac{1}{2\mu_1}\|f(t)\|^2_{Y^*}.
\end{aligned}$$

Integrating from t to $t+1$ we obtain

$$\begin{aligned}
&\frac{1}{2}\|u_{m,k}(t+1)\|^2 - \frac{1}{2}\|u_{m,k}(t)\|^2 + \frac{1}{4\mu_2}\|\tau_{m,k}(t+1)\|^2 - \frac{1}{4\mu_2}\|\tau_{m,k}(t)\|^2 \\
&\quad + \frac{\mu_1}{2}\int_t^{t+1}\|u_{m,k}(s)\|^2_Y\,ds + \frac{1}{2k\lambda_1\mu_2}\int_t^{t+1}\|\tau_{m,k}(s)\|^2_Y\,ds \leq \frac{1}{2\mu_1}\|f\|^2_{\mathfrak{X}}.
\end{aligned} \tag{6.8.5}$$

Adding (6.8.4) and (6.8.5) we get:

$$\begin{aligned}
&\frac{1}{2}\|u_{m,k}(t+1)\|^2 + \frac{1}{4\mu_2}\|\tau_{m,k}(t+1)\|^2 \\
&\qquad + \max_{s\in(t,t+1)}\left(\frac{1}{2}\|u_{m,k}(s)\|^2 + \frac{1}{4\mu_2}\|\tau_{m,k}(s)\|^2\right) \\
&\qquad + \frac{\mu_1}{2}\int_t^{t+1}\|u_{m,k}(s)\|^2_Y\,ds + \frac{1}{2k\lambda_1\mu_2}\int_t^{t+1}\|\tau_{m,k}(s)\|^2_Y\,ds \\
&\quad \leq e^{-2\gamma t}\left(\|a\|^2 + \frac{1}{2\mu_2}\|\tau_0\|^2\right) + \frac{2e^{4\gamma}+e^{2\gamma}-1}{2\mu_1(e^{2\gamma}-1)}\|f\|^2_{\mathfrak{X}},
\end{aligned} \tag{6.8.6}$$

$0 \leq t \leq T-1$.

Taking into account inequality (6.3.9) and ignoring the first two positive terms in the left-hand side we obtain

$$\begin{aligned}\frac{1}{4}\max_{s\in(t,t+1)}\|u_{m,k}(s)\|^2 &+ \frac{1}{8\mu_2}\max_{s\in(t,t+1)}\|\tau_{m,k}(s)\|^2\\ &+\frac{\mu_1}{2}\int_t^{t+1}\|u_{m,k}(s)\|_Y^2\,ds+\frac{1}{2k\lambda_1\mu_2}\int_t^{t+1}\|\tau_{m,k}(s)\|_Y^2\,ds\\ &\le e^{-2\gamma t}\Big(\|a\|^2+\frac{1}{2\mu_2}\|\tau_0\|^2\Big)+\frac{2e^{4\gamma}+e^{2\gamma}-1}{2\mu_1(e^{2\gamma}-1)}\|f\|_{\mathcal{X}}^2,\\ 0\le t\le T-1,&\end{aligned}\tag{6.8.7}$$

which yields (6.8.1). □

Proof of inequality (6.8.3). Without loss of generality we may assume that $\phi(s) = 0$ for $s > t$. Denote by $[t]$ the integer part of t. Then we have:

$$\begin{aligned}\int_0^t a^s\phi(s)\,ds &\le \sum_{i=1}^{[t]}\int_i^{i+1} a^s|\phi(s)|\,ds \le \sum_{i=0}^{[t]} a^{i+1}\int_i^{i+1}|\phi(s)|\,ds\\ &\le a(1+a+a^2+\cdots+a^{[t]})\sup_{s\in[0,t]}\int_s^{s+1}|\phi(\zeta)|\,d\zeta\\ &=\frac{a(a^{[t]+1}-1)}{a-1}\sup_{s\in[0,t-1]}\int_s^{s+1}|\phi(\zeta)|\,d\zeta\\ &\le\frac{a^{t+2}}{a-1}\sup_{s\in[0,t-1]}\int_s^{s+1}|\phi(\zeta)|\,d\zeta.\end{aligned}$$

□

Now we turn to the generalization of Lemma 6.7.1.

Lemma 6.8.1. *Let $f \in \mathcal{X}$. Assume that a pair (u, τ), where*

$$u\in L_2(0,T;V)\bigcap L_\infty(0,T;H),\ \ \tau\in L_\infty(0,T;L_2),\tag{6.8.8}$$

satisfies either identities (6.3.1), (6.3.2) or identities (6.3.3), (6.3.4) almost everywhere on $(0, T)$ for all $\varphi \in \mathcal{V}$ and $\Phi \in C_0^\infty$. Then for all $t \in [0, T]$ the following estimate is valid:

$$\begin{aligned}&\|u'\|_{L_{4/3}(t,T;Y^*)}+\|\tau'\|_{L_2(t,T;H^{-2})}\\ &\quad\le K_1(\|u\|_{L_\infty(t,T;H)},\|u\|_{L_2(t,T;Y)},\|\tau\|_{L_\infty(t,T;L_2)},\|f\|_{\mathcal{X}},T-t,\Omega).\end{aligned}\tag{6.8.9}$$

Proof. Let a pair (u,τ) from class (6.8.8) satisfy identities (6.3.3), (6.3.4) almost everywhere on $(0,T)$ for all $\varphi \in \mathcal{V}$ and $\Phi \in C_0^\infty$. Inequalities (2.1.23), (2.1.24) and (2.1.26) imply that there is a constant $K_2 = K_2(\Omega)$ such that

$$\|u(t)\|_{L_4} \leq K_2(\Omega)\|u(t)\|^{\frac{1}{4}}\|u(t)\|_Y^{\frac{3}{4}}, \quad n = 2,3. \tag{6.8.10}$$

Thus,

$$\|u\|_{L_{8/3}(t,T;L_4)} = \|\|u\|_{L_4(\Omega)}\|_{L_{8/3}(t,T)} \leq K_2(\Omega)\big\|\,\|u(t)\|^{\frac{1}{4}}\|u(t)\|_Y^{\frac{3}{4}}\big\|_{L_{8/3}(t,T)}.$$

By Hölder's inequality (2.1.1) with $\Omega = (0,T)$, $\psi_1 = \|u(t)\|^{\frac{1}{4}}$, $\psi_2 = \|u(t)\|_Y^{\frac{3}{4}}$, $p = p_2 = 8/3$, $p_1 = \infty$ the last expression does not exceed

$$K_2(\Omega)\|\,\|u(t)\|^{\frac{1}{4}}\|_{L_\infty(t,T)}\|\,\|u(t)\|_Y^{\frac{3}{4}}\|_{L_{8/3}(t,T)} = K_2(\Omega)\|u\|^{\frac{1}{4}}_{L_\infty(t,T;H)}\|u\|^{\frac{3}{4}}_{L_2(t,T;Y)}.$$

We have from (6.3.4):

$$\begin{aligned}
\|\langle u',\varphi\rangle\|_{L_{4/3}(t,T)} &= \left\|\frac{d}{dt}(u,\varphi)\right\|_{L_{4/3}(t,T)} \\
&\leq \sum_{i=1}^{n}\left\|\left(\frac{u_i u}{1+\frac{1}{m}\left(\frac{|\tau|^2}{2\mu_2}+|u|^2\right)}, \frac{\partial\varphi}{\partial x_i}\right)\right\|_{L_{4/3}(t,T)} \\
&\quad + \|\mu_1(\nabla u,\nabla\varphi)\|_{L_{4/3}(t,T)} + \|(\tau,\nabla\varphi)\|_{L_{4/3}(t,T)} \\
&\quad + \|\langle f,\varphi\rangle\|_{L_{4/3}(t,T)}.
\end{aligned}$$

Applying Hölder's inequality (2.1.1), Lemma 2.2.1 a), and the inequality

$$\frac{1}{1+\frac{1}{m}(\frac{|\tau|^2}{2\mu_2}+|u|^2)} \leq 1$$

we get that the right-hand side does not exceed

$$\begin{aligned}
&\|\varphi\|_Y\big(\|u\|^2_{L_{8/3}(t,T;L_4)} + \mu_1\|u\|_{L_{4/3}(t,T;Y)} + \|\tau\|_{L_{4/3}(t,T;L_2)} + \|f\|_{L_{4/3}(t,T;Y^*)}\big) \\
&\quad \leq \|\varphi\|_Y(K_2^2(\Omega)\|u\|^{1/2}_{L_\infty(t,T;H)}\|u\|^{\frac{3}{2}}_{L_2(t,T;Y)} + \mu_1\|u\|_{L_{4/3}(t,T;Y)} \\
&\qquad + \|\tau\|_{L_{4/3}(t,T;L_2)} + \|f\|_{L_{4/3}(t,T;Y^*)}) \\
&\quad \leq \|\varphi\|_Y K_3\big(\|u\|_{L_\infty(t,T;L_2)}, \|u\|_{L_2(t,T;Y)}, \|\tau\|_{L_\infty(t,T;L_2)}, \|f\|_{\mathfrak{X}}, T-t, \Omega\big),
\end{aligned}$$

and this gives the estimate of the first term in the left-hand side of (6.8.9).

Taking into account the embedding $H_0^2 \subset W_4^1$ for $n = 2, 3$, from (6.3.3) we obtain the estimate of the second term in the left-hand side of (6.8.9):

$$
\begin{aligned}
\|\langle \tau', \Phi\rangle\|_{L_2(t,T)} &= \Big\|\frac{d}{dt}(\tau, \Phi)\Big\|_{L_2(t,T)} \\
&\le \Big\|\frac{1}{\lambda_1}(\tau, \Phi)\Big\|_{L_2(t,T)} + \frac{1}{k\lambda_1}\|(\tau, \Delta\Phi)\|_{L_2(t,T)} \\
&\quad + \Big\|\sum_{i=1}^{n}\Big(\frac{u_i\tau}{1+\frac{1}{m}\Big(\frac{|\tau|^2}{2\mu_2}+|u|^2\Big)}, \frac{\partial\Phi}{\partial x_i}\Big)\Big\|_{L_2(t,T)} + \|2\mu_2(u, \operatorname{Div}\Phi)\|_{L_2(t,T)} \\
&\le \frac{1}{\lambda_1}\|\tau\|_{L_2(t,T;L_2)}\|\Phi\|_{L_2} + \frac{1}{k\lambda_1}\|\Phi\|_{H_0^2}\|\tau\|_{L_2(t,T;L_2)} \\
&\quad + \|u\|_{L_2(t,T;L_4)}\|\tau\|_{L_\infty(t,T;L_2)}\Big\|\frac{\partial\Phi}{\partial x_i}\Big\|_{L_4} + 2\mu_2\|u\|_{L_2(t,T;L_2)}\|\operatorname{Div}\Phi\|_{L_2} \\
&\le \|\Phi\|_{H_0^2} K_4\big(\|u\|_{L_\infty(t,T;L_2)}, \|u\|_{L_2(t,T;Y)}, \|\tau\|_{L_\infty(t,T;L_2)}, T-t, \Omega\big).
\end{aligned}
$$

If (u, τ) satisfies identities (6.3.1), (6.3.2) instead of (6.3.3), (6.3.4), then the proof of (6.8.9) is carried out in the same way. □

Remark 6.8.1. For arbitrary domain Ω (not necessarily bounded), using a similar way of proof with estimate (2.1.25) (which does not contain constants dependent on Ω) instead of (6.8.10), and with the space V instead Y, one obtains

$$
\begin{aligned}
&\|u'\|_{L_{4/3}(t,T;V^*)} + \|\tau'\|_{L_2(t,T;H^{-2})} \\
&\quad \le K_5\big(\|u\|_{L_\infty(t,T;H)}, \|u\|_{L_2(t,T;V)}, \|\tau\|_{L_\infty(t,T;L_2)}, \\
&\qquad \|f\|_{L_2(t,T;V^*)}, T-t\big)
\end{aligned}
\tag{6.8.11}
$$

where K_5 is independent of Ω.

6.8.2 Existence and structure of uniform attractors

Let us now construct the minimal uniform trajectory attractors and the uniform global attractors for problem (6.3.1) – (6.3.2). As in Section 6.7, we choose $H \times L_2(\Omega, \mathbb{R}_S^{n\times n})$ as the space E and the space $V_\delta^* \times H^{-\delta}(\Omega, \mathbb{R}_S^{n\times n})$ as the space E_0, where $\delta \in (0, 1]$ is a fixed number.

Fix some $f \in \mathfrak{X}$. We can choose the symbol space Σ, for instance, from the following variants:

a) $\{f\}$;

b) $\Sigma_0 = \{T(t)f \mid t \ge 0\}$;

c) the closure of Σ_0 in the strong topology of $\mathfrak{X}$;

d) the closure of Σ_0 in the weak topology of $\mathcal{X}$;

e) any other subset of $\mathcal{X}$ which contains Σ_0 and satisfies the property

$$\|\sigma\|_{\mathcal{X}} \leq \|f\|_{\mathcal{X}} \tag{6.8.12}$$

for all $\sigma \in \Sigma$;

f) any other subset of $\mathcal{X}$ which satisfies (6.8.12) and contains f.

As it was mentioned in Remark 4.3.1, the uniform attractors depend on the choice of the symbol space.

Remark 6.8.2. The sets b) – e) contain Σ_0 (cf. Remark 4.3.3).

Fix a symbol space Σ, given by one of the statements a) – f). Define the trajectory space $\mathcal{H}_\sigma^+$ for a symbol $\sigma \in \Sigma$ as the set of pairs of functions (u, τ) which

i) belong to class (6.7.9);

ii) satisfy identity (6.3.1) and the identity

$$\frac{d}{dt}(u(t), \varphi) - \sum_{i=1}^{n} \left(u_i(t)u(t), \frac{\partial \varphi}{\partial x_i}\right) + \mu_1(\nabla u(t), \nabla \varphi) + (\tau(t), \nabla \varphi) = \langle \sigma(t), \varphi \rangle \tag{6.8.13}$$

almost everywhere on $(0, +\infty)$ for all $\varphi \in \mathcal{V}$ and $\Phi \in C_0^\infty$;

iii) satisfy the energy inequality:

$$\begin{aligned} &\frac{1}{4}\|u\|^2_{L_\infty(t,t+1;H)} + \frac{1}{8\mu_2}\|\tau\|^2_{L_\infty(t,t+1;L_2)} + \frac{\mu_1}{2}\|u\|^2_{L_2(t,t+1;V)} \\ &\quad \leq e^{-2\gamma t}\left(\|u\|^2_{L_\infty(0,+\infty;H)} + \frac{1}{2\mu_2}\|\tau\|^2_{L_\infty(0,+\infty;L_2)}\right) \\ &\qquad + \frac{2e^{4\gamma} + e^{2\gamma} - 1}{2\mu_1(e^{2\gamma} - 1)}\|f\|^2_{\mathcal{X}} \end{aligned} \tag{6.8.14}$$

for all $t > 0$ where γ is as in Theorem 6.7.1.

Remark 6.8.3. The same arguments as in Remark 6.7.1 show that

$$\mathcal{H}_\sigma^+ \subset C([0, +\infty); E_0) \cap L_\infty(0, +\infty; E)$$

for all $\sigma \in \Sigma$.

Remark 6.8.4. It is clear that on account of inequality (6.8.14) the family of trajectory spaces $\mathcal{H}_\sigma^+$, $\sigma \in \Sigma$ is not translation-coordinated (cf. Remark 4.3.2).

Theorem 6.8.2 (cf. [77]). *Given $a \in H$, $\tau_0 \in L_2(\Omega, \mathbb{R}_S^{n\times n})$, $\sigma \in \Sigma$, there is a pair of functions (a trajectory) $(u, \tau) \in \mathcal{H}_\sigma^+$ satisfying initial condition (6.3.5).*

Proof. Take an increasing sequence of positive numbers $T_m \to \infty$. Put in Theorems 6.5.1 and 6.8.1 σ instead of f. Then, by Theorem 6.5.1, for every natural m there is a pair (u_m, τ_m) which belongs to class (6.5.1) with $T = T_m$, and satisfies identities (6.3.1), (6.3.2) for almost everywhere $t \in (0, T_m)$ and for all $\varphi \in \mathcal{V}$ and $\Phi \in C_0^\infty$. By Theorem 6.8.1, all pairs (u_m, τ_m) satisfy inequality (6.8.1) for $t \in [0, T_m - 1]$ with σ instead of f. Since $\|\sigma\|_{\mathcal{X}} \le \|f\|_{\mathcal{X}}$, they satisfy it also with f. Denote by $\widetilde{u}_m$ and $\widetilde{\tau}_m$ the functions which are equal to u_m and τ_m in $[0, T_m]$ and vanish on $(T_m, +\infty)$. Then, without loss of generality one may assume that there exist limits

$$u = \lim_{m\to\infty} \widetilde{u}_m, \text{ which is } *\text{-weak in } L_\infty(0, +\infty; H);$$

$$\tau = \lim_{m\to\infty} \widetilde{\tau}_m, \text{ which is } *\text{-weak in } L_\infty(0, +\infty; L_2).$$

Fix an arbitrary interval $[0, T]$. Then estimate (6.8.1) for all pairs (u_m, τ_m) yields uniform boundedness of the sequence $\{\widetilde{u}_m\}$ in $L_2(0, T; V)$. Therefore without loss of generality we may consider that $\widetilde{u}_m \rightharpoonup u$ weakly in $L_2(0, T; V)$. It is easy to see that the pair (u, τ) satisfies inequality (6.8.1) (and, hence, (6.8.14)) for $t \in [0, T-1]$, and that, for $T_m \ge T$, the functions $\widetilde{u}_m$ and $\widetilde{\tau}_m$ coincide with u_m and τ_m on the segment $[0, T]$.

Just as in the proof of Theorems 6.4.1, 6.5.1, 6.7.2, estimate (6.8.1) for all pairs (u_m, τ_m) and Lemma 6.8.1 ensure convergence of all terms in identities (6.3.1), (6.8.13) with (u_m, τ_m) substituted there, $T_m \ge T$, to the corresponding terms in (6.3.1), (6.8.13) with (u, τ) in the sense of scalar distributions on $(0, T)$. Therefore the pair (u, τ) satisfies identities (6.3.1), (6.8.13) on $(0, T)$ for all $\varphi \in \mathcal{V}$ and $\Phi \in C_0^\infty$, belongs to class (6.7.8) and satisfies condition (6.3.5). Since T was arbitrary, the pair (u, τ) satisfies conditions i) – iii) of the definition of $\mathcal{H}_\sigma^+$. □

The main results of this section are

Theorem 6.8.3 (cf. [77]). *There exists a minimal uniform (with respect to $\sigma \in \Sigma$) trajectory attractor $\mathfrak{U}_J$ for problem (6.3.1) – (6.3.2).*

Theorem 6.8.4 (cf. [77]). *In the space $H \times L_2$ there is a uniform (with respect to $\sigma \in \Sigma$) global attractor $\mathcal{A}_J$ (in $V_\delta^* \times H^{-\delta}$) for problem (6.3.1) – (6.3.2).*

Proof. Consider the set P which consists of pairs

$$(u, \tau) \in C([0, +\infty); E_0) \cap L_\infty(0, +\infty; E)$$

satisfying inequalities

$$\frac{1}{4}\|u\|^2_{L_\infty(t,t+1;H)} + \frac{1}{8\mu_2}\|\tau\|^2_{L_\infty(t,t+1;L_2)} + \frac{\mu_1}{2}\|u\|^2_{L_2(t,t+1;V)}$$
$$\leq \frac{2e^{4\gamma}+e^{2\gamma}-1}{\mu_1(e^{2\gamma}-1)}\|f\|^2_{\mathfrak{X}},$$
$$\|u'\|_{L_{4/3}(t,t+1;Y^*)} + \|\tau'\|_{L_2(t,t+1;H^{-2})}$$
$$\leq K_1\big(\|u\|_{L_\infty(t,t+1;H)}, \|u\|_{L_2(t,t+1;V)}, \|\tau\|_{L_\infty(t,t+1;L_2)}, \|f\|_{\mathfrak{X}}, 1, \Omega\big)$$

for all $t \geq 0$, where K_1 is the constant from estimate (6.8.9). Just as in the proof of Theorem 6.7.3, one shows that $\overline{P}$ (its closure in $C([0,+\infty); E_0)$) is a uniform trajectory semiattractor.

Now from Theorem 4.3.1 we get Theorem 6.8.3, and Theorems 4.3.3 and 6.8.3 imply Theorem 6.8.4. □

6.9 Stationary boundary-value problem for the Jeffreys model

6.9.1 Strong and weak statements of the stationary problem

One of the interesting and important problems in the theory of evolutionary equations, and of the equations from fluid mechanics, in particular, is the study of *stationary*, i.e. time-independent, solutions. Such solutions correspond to the *steady* flow regime.

The boundary value problem describing the steady flow of the Jeffreys medium with constitutive equation (1.3.12) easily results from problem (6.2.1) – (6.2.4):

$$\sum_{i=1}^{n} u_i \frac{\partial u}{\partial x_i} + \operatorname{grad} p = \operatorname{Div} \sigma + f, \tag{6.9.1}$$

$$\sigma + \lambda_1 \sum_{i=1}^{n} u_i \frac{\partial \sigma}{\partial x_i} = 2\eta\Big(\mathcal{E} + \lambda_2 \sum_{i=1}^{n} u_i \frac{\partial \mathcal{E}}{\partial x_i}\Big), \tag{6.9.2}$$

$$\operatorname{div} u = 0, \tag{6.9.3}$$

$$u\big|_{\partial\Omega} = 0. \tag{6.9.4}$$

Note that Ω here is an arbitrary (possibly unbounded) domain in $\mathbb{R}^n$, $n = 2, 3$.

Observe that stationary solutions may exist only for the autonomous problems (when f depends only on x). Let $f \in Y^*(\Omega)$.

Definition 6.9.1. A *weak solution* to problem (6.9.1) – (6.9.4) is a pair of functions $u \in Y$, $\sigma \in L_2(\Omega, \mathbb{R}_S^{n\times n})$ such that

$$(\sigma, \nabla\varphi) - \sum_{i=1}^{n} (u_i u, \frac{\partial\varphi}{\partial x_i}) = \langle f, \varphi\rangle_{Y^*\times Y}, \tag{6.9.5}$$

$$(\sigma, \Phi) - \lambda_1 \sum_{i=1}^{n} (u_i \sigma, \frac{\partial\Phi}{\partial x_i}) = -2\eta(u, \operatorname{Div}\Phi) - 2\eta\lambda_2 \sum_{i=1}^{n} (u_i \mathcal{E}(u), \frac{\partial\Phi}{\partial x_i}) \tag{6.9.6}$$

for all $\varphi \in \mathcal{V}$ and $\Phi \in C_0^\infty$.

This weak setting can by realized in the framework of the general scheme of Section 6.1.1 just as it was done for problem (6.2.1) – (6.2.5).

Theorem 6.9.1 (see [68]). *Let $f \in Y^*$. Then there exists a weak solution to problem (6.9.1) – (6.9.4).*

6.9.2 Auxiliary problem and a priori bound

As in the non-stationary case, we study an auxiliary problem first.

Using the notations $\mu_1 = \eta\frac{\lambda_2}{\lambda_1}$, $\mu_2 = \frac{\eta-\mu_1}{\lambda_1}$, $\tau = \sigma - 2\mu_1\mathcal{E}$, as in Section 6.3, we rewrite system (6.9.5) – (6.9.6) in the form:

$$\frac{1}{\lambda_1}(\tau, \Phi) - \sum_{i=1}^{n} (u_i \tau, \frac{\partial\Phi}{\partial x_i}) + 2\mu_2(u, \operatorname{Div}\Phi) = 0, \tag{6.9.7}$$

$$-\sum_{i=1}^{n} (u_i u, \frac{\partial\varphi}{\partial x_i}) + \mu_1(\nabla u, \nabla\varphi) + (\tau, \nabla\varphi) = \langle f, \varphi\rangle. \tag{6.9.8}$$

Now, consider the following auxiliary problem:

$$\frac{1}{\lambda_1}(\tau, \Phi) - \delta \sum_{i=1}^{n} (u_i \tau, \frac{\partial\Phi}{\partial x_i}) + 2\mu_2\delta(u, \operatorname{Div}\Phi) + \varepsilon(\nabla\tau, \nabla\Phi) = 0, \tag{6.9.9}$$

$$-\delta \sum_{i=1}^{n} (u_i u, \frac{\partial\varphi}{\partial x_i}) + \mu_1(\nabla u, \nabla\varphi) + \delta(\tau, \nabla\varphi) = \delta\langle f, \varphi\rangle \tag{6.9.10}$$

for all $\varphi \in Y$, $\Phi \in H_0^1$ (here $\varepsilon > 0$ and $0 \le \delta \le 1$ are parameters).

Let us prove the following a priori estimate:

Lemma 6.9.1. *Let Ω be bounded, and let a pair $(u \in Y, \tau \in H_0^1(\Omega, \mathbb{R}_S^{n\times n}))$ be a solution to (6.9.9), (6.9.10). Then the following a priori estimate takes place.*

$$\mu_1\|u\|_Y^2 + \frac{1}{2\lambda_1\mu_2}\|\tau\|^2 + \frac{\varepsilon}{2\mu_2}\|\tau\|_Y^2 \le \frac{1}{\mu_1}\|f\|_{Y^*}^2. \tag{6.9.11}$$

Proof. Put $\varphi = u$ in (6.9.10), and $\Phi = \dfrac{\tau}{2\mu_2}$ in (6.9.9), and add the obtained equalities. Taking into account (6.1.18) – (6.1.20), we get:

$$\mu_1(\nabla u, \nabla u) + \frac{1}{2\lambda_1\mu_2}(\tau, \tau) + \frac{\varepsilon}{2\mu_2}(\nabla\tau, \nabla\tau) = \delta\langle f, u\rangle. \tag{6.9.12}$$

Since

$$\delta\langle f, u\rangle \leq \|f\|_{Y^*}\|u\|_Y, \tag{6.9.13}$$

(6.9.12) yields

$$\mu_1\|u\|_Y^2 \leq \|f\|_{Y^*}\ \|u\|_Y,$$

so

$$\|u\|_Y \leq \frac{1}{\mu_1}\|f\|_{Y^*}. \tag{6.9.14}$$

Now (6.9.12), (6.9.13) and (6.9.14) imply (6.9.11). □

6.9.3 Solvability of the auxiliary problem

Theorem 6.9.2. *Let Ω be bounded and $f \in Y^*$. Then there is a solution ($u \in Y$, $\tau \in H_0^1(\Omega, \mathbb{R}_S^{n\times n})$) to problem (6.9.9), (6.9.10).*

Proof. Let us introduce auxiliary operators by the following formulas (in these formulas φ and Φ are arbitrary elements of Y and $H_0^1(\Omega, \mathbb{R}_S^{n\times n})$, respectively):

$$K : Y \to Y^*, \quad \langle K(u), \varphi\rangle = -\sum_{i=1}^{n}\left(u_i u, \frac{\partial\varphi}{\partial x_i}\right),$$

$$A : Y \to Y^*, \quad \langle A(u), \varphi\rangle = \mu_1(\nabla u, \nabla\varphi),$$

$$A_\varepsilon : H_0^1 \to H^{-1}, \quad \langle A_\varepsilon(\tau), \Phi\rangle = \varepsilon(\nabla\tau, \nabla\Phi) + \frac{1}{\lambda_1}(\tau, \Phi),$$

$$\tilde{A} : Y \times H_0^1 \to Y^* \times H^{-1}, \quad \tilde{A}(u, \tau) = (A(u), A_\varepsilon(\tau)),$$

$$N_1 : H_0^1 \to Y^*, \quad \langle N_1(\tau), \varphi\rangle = (\tau, \nabla\varphi),$$

$$N_2 : Y \to H^{-1}, \quad \langle N_2(u), \Phi\rangle = 2\mu_2(u, \operatorname{Div}\Phi),$$

$$\tilde{K} : Y \times H_0^1 \to H^{-1}, \quad \langle \tilde{K}(u, \tau), \Phi)\rangle = -\sum_{i=1}^{n}(u_i\tau, \frac{\partial\Phi}{\partial x_i}),$$

$$Q : Y \times H_0^1 \to Y^* \times H^{-1}, \quad Q(u, \tau) = (K(u) + N_1(\tau) - f,\ \tilde{K}(u, \tau) + N_2(u)).$$

Then system (6.9.9), (6.9.10) is equivalent to the operator equation

$$\tilde{A}(u, \tau) + \delta Q(u, \tau) = 0. \tag{6.9.15}$$

The linear operator N_1 is bounded as a map from L_2 into Y^*. But the embedding of H_0^1 into L_2 is compact, so the operator N_1 is compact as an operator from H_0^1 into Y^*. Similarly, the operator N_2 is compact since $Y \cong V$ (Ω is bounded now) is imbedded into L_2 compactly.

By Hölder's inequality (2.1.1), the operator K is continuous as map from L_4 into Y^*. But $Y \subset L_4$ compactly, so the operator K is compact (as map from Y into Y^*). Similarly, the operator $\tilde{K}$ is continuous as map from $L_4 \times L_4$ into H^{-1}, and the operator $\tilde{K} : Y \times H_0^1 \to H^{-1}$ is compact.

Hence, the operator Q is compact.

By Theorem 3.1.1 the operator $\tilde{A}$ is invertible. Rewrite equation (6.9.15) in the form

$$(u, \tau) + \delta \tilde{A}^{-1} Q(u, \tau) = 0. \tag{6.9.16}$$

Due to Lemma 6.9.1, equation (6.9.16) has no solutions on the boundary of a sufficiently large ball B in $Y \times H_0^1$ independent of δ. Then we can use the Leray–Schauder degree $\deg_{LS}(I + \delta \tilde{A} Q, B, 0)$ (see Section 3.2.2). By the homotopic invariance property of the degree,

$$\deg_{LS}(I + \delta \tilde{A} Q, B, 0) = \deg_{LS}(I, B, 0) = 1.$$

Hence, equation (6.9.16), and, therefore, system (6.9.9), (6.9.10), has a solution in B at every $\delta \in [0, 1]$. □

6.9.4 Proof of Theorem 6.9.1.

Denote by Ω_m the intersection of Ω with the ball B_m of radius m centered at the origin in the space $\mathbb{R}^n, m = 1, 2, \dots$.

Consider the "restriction" of f on Ω_m, $f|_{\Omega_m} \in Y^*(\Omega_m)$, which is defined by the formula

$$\langle f|_{\Omega_m}, \varphi \rangle = \langle f, \tilde{\varphi} \rangle$$

where φ is a function from $Y(\Omega_m)$, and $\tilde{\varphi}$ coincides with φ in Ω_m, and is identically zero in $\Omega \backslash \Omega_m$. Obviously, $\|f|_{\Omega_m}\|_{Y^*(\Omega_m)} \le \|f\|_{Y^*(\Omega)}$.

On each Ω_m, consider problem (6.9.9), (6.9.10) with $f = f|_{\Omega_m}$, $\varepsilon = \frac{1}{m}$, $\delta = 1$. By Theorem 6.9.2 these problems have at least one solution (u_m, τ_m). Denote by $(\tilde{u}_m, \tilde{\tau}_m)$ the functions which coincide with u_m and τ_m, respectively, in Ω_m, and are identically zero in $\Omega \backslash \Omega_m$. By Lemma 6.9.1, the norms $\|\tilde{u}_m\|_{Y(\Omega)} = \|u_m\|_{Y(\Omega_m)}$ and $\|\tilde{\tau}_m\|_{L_2(\Omega)} = \|\tau_m\|_{L_2(\Omega_m)}$ are uniformly bounded. Therefore, as $m \to \infty$, without loss of generality we may assume that $\tilde{u}_m \to \tilde{u}_0$ weakly in Y, $\tilde{\tau}_m \to \tilde{\tau}_0$ weakly in L_2. Let us show that $(\tilde{u}_0, \tilde{\tau}_0)$ is a solution to problem (6.9.7), (6.9.8).

Take arbitrary $\varphi \in \mathcal{V}$, $\Phi \in C_0^\infty$. At some k the supports of φ and Φ are contained in Ω_k. Denote by u_m^* the functions which coincide with $\tilde{u}_m$ in $\Omega \bigcap B_k$, and are identically zero in $B_k \backslash \Omega$. It is clear that $u_m^* \to u_0^*$ weakly in $W_2^1(B_k)$, that is, strongly in $L_4(B_k)$.

Therefore, all terms from (6.9.9), (6.9.10) with $\varepsilon = \frac{1}{m}$, $\delta = 1$, $u = u_m$, $\tau = \tau_m$ converge to the corresponding terms in (6.9.7), (6.9.8), and

$$|\varepsilon(\nabla\tau_m, \nabla\Phi)| = \Big|\frac{1}{m}(\tilde{\tau}_m, \Delta\Phi)\Big| \leq \frac{1}{m}\|\tilde{\tau}_m\|\,\|\Delta\Phi\| \to 0.$$

Thus, the pair $(\tilde{u}_0, \tilde{\tau}_0)$ satisfies identities (6.9.7), (6.9.8) for all $\varphi \in \mathcal{V}$, $\Phi \in C_0^\infty$. Let $\tilde{\sigma}_0 = \tilde{\tau}_0 + 2\mu_1\mathcal{E}(\tilde{u_0})$. It is clear that $\tilde{\sigma}_0 \in L_2$. Then $(\tilde{u_0}, \tilde{\sigma}_0)$ is a solution to problem (6.9.5), (6.9.6), i.e. a weak solution to problem (6.9.1) – (6.9.4).

Chapter 7

The regularized Jeffreys model

7.1 Formulation of the problem and the main results

Consider the motion of homogeneous incompressible Jeffreys' viscoelastic medium filling a domain Ω in $\mathbb{R}^n$, $2 \le n \le 4$, on a time interval $(0, T)$, $T > 0$. We recall (Section 1.3.3) that expressing σ from the constitutive relation (1.3.12) and substituting it in equation (1.1.12) one gets (1.3.14). In this chapter we are going to investigate the equation of motion (1.3.14) instead of system (6.2.1), (6.2.2). Let us point out that equation (1.3.14) seems to be more convenient for the applications and numerical schemes, since (in the three-dimensional case) it contains only four scalar function variables v_1, v_2, v_3, p, whereas system (6.2.1), (6.2.2) has three velocity variables, one pressure variable, and six stress variables $\sigma_{11}, \sigma_{22}, \sigma_{33}, \sigma_{12}, \sigma_{13}, \sigma_{23}$.

Hereafter we assume that Ω is a sufficiently regular bounded domain, and use Einstein's summation agreement.

Equation (1.3.14) contains the function z which describes the trajectories of particles of a medium. More precisely, $z(\varsigma, t, x)$ shows which position, at the moment ς, is occupied with the particle which at the moment t is at the point x (in our case, $x \in \Omega$). Rewrite the properties (1.3.4), (1.3.5) of z in the equivalent form

$$z(\varsigma, t, x) = x + \int_t^{\varsigma} v(s, z(s, t, x))\, ds. \tag{7.1.1}$$

Then the trajectories of the particles of the medium are determined by the velocity field v as the solution of this integral equation.

Combining (1.3.14) and (7.1.1) with the incompressibility condition (1.1.10), the no-slip condition (1.1.15), and the initial condition, we arrive at the following problem

$$\begin{aligned} &\frac{\partial v}{\partial t} + v_i \frac{\partial v}{\partial x_i} - \mu \operatorname{Div} \int_0^t e^{-\frac{t-s}{\lambda}} \mathcal{E}(v)(s, z(s, t, x))\, ds - \frac{\mu_0}{2} \Delta v \\ &\quad = -\operatorname{grad} p + f, \quad (t, x) \in (0, T) \times \Omega, \end{aligned} \tag{7.1.2}$$

$$z(\varsigma, t, x) = x + \int_t^{\varsigma} v(s, z(s, t, x))\, ds, \quad \varsigma \in [0, T], \ (t, x) \in (0, T) \times \Omega, \tag{7.1.3}$$

$$\operatorname{div} v = 0, \quad (t,x) \in (0,T) \times \Omega, \tag{7.1.4}$$

$$v\mid_{(0,T)\times\partial\Omega} = 0, \tag{7.1.5}$$

$$v(0,x) = a(x), \quad x \in \Omega. \tag{7.1.6}$$

Here we assumed for simplicity that

$$\sigma_0 - 2\eta\frac{\lambda_2}{\lambda_1}\mathcal{E}(a) = 0 \tag{7.1.7}$$

and used the notations

$$\mu_0 = 2\eta\frac{\lambda_2}{\lambda_1} > 0, \ \mu = 2\eta\frac{\lambda_1 - \lambda_2}{\lambda_1^2} > 0. \tag{7.1.8}$$

Let us introduce some additional notations required in the sequel.

First we point out that in the space $V(\Omega)$ we shall sometimes use the scalar product $(u,v)_V = \int_\Omega \mathcal{E}_{ij}(u)\cdot\mathcal{E}_{ij}(v)\,dx = \frac{1}{2}(u,v)_Y$ and the corresponding Euclid norm $\|v\|_V$.

We shall use the following notations for the vector function spaces:

$E = L_2(0,T;V)$	with the norm $\|v\|_E = \|v\|_{L_2(0,T;V)}$	for $v \in E$;
$E^* = L_2(0,T;V^*)$	with the norm $\|f\|_{E^*} = \|f\|_{L_2(0,T;V^*)}$	for $f \in E^*$;
$E_1^* = L_1(0,T;V^*)$	with the norm $\|f\|_{E_1^*} = \|f\|_{L_1(0,T;V^*)}$	for $f \in E_1^*$.

Denote by $C^1D(\overline{\Omega})$ the set of one-to-one maps $z : \overline{\Omega} \to \overline{\Omega}$ coinciding with the identity map on $\partial\Omega$ and having continuous partial derivatives of the first order such that $\det\left(\frac{\partial z}{\partial x}\right) = 1$ at each point of $\overline{\Omega}$. We shall use the norm of the space of continuous functions $C(\overline{\Omega})^n$ in this set.

Let us introduce the set $CG = C([0,T]\times[0,T], C^1D(\overline{\Omega}))$. Note that $CG \subset C([0,T]\times[0,T], C^1(\overline{\Omega})^n)$, and CG will be considered as a metric space with the metric of space $C([0,T]\times[0,T], C(\overline{\Omega})^n)$.

Let $Q_T = (0,T)\times\Omega$.

Equation (7.1.2) includes an integral along the trajectories of motion of particles of the medium (cf. Section 1.3.3). Therefore the trajectories should be unambiguously determined by the field of velocities $v(t,x)$. In other words, equation (7.1.3) should have a unique solution for a given velocity field $v(t,x)$. However, existence of solutions for equation (7.1.3) at fixed v is known only in the case $v \in L_1(0,T;C(\overline{\Omega})^n)$, and this solution is unique for $v \in L_1(0,T;C^1(\overline{\Omega})^n)$ such that $v\mid_{(0,T)\times\partial\Omega} = 0$ (see, for example, [46, 47], [48, Lemmas 1 and 2]). Therefore, for weak solutions, the motion trajectories cannot be determined unambiguously. A possible way out of the situation (see e.g. [39]) is a smoothing (a regularization) of the velocity field and determination of the trajectories $z = Z_\delta(v)$ for the smoothed velocity field $S_\delta(v)$.

Let us choose some linear regularization operator $S_\delta : H \to C^1(\overline{\Omega})^n \cap V$ for $\delta > 0$ such that the map $S_\delta : L_2(0,T;H) \to L_2(0,T;C^1(\overline{\Omega})^n \cap V)$ generated by the operator is continuous and the operators $S_\delta : L_2(0,T;H) \to L_2(0,T;H)$ converge strongly to the identity operator I as $\delta \to 0$. A construction of such an operator is given in Section 7.7.

Replace equation (7.1.3) in system (7.1.2) – (7.1.6) with the following equation:

$$z(\varsigma,t,x) = x + \int_t^\varsigma S_\delta v(s,z(s,t,x))\,ds, \quad \varsigma,t \in (0,T), \quad x \in \Omega. \tag{7.1.9}$$

For each $v \in L_2(0,T;V)$, this equation has a unique solution $z = Z_\delta(v)$ in the class CG. Substituting $Z_\delta(v)$ for z in equation (7.1.2), we arrive at the regularized problem

$$\begin{aligned}&\frac{\partial v}{\partial t} + v_i\frac{\partial v}{\partial x_i} - \mu\,\mathrm{Div}\int_0^t e^{-\frac{t-s}{\lambda}}\mathcal{E}(v)(s,Z_\delta(v)(s,t,x))\,ds - \frac{\mu_0}{2}\Delta v\\ &\quad= -\mathrm{grad}\,p + f, \quad (t,x) \in (0,T)\times\Omega,\end{aligned} \tag{7.1.10}$$

$$\mathrm{div}\; v = 0, \quad (t,x) \in (0,T)\times\Omega, \tag{7.1.11}$$

$$v\,|_{(0,T)\times\partial\Omega} = 0, \tag{7.1.12}$$

$$v(0,x) = a(x), \quad x \in \Omega. \tag{7.1.13}$$

In order to define weak solutions for this problem, let us introduce the maps

$$A : V \to V^*, \ \langle A(u),h\rangle = \mu_0(\mathcal{E}(u),\mathcal{E}(h)), \ u,h \in V,$$

$$K : V \to V^*, \ \langle K(u),h\rangle = (u_iu_j, \frac{\partial h_i}{\partial x_j}), \ u,h \in V,$$

$$\langle C(v,z)(t),h\rangle = \mu\left(\int_0^t e^{-\frac{t-s}{\lambda}}\mathcal{E}(v)(s,z(s,t,x))\,ds, \mathcal{E}(h)\right).$$

Here $v \in E$, $z \in CG$, $t \in (0,T)$, $C(v,z)(t) \in V^*$, $h \in V$. Below we show that $C : E \times CG \to E^*$.

Definition 7.1.1. Given $f \in L_1(0,T;V^*)$, $a \in H$, a *weak solution for the regularized problem* (7.1.10) – (7.1.13) is a function $v \in L_2(0,T;V) \cap C_w([0,T];H)$ with $v' \in L_1(0,T;V^*)$ which satisfies equalities

$$v' + A(v) - K(v) + C(v,Z_\delta(v)) = f, \tag{7.1.14}$$

$$v(0) = a. \tag{7.1.15}$$

We point out that a weak solution v belongs to the space

$$W_1 = \{v : \ v \in E, \ v' \in E_1^*\}$$

with the norm $\|v\|_{W_1} = \|v\|_E + \|v'\|_{E_1^*}$. By Lemma 2.2.3, $W_1 \subset C([0,T], V^*)$, so condition (7.1.15) makes sense.

As in the case of problems (6.1.5) – (6.1.8) and (6.2.1) – (6.2.5), using integration by parts one can check that this weak setting can by obtained in the framework of the general scheme of Section 6.1.1.

Note that if v_δ is a strong solution of the regularized problem, then passage to the limit in equality (7.1.9) (and then in (7.1.10)) as $\delta \to 0$ leads us to a solution of problem (7.1.2) – (7.1.6). In Section 7.6, we shall study this question for weak solutions as well.

The solvability of the Cauchy problem (7.1.14) – (7.1.15) will be established provided $f \in L_1(0,T;H^*)+L_2(0,T;V^*)$, i.e. $f = f_1+f_2$ where $f_1 \in L_1(0,T;H^*)$ and $f_2 \in L_2(0,T;V^*)$.

Let us construct the approximating equations. For this purpose, we make modifications in equation (7.1.14), so that all the terms belong to $L_2(0,T;V^*)$. Let $\varepsilon > 0$. Consider the operator

$$K_\varepsilon : V \to V^*, \ \langle K_\varepsilon(u), h\rangle = \left(\frac{u_i u_j}{1+\varepsilon|u|^2}, \frac{\partial h_i}{\partial x_j}\right),$$

and approximate the function f_1 from $L_1(0,T;H^*)$ by functions $f_{1,\varepsilon}$ from $L_2(0,T;H^*)$ so that

$$f_{1,\varepsilon} \to f_1 \text{ in } L_1(0,T;H^*) \text{ as } \varepsilon \to 0. \tag{7.1.16}$$

Let f_ε denote the function $f_\varepsilon = f_{1,\varepsilon} + f_2$.

Consider the following Cauchy problem

$$v' + A(v) - K_\varepsilon(v) + C(v, Z_\delta(v)) = f_\varepsilon, \tag{7.1.17}$$

$$v(0) = a, \tag{7.1.18}$$

in the space $W = \{v : v \in E,\ v' \in E^*\}$ with the norm $\|v\|_W = \|v\|_E + \|v'\|_{E^*}$. The space W is a Banach space. Since $W \subset C([0,T], H)$ (due to Lemma 2.2.7), (7.1.18) makes sense.

Introduce the maps $L, G, K_\varepsilon : W \to E^* \times H$ with the help of the following equalities:

$$L(v) = (v' + A(v), v|_{t=0}), \ \ G(v) = (C(v, Z_\delta(v)), 0), \ \ K_\varepsilon = (K_\varepsilon, 0).$$

Then problem (7.1.17) – (7.1.18) is equivalent to the operator equation

$$L(v) = K_\varepsilon - G(v) + (f_\varepsilon, a). \tag{7.1.19}$$

Now we can state the results on existence of weak solutions (see [88]).

Theorem 7.1.1. *Given $\varepsilon > 0$, $f_\varepsilon \in E^*$, $a \in H$, problem (7.1.17), (7.1.18) possesses at least one solution $v \in W$.*

Theorem 7.1.2. *Given $f \in L_1(0,T;H^*) + L_2(0,T;V^*)$ and $a \in H$, problem (7.1.14), (7.1.15) has a solution in $W_1 \cap C_w([0,T];H)$.*

The proof of these statements is given in the following three sections.

7.2 Properties of the operators

In this section we study properties of the operators which appear in equations (7.1.17), (7.1.18), (7.1.19).

Lemma 7.2.1. a) *For $v \in E$ one has $A(v) \in E^*$, the map $A : E \to E^*$ is continuous, and the following estimate is valid*

$$\|A(v)\|_{E^*} \leq C_0(1 + \|v\|_E). \tag{7.2.1}$$

b) *For $v \in E$ one has $K(v) \in L_1(0,T;V^*)$ and $K_\varepsilon(v) \in L_\infty(0,T;V^*)$. The maps $K : E \to L_1(0,T;V^*)$, $K_\varepsilon : E \to E^*$ are continuous, and*

$$\|K_\varepsilon(v)\|_{E^*} \leq \frac{C_1}{\varepsilon}, \quad \|K_\varepsilon(v)\|_{L_1(0,T;V^*)} \leq C_1\|v\|_E^2, \tag{7.2.2}$$

where the constants C_0, C_1 are independent of v. The second estimate holds also for $\varepsilon = 0$ (i.e. for $K_0 = K$). The map $K_\varepsilon : W \to E^*$ is compact for $\varepsilon > 0$.

These facts can be found in [16, Lemma 2.1 and Theorem 2.2] (cf. also the proof of Theorem 6.3.1, Section 6.3 of this book, and [88]).

As it was mentioned above, $W \subset E \cap C([0,T],H)$. For functions $v \in E \cap C([0,T],H)$, consider the norm

$$\|v\|_{EC} = \max_{0 \leq t \leq T} \|v(t)\| + \|v\|_E.$$

We shall also require the equivalent norms $\|v\|_{k,EC} = \|\bar{v}\|_{EC}$ where $\bar{v}(t) = \exp(-kt) \cdot v(t)$, $k \geq 0$.

Similarly, define equivalent norms $\|\cdot\|_{k,E}$, $\|\cdot\|_{k,E^* \times H}$, $\|\cdot\|_{k,L_2(Q_T)}$ in the spaces E, $E^* \times H$, $L_2(Q_T)$.

Theorem 7.2.1. *The map $L : W \to E^* \times H$ is invertible, and for all functions $u, v \in W$ the estimate*

$$\|v - u\|_{k,EC} \leq C_2\|L(v) - L(u)\|_{k,E^* \times H} \tag{7.2.3}$$

holds for any $k \geq 0$. Here the constant C_2 is independent of u, v, k.

This statement is a particular case of Theorem 2.1 from [16], but as a matter of fact invertibility of L follows easily from Lemma 3.1.3.

Lemma 7.2.2. *For any $v \in E$, $z \in CG$, one has $C(v,z) \in E^*$, and the map*

$$C : E \times CG \to E^*$$

is continuous and bounded.

Proof. By definition of the map C, for $v \in E, z \in CG, h \in E$ and $t \in (0,T)$:

$$\langle C(v(t), z(\cdot,t,\cdot)), h(t)\rangle = \mu\left(\int_0^t e^{-\frac{t-s}{\lambda}} \mathcal{E}(v)(s, z(s,t,x))\, ds, \mathcal{E}(h(t))\right).$$

Let

$$B : (v,z) \mapsto \int_0^t e^{-\frac{t-s}{\lambda}} \mathcal{E}(s, z(s,t,x))\, ds.$$

It suffices to show that

$$B : E \times CG \to L_2([0,T], L_2(\Omega, \mathbb{R}^{n\times n})),$$

as well as continuity and boundedness of B.

But B is a superposition of an integral operator with the map

$$\begin{aligned} &\Phi : E \times CG \to L_2([0,T]\times[0,T], L_2(\Omega, \mathbb{R}^{n\times n})),\\ &\Phi : (v,z) \mapsto \mathcal{E}(v)(s, z(s,t,x)). \end{aligned} \tag{7.2.4}$$

At any fixed $z \in CG$,

$$\begin{aligned} &\int_0^T\int_0^T \|\mathcal{E}(v)(s,z(s,t,x)) - \mathcal{E}(u)(s,z(s,t,x))\|^2_{L_2(\Omega,\mathbb{R}^{n\times n})}\, ds\, dt\\ &\quad = \int_0^T\int_0^T\int_\Omega |\mathcal{E}|^2(v-u)(s,z(s,t,x))\, dx\, ds\, dt\\ &\quad = T\int_0^T\int_\Omega |\mathcal{E}|^2(v-u)(s,z)\, dz\, ds\\ &\quad = T\int_0^T \|\mathcal{E}(v-u)(s,\cdot)\|^2_{L_2(\Omega,\mathbb{R}^{n\times n})}\, ds = T\|v-u\|^2_E. \end{aligned} \tag{7.2.5}$$

Note that the change of variables $x \to z = z(s,t,x)$ at fixed s,t does not change the norm because $\det\left(\frac{\partial z}{\partial x}\right) = 1$ for all $(s,t,x) \in [0,T]\times[0,T]\times\overline{\Omega}$. Therefore the map Φ is continuous in v uniformly with respect to z.

Now, to prove continuity of Φ, it remains to check its continuity in z at a fixed value of v.

Let z_l be an arbitrary sequence from CG, converging to $z_0 \in CG$ in the space $C([0,T]\times[0,T], C(\overline{\Omega})^n)$, and let $\varepsilon > 0$ be an arbitrary number. Since the set of continuous functions $C(\overline{Q_T}, \mathbb{R}^{n\times n})$ is dense in $L_2(Q_T, \mathbb{R}^{n\times n})$, there exists a continuous $\varepsilon/(3\sqrt{T})$ - approximation of the function $\mathcal{E}(v)$ (denote it by ξ), i.e.

$$\|\mathcal{E}(v) - \xi\|_{L_2(Q_T,\mathbb{R}^{n\times n})} < \frac{\varepsilon}{3\sqrt{T}}.$$

Then

$$\begin{aligned}
&\|\mathcal{E}(v)(s, z_l(s,t,x)) - \mathcal{E}(v)(s, z_0(s,t,x))\|_{L_2([0,T]\times[0,T],L_2(\Omega,\mathbb{R}^{n\times n}))} \\
&\quad\le \|\mathcal{E}(v)(s, z_l(s,t,x)) - \xi(s, z_l(s,t,x))\|_{L_2([0,T]\times[0,T],L_2(\Omega,\mathbb{R}^{n\times n}))} \\
&\qquad + \|\xi(s, z_l(s,t,x)) - \xi(s, z_0(s,t,x))\|_{L_2([0,T]\times[0,T],L_2(\Omega,\mathbb{R}^{n\times n}))} \\
&\qquad + \|\mathcal{E}(v)(s, z_0(s,t,x)) - \xi(s, z_0(s,t,x))\|_{L_2([0,T]\times[0,T],L_2(\Omega,\mathbb{R}^{n\times n}))}.
\end{aligned}$$

Due to the choice of ξ, the first and last terms are less than $\varepsilon/3$. The function ξ is uniformly continuous on Q_T, so the operator of superposition $z \mapsto \xi(\cdot, z)$ is continuous from CG to $C([0,T]\times[0,T], C(\Omega,\mathbb{R}^{n\times n}))$. Therefore,

$$\|\xi(s, z_l(s,t,x)) - \xi(s, z_0(s,t,x))\|_{C([0,T]\times[0,T],C(\Omega,\mathbb{R}^{n\times n}))} \to 0 \ \text{ as } \ l \to \infty.$$

Choosing l large enough, $l > l_0$, we obtain

$$\|\xi(s, z_l(s,t,x)) - \xi(s, z_0(s,t,x))\|_{L_2([0,T]\times[0,T],L_2(\Omega,\mathbb{R}^{n\times n}))} < \varepsilon/3.$$

Hence,

$$\|\mathcal{E}(v)(s, z_l(s,t,x)) - \mathcal{E}(v)(s, z_0(s,t,x))\|_{L_2([0,T]\times[0,T],L_2(\Omega,\mathbb{R}^{n\times n}))} < \varepsilon.$$

Thus, the map Φ is continuous in z.

Estimate (7.2.5) with $u = 0$, continuity of Φ, and boundedness of the integral operator yield boundedness and continuity of the map B. □

Lemma 7.2.3. *The map $Z_\delta : L_2(0,T;H) \to CG$ is continuous.*

Proof. The operator S_δ is continuous from $L_2(0,T;H)$ to $L_2(0,T;C^1(\overline{\Omega})^n \cap V)$. The map Z_δ is a superposition of S_δ with the resolving operator for equation (7.1.3), $Z_0(v) = z$. It remains to show that $Z_0 : L_2(0,T;C^1(\Omega)^n \cap V) \to CG$ is continuous.

Choose an arbitrary sequence

$$\{v_l\},\ \ v_l \in L_2(0,T;C^1(\Omega)^n \cap V),\ \ v_l \to v_0 \ \text{as}\ l \to \infty.$$

By [48, Lemmas 1 and 3], for every $v = v_l$, $l = 0, 1, 2, \dots$, there is a unique solution z_l in CG to equation (7.1.3), and one has the estimate

$$\|z_l(\varsigma,t,\cdot) - z_0(\varsigma,t,\cdot)\|_{C(\overline{\Omega})^n} \le M \left| \int_t^{\varsigma} \|v_l(s,\cdot) - v_0(s,\cdot)\|_{C(\overline{\Omega})^n}\, ds \right|$$

with some constant M independent of l, ς, t. Then

$$\|z_l - z_0\|_{C([0,T]\times[0,T],C(\overline{\Omega})^n)} \le M\|v_l - v_0\|_{L_1(0,T;C^1(\Omega)^n)} \le C\|v_l - v_0\|_{L_2(0,T;C^1(\Omega)^n)}.$$

Since $v_l \to v_0$ as $l \to \infty$ in the space $L_2(0,T;C^1(\Omega)^n)$, we get $z_l = Z_0(v_l) \to Z_0(v_0) = z_0$ strongly in the space $C([0,T]\times[0,T], C(\overline{\Omega})^n)$. □

Lemma 7.2.4. *For $z \in CG$, $u, v \in E$ there is the estimate*

$$\|C(v,z) - C(u,z)\|_{k,E^*} \le \mu\sqrt{\frac{T}{2k}}\,\|u - v\|_{k,E}. \tag{7.2.6}$$

Proof. Let $\overline{v}(t) = \exp(-kt)v(t)$, $\overline{u}(t) = \exp(-kt)u(t)$. By the definition of C, for $h \in E$ we have

$$\big\langle \exp(-kt)C(v(t), z(\cdot,t,\cdot)) - \exp(-kt)C(u(t), z(\cdot,t,\cdot)), h(t)\big\rangle_{E^*\times E}$$
$$= \mu\int_0^T\int_\Omega\int_0^t e^{-(t-s)(\frac{1}{\lambda}+k)}\mathcal{E}_{ij}(\overline{v}-\overline{u})(s, z(s,t,x))\,ds\,\mathcal{E}_{ij}(h)(t,x)\,dx\,dt$$

Then with the help of Hölder's inequality we get

$$\big|\big\langle \exp(-kt)C(v(t), z(\cdot,t,\cdot)) - \exp(-kt)C(u(t), z(\cdot,t,\cdot)), h(t)\big\rangle\big|$$
$$\le \mu\int_0^T\int_0^t e^{-(t-s)(\frac{1}{\lambda}+k)}\Big(\int_\Omega |\mathcal{E}|^2(\overline{v}-\overline{u})(s, z(s,t,x))\,dx\Big)^{1/2} \cdot\Big(\int_\Omega |\mathcal{E}|^2(h)(t,x)\,dx\Big)^{1/2} ds\,dt$$

$$= \mu\int_0^T\int_0^t e^{-(t-s)(\frac{1}{\lambda}+k)}\Big(\int_\Omega |\mathcal{E}|^2(\overline{v}-\overline{u})(s,z)\,dz\Big)^{1/2}\cdot\|h(t,\cdot)\|_V\,ds\,dt$$
$$= \mu\int_0^T\int_0^t e^{-(t-s)(\frac{1}{\lambda}+k)}\|(\overline{v}-\overline{u})(s,\cdot)\|_V\cdot\|h(t,\cdot)\|_V\,ds\,dt$$
$$\le \mu\int_0^T\Big(\int_0^t e^{-2(t-s)(\frac{1}{\lambda}+k)}\,ds\Big)^{1/2}\Big(\int_0^t \|(\overline{v}-\overline{u})(s,\cdot)\|_V^2\,ds\Big)^{1/2} \cdot\|h(t,\cdot)\|_V\,dt$$
$$\le \mu\|\overline{v}-\overline{u}\|_E\cdot\|h\|_E\cdot\Big(\int_0^T\int_0^t e^{-2(t-s)(\frac{1}{\lambda}+k)}\,ds\,dt\Big)^{1/2}.$$

Let us estimate the remaining integral. Denote $\lambda_0 = 2(\frac{1}{\lambda}+k)$, then

$$\int_0^T\int_0^t e^{-(t-s)\lambda_0}\,ds\,dt = \frac{1}{\lambda_0}\int_0^T(1-e^{-t\lambda_0})\,dt \le \frac{1}{\lambda_0}\int_0^T dt = \frac{T}{\lambda_0}.$$

Thus,

$$\left|\langle e^{-kt}C(v,z)-e^{-kt}C(u,z),h\rangle\right| \le \mu\sqrt{\frac{T\lambda}{2+2\lambda k}}\,\|\overline{v}-\overline{u}\|_E\cdot\|h\|_E,$$

whence

$$\begin{aligned}\|C(v,z)-C(u,z)\|_{k,E^*} &= \|e^{-kt}(C(v,z)-C(u,z))\|_{E^*}\\ &\le \mu\sqrt{\frac{T\lambda}{2+2\lambda k}}\,\|\overline{v}-\overline{u}\|_E \le \mu\sqrt{\frac{T}{2k}}\,\|v-u\|_{k,E}.\quad\square\end{aligned}$$

These lemmas allow us to investigate properties of the map G.

We recall [86] that, for given bounded maps $F, \mathcal{A} : X \to Y$ (X and Y are Banach spaces), the map F is called $\mathcal{A}$*-condensing* with respect to a measure of noncompactness [5] γ in Y if

$$\gamma(F(M)) \le q\gamma(\mathcal{A}(M))$$

for any bounded set M in X with some constant $q < 1$.

Let γ_k denote the *Kuratowski measure of noncompactness* [5] in the space E^* with the norm $\|\cdot\|_{k,E^*}$.

Theorem 7.2.2. *The map $G : W \to E^* \times H$ is L-condensing with respect to γ_k for k large enough.*

Proof. Let $M \subset W$ be an arbitrary bounded set. Then it is relatively compact in $L_2(0,T;H)$ by Theorem 2.2.6. Lemma 7.2.3 implies that the set $Z_\delta(M)$ is relatively compact in CG. Then the set $C(v, Z_\delta(M))$ is relatively compact in E^* for any fixed $v \in W$. Besides, for any $z \in Z_\delta(M)$ the map $C(\cdot,z)$ satisfies the Lipschitz condition (7.2.6) with constant $\mu\sqrt{\frac{T}{2k}}$. Then by Theorem 1.5.7 from [5] the map $C(v, Z_\delta(v))$ (and hence G) is $\mu\sqrt{\frac{T}{2k}}$*-bounded* with respect to the *Hausdorff measure of noncompactness* χ_k, i.e.

$$\chi_k(G(M)) \le \mu\sqrt{\frac{T}{2k}}\,\chi_k(M).$$

Due to [5, Theorem 1.1.7],

$$\chi_k(M_*) \le \gamma_k(M_*) \le 2\chi_k(M_*)$$

for any bounded set M_*. Therefore

$$\gamma_k(G(M)) \le 2\chi_k(G(M)) \le 2\mu\sqrt{\frac{T}{2k}}\,\gamma_k(M). \tag{7.2.7}$$

Estimate (7.2.3) yields

$$\gamma_k(M) \le C_2\, \gamma_k(L(M)). \tag{7.2.8}$$

Estimates (7.2.7) and (7.2.8) imply

$$\gamma_k(G(M)) \le 2C_2\, \mu \sqrt{\frac{T}{2k}}\, \gamma_k(L(M)).$$

Choosing k so that $2C_2\, \mu \sqrt{\frac{T}{2k}} < 1$, we conclude. □

7.3 A priori estimates of solutions and solvability of the approximating equations

Consider the following auxiliary set of operator equations:

$$v' + A(v) - \lambda K_\varepsilon(v) + \lambda C(v, Z_\delta(v)) = f_\varepsilon, \quad \lambda \in [0,1]. \tag{7.3.1$_\lambda$}$$

At $\lambda = 1$ this equation coincides with (7.1.17).

Let us derive some a priori estimates of solutions for this set.

Theorem 7.3.1. *For any solution $v \in W$ to the Cauchy problem (7.3.1$_\lambda$), (7.1.18), $\lambda \in [0,1]$, one has*

$$\|v\|_{EC} \le C\,(1 + \|f_\varepsilon\|_{E^*} + \|a\|), \tag{7.3.2}$$

$$\|v'\|_{E^*} \le C\,(1 + \|f_\varepsilon\|_{E^*} + \|a\|), \tag{7.3.3}$$

where the constant C depends on ε, but is independent of v and $\lambda \in [0,1]$.

Proof. Let $v \in W$ be an arbitrary solution to the Cauchy problem (7.3.1$_\lambda$), (7.1.18) for some $\lambda \in [0,1]$. Then

$$L(v) = \lambda K_\varepsilon - \lambda G(v) + (f_\varepsilon, a). \tag{7.3.4}$$

As $L(0) = 0$, estimate (7.2.3) yields

$$\|v\|_{k,EC} \le C_2 \|L(v)\|_{k,E^*\times H}. \tag{7.3.5}$$

Similarly, $C(0, Z_\delta(v)) = 0$, and estimate (7.2.6) implies the following inequality

$$\|C(v, Z_\delta(v))\|_{k,E^*} = \|C(v, Z_\delta(v)) - C(0, Z_\delta(v))\|_{k,E^*} \le \mu \sqrt{\frac{T}{2k}}\, \|v\|_{k,E}. \tag{7.3.6}$$

Taking into account estimate (7.2.2), we conclude from equality (7.3.4) and estimates (7.3.5), (7.3.6) that

$$\|v\|_{k,EC} \le C\Big(\frac{1}{\varepsilon} + \mu \sqrt{\frac{T}{2k}}\, \|v\|_{k,E} + \|f_\varepsilon\|_{k,E^*} + \|a\|\Big).$$

Since $\|v\|_{k,E} \leq \|v\|_{k,EC}$, for k large enough (such that $C\mu\sqrt{\frac{T}{2k}} < \frac{1}{2}$) we obtain

$$\|v\|_{k,EC} \leq 2C\left(\frac{1}{\varepsilon} + \|f_\varepsilon\|_{k,E^*} + \|a\|\right).$$

Taking into account the equivalence of norms $\|\cdot\|_{k,EC}$ and $\|\cdot\|_{EC}$, $\|\cdot\|_{k,E^*}$ and $\|\cdot\|_{E^*}$, we get estimate (7.3.2).

Now express v' from equation (7.3.1$_\lambda$):

$$v' = -(A(v) - \lambda K_\varepsilon(v) + \lambda C(v, Z_\delta(v)) - f_\varepsilon).$$

Estimate (7.3.3) follows from estimate (7.3.2) and the properties of the maps A, K_ε, C. □

Let us show the solvability of the Cauchy problem (7.1.17), (7.1.18) via the degree theory for the class of $\mathcal{A}$-condensing perturbations of invertible maps (see [86]).

Proof of Theorem 7.1.1. As it was already mentioned, problem (7.1.17), (7.1.18) is equivalent to the operator equation (7.1.19).

Similarly, the problem (7.3.1$_\lambda$), (7.1.18) is equivalent to the operator equation

$$L(v) - \lambda(K_\varepsilon(v) - G(v)) = (f_\varepsilon, a), \tag{7.3.7$_\lambda$}$$

so it suffices to show solvability of (7.3.7$_\lambda$).

Due to Theorem 7.2.2 and Lemma 7.2.1, the map $\lambda(K_\varepsilon - G) : W \times [0,1] \to E^*$ is L-condensing with respect to the Kuratowski measure of noncompactness γ_k. Moreover, a priori estimates (7.3.2), (7.3.3) imply that equations (7.3.7$_\lambda$) have no solutions on the boundary of a ball $B_R \subset W$ of large radius R centered at the origin. Hence, for every $\lambda \in [0,1]$ the topological degree (see [86])

$$\deg_2(L - \lambda(K_\varepsilon - G), \overline{B}_R, (f_\varepsilon, a))$$

is defined. Since the degree of a map is conserved with the change of λ (homotopy invariance), we have

$$\deg_2(L - K_\varepsilon + G, \overline{B}_R, (f_\varepsilon, a)) = \deg_2(L, \overline{B}_R, (f_\varepsilon, a)).$$

The map L is invertible, so the equation

$$L(v) = (f, a)$$

has a unique solution u_0 in W, and for u_0 a priori estimates (7.3.2), (7.3.3) hold. Then $u_0 \in B_R$ and $\deg_2(L, \overline{B}_R, (f_\varepsilon, a)) = 1$ [86]. Therefore

$$\deg_2(L - K_\varepsilon + G, \overline{B}_R, (f_\varepsilon, a)) = 1.$$

Since this degree is nonzero, we have existence of solutions for the operator equation (7.3.7$_\lambda$). □

7.4 A priori estimate and existence of solutions for the regularized problem

In this section we show that the solutions of approximating problems (7.1.17), (7.1.18) converge in some sense to a solution to problem (7.1.14), (7.1.15) as $\varepsilon \to 0$.

For the functions $v \in E \cap C_w([0,T];H)$, consider the norm

$$\|v\|_{EL} = \|v(t)\|_{L_\infty(0,T;H)} + \|v\|_E$$

and the equivalent norms $\|v\|_{k,EL} = \|\bar{v}\|_{EL}$ where $\bar{v}(t) = e^{-kt}v(t)$, $k \geq 0$.

Theorem 7.4.1. *For any solution $v \in W_1 \cap C_w([0,T];H)$ of problem (7.1.17), (7.1.18) with $\varepsilon \geq 0$, one has the estimate*

$$\|v\|_{EL} \leq C\left(1 + \|f_{1,\varepsilon}\|_{L_1(0,T;H^*)} + \|f_2\|_{L_2(0,T;V^*)} + \|a\|\right), \tag{7.4.1}$$

with a constant C independent of ε and δ.

Proof. Let $v \in W_1 \cap C_w([0,T];H)$ be a solution of problem (7.1.17), (7.1.18) for some $\varepsilon \geq 0$. Then

$$v' + A(v) - K_\varepsilon(v) + C(v, Z_\delta(v)) = f_{1,\varepsilon} + f_2. \tag{7.4.2}$$

Let $v(t) = e^{kt}\overline{v}(t)$, $\overline{K}_\varepsilon(\overline{v}) = e^{-kt}K_\varepsilon(e^{kt}\overline{v})$, $\overline{C}(\overline{v}, Z_\delta(e^{kt}\overline{v})) = e^{-kt}C(e^{kt}\overline{v}, Z_\delta(e^{kt}\overline{v}))$, $\overline{f}_{1,\varepsilon} = e^{-kt}f_{1,\varepsilon}$, $\overline{f}_2 = e^{-kt}f_2$. Multiplying (7.4.2) by e^{-kt}, we obtain

$$\overline{v}' + k\overline{v} + A(\overline{v}) - \overline{K}_\varepsilon(\overline{v}) + \overline{C}(\overline{v}, Z_\delta(e^{kt}\overline{v})) = \overline{f}_{1,\varepsilon} + \overline{f}_2. \tag{7.4.3}$$

Consider the values of the functionals in the left-hand and right-hand members of (7.4.3) on the function $\overline{v}$:

$$\begin{aligned}\frac{1}{2}\frac{d}{dt}\|\overline{v}(t)\|^2 &+ k\|\overline{v}(t)\|^2 + (A(\overline{v}(t)), \overline{v}(t)) - (\overline{K}_\varepsilon(\overline{v}(t)), \overline{v}(t))\\ &= -(\overline{C}(\overline{v}, Z_\delta(e^{kt}\overline{v}))(t), \overline{v}(t)) + (\overline{f}_{1,\varepsilon}, \overline{v}(t)) + (\overline{f}_2, \overline{v}(t)).\end{aligned} \tag{7.4.4}$$

Formula (6.1.21) with $\tau \equiv 0$ yields $(\overline{K}_\varepsilon(\overline{v}(t)), \overline{v}(t)) = 0$ for all $t \in [0,T]$. Therefore, integrating both parts of (7.4.4) along a segment $[0,t]$, we get

$$\begin{aligned}\frac{1}{2}\|\overline{v}(t)\|^2 &+ k\|\overline{v}\|^2_{L_2(0,T;H)} + \mu_0\|\overline{v}\|^2_{L_2(0,T;V)}\\ &= \frac{1}{2}\|\overline{a}\|^2 - \int_0^t (\overline{C}(\overline{v}, Z_\delta(e^{k\varsigma}\overline{v}))(\varsigma), \overline{v}(\varsigma))\,d\varsigma + \int_0^t (\overline{f}_{1,\varepsilon}(\varsigma), \overline{v}(\varsigma))\,d\varsigma\\ &\quad + \int_0^t (\overline{f}_2(\varsigma), \overline{v}(\varsigma))\,d\varsigma.\end{aligned}$$

From here and from estimate (7.2.6) for $u = 0$ with the help of Cauchy's inequality we arrive at

$$\frac{1}{2}\,\|\overline{v}(t)\|^2 + k\|\overline{v}\|^2_{L_2(0,T;H)} + \mu_0\|\overline{v}\|^2_{L_2(0,T;V)}$$
$$\leq \frac{1}{2}\,\|\overline{a}\|^2 - \mu\sqrt{\frac{T}{2k}}\|\overline{v}\|^2_E + \|\overline{f}_{1,\varepsilon}\|_{L_1(0,T;H^*)}\cdot\|\overline{v}\|_{L_\infty(0,T;H)}$$
$$+ \|\overline{f}_2\|_{L_2(0,T;V^*)}\cdot\|\overline{v}\|_{L_2(0,T;V)}.$$

Considering k large enough so that $\mu\sqrt{\frac{T}{2k}} < \frac{\mu_0}{2}$, and using Cauchy's inequality, we get the estimate

$$\|\overline{v}\|^2_{L_\infty(0,T;H)} + 2k\|\overline{v}\|^2_{L_2(Q_T)} + \mu_0\|\overline{v}\|_E$$
$$\leq \|a\|^2_H + \frac{1}{2}\,\|\overline{v}\|^2_{L_\infty(0,T;H)} + \frac{1}{2}\,\mu_0\,\|\overline{v}\|^2_E + 2\|\overline{f}_{1,\varepsilon}\|^2_{L_1(0,T;H^*)}$$
$$+ \frac{2}{\mu_0}\|\overline{f}_2\|^2_{L_2(0,T;V^*)}.$$

Hence,

$$\frac{1}{2}\,\|\overline{v}\|^2_{L_\infty(0,T;H)} + 2k\|\overline{v}\|^2_{L_2(Q_T)} + \frac{1}{2}\,\mu_0\|\overline{v}\|_E$$
$$\leq \|\overline{a}\|^2 + 2\|\overline{f}_{1,\varepsilon}\|^2_{L_1(0,T;H^*)} + \frac{2}{\mu_0}\|\overline{f}_2\|^2_{L_2(0,T;V^*)}.$$

This estimate implies (7.4.1). □

Theorem 7.4.2. *For any solution $v \in W_1 \cap C_w([0,T];H)$ of problem (7.1.17), (7.1.18) with $\varepsilon \geq 0$, there is the estimate*

$$\|v'\|_{L_1(0,T;V^*)} \leq C\,(1 + \|f_{1,\varepsilon}\|_{L_1(0,T;H^*)} + \|f_2\|_{L_2(0,T;V^*)} + \|a\|)^2, \quad (7.4.5)$$

with a constant C independent of ε and δ.

Proof. We repeat the arguments used in the proof of estimate (7.3.3). Express v' from equality (7.1.17): $v' = -(A(v) - K_\varepsilon(v) + C(v, Z_\delta(v)) - f_\varepsilon)$. Thus,

$$\|v'\|_{L_1(0,T;V^*)} \leq C\,\big(\|A(v)\|_{E^*} + \|K_\varepsilon(v)\|_{L_1(0,T;V^*)}$$
$$+ \|C(v, Z_\delta(v))\|_{E^*} + \|f_{1,\varepsilon}\|_{L_1(0,T;H^*)} \quad (7.4.6)$$
$$+ \|f_2\|_{L_2(0,T;V^*)}\big).$$

Let us estimate $\|K_\varepsilon(v)\|_{L_1(0,T;V^*)}$. Since, for $n \leq 4$, the embedding $V \subset L_4(\Omega)^n$ is continuous, for $u \in V$ we have

$$\|K_\varepsilon(u)\|_{V^*} \leq \max_{i,j}\Big\|\frac{u_i u_j}{1+\varepsilon|u|^2}\Big\|_{L_2(\Omega)} \leq \max_{i,j}\|u_i u_j\|_{L_2(\Omega)} \leq C\|u\|^2_{L_4(\Omega)^n}.$$

Hence, $\|K_\varepsilon(v)\|_{L_1(0,T;V^*)} \le C\,\|v\|^2_{L_2(0,T;L_4(\Omega)^n)}$. Furthermore, the embedding $E = L_2(0,T;V) \subset L_2(0,T;L_4(\Omega)^n)$ is also continuous, so $\|v\|_{L_2(0,T;L_4(\Omega)^n)} \le C\,\|v\|_E$, and $\|K_\varepsilon(v)\|_{L_1(0,T;V^*)} \le C\,\|v\|^2_E$. This estimate and the properties of the maps A and C allow us to derive (7.4.5) from (7.4.1) and (7.4.6). □

Proof of Theorem 7.1.2. Take a sequence of positive numbers $\{\varepsilon_l\}$ converging to zero. For every number ε_l, the corresponding problem (7.1.17), (7.1.18) has at least one solution $v_l \in W$. Due to (7.1.16),

$$f_{1,\varepsilon_l} \to f_1 \ \text{ in } \ L_1(0,T;H^*) \ \text{ as } \ l \to \infty.$$

Hence, $\|f_{1,\varepsilon_l}\|_{L_1(0,T;H^*)}$ are uniformly bounded. Due to estimate (7.4.1), the sequence $\{\|v_l\|_{EL}\}$ is bounded. From estimate (7.4.5) it follows that the sequence of derivatives $\{v_l'\}$ is bounded in the space $L_1(0,T;V^*)$. Then without loss of generality we may assume that

$$\begin{aligned} v_l &\rightharpoonup v^* && \text{weakly in } E;\\ v_l &\rightharpoonup v^* && * \text{ - weakly in } L_\infty(0,T;H);\\ v_l &\to v^* && \text{strongly in } L_2(Q_T)^n;\\ v_l' &\rightharpoonup v^{*\prime} && \text{in the sense of distributions.} \end{aligned}$$

Since linear operators are weakly continuous, $A(v_l) \rightharpoonup A(v^*)$ weakly in E^*, and $\mathcal{E}(v_l)(s,x) \rightharpoonup \mathcal{E}(v^*)(s,x)$ weakly in $L_2(Q_T,\mathbb{R}^{n\times n})$.

Let us show that

$$C(v_l, Z_\delta(v_l)) \rightharpoonup C(v^*, Z_\delta(v^*)) \quad \text{weakly in } E^*. \tag{7.4.7}$$

Due to Lemma 7.2.3,

$$Z_\delta(v_l) \to Z_\delta(v^*) \ \text{ in } \ C([0,T]\times[0,T], C(\overline{\Omega})^n). \tag{7.4.8}$$

Let $h \in E$ be an arbitrary function. Then

$$\begin{aligned} &\big\langle C(v_l, Z_\delta(v_l)) - C(v^*, Z_\delta(v^*)), h\big\rangle \\ &\quad = \big\langle C(v_l, Z_\delta(v_l)) - C(v^*, Z_\delta(v_l)), h\big\rangle + \langle C(v^*, Z_\delta(v_l)) - C(v^*, Z_\delta(v^*)), h\big\rangle. \end{aligned}$$

The second term converges to zero due to (7.4.8) and continuity of the map C in z. In the first term, let us change the variable x for $z = Z_\delta(v_l)(s,t,x)$. The inverse change of variable looks like $x = Z_\delta(v_l)(t,s,z)$.

$$\begin{aligned} &\big\langle C(v_l, Z_\delta(v_l)) - C(v^*, Z_\delta(v_l)), h\big\rangle \\ &\quad = \int_0^T \int_\Omega \int_0^t \big[\mathcal{E}(v_l)(s, Z_\delta(v_l)(s,t,x)) - \mathcal{E}(v^*)(s, Z_\delta(v_l)(s,t,x))\big]\, ds \\ &\qquad\quad \cdot \mathcal{E}(h)(t,x)\, dx\, dt \\ &\quad = \int_0^T \int_\Omega \int_0^t \big[\mathcal{E}(v_l)(s,z) - \mathcal{E}(v^*)(s,z)\big]\, ds \cdot \big[\mathcal{E}(h)(t, Z_\delta(v_l)(t,s,z))\big]\, dz\, dt. \end{aligned}$$

The first expression in square brackets converges to zero weakly in $L_2(Q_T, \mathbb{R}^{n\times n})$. Using continuity of the map Φ defined by equality (7.2.4), from the proof of Lemma 7.2.2 and (7.4.8) we conclude that the second expression converges strongly in $L_2([0,T]\times[0,T], L_2(\Omega, \mathbb{R}^{n\times n}))$. Then the triple integral converges to zero as $l \to \infty$, and this finishes the proof of (7.4.7).

Note that $K_{\varepsilon_l}(v_l) \to K(v^*)$ in the sense of distributions (cf. (6.4.8) with $\tau \equiv 0$). Passing to the limit in the sense of distributions as $l \to \infty$ in the equality

$$v_l' + A(v_l) - K_\varepsilon(v_l) + C(v_l, Z_\delta(v_l)) = f_{1,\varepsilon_l} + f_2,$$

we get equality (7.1.14) for the function v^*. Hence, v^* is a solution of equation (7.1.14). Note that as $v^* \in E$, equality (7.1.14) implies $v^{*'} \in E_1^*$ (cf. the proof of Theorem 7.4.2). Therefore, $v^* \in W_1$. Since $v^* \in L_\infty(0,T;H)$, Lemmas 2.2.3 and 2.2.6 yield $v^* \in C_w([0,T];H)$. Arguments as in the proof of Theorem 6.4.1 show that v^* satisfies the initial condition. □

7.5 Another weak formulation for the regularized Jeffreys model

Now we are going to study the problem of convergence of weak solutions of the initial-boundary value problems for the regularized model (7.1.10) – (7.1.13) to weak solutions of the initial-boundary value problem for system (6.2.1) – (6.2.5) as δ tends to zero.

Hereafter we assume that Ω is a sufficiently regular bounded domain in $\mathbb{R}^n$, $2 \le n \le 3$.

Let $a \in H$, $\sigma_0 \in W_2^{-1}(\Omega, \mathbb{R}_S^{n\times n})$, $f \in L_2(0,T;V^*)$. We recall (Definition 6.2.1) that a weak solution of problem (6.2.1) – (6.2.5) is a pair of functions (u,σ),

$$u \in L_2(0,T;V)\bigcap C_w([0,T];H), \ \frac{du}{dt} \in L_1(0,T;V^*), \tag{7.5.1}$$

$$\sigma \in L_2(0,T;L_2(\Omega,\mathbb{R}_S^{N\times N}))\bigcap C_w([0,T];H^{-1}(\Omega,\mathbb{R}_S^{N\times N})) \tag{7.5.2}$$

satisfying the initial condition

$$u|_{t=0} = a, \ \sigma|_{t=0} = \sigma_0 \tag{7.5.3}$$

and the identities

$$\frac{d}{dt}(u,\varphi) + (\sigma, \nabla\varphi) - \sum_{i=1}^{n} \Big(u_i u, \frac{\partial \varphi}{\partial x_i}\Big) = \langle f, \varphi \rangle, \tag{7.5.4}$$

$$\begin{aligned}(\sigma,\Phi) + \lambda_1 \frac{d}{dt}(\sigma,\Phi) - \lambda_1 \sum_{i=1}^{n} \Big(u_i\sigma, \frac{\partial \Phi}{\partial x_i}\Big)\\ = 2\eta(\mathcal{E}(u),\Phi) + 2\eta\lambda_2\Big(\frac{d}{dt}(\mathcal{E}(u),\Phi) - \sum_{i=1}^{n}\Big(u_i\mathcal{E}(u), \frac{\partial \Phi}{\partial x_i}\Big)\Big)\end{aligned} \tag{7.5.5}$$

for all $\varphi \in \mathcal{V}$ and $\Phi \in C_0^\infty$ in the sense of distributions on $(0, T)$. As in Chapter 6, C_0^∞ stands for the space $C_0^\infty(\Omega, \mathbb{R}_S^{n\times n})$.

Hereafter let us assume (7.1.7) again. Definition 7.1.1 implies that a weak solution of the regularized problem (7.1.10) – (7.1.13) is a function (let us denote it u here) from class (7.5.1) satisfying the initial condition

$$u|_{t=0} = a \tag{7.5.6}$$

and the identity

$$\begin{aligned}\frac{d}{dt}(u(t),\varphi) - \sum_{i=1}^{n}\Big(u_i(t)u(t), \frac{\partial \varphi}{\partial x_i}\Big) + \frac{1}{2}\mu_0(\nabla u(t), \nabla\varphi)\\ + \mu\Big(\int_0^t e^{\frac{s-t}{\lambda_1}} \mathcal{E}(s, Z_\delta(u)(s,t,\cdot))\,ds, \mathcal{E}(\varphi)\Big) = \langle f(t), \varphi\rangle\end{aligned} \tag{7.5.7}$$

for all $\varphi \in V$ on $(0, T)$.

Let us also consider another problem which is obtained as an immediate regularization of problem (6.2.1) – (6.2.5). Let us define a weak solution of the problem as a pair of functions (u, σ) satisfying conditions (7.5.3), (7.5.4) and the equality

$$\begin{aligned}(\sigma,\Phi) + \lambda_1 \frac{d}{dt}(\sigma,\Phi) - \lambda_1 \sum_{i=1}^{n} \Big((S_\delta u)_i\sigma, \frac{\partial \Phi}{\partial x_i}\Big)\\ = 2\eta(\mathcal{E}(u),\Phi) + 2\eta\lambda_2\Big(\frac{d}{dt}(\mathcal{E}(u),\Phi) - \sum_{i=1}^{n}\Big((S_\delta u)_i\mathcal{E}(u), \frac{\partial \Phi}{\partial x_i}\Big)\Big)\end{aligned} \tag{7.5.8}$$

for all $\varphi \in \mathcal{V}$ and $\Phi \in C_0^\infty$ in the sense of distributions on $(0, T)$.

It turns out that problems (7.5.6), (7.5.7) and (7.5.3), (7.5.4), (7.5.8) are associated as follows:

Theorem 7.5.1 (see [74]). *Let a function u from class (7.5.1) be a solution of problem (7.5.6), (7.5.7). Then the pair (u, σ) where*

$$\sigma(t) = \mu_0\mathcal{E}(u)(t) + \mu\int_0^t e^{-\frac{t-s}{\lambda_1}} \mathcal{E}(u)(s, Z_\delta(u)(s,t,\cdot))\,ds, \tag{7.5.9}$$

is a solution of problem (7.5.3), (7.5.4), (7.5.8) in class (7.5.1), (7.5.2).

For the proof of Theorem 7.5.1 we will need two technical lemmas.
Denote by I_1 the second term in the right-hand side of (7.5.9).

Lemma 7.5.1. *The following estimate is valid:*

$$\|I_1\|_{L_\infty(0,T;L_2)} \leq K(\|u\|_{L_2(0,T;V)}), \tag{7.5.10}$$

where the constant K does not depend on δ.

Proof. Note that the expression I_1 can be rewritten as $\mu B(u, Z_\delta(u))$. We recall that the operator B is defined by the formula

$$B(u,z)(t) = \int_0^t e^{-\frac{t-s}{\lambda_1}} \mathcal{E}(u)(s, z(s,t,\cdot))\, ds, \quad u \in L_2(0,T;V), \quad z \in CG.$$

We have:

$$\begin{aligned}
\|B(u,z)\|^2_{L_\infty(0,T;L_2)} &= \sup_{t\in(0,T)} \int_\Omega \int_0^t \left[e^{-\frac{t-s}{\lambda_1}} |\mathcal{E}(u)(s,z(s,t,x))|\, ds\right]^2 dx \\
&\leq \sup_{t\in(0,T)} \int_0^T \int_\Omega |\mathcal{E}(u)(s,z(s,t,x))|^2\, dx ds \\
&= \int_0^T \int_\Omega |\mathcal{E}(u)(s,z)|^2\, dz ds \\
&= \int_0^T \|\mathcal{E}(u)(s,\cdot)\|^2 ds \leq K\|u\|^2_{L_2(0,T;V)}.
\end{aligned}$$

We have taken into account the identity $\det(\partial z/\partial x) = 1$. From these equalities and inequalities estimate (7.5.10) immediately follows. □

Lemma 7.5.2. *Let $v \in L_2(0,T; C^1(\overline{\Omega})^n \cap V)$, and z_v be a solution of equation (7.1.1). Then for every function $\zeta \in L_2(0,T;L_2(\Omega,\mathbb{R}))$ the integral*

$$\xi(t) = \int_0^t \zeta(s, z_v(s,t,\cdot))\, ds$$

belongs to $L_2(0,T;L_2)$ and the identity

$$\frac{d}{dt}(\xi,\psi) - \sum_{i=1}^n \left(v_i\xi, \frac{\partial\psi}{\partial x_i}\right) = (\zeta,\psi) \tag{7.5.11}$$

is valid for all $\psi \in C_0^\infty(\Omega,\mathbb{R})$ in the sense of distributions on $(0,T)$. Furthermore, $\frac{d\xi}{dt} \in L_1(0,T;H^{-2})$.

Proof. Take a sequence of functions $\{\zeta_m\}$ smooth enough and converging to ζ in $L_2(0,T;L_2)$. Then the corresponding ξ_m converge to ξ in $L_2(0,T;L_2)$. Really,

$$\begin{aligned}
&\|\xi_m-\xi\|^2_{L_2(0,T;L_2)}\\
&\quad=\int_0^T\Big\|\int_0^t \zeta_m(s,z_v(s,t,x))-\zeta(s,z_v(s,t,x))\,ds\Big\|^2_{L_2(\Omega,\mathbb{R}^{n\times n})}dt\\
&\quad\le\int_0^T\int_0^T\int_\Omega |(\zeta_m-\zeta)(s,z_v(s,t,x))|^2\,dx\,ds\,dt\\
&\quad=T\int_0^T\int_\Omega |(\zeta_m-\zeta)(s,z)|^2\,dz\,ds=T\|\zeta_m-\zeta\|^2_{L_2(0,T;L_2)}\\
&\quad\to 0.
\end{aligned}$$

But for the smooth functions ξ_m and ζ_m formula (1.3.9) gives

$$\frac{d}{dt}\xi_m=\zeta_m.$$

Here (only here!) $\frac{d}{dt}$ stands for the substantial derivative, i.e.

$$\frac{\partial\xi_m(t)(x)}{\partial t}+\sum_{i=1}^n v_i\frac{\partial\xi_m(t)(x)}{\partial x_i}=\zeta_m(t)(x).$$

Taking the $L_2(\Omega,\mathbb{R})$-scalar product of this equality with an arbitrary function $\psi\in C_0^\infty(\Omega,\mathbb{R})$ at almost all $t\in(0,T)$, and integrating by parts in the second term of the left-hand side, we arrive at

$$\frac{d}{dt}(\xi_m,\psi)-\sum_{i=1}^n\Big(v_i\xi_m,\frac{\partial\psi}{\partial x_i}\Big)=(\zeta_m,\psi).$$

Passing to the limit as $m\to\infty$ in the sense of distributions on $(0,T)$ we obtain formula (7.5.11).

We have from (7.5.11), using Hölder's inequality:

$$\begin{aligned}
\|\langle\xi',\psi\rangle\|_{L_1(0,T)}&=\Big\|\frac{d}{dt}(\xi,\psi)\Big\|_{L_1(0,T)}\le\sum_{i=1}^n\Big\|\Big(v_i\xi,\frac{\partial\psi}{\partial x_i}\Big)\Big\|_{L_1(0,T)}+\|(\zeta,\psi)\|_{L_1(0,T)}\\
&\le\sum_{i=1}^n\|v_i\|_{L_2(0,T;L_4)}\|\xi\|_{L_2(0,T;L_2)}\Big\|\frac{\partial\psi}{\partial x_i}\Big\|_{L_4}+\|\zeta\|_{L_2(0,T;L_2)}\|\psi\|_{L_2}\\
&\le\|v\|_{L_2(0,T;V)}\|\xi\|_{L_2(0,T;L_2)}\|\psi\|_{H_0^2}+\|\zeta\|_{L_2(0,T;L_2)}\|\psi\|_{L_2}\le C\|\psi\|_{H_0^2}.
\end{aligned}$$

Here we have used the continuity of embedding of $H_0^1(\Omega)$ and V into L_4. Thus, $\frac{d\xi}{dt}\in L_1(0,T;H^{-2})$. □

Proof of Theorem 7.5.1. Let us show first that the defined by formula (7.5.9) function σ belongs to class (7.5.2). Really, since $u \in L_2(0,T;V)\bigcap C_w([0,T];H)$, the first term in (7.5.9) $\mu_0\mathcal{E}(u) \in L_2(0,T;L_2)\bigcap C_w([0,T];H^{-1})$. By Lemma 7.5.1 the second term in (7.5.9) $I_1 \in L_\infty(0,T;H)$. And from Lemma 7.5.2 it follows that $\frac{d}{dt}I_1 \in L_1(0,T;H^{-2})$. Then, by Lemma 2.2.3, $I_1 \in C([0,T];H^{-2})$, so, by Lemma 2.2.6, $I_1 \in C_w([0,T];H)$. Hence, σ belongs to class (7.5.2).

From (7.5.7) and (7.5.9) it immediately follows that (7.5.4) is fulfilled. And (7.5.9) and (7.1.7) yield

$$\sigma|_{t=0} = \mu_0\mathcal{E}(u)|_{t=0} = \mu_0\mathcal{E}(a) = \sigma_0.$$

Thus, the initial condition (7.5.3) is fulfilled.

It remains to show (7.5.8). Take an arbitrary $\Phi \in C_0^\infty$. We have from (7.5.9):

$$e^{\frac{t}{\lambda_1}}(\sigma - \mu_0\mathcal{E}(u)) = \mu\int_0^t e^{\frac{s}{\lambda_1}}\mathcal{E}(u)(s, Z_\delta(u)(s,t,\cdot))\,ds.$$

By Lemma 7.5.2 we have

$$\frac{d}{dt}(e^{\frac{t}{\lambda_1}}(\sigma-\mu_0\mathcal{E}(u)),\Phi)-e^{\frac{t}{\lambda_1}}\sum_{i=1}^n((S_\delta(u))_i(\sigma-\mu_0\mathcal{E}(u)),\frac{\partial\Phi}{\partial x_i}) = \mu e^{\frac{t}{\lambda_1}}(\mathcal{E}(u),\Phi),$$

whence

$$\frac{1}{\lambda_1}e^{\frac{t}{\lambda_1}}(\sigma - \mu_0\mathcal{E}(u),\Phi) + e^{\frac{t}{\lambda_1}}\frac{d}{dt}(\sigma - \mu_0\mathcal{E}(u),\Phi)$$
$$- e^{\frac{t}{\lambda_1}}\sum_{i=1}^n((S_\delta(u))_i(\sigma - \mu_0\mathcal{E}(u)),\frac{\partial\Phi}{\partial x_i}) = \mu e^{\frac{t}{\lambda_1}}(\mathcal{E}(u),\Phi).$$

Dividing by $\frac{1}{\lambda_1}e^{\frac{t}{\lambda_1}}$ and rearranging the terms, we obtain:

$$(\sigma,\Phi) + \lambda_1\frac{d}{dt}(\sigma,\Phi) - \lambda_1\sum_{i=1}^n((S_\delta u)_i\sigma,\frac{\partial\Phi}{\partial x_i})$$
$$= (\mu_0 + \mu\lambda_1)(\mathcal{E}(u),\Phi) + \mu_0\lambda_1\frac{d}{dt}(\mathcal{E}(u),\Phi) - \mu_0\lambda_1\sum_{i=1}^n((S_\delta u)_i\mathcal{E}(u),\frac{\partial\Phi}{\partial x_i}).$$

This immediately yields (7.5.8). □

7.6 Behaviour of solutions of regularized problems as $\delta \to 0$

In this section we assume $a \in H$, $\sigma_0 \in W_2^{-1}(\Omega, \mathbb{R}_S^{n\times n})$, $f \in L_2(0,T;V^*)$ and condition (7.1.7) to be fulfilled. Theorems 7.1.2, 7.4.1, 7.4.2 guarantee existence of solutions for problem (7.5.6), (7.5.7) which satisfy the estimate

$$\|u\|_{L_2(0,T;V)} + \|u\|_{L_\infty(0,T;H)} + \left\|\frac{du}{dt}\right\|_{L_1(0,T;V^*)} \leq K(\|a\|, \|f\|_{L_2(0,T;V^*)}), \tag{7.6.1}$$

where the constant K does not depend on δ. By Theorem 7.5.1 these solutions generate pairs of solutions for problem (7.5.3), (7.5.4), (7.5.8). Now we are interested in behavior of these solutions as $\delta \to 0$. The following theorem gives an answer to this question.

Theorem 7.6.1 (see [74]). *Let $\delta_k \to 0$ for $k \to \infty$ be a sequence of numbers, u_k be some solutions of problem (7.5.6), (7.5.7) at corresponding δ_k, satisfying estimate (7.6.1), and let σ_k be the corresponding deviatoric stress tensors given by formula (7.5.9). Then it is possible to select a subsequence $\delta_{k_m} \to 0$ for $m \to \infty$ and a pair (u_*, σ_*) so that*

the pair (u_*, σ_*) is a weak solution for problem (6.2.1) – (6.2.5);

$$u_{k_m} \to u_* \text{*-weakly in } L_\infty(0,T;H), \text{weakly in } L_2(0,T;V),$$
$$\text{strongly in } L_2(0,T;H);$$
$$\sigma_{k_m} \to \sigma_* \text{*-weakly in } L_\infty(0,T;H^{-1}), \text{weakly in } L_2(0,T;L_2).$$

Proof. From estimate (7.6.1) it follows that there exist a subsequence δ_{k_m} of sequence δ_k and a function u_* so that

$$u_{k_m} \to u_* \text{ weakly in } L_2(0,T;V),$$
$$u_{k_m} \to u_* \text{ *-weakly in } L_\infty(0,T;H)$$

and the sequence $\|\frac{du_{k_m}}{dt}\|_{L_1(0,T;V^*)}$ is bounded. Due to Theorem 2.2.6, without loss of generality we may assume that

$$u_{k_m} \to u_* \text{ strongly in } L_2(0,T;H).$$

So we have:

$$\mathcal{E}(u_{k_m}) \to \mathcal{E}(u_*) \text{ weakly in } L_2(0,T;L_2),$$
$$\mathcal{E}(u_{k_m}) \to \mathcal{E}(u_*) \text{ *-weakly in } L_\infty(0,T;H^{-1}).$$

By Lemma 7.5.1 the expressions

$$\mu \int_0^t e^{-\frac{t-s}{\lambda_1}} \mathcal{E}(u_{k_m})(s, Z_{\delta_{k_m}}(u_{k_m})(s,t,x))\,ds$$

are bounded in $L_\infty(0,T;L_2)$ and, therefore, in $L_\infty(0,T;H^{-1})$ and $L_2(0,T;L_2)$.

Then from (7.5.9) it follows that without loss of generality there is σ_* such that

$$\sigma_{k_m} \to \sigma_* \text{ weakly in } L_2(0,T;L_2),$$
$$\sigma_{k_m} \to \sigma_* \text{ *-weakly in } L_\infty(0,T;H^{-1}).$$

The operators S_δ converge strongly to the identity operator in $L_2(0,T;H)$ as $\delta \to 0$; therefore $\|S_\delta\| \le K$, where K does not depend on δ.

We have:

$$\begin{aligned}&\|S_{\delta_{k_m}} u_{k_m} - u_*\|_{L_2(0,T;H)}\\ &\quad\le \|S_{\delta_{k_m}}\|\, \|u_{k_m} - u_*\|_{L_2(0,T;H)} + \|(S_{\delta_{k_m}} - I)u_*\|_{L_2(0,T;H)}.\end{aligned}$$

Both terms vanish as $m \to \infty$. Thus,

$$S_{\delta_{k_m}} u_{k_m} \to u_* \text{ strongly in } L_2(0,T;H).$$

Fix an arbitrary function $\varphi \in \mathcal{V}$ and $\Phi \in C_0^\infty$. We have:

$$\begin{aligned}&\int_0^T \Big|\sum_{i=1}^n \Big[\Big((S_{\delta_{k_m}} u_{k_m})_i \sigma_{k_m}, \frac{\partial \Phi}{\partial x_i}\Big) - \Big(u_{*i}\sigma_*, \frac{\partial \Phi}{\partial x_i}\Big)\Big]\Big|\,dt\\
&\quad\le \int_0^T \sum_{i=1}^n \Big[\Big|\Big([(S_{\delta_{k_m}} u_{k_m})_i - u_{*i}]\sigma_{k_m}, \frac{\partial \Phi}{\partial x_i}\Big)\Big| + \Big|\Big(u_{*i}[\sigma_{k_m} - \sigma_*], \frac{\partial \Phi}{\partial x_i}\Big)\Big|\Big]\,dt\\
&\quad\le \sum_{i=1}^n \Big[\|S_{\delta_{k_m}} u_{k_m} - u_*\|_{L_2(0,T;H)} \|\sigma_{k_m}\|_{L_2(0,T;L_2)} \Big\|\frac{\partial \Phi}{\partial x_i}\Big\|_{L_\infty}\\
&\qquad + \Big\langle \sigma_{k_m} - \sigma_*, u_{*i}\frac{\partial \Phi}{\partial x_i}\Big\rangle_{L_\infty(0,T;L_2)\times L_1(0,T;L_2)}\Big].\end{aligned}$$

All terms converge to zero. Therefore

$$\sum_{i=1}^n \Big((S_{\delta_{k_m}} u_{k_m})_i \sigma_{k_m}, \frac{\partial \Phi}{\partial x_i}\Big) \to \sum_{i=1}^n \Big(u_{*i}\sigma_*, \frac{\partial \Phi}{\partial x_i}\Big) \text{ in } L_1(0,T).$$

Similarly

$$\sum_{i=1}^n \Big((S_{\delta_{k_m}} u_{k_m})_i \mathcal{E}(u_{k_m}), \frac{\partial \Phi}{\partial x_i}\Big) \to \sum_{i=1}^n \Big(u_{*i}\mathcal{E}(u_*), \frac{\partial \Phi}{\partial x_i}\Big) \text{ in } L_1(0,T).$$

It is easy to see that the remaining terms in (7.5.4) and (7.5.8) with $u_{k_m}, \sigma_{k_m}, \delta_{k_m}$ substituted into these equalities converge to corresponding terms of (7.5.4) and (7.5.5) with u_*, σ_* in the sense of distributions on $(0, T)$. Therefore the pair (u_*, σ_*) satisfies (7.5.4), (7.5.5). As in the proof of Theorem 6.4.1, one can check that the initial condition (7.5.3) is fulfilled. Furthermore, as in the proof of estimates (6.3.14) and (6.5.2) one can establish that

$$\frac{du_*}{dt} \in L_1(0, T; V^*), \frac{d}{dt}(\sigma_* - \mu_0 \mathcal{E}(u_*)) \in L_2(0, T; H^{-2}).$$

By Lemma 2.2.3, $u_* \in C([0, T]; V^*), \sigma_* - \mu_0 \mathcal{E}(u_*) \in C([0, T]; H^{-2})$. Since $u_* \in L_\infty(0, T; H)$ and $\sigma_* - \mu_0 \mathcal{E}(u_*) \in L_\infty(0, T; H^{-1})$, Lemma 2.2.6 implies that $u_* \in C_w([0, T]; H)$, $\sigma_* - \mu_0 \mathcal{E}(u_*) \in C_w([0, T]; H^{-1})$. This yields $\mathcal{E}(u_*) \in C_w([0, T]; H^{-1})$ and $\sigma_* \in C_w([0, T]; H^{-1})$. □

7.7 Two constructions of regularization operator

7.7.1 The first construction

Let Ω be a sufficiently regular bounded domain in $\mathbb{R}^n$.

For $1 \leq q < \infty$, denote the closure of the set $\mathcal{V}(\Omega)$ in the norm of the space $L_q(\Omega)^n$ by H_q, and the closure of $\mathcal{V}(\Omega)$ in the norm of the space $W_q^1(\Omega)^n$ by V_q.

In this section we construct the regularization operator S_δ (see [17, 18]). We recall (Section 7.1) that it should satisfy the properties: i) the operator $S_\delta : H \to C^1(\overline{\Omega})^n \cap V$ for $\delta > 0$; ii) the map $S_\delta : L_2(0, T; H) \to L_2(0, T; C^1(\overline{\Omega})^n \cap V)$ generated by this operator is continuous; iii) the operators $S_\delta : L_2(0, T; H) \to L_2(0, T; H)$ converge strongly to the identity operator I as $\delta \to 0$. However, we are going to present more general constructions which are suitable not only for L_2-case but also for L_q-case. Namely, we are constructing an operator $S_\delta : H_q \to C^1(\overline{\Omega})^n \cap V_q$ such that the map $S_\delta : L_q(0, T; H_q) \to L_q(0, T; C^1(\overline{\Omega})^n \cap V_q)$ is continuous, and $S_\delta(v)$ converges to v in $L_q(0, T; H_q)$ as $\delta \to 0$ for all $v \in L_q(0, T; V_q)$ (and even for $v \in L_2(0, T; H)$ if $q = 2$).

We point out that classical averaging procedures are not directly applicable here since they do not conserve the condition $v(t) \in V_q$.

Choose a finite covering of the domain Ω by sets $U, U_1, \dots, U_k$ such that i) $U \subset \overline{U} \subset \Omega$, ii) each of the sets $\partial\Omega \cap U_j$, $j = 1, 2, \dots, k$, is not empty and is a graph of a Lipschitz function, iii) each of the sets $\Omega \cap U_j$ is star-shaped with respect to one of its points x_j^*.

Choose a C^∞-smooth partition of unity subordinated to the covering $U, U_1, \dots, U_k$ of the set Ω, i.e. scalar functions ϕ, ϕ_j, $j = 1, 2, \dots, k$, on Ω such that

$$\phi + \sum_{j=1}^{k} \phi_j = 1, \quad \text{where} \quad \operatorname{supp} \phi \subset U, \quad \operatorname{supp} \phi_j \subset U_j, \quad j = 1, 2, \dots, k.$$

Then for a function $v \in H_q$ we have

$$v = \phi v + \sum_{j=1}^{k} \phi_j v, \tag{7.7.1}$$

and each term belongs to $L_q(\Omega)^n$.

Let $\sigma_\varepsilon(x_j^*)$, $\varepsilon \neq 0$, denote the homothetic transformation with coefficient $1 - \varepsilon$:

$$x \mapsto x_j^* \cdot \varepsilon + x \cdot (1 - \varepsilon).$$

For this transformation $\sigma_\varepsilon(\Omega \cap U_j) \subset \sigma_\varepsilon(\overline{\Omega} \cap U_j) \subset \Omega \cap U_j$ for every j. Denote by u_j the function $u_j = \phi_j v$ and by $\sigma_\varepsilon[u_j]$ the function $\sigma_\varepsilon[u_j](x) = u_j(\sigma_\varepsilon(x))$. We assume that each of the functions u_j is continued by zero onto the whole space $\mathbb{R}^n$. Due to [61, Chapter I, Lemma 1.1], $\sigma_\varepsilon[u_j]$ converges to u_j in $L_q(\Omega)^n$ as $\varepsilon \to 0$. Note also that for fixed $\varepsilon > 0$ the support of each function $\sigma_\varepsilon[u_j]$, $j = 1, 2, \dots, k$, and its 2δ-neighbourhood is contained in $U_j \cap \Omega$ for $\delta > 0$ small enough. The same is true for the function $u = \phi v$, since the function ϕ has a compact support in Ω. Thus, the function

$$\overline{v} = u + \sum_{j=1}^{k} \sigma_\varepsilon[u_j]$$

has a compact support in Ω.

Take the subdomain of the domain Ω of the form

$$\Omega_\delta = \{x \in \Omega : \operatorname{dist}(x, \partial\Omega) > \delta\},$$

containing the support of the function $\overline{v}$.

It is known (see [6, 36]) that the space $L_q(\Omega_\delta)^n$ may be decomposed in the following direct sum: $L_q(\Omega_\delta)^n = H_q(\Omega_\delta) \oplus G_q(\Omega_\delta)$, where $G_q(\Omega_\delta) = \{\nabla p : p \in W_q^1(\Omega_\delta)\}$. Denote by P_δ the operator of projection from $L_q(\Omega_\delta)^n$ onto $H_q(\Omega_\delta)$. Applying this operator to the function $\overline{v}$, one obtains $v_\delta = P_\delta(\overline{v})$. The function v_δ belongs to space $H_q(\Omega_\delta)$, therefore its continuation by zero onto the domain Ω has the property $\operatorname{div} v_\delta = 0$ in every point of the domain. Hence, $v_\delta \in H_q = H_q(\Omega)$.

Let ρ be a function of the class C^∞ with a compact support in the ball $B_1(0)$ of radius 1 centered at the origin, such that $\rho \geq 0$, $\int_{\mathbb{R}^n} \rho(x)dx = \int_{B_1(0)} \rho(x)dx = 1$. Denote by ρ_δ the function $\frac{1}{\delta^n}\rho(\frac{x}{\delta})$. As $\delta \to 0$ the functions ρ_δ converge in the sense of distributions to Dirac function and $\rho_\delta * v \to v$ in $L_q(\mathbb{R}^n)^n$ for any function $v \in L_q(\mathbb{R}^n)^n$, where $*$ is the convolution of functions.

Applying the operation of Steklov averaging to this function, we obtain $\widetilde{v} = \rho_\delta * v_\delta$. The choice of δ ensures the condition that the support of $\widetilde{v}$ is a compact set in Ω. Since

$$\operatorname{div} \widetilde{v} = \operatorname{div}(\rho_\delta * v_\delta) = \rho_\delta * \operatorname{div} v_\delta = 0,$$

one has $\widetilde{v} \in V \cap C^\infty(\overline{\Omega})^n$.

Each of the transformations used at the construction of $\widetilde{v}$ determines a linear bounded map in the corresponding spaces. Therefore the map $S_\delta : v \to \widetilde{v} = \rho_\delta * v_\delta$ satisfies the conditions on a regularization operator. Namely, $S_\delta : H_q \to V_q \cap C^1(\overline{\Omega})^n$ is continuous. Since the construction of S_δ does not depend on t, the map $S_\delta : L_q(0, T; H_q) \to L_q(0, T; V_q \cap C^1(\overline{\Omega})^n)$ is also continuous.

Besides, $S_\delta v \to v$ in $L_q(0, T; H_q)$ as $\delta \to 0$. The check of this fact is rather long, so we shall consider another construction of a regularization operator and present a proof of the convergence only for the second construction.

7.7.2 The second construction

Let Ω be a sufficiently regular bounded domain in $\mathbb{R}^n$. For $\delta > 0$, we denote by Ω_δ the set

$$\Omega_\delta = \{x \in \Omega : \operatorname{dist}(x, \partial\Omega) > \delta\}.$$

Let $v \in H_q$ be an arbitrary function. Denote by v_δ the restriction of the function v to the domain Ω_δ: $v_\delta = v \mid_{\Omega_\delta}$. We suppose that the function v_δ is continued by zero onto $\Omega \backslash \Omega_\delta$.

We recall that by P_δ we denote the operator of projection from $L_q(\Omega_\delta)^n$ onto $H_q(\Omega_\delta)$. Applying this operator to the function v_δ, we obtain $\overline{v} = P_\delta(v_\delta)$. The function $\overline{v}$ belongs to the space $H_q(\Omega_\delta)$, therefore its continuation by zero onto the domain Ω has the property $\overline{v} \in H_q = H_q(\Omega)$. Hence, $\operatorname{div} \overline{v} = 0$ in every point of the domain. Really, by definition of the space $H_q(\Omega_\delta)$ the function $\overline{v}$ is a limit of a sequence of functions $w_k \in \mathcal{V}(\Omega_\delta)$ in the norm of the space $L_q(\Omega_\delta)$. Continuing the functions w_k by zero on the complement of the domain Ω_δ, we obtain functions $\overline{w}_k \in \mathcal{V}(\Omega)$. It is easy to see that the function $\overline{v}$ is a limit of the sequence of functions $\overline{w}_k$ in the norm of the space $L_q(\Omega)$. Therefore this function belongs to the space $H_q(\Omega)$.

Again, let ρ be a function of the class C^∞ with a compact support in the ball of radius 1 centered at the origin, such that $\rho \geq 0$, $\int_{\mathbb{R}^n} \rho(x)dx = \int_{B_1(0)} \rho(x)dx = 1$. Denote by ρ_δ the function $\frac{2^n}{\delta^n}\rho(\frac{2x}{\delta})$.

Applying the operation of Steklov averaging to the function $\overline{v}$ we obtain $\widetilde{v} = \rho_\delta * \overline{v}$.

The choice of δ ensures the condition that the support of $\widetilde{v}$ is a compact set in Ω. Since

$$\operatorname{div} \widetilde{v} = \operatorname{div}(\rho_\delta * \overline{v}) = \rho_\delta * \operatorname{div} \overline{v} = 0,$$

one has $\widetilde{v} \in V \cap C^\infty(\overline{\Omega})^n$.

Each of the transformations used at the construction of $\widetilde{v}$ determines a linear bounded map in the corresponding spaces. Therefore the map $S_\delta : v \to \widetilde{v} = \rho_\delta * v_\delta$ satisfies the conditions on a regularization operator. Namely, $S_\delta : H_q \to V_q \cap C^1(\overline{\Omega})^n$ is continuous. Since the construction of S_δ does not depend on t, the map $S_\delta : L_q(0, T; H_q) \to L_q(0, T; V_q \cap C^1(\overline{\Omega})^n)$ is also continuous.

It is easy to see that the maps $S_\delta : V_q \to V_q \cap C^1(\overline{\Omega})^n$ and $S_\delta : L_q(0,T;V_q) \to L_q(0,T;V_q \cap C^1(\overline{\Omega})^n)$ are correctly defined and continuous.

Let us check that for any function $v \in V_q$ (or $v \in H$ for $q = 2$) one has

$$S_\delta v \to v \quad \text{in} \quad H_q \quad \text{as} \quad \delta \to 0.$$

First of all, it is easy to see that $\|v\ |_{\Omega\setminus\Omega_\delta}\ \|_{L_q(\Omega\setminus\Omega_\delta)^n} \to 0$ as $\delta \to 0$. It follows from the property of absolute continuity of the Lebesgue integral.

Let p be a solution of the boundary value problem

$$\begin{cases} \Delta p = 0, \\ \frac{\partial p}{\partial \nu_\delta}|_{\partial\Omega_\delta} = \gamma_{\nu_\delta}(v_\delta), \end{cases}$$

such that $\int_\Omega p(x)dx = 0$, where ν_δ is the exterior normal vector to the boundary $\partial\Omega_\delta$ of the domain Ω_δ and $\gamma_{\nu_\delta}(v_\delta) = v_\delta\ |_{\partial\Omega_\delta} \cdot \nu_\delta$ (this has a special meaning for $q = 2$ and $v \in H$, see [61]). Then $\overline{v} = P_\delta v_\delta = v_\delta - \operatorname{grad} p$ (cf. [61]).

Since

$$\|p\|_{W_q^1(\Omega_\delta)^n} \le C_0 \|\gamma_{\nu_\delta}(v_\delta)\|_{B_q^{1-1/q}(\partial\Omega_\delta)^n}$$

with some constant C_0 independent from δ, one has $\|p\|_{W_q^1(\Omega_\delta)^n} \to 0$ as $\delta \to 0$, if $\|\gamma_{\nu_\delta}(v_\delta)\|_{B_q^{1-1/q}(\partial\Omega_\delta)^n} \to 0$ as $\delta \to 0$ ($B_q^{1-1/q}$ is the Besov space [11]; it must be replaced by $H^{-1/2}$ if $q = 2$ and $v \in H$).

The concept of trace of a function assumes "continuous dependence" of values of a function on the variation of the manifold $\partial\Omega$. For small δ the manifolds $\partial\Omega_\delta$ and $\partial\Omega$ are close and there exists a one-to-one map of one manifold onto another, therefore the traces of a function on these manifolds $\gamma_{\nu_\delta}(v)$ and $\gamma_\nu(v)$ are also close. Since $\gamma_\nu(v) = 0$, one has $\|\gamma_{\nu_\delta}(v_\delta)\|_{B_q^{1-1/q}(\partial\Omega_\delta)^n}$ is close to zero, i.e. $\|\gamma_{\nu_\delta}(v_\delta)\|_{B_q^{1-1/q}(\partial\Omega_\delta)^n} \to 0$ as $\delta \to 0$. Thus, p tends to zero in W_q^1 as $\delta \to 0$.

Finally, the operation of Steklov averaging has the property

$$\rho_\delta * v \ \to \ v \ \text{in} \ L_q(\mathbb{R}^n)^n$$

for any function $v \in L_q(\mathbb{R}^n)^n$. Therefore

$$\|\widetilde{v} - v\|_{L_q(\Omega)^n} \le \|\rho_\delta * \overline{v} - \rho_\delta * v\|_{L_q(\Omega)^n} + \|\rho_\delta * v - v\|_{L_q(\Omega)^n},$$

and the second term tends to zero. For the first term we have the estimate

$$\begin{aligned} \|\rho_\delta * \overline{v} - \rho_\delta * v\|_{L_q(\Omega)^n} &\le C_1 \|\overline{v} - v\|_{L_q(\Omega)^n} \\ &\le C_1(\|\operatorname{grad} p\|_{L_q} + \|v\ |_{\Omega\setminus\Omega_\delta}\ \|_{L_q(\Omega\setminus\Omega_\delta)^n}) \\ &\le C_1(\|p\|_{W_q^1} + \|v\ |_{\Omega\setminus\Omega_\delta}\ \|_{L_q(\Omega\setminus\Omega_\delta)^n}) \end{aligned}$$

with C_1 independent from δ. Both terms in the parentheses vanish as $\delta \to 0$. Thus, $\|\widetilde{v} - v\|_{L_q(\Omega)^n} \to 0$ as $\delta \to 0$.

Since $S_\delta v = \widetilde{v}$, for every function $v \in V_q$ (or H for $q = 2$) we have

$$S_\delta v \to v \quad \text{in} \quad H_q \quad \text{as} \quad \delta \to 0.$$

Moreover, if $v \in L_q(0, T; V_q)$ (or $v \in L_2(0, T; H)$ for $q = 2$), a similar reasoning shows that

$$S_\delta v \to v \quad \text{in} \quad L_q(0, T; H_q) \quad \text{as} \quad \delta \to 0.$$

References

[1] R. A. Adams, Sobolev spaces, Acad. Press, New York–San Francisco–London, 1975.

[2] Ju. Ja. Agranovich, P. E. Sobolevskii, Motion of a nonlinear viscoelastic fluid, DAN SSSR, 1990, V. 314, no. 3, 521–525.

[3] Yu. Ya. Agranovich, P. E. Sobolevskii, Motion of nonlinear viscoelastic fluid, Nonlinear Anal. TMA, #6 (1998), V. 32, 755–760.

[4] Y. Y. Agranovich, V. G. Zvyagin, Investigations of Properties of Attractors for a Regularized Model of the Motion of a Nonlinear-Viscous Fluid, International Journal of Mathematics, Game Theory and Algebra, 12(1) (2001), 71–87.

[5] R. R. Akhmerov, M. I. Kamenskii, A. S. Potapov, A. E. Rodkina, B. N. Sadovskii, Measures of Noncompactness and Condensing Operators, Birkhäuser, Boston, Basel, Berlin, 1992.

[6] H. Amann, On the strong solvability of the Navier–Stokes equations, J. Math. Fluid Mech., 2 (2000), 16–98.

[7] G. Astarita, G. Marucci, Principles of non-Newtonian fluid hydromechanics, McGraw Hill, New York, 1974.

[8] J. P. Aubin, Un théorème de compacité, C.R. Acad. Sci., V. 256 (1963), 5042–5044.

[9] A. V. Babin, M. I. Vishik, Attractors of evolutionary equations, Nauka, Moscow, 1989.

[10] G. K. Batchelor, An Introduction to Fluid Dynamics, Cambridge University Press, 2000.

[11] O. V. Besov, V. P. Il'in, S. M. Nikol'skii, Integral representations of functions and imbedding theorems, Scripta Series in Mathematics, Edited by Mitchell H. Taibleson, V. H. Winston and Sons, Washington, D.C., Halsted Press [John Wiley and Sons], New York–Toronto, Ont.–London, 1978–1979.

[12] D. Cheban, Global Attractors of Nonautonomous Dissipative Dynamical Systems, Interdisciplinary Mathematical Sciences 1. River Edge, NJ, World Scientific, 2004.

[13] J.-Y. Chemin, N. Masmoudi, About lifespan of regular solutions of equations related to viscoelastic fluids, SIAM J. Math. Anal. 33 (2001), no. 1, 84–112.

[14] V. V. Chepyzhov, M. I. Vishik, Attractors of non-autonomous dynamical systems and their dimension, J. Math Pures Appl. 73 (1994), 279–333.

[15] V. V. Chepyzhov, M. I. Vishik, Evolution equations and their trajectory attractors, J. Math Pures Appl. 76 (1997), 913–964.

[16] V. T. Dmitrienko, V. G. Zvyagin, The topological degree method for equations of the Navier–Stokes type, Abstract and Applied Analysis, V. 2 (1997), No. 1–2, 1–45.

[17] V. T. Dmitrienko, V. G. Zvyagin, Investigation of a regularized model of motion of a viscoelastic medium, in: Analytical Approaches to Multidimensional Balance Laws, O. Rozanova (Ed.), Nova Science, New York, 2004.

[18] V. T. Dmitrienko, V. G. Zvyagin, Construction of a regularization operator in models of viscoelastic media motion, Vestnik VSU, Ser. phys.-math., no.2 (2004), 148–153 (in Russian).

[19] Yu A. Dubinskii, Weak convergence for nonlinear elliptic and parabolic equations, Mat. Sb., 67 (109) (1965), No. 4, 609–642.

[20] H. Gajewski, K. Groeger, K. Zacharias, Nichtlineare Operatorgleichungen und Operatordifferentialgleichungen, Akademie-Verlag, Berlin, 1974.

[21] C. Guillopé, J. C. Saut, Existence results for the flow of viscoelastic fluids with a differential constitutive law, Nonlinear Anal. TMA, # 9 (1990), V.15, 849–869.

[22] C. Guillopé, J.-C. Saut, Mathematical problems arising in differential models for viscoelastic fluids, Mathematical topics in fluid mechanics (Lisbon, 1991), 64–92, Pitman Res. Notes Math. Ser., 274, Longman Sci. Tech., Harlow, 1992.

[23] I. Gyarmati Non-equilibrium thermodynamics. Field theory and variational principles, Springer, Heidelberg, 1970.

[24] E. Fernández Cara, F. Guillén, R. R. Ortega, Existence et unicité de solution forte locale en temps pour des fluides non Newtoniens de type Oldroyd (version L_r - L_s), C.R. Academie des Sciences Paris, tome 319, serie I, pp. 411–416.

[25] E. Fernández Cara, F. Guillén, R. R. Ortega, Some theoretical results concerning non-Newtonian fluids of the Oldroyd kind, Ann. Scuola Norm. Sup. Pisa Cl. Sci., V. 26 (1998), 1–29.

[26] A. Hakim, Mathematical analysis of viscoelastic fluids of White-Metzner type, J. Math. Anal. Appl. #3 (1994), V. 185, 675–705.

[27] J. Hale, Asymptotic behavior of dissipative systems, Mathematical surveys and Monographs, 25, American Math. Soc., Providence, R.I., 1988.

[28] A. Haraux, Systèmes dynamiques dissipatifs et applications, Masson, Paris, 1991.

[29] R. H. W. Hoppe, M. Y. Kuzmin, W. G. Litvinov, and V. G. Zvyagin, Flow of electrorheological fluid under conditions of slip on the boundary, Abstract and Applied Analysis, vol. 2006 (2006), Article ID 43560, 14 pages.

[30] A. A. Kotsiolis, A. P. Oskolkov, On the limiting conditions and an attractor for the motion equations of Oldroyd's fluid, Zap. nauch. sem. LOMI, V. 152 (1986), 67–71.

[31] M. Krasnoselskii, Topological methods in the theory of nonlinear integral equations, Gostehizdat, 1956, (Russian); Engl. transl., Macmillan, 1964.

[32] M. A. Krasnosel'skii, P. P. Zabreiko, Geometrical methods of nonlinear analysis, Grundlehren der Mathematischen Wissenschaften [Fundamental Principles of Mathematical Sciences], 263. Springer-Verlag, Berlin, 1984.

[33] M. A. Krasnosel'skii, P. P. Zabreiko, E. I. Pustyl'nik, P. E. Sobolevskii, Integral operators in spaces of summable functions, Monographs and Textbooks on Mechanics of Solids and Fluids, Mechanics: Analysis, Noordhoff International Publishing, Leiden, 1976.

[34] S. G. Krein and Y. I. Petunin, Scales of Banach spaces, Russ. Math. Surv., V. 21 (1966), no. 2, 85–159.

[35] O. A. Ladyzenskaya, On the dynamic system generated by Navier–Stokes equations, Zap. nauch. sem. LOMI, V. 27 (1972), 91–115.

[36] O. A. Ladyzenskaya, The Mathematical Theory of Viscous Incompressible Flow, New York: Gordon and Breach, 2nd edition (1963).

[37] J.-L. Lions, Quelques méthodes de résolution des problèmes aux limites non linéaires, Dunod, Paris, 1969.

[38] V. G. Litvinov, Motion of the nonlinear-viscous fluid, Moscow, Nauka, 1982, (Russian).

[39] W. G. Litvinov, A model and general problem on plastic flow under deformations, Universität Stuttgart, Bericht, 1999, 99/07.

[40] N. G. Lloyd, Degree theory, Cambridge University Press, 1978.

[41] E. Magenes and G. Stampacchia, I problemi al contorno per le equazioni differenziali di tipo elliptico, Ann. Scuola Norm. Sup. Pisa, # 3 (1958), V. 12, 247–358.

[42] A. Miranville and X. Wang, Attractors for nonautonomous nonhomogeneous Navier–Stokes equations, Nonlinearity 10 (1997), 1047–1061.

[43] S. M. Nikol'skii, Approximation of functions of several variables and imbedding theorems, Die Grundlehren der Mathematischen Wissenschaften, Band 205, Springer-Verlag, New York–Heidelberg, 1975.

[44] J. G. Oldroyd, Non-Newtonian flow of fluids and solids, in: Rheology: theory and applications, F. R. Eirich (Ed.), Academic Press Inc., Publishers, 1956.

[45] J. G. Oldroyd, On the formation of rheological equations of state, Proc. R. Soc. Lond, A200, 523–541, 1950.

[46] V. P. Orlov, P. E. Sobolevskii, On mathematical models of viscoelasticity with memory, Differential and Integral Equations, V.4, no. 1 (1991), 103–115.

[47] V. P. Orlov, P. E. Sobolevskii, Investigation of mathematical models of multidimensional viscoelastic media, Dokl. Akad. Nauk Ukrain. SSR Ser. A, no. 10 (1989), 31–35, 87 (Russian).

[48] V. P. Orlov, P. E. Sobolevskii, Investigation of mathematical models of multidimensional viscoelastic media, Preprint Akad. Nauk Ukrain. SSR, No. 27–29, Institute of applied problems of mechanics and mathematics, Lvov, 1989 (Russian).

[49] M. Reiner, Rheology, in: Handbuch der Physik, S. Flugge (Ed.), Bd. VI, Springer, 1958.

[50] M. Renardy, Existence of slow steady flows of viscoelastic fluids with differential constitutive equations, Z. angew. Math. Mech. V. 65 (1985), 449–451.

[51] M. Renardy, W. J. Hrusa, J. A. Nohel, Mathematical Problems in Viscoelasticity, Longman, London, 1987.

[52] G. de Rham, Variétés différentiables, Hermann, Paris, 1960.

[53] G. Sell, Global attractors for the three-dimensional Navier–Stokes equations, J. Dyn. Diff. Eq., V. 8 (1996), No. 1, 1–33.

[54] J. Simon, Compact sets in the space $L^p(0, T; B)$, Ann. Mat. Pura Appl., V. 146 (1987), 65–96.

[55] J. Simon, On the existence of the pressure for solutions of the variational Navier–Stokes equation, J. Mathematical Fluid Mech., 3 (1999), 225–234.

[56] J. Simon, Primitives de distributions et applications, Séminaire Analyse, Univ. Blaise Pascal, V. 6–7 (1991).

[57] G. F. Smith, On a Fundamental Error in Two Papers of C.-C. Wang "On Representations for Isotropic Functions, Parts I and II", Arch. Rational Mech. Anal., V. 36 (1970), 161–165.

[58] P. E. Sobolevskii, On the equations of parabolic type in a Banach space, Trudy Mos. Mat. Obsch., V.10 (1961), 297–350.

[59] E. M. Stein, Singular Integrals and Differentiability Properties of Functions, Princeton University Press, Princeton, New Jersey, 1970.

[60] R. Talhouk, Existence locale et unicité d'écoulements de fluides viscoélastiques dans des domaines non bornés, C.R. Acad. Sci. Paris. Serie I., t.328 (1999), 87–92.

[61] R. Temam, Navier–Stokes equations, Theory and numerical analysis, Studies in Mathematics and its Applications, Vol. 2, North-Holland Publishing Co., Amsterdam–New York–Oxford, 1977.

[62] R. Temam, Infinite-dimensional dynamical systems in mechanics and physics, 2nd edition, Springer-Verlag, 1997.

[63] C. Truesdell, A first course in rational continuum mechanics, The John Hopkins University, Baltimore, 1972.

[64] M. V. Turbin, Research of a mathematical model of low-concentrated aqueous polymer solutions, Abstr. Appl. Anal. 2006, Art. ID 12497, 27 pp.

[65] E. M. Turganbaev, Initial-boundary value problems for the equations of a viscoelastic fluid of Oldroyd type, Siberian Mathematical Journal, Vol. 36 (1995), No. 2, 389–403.

[66] M. I. Vishik, V. V. Chepyzhov, Trajectory and global attractors for 3D Navier–Stokes system, Mat. Zametki, V. 71 (2002), 194–213.

[67] D. A. Vorotnikov, On the motion of a nonlinear-viscous fluid in R^n, Vestnik VSU, Ser. phys.–math., no.1 (2002), p. 102–120 (in Russian).

[68] D. A. Vorotnikov, On the existence of weak stationary solutions of the boundary value problem in the Jeffreys model of motion of a viscoelastic medium, Izvestiya Vuzov, Ser. Math., no. 9 (2004), 13–17.

[69] D. A. Vorotnikov, Energy inequality and uniqueness of a weak solution of the initial-boundary value problem for the motion equations of viscoelastic medium, Vestnik VSU, Ser. phys.–math., no. 1 (2004), 96–102 (in Russian).

[70] D. A. Vorotnikov, Continuous dependence of solutions on data of initial-value problem for motion equations of nonlinear visoelastic medium, Vestnik VSU, Ser. phys.–math., no. 1 (2005), 148–151 (in Russian).

[71] D. A. Vorotnikov, Dissipative solutions for equations of viscoelastic diffusion in polymers, J. Math. Anal. Appl., Volume 339 (2008), 876–888.

[72] D. A. Vorotnikov, V. G. Zvyagin, On the existence of weak solutions for the initial-boundary value problem in the Jeffreys model of motion of a viscoelastic medium, Abstr. Appl. Anal., # 10 (2004), V. 2004, 815–829.

[73] D. A. Vorotnikov, V. G. Zvyagin, On the solvability of the initial-boundary value problem for the motion equations of nonlinear viscoelastic medium in the whole space, Nonlinear Anal. TMA, V. 58 (2004), 631–656.

[74] D. A. Vorotnikov, V. G. Zvyagin, On convergence of solutions of the regularized problem for motion equations of Jeffreys viscoelastic medium to solutions of the original problem, Journal of Mathematical Sciences, V.144 (2007), No.5, 4398–4408. Translated from Fundamentalnaya i prikladnaya matematika (FPM), V. 11 (2005), no. 4, 49–63.

[75] D. A. Vorotnikov, V. G. Zvyagin, On the trajectory and global attractors for the equations of motion of a visco-elastic medium, Russ. Math. Surv., V. 61 (2006), No. 2, 368–370.

[76] D. A. Vorotnikov, V. G. Zvyagin, Trajectory and global attractors of the boundary value problem for autonomous motion equations of viscoelastic medium, J. Math. Fluid Mech., V. 10 (2008), 19–44.

[77] D. A. Vorotnikov, V. G. Zvyagin, Uniform attractors for non-autonomous motion equations of viscoelastic medium, J. Math. Anal. Appl., Volume 325 (2007), Issue 1, 438–458.

[78] W. von Wahl, The equations of Navier–Stokes and abstract parabolic equations, Friedr. Vieweg und Sohn, Braunschweig, 1985.

[79] C.-C. Wang, A New Representation Theorem for Isotropic Functions: An Answer to Professor G. F. Smith's Criticism of my Papers on Representations for Isotropic Functions. Part 1. Scalar-valued Isotropic Functions, Arch. Rational Mech. Anal., V. 36 (1970), 166–197.

[80] C.-C. Wang, A New Representation Theorem for Isotropic Functions: An Answer to Professor G. F. Smith's Criticism of my Papers on Representations for Isotropic Functions. Part 2. Vector-valued Isotropic Functions, Symmetric Tensor-valued Isotropic Functions, and Skew-symmetric Tensor-valued Isotropic Functions, Arch. Rational Mech. Anal., V. 36 (1970), 198–223.

[81] C.-C. Wang, Corrigendum to My Recent Papers on "Representations for Isotropic Functions", Arch. Rational Mech. Anal., 1971, V. 43, 392–395.

[82] X. Wang. A remark on the characterization of the gradient of a distribution, Appl. Anal., V. 51 (1993), # 1–4, 35–40.

[83] W. L. Wilkinson, Non-Newtonian fluids, Pergamon Press, 1960.

[84] K. Yosida, Functional analysis, Sixth edition, Grundlehren der Mathematischen Wissenschaften [Fundamental Principles of Mathematical Sciences], 123, Springer-Verlag, Berlin–New York, 1980.

[85] V. G. Zvyagin, On the solvability of some initial-boundary value problems for mathematical models of the motion of nonlinearly viscous and viscoelastic fluids, Sovrem. Mat. Fundam. Napravl., 2 (2003), 57–69 (electronic) (Russian); translation in J. Math. Sci. (N. Y.), 124 (2004), no. 5, 5321–5334.

[86] V. T. Dmitrienko, V. G. Zvyagin, Homotopy classification of a class of continuous mappings, Mat. Zametki 31 (1982), no. 5, 801–812, 816 (Russian).

[87] V. G. Zvyagin, V. T. Dmitrienko, Approximating–topological approach to investigation of problems of hydrodynamics. Navier–Stokes system, Editorial URSS, Moscow, 2004.

[88] V. G. Zvyagin, V. T. Dmitrienko, On the weak solutions of the regularized model of a viscoelastic fluid, Differencial'nye Uravnenija, V. 38 (2002), no. 12, 1633–1645.

Index

Bei Fragen zur Produktsicherheit wenden Sie sich bitte an:
If you have any questions regarding product safety,
please contact:

Walter de Gruyter GmbH
Genthiner Straße 13
10785 Berlin
productsafety@degruyterbrill.com

पर्यटन
एवं
आर्थिक विकास

विनोद अग्रवाल

अर्जुन पब्लिशिंग हाउस

पर्यटन एवं आर्थिक विकास

संस्करण, 2012

ISBN 978-81-8330-401-6

प्रकाशक :

अर्जुन पब्लिशिंग हाऊस

4831/24, प्रह्लाद गली,

अंसारी रोड, दरियागंज,

नई दिल्ली–110002

फोन : 23272541, 23257835

फैक्स : 91-011-23257835

e-mail : arjunbooks2001@gmail.com

टाइपसेटिंग :

डिम्पल कम्प्यूटर्स

दिल्ली–110092

प्रिन्टर्स :

रोशन आफसेट प्रिंटर्स

दिल्ली - 53

प्रस्तावना

राष्ट्र की उन्नति व विकास करने में पर्यटन न केवल एक विशिष्ट प्रक्रिया के रूप में बल्कि संसार के तमाम राष्ट्रों में सभ्यता व सांस्कृतिक एवं भाषा परिवर्तन लाने का एक अत्यधिक उपयोगी साधन भी है। राष्ट्र के आर्थिक क्षेत्र में महत्व, विदेशी मुद्रा अर्जित करने में योगदान एवं रोजगार के विभिन्न स्रोत उपलब्ध कराने में पर्यटन का अपना एक विशिष्ट स्थान रहा है। पर्यटन के महत्त्व को देश की सामाजिक शिक्षा एवं सामाजिक गतिशीलता का माध्यम समझने तथा साथ ही पर्यटन के द्वारा विश्व के विभिन्न देशों में आपसी अच्छे संबंध स्थापित करने, घनिष्ठ सांस्कृतिक एवं व्यापारिक संबंध बनाने तथा विश्व शान्ति स्थापित करने का यह उचित समय है। राष्ट्रीय अर्थव्यवस्था में पर्यटन का पूर्ण प्रभाव सिर्फ पर्यटक व्यापार को अग्रगामी सीमा पर व्यय करने से प्राप्त नहीं होता है। इसमें उन समस्त उद्योगों को शामिल किया जाता है जो विभिन्न स्तरों पर माल एवं सेवाओं की पूर्ति करने के लिए जरूरी होते हैं। प्रत्यक्ष पर्यटन क्षेत्र व्यवहार उद्योगों द्वारा निर्मित उत्पादन वास्तविक अर्थ में पर्यटकों की आवश्यकतानुसार है जो सर्वप्रथम धन व्यय करते हैं तथा फिर समस्त क्रिया को जारी रखते हैं।

राष्ट्र के आर्थिक विकास में पर्यटन उद्योग की महत्त्वपूर्ण भूमिका होती है क्योंकि वर्तमान समय में पर्यटन उद्योग का लगातार विकास हो रहा है जो राष्ट्रीय आय का प्रमुख स्रोत है। राष्ट्रीय और अन्तर्राष्ट्रीय दोनों ही स्तर पर होने वाली आय से राष्ट्रीय आय में वृद्धि होती है। हालांकि स्वदेशी पर्यटन के क्षेत्र में यह तथ्य सत्य नहीं उतरता है, क्योंकि जहाँ तक खान-पान के व्यय का प्रश्न है यह अर्थव्यवस्था में एक स्थान से दूसरे स्थान में परिवर्तित होता रहता है।

पर्यटन के तहत सेवाओं व उपभोक्ताओं की वस्तुओं की मांग सम्मिलित है, अन्यथा इनका उत्पादन ही नहीं होता। इससे उत्पादन के क्षेत्रों को बढ़ावा मिलेगा, जिससे स्वयं राष्ट्रीय उपभोग में वृद्धि होगी। क्योंकि पर्यटन व्यय के आकार का प्रभाव उत्पादन की वृद्धि में परिलक्षित होता है, अतः विभिन्न क्षेत्रीय कच्चे माल के माध्यम से उस क्षेत्र व अन्य क्षेत्रों में आय वृद्धि की क्रिया से देखा जा सकता है। यह विकास उपभोग से बिल्कुल स्वतंत्र है एवं इससे समस्त अर्थव्यवस्था पर धारावाहिक प्रतिक्रिया के रूप में प्रभाव पड़ता है। आवास का निर्माण, पर्यटकों हेतु शहरीकरण, आधारिक संरचना की संस्थापना, बढ़ते हुए पर्यटन के परिवहन व यन्त्र आदि कुछ उदाहरण हैं, जिनसे जाहिर होता है कि पर्यटकों की माँग से किस तरह विनियोजन को बढ़ावा मिलता है। इस तरह यहाँ एक वेग वृद्धि सिद्धान्त लागू होता है। साधारण अर्थव्यवस्था में जब कभी उपभोग में संचित दर से वृद्धि होती है तो यह सिद्धान्त लागू होता है परन्तु यदि प्रति समय यह कम दर से बढ़ रही हो तो यह बिल्कुल लागू नहीं होगा एवं समान रूप में पर्यटन में भी लागू होता है। पर्यटन के क्षेत्र में उन्नति का द्विमुखी विकास व संचित विकास कायम रहना चाहिए।

प्रस्तुत पुस्तक 'पर्यटन एवं आर्थिक विकास' की रचना सरल व सुबोध भाषा में की गई है। इसकी रचना में कुछ लिखित व अलिखित स्रोतों से मदद ली गई है, जिनके लेखकों व प्रकाशकों के प्रति मैं अपना विनम्र आभार प्रकट करता हूँ और आशा करता हूँ कि पुस्तक पाठकों का सही मार्गदर्शन कर सकेगी।

—लेखक

अनुक्रम

1

पर्यटन की अवधारणा और परिभाषाएँ

आज से लगभग कुछ सालों पहले पर्यटन अर्थात् देश-विदेश की यात्रा करना उच्च वर्ग तक के लिए सीमित था। मगर अब सरकार, जनता तथा प्रशासन के द्वारा मिले समर्थन से यह अब उद्योग के रूप में दिखने लगा। राष्ट्रमंडल खेलों के समय दिल्ली में आए पर्यटकों के लिए विशेष रूप से पर्यटन से संबंधित बसों की व्यवस्था करना एक प्रकार से पर्यटन उद्योग का बिम्ब प्रस्तुत करता है तथा पर्यटकों की सुरक्षा, पर्यटन स्थलों की जानकारी, रहन-सहन का प्रबंध आदि व्यवस्था से पर्यटन उद्योग का विकास तेजी से हो रहा है।

आज के समय में विकसित एवं विकासशील देशों में आर्थिक वृद्धि दर उच्च होने से तथा यातायात के साधनों में क्रांतिकारी परिवर्तन होने से अब यह मध्यम तथा निम्न वर्ग के लोगों के लिए भी विदेशी यात्रा करना सुलभ हो गया है। इसलिए जगह-जगह पर हमें पर्यटन से संबंधित केन्द्र देखने को मिल जाते हैं तथा इसमें बैंक भी वित्तीय सहायता आसानी से प्रदान कर देते हैं।

राष्ट्र की उन्नति व विकास करने में पर्यटन न केवल एक विशिष्ट प्रक्रिया के रूप में बल्कि संसार के तमाम राष्ट्रों में सभ्यता व सांस्कृतिक एवं भाषा परिवर्तन लाने का एक अत्यधिक उपयोगी साधन भी है। राष्ट्र के आर्थिक क्षेत्र में महत्व, विदेशी मुद्रा अर्जित करने में योगदान एवं रोजगार के विभिन्न स्रोत उपलब्ध कराने में पर्यटन का अपना एक विशिष्ट स्थान रहा है। पर्यटन के महत्व को देश की सामाजिक शिक्षा एवं सामाजिक गतिशीलता का माध्यम समझने तथा साथ ही पर्यटन के द्वारा विश्व के विभिन्न देशों में आपसी अच्छे संबंध स्थापित करने, घनिष्ठ सांस्कृतिक एवं व्यापारिक संबंध बनाने तथा विश्व शान्ति स्थापित करने का यह उचित समय है, राष्ट्रीय अर्थव्यवस्था में पर्यटन का पूर्ण प्रभाव सिर्फ पर्यटक व्यापार को अग्रगामी सीमा पर व्यय करने से प्राप्त नहीं होता है। इसमें उन समस्त उद्योगों को शामिल किया जाता है जो विभिन्न स्तरों पर माल एवं सेवाओं की पूर्ति करने के लिए जरूरी होते हैं। प्रत्यक्ष पर्यटन क्षेत्र व्यवहार उद्योगों द्वारा निर्मित उत्पादन वास्तविक अर्थ में पर्यटकों की आवश्यकतानुसार हैं जो सर्वप्रथम धन व्यय करते हैं तथा फिर समस्त क्रिया को जारी रखते हैं। श्री **रिचर्ड्स** के मतानुसार, पर्यटन के व्यय करने एवं राष्ट्रीय अर्थव्यवस्था में दो मुख्य संबंध हैं। प्रथम, उपभोक्ता से संबंधित वस्तुएं जिनका सीधा संबंध है। इसके तहत पर्यटन के लिए किए गए व्यय जैसे परिवहन, आवास, खान-पान व अन्य संबंधित खर्चे व सेवाएं हैं। द्वितीय, पर्यटन से प्रत्यक्ष जुड़े हुए व्यवसाय व अन्य व्यवसाय व उद्योग जो पर्यटन उद्योग की वस्तुओं एवं सेवाओं की पूर्ति करते हैं।

विलासितापूर्ण क्रियाओं के अतिरिक्त पर्यटन के असंख्य रूप—सांस्कृतिक पर्यटन, स्वास्थ्यलाभ पर्यटन, क्रीड़ा पर्यटन, सम्मेलन/महासम्मेलन पर्यटन, धार्मिक सम्मेलन आदि-आदि हैं। इसी प्रकार, सामूहिक पर्यटन पर भी विशेष तौर पर ध्यान देने की जरूरत है ताकि व्यक्तिगत पर्यटक एवं सामूहिक पर्यटक को ठीक किस्म की वस्तुएं एवं सेवाएँ प्राप्त हो सकें।

पर्यटन अंग्रेजी शब्द 'Tourism' का संबंध 'Tour' से है जो लैटिन भाषा से लिया गया है। 'Tornos' का मतलब एक औजार से है जो एक पहिये की तरह गोलाकार होता है। यह एक गोलाकार पिन है, इसी 'Tornos' शब्द से यात्रा-चक्र या पैकेज टूर/एकमुश्त यात्रा का विचार सृजित हुआ है जो कि वर्तमान यात्रा व पर्यटन का मुख्य आवलंबन है। इस शब्द का प्रयोग करीब 1643 ई. में कई स्थानों के टूर, मनोविनोद भ्रमण यात्रा करने के लिये हुआ था। 'Tour' एक हिब्रू (Hebrew) शब्द है जो कि हिब्रू शब्द 'Torah' से लिया गया है। यहूदियों के

विधान से संबंधित है। यहूदी विधान, यहूदियों के रहन-सहन की परिभाषा वर्णित करता है। 'Tour' का अर्थ है कि यात्री किसी विशेष स्थान पर जाकर खोज करता है। ठीक इसी प्रकार एक पर्यटक की जिज्ञासा होती है कि वह उस स्थान के बावत पूर्ण ज्ञान अर्जित करे जिसकी बावत उसने सुना तथा साथ ही उसको व्यापार व रोजगार की संभावनाओं के बावत जानकारी, स्वास्थ्य एवं शिक्षा लाभ, पर्यावरण व मनोरंजन से संबंधित सम्पत्ति के विषय में पूर्ण जानकारी प्राप्त करने की अभिलाषा रहती है।

फ्रांसीसी साहित्य में 'grand tour' का अभिप्राय 'great tour', जो 'Tour' शब्द से लिया गया है। 'Tour' शब्द का अर्थ है; फ्रांस, जर्मनी, स्विट्जरलैन्ड, इटली आदि राष्ट्रों की यात्रा करना है।

पर्यटन शब्द के अन्तर्गत वे सभी निर्याती प्रक्रियाएं जुड़ी हैं जो पर्यटकों की आवश्यकताओं को पूरा करती हैं। किसी की गन्तव्य यात्रा का पर्यटन से मिलाप व सम्पर्क हो जाता है अगर वह निम्नलिखित शर्तें पूरी करने में सक्षम होती है–

– यात्रा ऐच्छिक हो।
– यात्रा का आशय किसी प्रकार का पारिश्रमिक अर्जित करना नहीं होना चाहिए।
– यात्रा अस्थायी हो।

पर्यटन हेतु संस्कृत साहित्य में तीन शब्दों का उपयोग किया गया है, जिनकी उत्पत्ति 'पर्यटन' शब्द से हुई है। इसका अर्थ अपने निवासित स्थान से कहीं दूर प्रस्थान करना है। ये तीन शब्द हैं–

पर्यटन–इसका अर्थ है–आराम एवं ज्ञान प्राप्ति के लिए यात्रा करना।

देशाटन–विदेशों में मुख्यतः आर्थिक फायदे के लिए यात्रा करना।

तीर्थाटन–धार्मिक लाभों के कारण यात्रा करना।

री जोस इगनेसी डी. एग्रील्ला के मतानुसार, पहले पर्यटन एक सैलानी के रूप में मोटर द्वारा यात्रा करने, साइकिलिंग, ऊँचे पर्वतों पर आरोहण, पड़ाव डालना, साहसिक भ्रमण तथा केलिपोत के संकलन के रूप में जाना जाता था। स्विस प्रोफेसर हुन्जीकर एवं क्राफ्ट के मतानुसार, "पर्यटन उन प्रवृत्तियों एवं संबंधों का गठजोड़ है जो यात्रा एवं प्रवासी के ठहरने से उत्पन्न होती है बशर्ते कि यह स्थायी निवास की क्रिया से संबंधित नहीं है।"

उपरोक्त परिभाषा को पर्यटन वैज्ञानिक विशेषज्ञों के आर्थिक संगठन ने भी मान्यता दी है। एक अन्य स्थान पर प्रोफेसर हुन्जीकर के मतानुसार, "पर्यटन उन

संबंधों एवं तथ्यों का गठजोड़ है जो यात्रा एवं व्यक्तियों के अल्पकालीन निवास के परिणामस्वरूप उत्पन्न होती है बशर्ते कि यह स्वयं के साधारण निवास स्थान से दूर हो तथा किसी किस्म के धन कमाने की इच्छा या इससे संबंधित न हो।"

एक अन्य स्थान पर पर्यटन की एक सामान्य परिभाषा दी गयी है, जिसके तहत लगभग सम्पूर्ण पहलू एवं परिणाम शामिल करने की कोशिश की गयी है। उनके लिए पर्यटन वह संबंध स्थापित करता है जो विदेशी व्यक्तियों के आवास (ठहरने) से संबद्ध है बशर्ते कि इस ठहरने की क्रिया का स्थायी व अस्थायी रूप में धन अर्जन करने से कोई संबंध न हो।

उपरोक्त वर्णित परिभाषा में यात्रा के सम्पूर्ण पक्षों की उपस्थिति है, परिणामस्वरूप पर्यटन एक मिश्रित प्रवृत्ति के रूप में निखारता है। इसके तहत विदेशी पर्यटक एवं पर्यटन के आर्थिक पहलू पर जोर दिया गया है जो कि विकास के लिए बहुत ही जरूरी है। यदि इसका अध्ययन आर्थिक दृष्टि से किया जाए तो यह एक बहुत ही विस्तृत एवं संतोषजनक परिभाषा है।

इस परिभाषा की मुख्य आलोचना इसलिए की जाती है कि यह यात्रा के सामाजिक पक्ष पर बिल्कुल ध्यान नहीं देती है। एक समाजशास्त्री यह जानकारी प्राप्त करना चाहेगा कि कहां यात्रा की गयी? कहां रुका गया? एवं किस प्रकार ठहरा गया? किस कारणवश यात्रा की गयी है? अतएव पर्यटक आवास काल को व्यतीत करने वाला वह व्यक्ति है जो कि ज्ञान, आराम एवं विलासिता इत्यादि वस्तुओं एवं सेवाओं की खोज करता है।

सम्पूर्ण प्रदेशों की राष्ट्रीय यात्रा की विधि को तय करने में पर्यटन के शैक्षिक, मनोविनोद सामाजिक पहलुओं पर दृष्टिगोचर करना परम आवश्यक है। साथ ही राज्य सरकारों का यह कर्त्तव्य है कि सामाजिक नीति की दृष्टि से भी पर्यटकों के आवास से संबंधित सेवाओं के विषय में आवश्यक कदम उठायें।

सैद्धान्तिक रूप में पर्यटन का आशय आराम व मनोरंजन तथा यात्रा व देशान्तर गमन से अलग है। स्पष्ट रूप में देखा जाए तो पर्यटन एक आराम एवं विश्राम की क्रिया है; जिसमें स्वेदश में अर्जित धन को पर्यटन क्षेत्रों में खर्च किया जाता है। इसके तहत पर्यटक स्वेच्छानुसार धन एवं समय का उपयोग करते हैं। डॉक्टर जेभिड्डिन के कथनानुसार, "यह एक सामाजिक आन्दोलन है ताकि आराम, विनोद, क्रीड़ा व सांस्कृतिक जरूरतों की संतुष्टि हो सके।"

प्रिमाउन्ट के विचारों में पर्यटन का अर्थ, "समस्त मानवीय क्रियाओं के क्षेत्र तथा समस्त प्राकृतिक पहलुओं में अनभिज्ञता की जिज्ञासा/खोज करना है।"

प्रौद्योगिकी दृष्टिकोण यदि देखा जाए तो पर्यटन के तहत विभिन्न तरह की दर्शन यात्राएं शामिल हैं तथा इसको एक विशेष यात्रा के उद्देश्य, समय व अन्य संबंधित कारणों के अनुसार परिभाषित किया गया है।

पर्यटक शब्द की परिभाषा सांख्यिकीयविदों एवं अर्थशास्त्रियों के लिए जरूरी है। साथ ही स्वेदशी एवं विदेशी पर्यटकों के मध्य अन्तर को स्पष्ट करने में मददगार होती है।

एक तकनीकी परिभाषा के अंतर्गत यात्रा व लघुकालीन यात्राओं की श्रेणियां, समय व अन्य संबंधित दशाओं का स्पष्ट ज्ञान होना चाहिए। यात्रा एवं यात्रियों के विभिन्न वर्गों, जो चाहे वे किसी भी विशेष उद्देश्यों की प्राप्ति हेतु हो तथा पर्यटन की सीमा के अन्तर्गत भी हो, का उल्लेख होना अनिवार्य है।

शब्दकोश के अनुसार पर्यटक शब्द की परिभाषा इस प्रकार है—"एक दृश्यावलोकन करने वाला व्यक्ति जो कि यात्रा करता है।" यूनीवर्सल शब्दकोश के अनुसार पर्यटक शब्द 1876 से भी पूर्व उपयोग किया जाता था। इसके अनुसार पर्यटक वह व्यक्ति है जो जानकारी हासिल करने तथा मनोरंजन करने के लिए यात्रा करता है ताकि अन्य व्यक्तियों को अपनी यात्रा का वर्णन दे सके।

एक अर्थशास्त्री एवं पर्यटक लेखक श्री नोरमल ने पर्यटक की परिभाषा इस प्रकार दी है—"प्रत्येक व्यक्ति जो विदेशों में स्थायी रूप में बसने या रोजगार की दृष्टि के अतिरिक्त अन्य कारणों से प्रवेश करता है तथा जो अपने अस्थायी ठहरने के दौरान इस देश में अर्जित धन को व्यय करना चाहता है।"

श्री लिकोरिश के मतानुसार, वे सभी व्यक्ति जो एक महीने से अधिक व चौबीस घंटे से कम अवधि के लिए किसी स्थान पर ठहरते हैं, यात्रियों की श्रेणी में नहीं आते हैं। उनके अनुसार चौबीस घंटे से कम ठहरने वाले व्यक्तियों को मनोरंजन भ्रमणकारी या यात्री, आगन्तुक के तहत सम्मिलित किया जा सकता है। वे लिखते हैं कि सैद्धान्तिक रूप में यात्री का अर्थ है, वे समस्त व्यक्ति जो स्वेदश से किसी अन्य देश में कम-से-कम चौबीस घंटे के आराम व मनोरंजन के लिए यात्रा करते हैं तथा किसी प्रकार का लाभान्वित व्यवसाय नहीं करते हैं। अल्पकालीन आगन्तुकों के तहत उन व्यक्तियों को शामिल किया जाता है जो स्वेदश में ही कम-से-कम चौबीस घंटे की यात्रा करते हैं तथा अल्पकाल के लिए यात्रा के उद्देश्य से ठहरते हैं।

आर्थिक दृष्टि से नोरमल की परिभाषा में पर्यटन के धन एवं वित्त के पक्ष पर प्रकाश डाला गया है। शिक्षा, सांस्कृतिक एवं सामाजिक पक्षों के अतिरिक्त आर्थिक पक्ष की काफी महत्ता है। पर्यटन मार्केट में पर्यटक बहुत ही महत्त्वपूर्ण स्थान रखते

हैं। पर्यटक ही हर क्रिया पर्यटन का आधार है। ये उपभोक्ता हैं तथा अर्थव्यवस्था में विदेशी पूंजी लगाते हैं। पर्यटक स्वयं पर्यटन देश में कोई उत्पादन क्रिया नहीं करते हैं तथा न ही वे वेतन से संबंधित कोई रोजगार करते हैं। जो धन वे व्यय करते हैं वह विदेशों में अर्जित किया हुआ होता है।

प्रत्येक राष्ट्र को विदेशीय यात्रा व पर्यटन से जुड़े विषयों के बावजूद सांख्यकीय सूचनाओं के लिए पर्यटक शब्द की परिभाषा प्रयोग करनी होती है, क्योंकि विभिन्न परिभाषाओं के कारण बहुत ही अव्यवस्था रहती है।

भारत में 1970 में पर्यटक संबंधी आंकड़े निम्न परिभाषा के आधार पर इकट्ठा किए जाते थे :

"एक व्यक्ति जो विदेशी पार-पत्र पर भारत की यात्रा कम-से-कम 24 घंटे तक करे तथा उसका समय छह महीने से ज्यादा न हो बशर्ते यह यात्रा देशान्तरवास व रोजगार दोनों से विहीन हो।"

1971 के उपरांत पर्यटक शब्द की एक नई परिभाषा अपनायी गई जो निम्नवत् है–

"एक व्यक्ति जो कम-से-कम 24 घंटे के अर्से के लिए विदेशी पार-पत्र पर बिना देशान्तरवास व बिना रोजगार के उद्देश्य से यात्रा करे।"

उपर्युक्त परिभाषा राष्ट्रसंघ के अन्तर्राष्ट्रीय यात्रा एवं पर्यटन सम्मेलन 1963 के अनुसार है तथा इसकी विशेषता यह है कि इसमें समय की ऊपरी सीमा पर कोई प्रतिबन्ध नहीं है।

आई. ओ. टी. ओ. ने सांख्यिकीय सूचना की महत्ता समझते हुए तथा सूचना को आसान एवं तुलनात्मक बनाने हेतु अपने सदस्यों के लिए एक नई परिभाषा बनाने का सुझाव दिया था, जो निम्न प्रकार है–

"सांख्यिकीय उद्देश्य से आगन्तुक शब्द उस व्यक्ति विशेष का वर्णन करता है जो कि स्वयं के राष्ट्र के अतिरिक्त किसी विदेश की यात्रा करता है तथा उसकी यात्रा का उद्देश्य पुरस्कृत-व्यवसाय प्राप्त करने से भिन्न है।"

उपरोक्त वर्णित परिभाषा में निम्नलिखित तथ्यों का समावेशन है–

पर्यटक–पर्यटक अर्थात् अस्थायी यात्री जो कि दर्शनीय स्थान पर कम-से-कम चौबीस घंटे निवास करते हैं और जिनकी यात्रा का उद्देश्य निम्नांकित शीर्षकों के अन्तर्गत आता है–

(1) सैलानी–अस्थायी यात्री जो पर्यटन देश में चौबीस घंटे से कम ठहरते हैं। सामुद्रिक भ्रमणकर्ता भी इसके तहत आते हैं।

(2) विलासिता—मनोरंजन, अवकाश, स्वास्थ्य अध्ययन, धर्म व क्रीड़ा।

(3) व्यापार—पारिवारिक, धर्मार्थ कार्य, सम्मेलन।

भ्रमणकर्ता जो वैधानिक रूप से देश के अन्दर प्रवेश नहीं करते हैं और वायुयान द्वारा एवं हवाई अड्डों के संक्रमक क्षेत्र से बाहर नहीं आते हैं, उन्हें सांख्यिकीय सूचनाओं में सम्मिलित नहीं किया जाता है।

1967 में एक अन्तर्राष्ट्रीय सांख्यिकीय यात्रा के विशेषज्ञ संगठन, जो संयुक्त राष्ट्र ने आमन्त्रित की थी, ने पूर्व वर्णित परिभाषा में कुछ मामूली बदलाव किए थे। बहुत विचार-विमर्श के उपरांत यह समझा गया कि यात्रियों की परिभाषा में कुछ यात्रियों को अलग वर्ग में रखा जाए तथा उन्हें दैनिक सैर पर देश की सीमा के भीतर बिना रोजगार के उद्देश्य से आते हैं तथा सामुद्रिक यात्री एवं भ्रमणकर्ता जो देश के भीतर एक रात्रि से अधिक उपलब्ध आवास सुविधाओं में नहीं ठहरते हैं।

वस्तुतः मान्य की गयी परिभाषा के द्वारा, "पर्यटक वे अस्थायी आगन्तुक हैं जो कम-से-कम एक रात्रि के लिए यात्रा करने वाले देश में ठहरते हैं तथा जिनकी यात्रा का उद्देश्य विलासिता (मनोरंजन अवकाश, स्वास्थ्य, अध्ययन व क्रीड़ा), व्यापार व पारिवारिक मेल-मिलाप से संबंधित है।

इस तरह हम देखते हैं कि पर्यटन का अर्थ सिर्फ आराम, मनोरंजन तथा आनन्द तक ही सीमित नहीं है, बल्कि इसके तहत व्यापार संबंधी यात्राएं व अन्य सभी प्रकार की यात्राएं शामिल हैं जिनका वेतनयुक्त रोजगार से कोई संबंध नहीं है। सम्पूर्ण विश्व में आय स्तर में वृद्धि होने से पर्यटन की संभाव्यता में महत्वपूर्ण वृद्धि होने लगी है, अतएव पर्यटन उद्योग संबंधी आंकड़े एकत्रित करने तथा उन्हें अन्तर्राष्ट्रीय स्तर पर विश्लेषण करने के लिए इन सूचनाओं एवं आंकड़ों की महत्ता बढ़ती जा रही है।

विशेषताएं—पर्यटन के अपने कुछ विशेष गुण हैं जिनका अध्ययन एक उपयोगी भाषा निर्मित करने के लिए अत्यंत जरूरी है। वर्तमान समय में लोग किसी कार्य को करने या आनन्द लेने के लिए देश के भीतर या विदेशों की यात्रा करते हैं। पर्यटन की एक उचित परिभाषा देना बड़ा ही कठिन कार्य है। क्योंकि इसके तहत वे सभी व्यवसाय आते हैं, जो पर्यटकों की आवश्यकता पूर्ति में तल्लीन हैं। पर्यटकों द्वारा किया गया कोई भी व्यय स्वयं इस उद्योग की समृद्धि एवं विकास में सहायक होता है। हालांकि यात्रा अभिप्राय कुछ अवधि के लिए अपने निर्वासित स्थान को छोड़कर बाहर जाना है, परन्तु इस अस्थायी समय में अपने स्थान में रहने की तुलना में पर्यटक को अपने व बाह्य जीवन में पूर्णतः संतुलन का अहसास होने लगता है।

साथ ही बहुत से माध्यमों की मदद लेनी होती है, ताकि यात्रा का उद्देश्य पूरा हो सके। एक पर्यटक जिस समय अपने स्थान को छोड़ता है, यात्रा करने के लिए तुरन्त ही व्यय प्रारंभ करने लगता है। उसको निम्न प्रकार की सेवाएं प्राप्त होती हैं जिन पर उसे धन खर्च करना होता है।

पर्यटन का स्वरूप तथा उत्पत्ति

भारत में पर्यटन का क्षेत्र—इस मामले में भारत का बहुत बड़ा क्षेत्र है। इसका मुख्य कारण भिन्न-भिन्न संस्कृतियां और भव्य एवं रमणीय प्राकृतिक आकर्षण हैं। ये पर्यटन उद्योग के मुख्य स्रोत हैं। भारत की अपनी एक प्राचीन सभ्यता रही है। यह सभ्यता हमारे धर्म, रीति-रिवाज, रहन-सहन, वास्तुकला इत्यादि में आज भी मौजूद है। हमारी एक अद्वितीय संस्कृति एवं रहन-सहन का तरीका है। हमारे मन्दिरों में लुप्त यह सभ्यता व सांस्कृतिक धरोहर आज भी आगन्तुकों के लिए प्रमुख आकर्षण का केन्द्र हैं। हमारे रीति-रिवाज, नृत्य एवं संगीत समस्त लोगों द्वारा पसंद किए जाते हैं। विभिन्न प्रकार के प्राकृतिक सौन्दर्य जैसे—हिमाच्छादित पहाड़ियां, हिमखण्ड, गर्म जल के फव्वारे, उष्ण जल स्रोत, विचित्र चमत्कारिक गर्म गुफाएं, शानदार झीलें, सुन्दर समुद्र तट व आकर्षणयुक्त पवर्तमालाएं, रोमांचक वन्य प्राणी, मनभावन मरुस्थल आदि। 5000 वर्ष पहले की हमारी ऐतिहासिक परम्परा वर्तमान में भी विद्यमान है। साथ ही हमें बड़ी-बड़ी ऐतिहासिक इमारतें जैसे—साँची के स्तूप, प्रेम का प्रतीक ताजमहल व अन्य प्रकार के उदाहरण कला व वास्तुकला के बेशुमार मन्दिरों, मस्जिदों एवं गिरजाघरों में मिलते हैं। हमें इस बात पर गर्व है कि हिन्दू, बौद्ध और इस्लाम जो कि दुनिया के महान् धर्म समझे जाते हैं, भारत में अपना अस्तित्व रखते हैं। इसके अलावा पर्यटकों के लिए यह क्षेत्र खान-पान, रहन-सहन एवं त्योहारों के लिए भी आकर्षण का केन्द्र है। पर्यटन उद्योग की एक बहुत ही विचित्र सम्पत्ति है—असीमित कच्चा माल। अद्भुत विभिन्न भारतीय संस्कृतियां एवं प्राकृतिक आकर्षण इस उद्योग के प्रधान स्रोत हैं।

भारत का आबादी के नजरिए से विश्व के राष्ट्रों में द्वितीय स्थान तथा क्षेत्रफल के नजरिए से 7वां स्थान है। उत्तरी गोलार्द्ध में स्थिति यह देश 32,87,000 वर्ग किलोमीटर क्षेत्र से घिरा हुआ है। भारत की यह असीमित व दूर-दूर तक फैली हुई भूमि, जो उत्तरी अक्षांश में 8° 4' से 37° 6' तथा देशांतर में 68° 7' से 97° 25' तक बिखरी हुई है। अपने 5000 वर्ष के इतिहास के साथ यह विश्व के पर्यटकों को बहुत कुछ प्रदान करती है। उत्तर से दक्षिण एवं पूर्व से पश्चिम तक इस देश

की लम्बाई व चौड़ाई क्रमशः 3214 किलोमीटर व 2933 किलोमीटर तक है। इसके अलावा भारत की भू-सीमा 15,200 किलोमीटर व समुद्र तट करीब 6100 किलोमीटर है।

भारत पूर्वी, पश्चिमी और दक्षिणी इन तीनों दिशाओं से समुद्र से घिरा हुआ है और उत्तरी सीमा में चीन और नेपाल राष्ट्र से जुड़ा है तथा असंख्य पर्वत श्रृंखलाएं इसको बर्मा से पृथक् करती हैं। पूर्व दिशा में बंग्लादेश तथा उत्तर-पश्चिम में पाकिस्तान एवं अफगानिस्तान, दक्षिण दिशा में हिन्द महासागर, पश्चिम में अरब सागर व पूर्व में स्थित बंगाल की खाड़ी, मन्नार की खाड़ी व पाक स्ट्रीट भारत को श्रीलंका राष्ट्र से पृथक् करती हैं। अरब सागर में स्थित मोनीकाइ, दमनद्वीप-जल-द्वीप और बंगाल में स्थित अण्डमान व निकोबार द्वीप भारत के हिस्से हैं। इस प्रकार भारत देश एशिया महाद्वीप से विश्व की सबसे महान् हिमालय पर्वत श्रृंखलाओं से अलग हो जाता है। पश्चिम से पूर्व दिशा की ओर भारत की भूमि सीमा पर पाकिस्तान, बंग्लादेश, भूटान, चीन, नेपाल व बर्मा स्थित हैं। दक्षिण में यह देश अरब सागर व हिन्द महासागर व पूर्व में बंगाल की खाड़ी से घिरा हुआ है। भारत को भौगोलिक दृष्टि से अधोलिखित भागों में विभाजित किया जा सकता है।

1. दक्षिण का पठार या दक्षिण प्रायद्वीप जो इस उपद्वीप के ऊपरी व दक्षिणी द्वीप को घेरे हुए है।
2. गंगा का मैदानी भाग जो कि दक्षिण में स्थित पर्वत श्रृंखला के समानान्तर है।
3. हिमालय एवं इससे संबंधित सामूहिक पर्वत श्रृंखलाएं।

भारत की मुख्य भूमि की विशेषता यह है कि इसकी हिमालय पर्वत श्रृंखला नाटकीय रूप में ऊँची है तथा गंगा का मैदान सुन्दर एवं दर्शनीय तथा समान ऊँचाई का मैदानी क्षेत्र है। इसके अलावा दक्षिणी प्रायद्वीप में ऊँचे पठार एवं उबड़-खाबड़ तटबन्ध हैं।

भारतवर्ष में प्रमुख रूप से चार ऋतुएं होती हैं, जो इस प्रकार हैं–

1. शिशिर ऋतु।
2. ग्रीष्म ऋतु।
3. वर्षा ऋतु।
4. मानसून के उपरांत का मौसम।

उत्तर में हिमालय से लेकर दक्षिण में कन्याकुमारी तक विभिन्न जलवायु व भिन्न-भिन्न तरह के निवासी हैं जिनकी सभ्यता एवं संस्कृति भी पृथक्-पृथक् है, नवम्बर से फरवरी तक पूरे देश में शीतऋतु का असर रहता है तथा इसके उपरांत

बसन्त ऋतु का आगमन, फिर ग्रीष्म ऋतु, तदुपरांत जून, जुलाई एवं अगस्त में मानसून का मौसम और फिर बहुत ठण्डा मौसम जो कि शनैः-शनैः पूर्व से दक्षिण क्षेत्र की तरफ आगे बढ़ता है। देश का उत्तरी भाग समशीतोष्ण कटिबन्ध के तहत आता है, जिसके कारण मौसम में ज्यादातर परिवर्तन होते रहते हैं तथा शिशिर ऋतु के कारण अत्यधिक ठण्ड रहती है। देश के अन्य भागों में तापमान सामान्यतः गर्म रहता है। अप्रैल एवं मई के महीने बहुत ही गर्म हैं। यहां तक कि राजस्थान जैसे क्षेत्रों में तापमान इन महीनों में 120° F तक पहुंच जाता है।

पूरे भारत देश में बारिश भी पृथक्-पृथक् रूप से होती है। पश्चिम बंगाल में और कोलकाता के आस-पास के क्षेत्रों में अत्यधिक बारिश होती है, जिससे यहां नमीयुक्त जलवायु पाई जाती है। कुछ भागों में तो वर्ष भर में 80 से.मी. तक की वर्षा होती है। मानूसनी हवाएं ही इस वर्षाकालीन व शुष्क मौसम की वजह हैं। ग्रीष्मकाल में यह मानसून हिन्द महासागर से नमी लेकर उत्तर दिशा को प्रवाहित होता है जबकि शिशिर ऋतु में साइबेरिया से प्रारंभ होकर दक्षिण दिशा को चलते-चलते भारत पहुंचकर पूर्णतः शुष्क हो जाता है।

भारत में प्रमुख रूप से बहने वाली नदियों को निम्नलिखित तीन भागों में बांटा गया है—

1. दक्षिण की नदियां।
2. समुद्रतटीय नदियां।
3. हिमालय की नदियां।

दक्षिण भारत की नदियों में वर्षा के बहाव में अंतर हो जाता है। हिमालय से निकले वाली नदियां अधिकतर बर्फ से जल उपलब्ध के कारण सालभर अर्थात् साल के बारहों मास बहती रहती हैं। समुद्र तटीय नदियां मुख्यतः पश्चिमी तट पर लम्बाई में छोटी हैं तथा उनका जल ग्रहण स्थान भी सीमित है। पश्चिमी राजस्थान की मुख्य भूमि से संबंधित नदियां बहुत कम हैं। गंगा का प्रवहण क्षेत्र देश में सर्वाधिक विशाल है। हिमालय में गंगा के पानी के दो मुख्य स्रोत हैं—भागीरथी जो कि गोमुख में गंगोत्री से निकलती है तथा अलकनन्दा। हिमालय से निकलने वाली ज्यादातर नदियां जैसे—यमुना, गन्डक, घाघरा एवं कोसी आदि गंगा में आकर मिलती हैं। दक्षिण क्षेत्र की मुख्य नदियां—महानदी, गोदावरी, कृष्णा, तुंगभद्रा तथा कावेरी हैं।

भौतिक लक्षण की दृष्टि से भी भारत की कई विशेषताएं हैं। जनसंख्या की दृष्टि से भी यहां पर विश्व की हर जाति व धर्म के लोग निवास करते हैं, जिनकी स्वयं की अपनी कुछ विशेषताएं हैं। इस शताब्दी के शुरू में हड़प्पा संस्कृति की खोज

ने भारत को विश्व की प्राचीन संस्कृतियों में एक महत्वपूर्ण स्थान दिया तथा तत्कालीन मिस्र एवं सुमेर की सभ्यता की विशेषताओं के बराबरी का दर्जा प्राप्त किया। इस सभ्यता का पतन व आर्य सभ्यता का उत्थान भारतीय इतिहास में एक अद्वितीय युग था। प्राकृतिक एवं ऐतिहासिक दृष्टि से भारत संसार के देशों में पर्यटन के लिए महत्वपूर्ण स्थान रखता है। यह देश पर्यटकों को विभिन्न पर्वत शृंखलाओं व पहाड़ियों, उच्च शृंखलाओं, समुद्री किनारों, प्राकृतिक दृश्य सौन्दर्य, स्मारकों एवं संग्रहालयों, प्राचीन कला एवं नृत्य, मेले एवं त्योहार व अन्य अनेक मूल्यवान वस्तुओं से आकर्षित करता है, जिनके कारण यह देश चित्रकारी का एक जीता-जागता अनुपम उदाहरण बन चुका है।

पर्यटन उद्योग का मुख्य आधार विविधता होती है। भारत में यह विविधता प्रत्येक पर्यटक के लिए कुतूहल का विषय है। एक सपेरे से लेकर हिमाच्छादित हिमालय पर्वत की शृंखलाएं, साधारण नृत्य से लेकर बड़े-बड़े पर्वतीय शिकारगाह, पर्यटन के मुख्य आकर्षण इस देश में मौजूद हैं। अन्य निर्यात करने वाले उद्योगों की तरह पर्यटन उद्योग के विकास में अधिक नियोजन की जरूरत नहीं है। इस उद्योग के विस्तार की कोई सीमा नहीं है तथा आधुनिक पर्यटन में तो अलग-अलग तरह की चीजें स्वयं अपना स्थान बनाने लगी हैं।

यात्रा, आराम, विश्राम व दूसरे क्षेत्रों में संस्कृति की व्याख्या को ही पर्यटन कहा जा सकता है। यह एक मूलभूत उद्योग है जो कि आय एक प्रमुख स्रोत है। विदेशी मुद्रा अर्जन करने का साधन है तथा राष्ट्र की छवि को सुदृढ़ बनाने में सहायक है। इस देश में पर्यटन के बहुत ही आवश्यक तत्व मौजूद हैं। विशिष्ट तौर पर विदेशी पर्यटक भारत में अपने पर्यटन का उद्देश्य कतिपय प्राचीन इमारतों व दर्शनीय आश्चर्यजनक स्थानों तक ही सीमित रखते हैं। भारत को पर्यटन की दृष्टि से खजुराहो, गुलमर्ग, ताजमहल, महाबलीपुरम् व कुछ ठहरने के स्थानों तक ही सीमित समझा जाता है। अतः इस प्रवृत्ति को परिवर्तित करने के लिए जरूरी है कि आकर्षक स्थानों, पर्यटक स्थलों तथा उनसे संबंधित अन्य सुविधाओं व सेवाओं का भी विकास किया जाए। ऐतिहासिक मन्दिरों, प्राचीन इमारतों, हिमालय, गंगा, ताजमहल, सुन्दर समुद्री तटों व अन्य स्थान व सुविधाएं जो विदेशी पर्यटकों के लिए मुख्य आकर्षण हैं, भारत में पहले से ही मौजूद हैं।

भारत सौंदर्य व आकर्षण स्थलों में धनी है यही कारण है कि अन्य देशों की अपेक्षा विदेशी पर्यटक यहाँ पर अधिक संख्या में आकर्षित होते हैं। विदेशी पर्यटक स्थानों में बहुत ही सीमित पर्यटन के तत्व विद्यमान होने पर भी एक माह की अवधि

में इतनी संख्या में पर्यटक आते-जाते हैं, जितने कि पूरे वर्ष भर में भारत में नहीं आते। यह एक जटिल सवाल है जो विचार करने योग्य है।

पर्यटन उद्योग का आकार—पर्यटन उद्योग सम्पूर्ण विलासितापूर्ण उद्योग का एक अभिन्न भाग है जिसके अन्तर्गत पर्यटन के अतिरिक्त टिकट संग्रह, पाल नौका, केलिपोत क्रीड़ा, गोल्फ, बागबानी आदि विभिन्न तरह की क्रियाएं आती हैं। पर्यटन को इस विस्तृत उद्योग का एक लघु उद्योग कहा जा सकता है। पर्यटन बहुत से अंगों के सम्मिश्रण का उद्योग है, जिसके तहत बहुत बड़ी विमान सेवाओं और बहुत बड़े-बड़े होटल उद्योग से लेकर दो व्यक्ति वाली पति-पत्नी की पथ-प्रदर्शक सेवाएं भी सम्मिलित की जा सकती हैं। सुविधाजनक एवं आरामदायक पर्यटन को कामयाब बनाने के लिए अति उत्तम होगा कि लघु उद्योगों को प्रोत्साहन दिया जाए। आवास, परिवहन एवं यात्रा अभिकर्ता पर्यटन क्षेत्र तक ही सीमित था तथा धर्म के नाम पर यात्रा करने में कठिनाइयों का सामना करना जरूरी समझा जाता था।

अतः आवास की सुविधाओं व परिवहन व्यवस्था का कोई अस्तित्व ही नहीं था। लोग मन्दिरों एवं धार्मिक स्थानों के दर्शन के लिए हजारों मील की यात्रा पैदल ही किया करते थे। पर्यटन पथ-प्रदर्शक व यात्रा-अभिकर्ता का अस्तित्व निर्माता था। पर्यटन के तीन मुख्य अंग—स्थान, परिवहन एवं आवास हैं। पर्यटक को किसी भी गंतव्य स्थान तक पहुंचने के लिए यात्रा-अभिकर्ता व यात्रा-संचालकों, सामान-वाहक आदि की सेवाओं की जरूरत पड़ती है। पर्यटन स्थल पर पहुंचने पर पर्यटक के आवास से संबंधित व अन्य आवास सुविधाओं एवं सेवाओं की पूर्ति होटल एवं खान-पान उद्योग द्वारा की जाती है।

पर्यटन उद्योग के विभिन्न आयाम

पर्यटन उद्योग के विभिन्न आयामों को निम्न शीर्षकों के द्वारा बताया जा सकता है—

(1) पर्यटन के आकर्षण—भारतीय पर्यटन मुख्यतः सांस्कृतिक क्षेत्र तक ही सीमित रहा है। पर्यटकों के स्वभाव में बदलाव के कारण आराम व मनोविनोद संबंधी क्रियाओं का पर्यटन के क्षेत्र में महत्व बढ़ता जा रहा है, इसलिए आधुनिक पर्यटन की मांग को पूरा करने के लिए आवासकालीन पर्यटन स्थल, दर्शनीय सौन्दर्यता, पर्वत एवं समुद्रतट, विलासिता व मनोरंजन की काफी जरूरत है।

(2) मनोरंजन—पर्यटन उद्योग के विभिन्न रूपों जैसे—रेस्टोरेन्ट, नृत्य एवं संगीत, नाट्यशाला एवं खरीददारी आदि का विकास बहुत ही आवश्यक है।

स्मारकों व संग्रहालयों, कला व हस्तकारिता, मेले व त्योहारों, पर्व, प्रदर्शनियों इत्यादि के अतिरिक्त पर्यटक बहुत-सी अन्य मनोविनोद की क्रियाओं की अभिलाषा रखते हैं। विशेषकर सन्ध्यकालीन समय को खुशी से व्यतीत करने के लिए मनोविनोद के साधन तो बहुत ही जरूरी हो जाते हैं। ताकि पर्यटकों के मनोविनोद एवं आकर्षण के मुख्य साधन–कैबरे, नृत्य व गान, नाट्यशाला व लोगों को एक शानदार आरामदायक वातावरण में मिलने-जुलने व वार्तालाप करने का मौका प्राप्त हो।

(3) विश्रामालय एवं खान-पान–पर्यटन के गंतव्य स्थान पर आवास एवं खान-पान की सुविधाएं, पर्यटक सेवाओं से संबंधित बहुत ही अनिवार्य अंग हैं। पर्यटकों द्वारा प्रयोग की जाने वाली भिन्न-भिन्न आवास सुविधाएं–होटल, सहायक आवास सुविधाएं, मित्र एवं संबंधियों के साथ ठहरना, निजी आवास तथा काफिला व टेन्ट आदि बहुत ही महत्वपूर्ण भूमिका अदा करते हैं।

(4) आवागमन के अवांछित नियमों का पालन–सभी यात्रियों के आने जाने की अवांछित विधियों का पालन करना पड़ता है। इसके साथ-साथ राष्ट्र में कुछ ऐसे नियम विद्यमान हैं जो यात्रा में मदद और बाधा होते हैं। जितने अधिक सरल एवं सुविधाजनक आवागमन से संबंधित नियम होंगे तथा पर्यटकों को स्वतन्त्र रूप से भ्रमण करने में मदद मिलेगी तथा आगमन प्रस्थान व्यवस्था सरल होगी उतने ही ज्यादा पर्यटक आकर्षित होंगे। पर्यटन स्थलों का पुनः दर्शन करने के लिए समय-समय पर किए गए सर्वेक्षणों से ज्ञात होता है कि भारत में पर्यटन से संबंधित नियम व सुविधाओं में बहुत ही कमियां खास तौर पर पर्यटकों की शिकायतें–बहुत-से फार्म भरने, तर्कहीन व अवरोधक नियम, अनावश्यक लालफीताशाही, विशेषकर सामान की छँटनी एवं ढूँढने में अव्यवस्था तथा अदक्ष प्रबंध के बावत होती हैं। हमें इन समस्त खामियों पर ध्यान देना होगा ताकि पर्यटन के क्षेत्र में चहुंमुखी विकास हो सके।

(5) यातायात–तमाम पर्यटन क्षेत्रों में घूमने, टहलने व देखने के लिए यातायात सुविधा का होना अति आवश्यक है। पर्यटक के लिए कार व उच्च कोटि की गाड़ियों की आवश्यकता होती है, साथ ही नवीन सड़कों का निर्माण तथा चुंगी कर व सीमाकर जो इस क्षेत्र में अवरोधक हैं को समाप्त करने की जरूरत है। वायु, रेल एवं समुद्री परिवहन व्यवस्था के क्षेत्र में विकास की आवश्यकता है; क्योंकि यह प्राथमिक सेवा है, जिसके अभाव में पर्यटक को अपने गंतव्य स्थान पर पहुंचने में परेशानी होती है। यह साधन इधर-उधर यात्रा करने में सहायक होता है। यात्रा

परिवहन को भिन्न-भिन्न रूपों में बांटा जा सकता है–सार्वजनिक एवं निजी, अन्तर्देशीय व अन्तर्राष्ट्रीय तथा हवाई, जल एवं भूमिगत।

(i) सड़क-यातायात–बसें, आलीशान मोटरगड़ियां किसी भी निश्चित स्थान की सीमित या मध्यम दूरी तथा गंतव्य स्थान भ्रमण करने हेतु प्रत्यक्ष एवं सतत् रूप में उपयोग में लायी जाती हैं। बहुत से स्थानों में यह रेल परिवहन सेवाओं के विकल्प का कार्य करती हैं। निजी मोटर व्यवस्था के शुरू होने से सड़क परिवहन का महत्व दिन-प्रतिदिन बढ़ता जा रहा है।

(ii) रेल-यातायात–देश के अन्दर व भिन्न-भिन्न देशों के अन्दर व बीच की सीमित मध्यम दूरी की यात्रा करने का यह बहुत ही उत्तम साधन है। यह गंतव्य स्थान पर पहुंचने एवं वहां पर भ्रमण करने में सहायक होता है।

(iii) जल-यातायात–सीमित सामुद्रिक मार्ग व जल के रास्तों एवं समुद्र द्वारा यात्रा करने में जल परिवहन एक महत्वपूर्ण साधन है।

(iv) वायु-यातायात–सुदूर पर्यटन के लिए वायु परिवहन तीव्र गति से विकसित होता हुआ माध्यम है। यह चहुं दिशाओं के परिवहन का प्राथमिक साधन है।

(1) पर्यटन का उद्‌भव एवं विकास–प्राचीन काल में सामाजिक सीमितता पर्यटन का आधार थी, बुद्धिजीवियों ने अपने विचारों का इस्तेमाल किया, सामान्य व्यक्तियों ने उनका अनुसरण किया तथा साथ ही आय में वृद्धि के कारण जनसाधारण ने पर्यटन को अपने जीवन का एक अभिन्न अंग बना लिया है।

पर्यटन के उद्‌भव का दूसरी वजह धार्मिक स्थानों को जाता है जहां धर्म से संबंधित स्थलों को देखा जा सके। वर्तमान समय में तीव्रगामी और सरल यात्रा सुविधाओं की प्राप्ति, शिक्षा तकनीकियों में संशोधन से परिवर्तन, सांस्कृतिक जिज्ञासा आदि के कारण यात्रा का विचार भिन्न-भिन्न क्षेत्रों में पदार्पण कर चुका है, विशेषकर शिक्षा के क्षेत्र में। साहित्य, इतिहास एवं कला के अध्ययन से यात्रा के नये-नये विचारों का आविर्भाव हुआ है।

इलैक्ट्रॉनिक मीडिया एवं प्रिन्ट मीडिया में तीव्रगामी उन्नति होने से यात्रा के विचारों में नई तेजी आई है। अखबार, रेडियो, टेलीविजन, चलचित्र, पत्रिकाएं आदि यात्रा के विचारों को अभिव्यक्त करने में महत्वपूर्ण भूमिका निभाते हैं। यात्रा के विकास पर इन साधनों का बुरा असर भी पड़ सकता है। अशान्त एवं बिगड़े हुए राजनैतिक एवं आर्थिक समाचार किसी भी क्षेत्र की अति पुरानी सुव्यवस्थित परम्परा को नष्ट कर सकते हैं। यात्रा के विचार को प्रभावित्त करने में सामाजिक, सांस्कृतिक

एवं राजनैतिक पक्षों का बहुत ही मानवीय स्थान है, अतएव राष्ट्रीय, क्षेत्रीय एवं उप-क्षेत्रीय नजरिए से यात्रा के विचार को विकसित करने हेतु एक परिपक्व, व्यवस्थित व सुनियोजित पर्यटन व्यवस्था की जरूरत होती है।

(2) पर्यटन का रूप—पर्यटन के अनगिनत रूप हैं। वस्तुतः यह राष्ट्रीय जीवन सम्पूर्ण पहलुओं को आकर्षित करता है। राष्ट्रीय एवं अन्तर्राष्ट्रीय पर्यटन के मौलिक सिद्धान्त हैं—निरन्तर एवं तीव्रगति से विकास की क्रियाएं। आज का पर्यटन विस्तृत सामाजिक भावनाओं से ओत-प्रोत होने के कारण उसकी इच्छाएं व रुचियां भिन्न-भिन्न तरह की हैं तथा साथ ही साथ उसके विश्राम का समय भी अत्यधिक सीमित है। अति प्रेमी क्रियाएं व विलासिता का व्यापारीकरण होने के कारण ये सुविधाएं आम आदमी को उपलब्ध होने लगी हैं। असंख्य लोग आज उत्तेजित एवं आश्चर्यजनक पवर्तारोहण, जल-क्रीड़ा, तरण-तारण क्रियाएं, घुड़सवारी जैसी क्रियाओं के लिए विदेश भ्रमण करते हैं। इन पर्यटकों को कम खर्चीले आवास तथा बहुसंख्यक पर्यटकों के लिए संगठित घुड़सवारी, नाविक व शिकार खेलों (क्रीड़ाओं) की प्राप्ति की इच्छा होती है। साथ ही इस तरह का वर्ग स्वयं के लिए मनोविनोद के साधन व अन्य सुविधाएं एकत्रित करता है। इसी वजह से सामाजिक पर्यटन काफी लोगों को उपलब्ध है तथा साथ ही अविकसित क्षेत्रों तक पहुंचने में सफल हुआ है।

(३) पर्यटन व्यवस्थापक व संचालन—ट्रेवल एजेन्ट्स एक ओर पर्यटक एवं दूसरी ओर परिवहन एवं आवास मालिकों के कार्यों की मध्यस्थता की महत्वपूर्ण कड़ी हैं। यात्रा-अभिकर्ता पर्यटन के भिन्न-भिन्न अंगों का इस तरह संगठन करते हैं कि यह वस्तु या सेवा के रूप में पर्यटकों को प्राप्त हो सके। यह पर्यटन वस्तु और सेवा उन्हीं के जोखिम के नाम पर बाजार में बेची जाती हैं। इस तरह पर्यटक निर्मित वस्तुओं और सेवाओं की मांग करते हैं तथा यह पर्यटक वस्तुएं एवं सेवा पूर्ति के रूप में बाजार में प्रस्तुत होती हैं। पर्यटन की सुविधाओं के तहत हमारी अर्थव्यवस्था के वे सब हिस्से आ जाते हैं जिनका समान उद्देश्य पर्यटक सेवाओं की पूर्ति करना है। इन सभी सेवाओं की अपनी एक विशेष प्रकृति होने एवं एक विशद् पर्यटन वस्तु से संबंधित होने की वजह से उन्हें पर्यटन उद्योग के रूप में परिभाषित किया जा सकता है।

पर्यटन के आधार तथा विशेषताएँ

पर्यटन आयोजन का मुख्य लक्ष्य स्वदेशीय तथा अंतर्राष्ट्रीय पर्यटन के क्षेत्रों में संवर्धानात्मक कार्य और पर्यटन उद्योग की आधार संरचना का विकास करना है। इन

प्रक्रियाओं में यातायात प्रबन्ध, यात्रा सुविधाओं की उन्नति, रेस्टोरेन्टों, पर्यटन संचालकों, यातायात चालकों, वन्य-जीवन, साज-समान, स्मारकों, मार्गदर्शकों, पर्यटन सूचना का प्रचार इत्यादि जैसे विभिन्न संघटकों के कार्यकलापों का राष्ट्रीय एवं अन्तर्राष्ट्रीय स्तर पर विनियमन होता है। इस संबंध में पर्यटन के विभिन्न पहलुओं के विषय में सूचनाओं के एकीकरण, सारणीयन और प्रचार के लिए 'पर्यटन' शब्द को परिभाषित करना अत्यंत ही जरूरी है। परिभाषा पर एकत्रित आंकड़े अन्तर्राष्ट्रीय एवं स्वदेशी पर्यटक आगमन, अनुमोदित होटलों में अधियोग, चुनिन्दा स्मारकों की यात्रा करने वाले पर्यटकों की संख्या से भी संबंधित होते हैं। इसके अतिरिक्त आधार-संरचनात्मक सुविधाओं की उपलब्धता एवं पर्याप्तता, मांग संबंधी विशिष्टताओं और पर्यटकों की प्रतिक्रिया आदि का जायजा लेने के लिए समय-समय पर पर्यटन संबंधी संस्थाएं विभिन्न सर्वेक्षण एवं अध्ययन करती रहती हैं।

अतएव सांख्यकीय एवं वैधानिक अध्ययन के नजरिए व औद्योगिक उद्देश्य से पर्यटन शब्द को परिभाषित करना अत्यधिक जरूरी है। पर्यटन का पूर्ण अध्ययन तभी सम्भव है जबकि इस शब्द की एक स्पष्ट एवं संक्षिप्त परिभाषा हो। किसी भी प्रवृत्ति का यथार्थ अवलोकन तभी किया जा सकता है, बशर्ते कि परिभाषा के तहत आने वाले सभी क्षेत्रों एवं प्रवृत्तियों का स्पष्टीकरण किया जाए। पर्यटन से संबंधित महत्वपूर्ण सांख्यिकीय सूचनाओं की जरूरत होती है ताकि हर प्रवृत्ति का गहन अध्ययन किया जा सके। इसीलिए एक उचित परिभाषा की आवश्यकता अनुभव होती है।

प्रत्यक्ष एवं अप्रत्यक्ष रूप से विभिन्न कानूनों, नियमों व विधानों का पर्यटन से संबंध है, जिनमें कुछ पर्यटन की क्रियाओं से संबंधित हैं तथा कुछ इनसे संबंधित नहीं हैं।

इसके अतिरिक्त आर्थिक प्रक्रियाओं के कारण पर्यटन को बाजारीकरण के शोध की आवश्यकता होती है जो उद्योग संस्था के विकास का आलंबन है। किसी भी प्रवृत्ति का उचित रूप से परीक्षण करने के लिए एक साधारणतया धारणायुक्त परिभाषा का होना परमावश्यक है। पर्यटन का एक उचित विचार एवं तत्व इससे संबंधित सैद्धान्तिक व प्रसिद्ध स्वरूप का ज्ञान करवाने में मददगार साबित होता है। यह पर्यटन की मुख्य विशेषता से अवगत कराता है तथा साथ ही संबंधित व अन्य विभिन्न प्रवृत्तियों की भिन्नता को प्रकट करता है।

वस्तुतः पर्यटन शब्द अर्थशास्त्र से जुड़ा है जोकि व्यवसाय एवं उद्योग का परिचायक है। पर्यटन एक आर्थिक प्रक्रिया के रूप में है और आर्थिक प्रक्रियाओं की

तरह इसकी उपलब्धता व मांग का विकास होता है तथा अन्य उद्योगों के लिए बाजार प्रदान करता है। श्री जी. जनता के मतानुसार पर्यटन को दो मुख्य क्षेत्रों में विभक्त किया जा सकता है—स्थिर क्षेत्र एवं अस्थिर क्षेत्र। स्थिर क्षेत्र के तहत वे सभी आर्थिक क्रियाएं आती हैं जो कि समुदाय के निर्माण, मांग को आकर्षित करने एवं परिवहन व्यवस्था से जुड़े हैं। इसके तहत परिवहन प्रचालकों, यात्रा-अभिकर्ताओं, परिवहन संस्थाओं व अन्य सहायक सेवा की क्रियाएं शामिल हैं। द्वितीय, अस्थिर क्षेत्र के अन्तर्गत अल्पकालीन प्रवास, आवास सुविधा, खान-पान की मांग आदि क्रियाएं सम्मिलित की जा सकती हैं। एक सीमित अर्थ में पर्यटन को उद्योग नहीं कहा जा सकता है। परन्तु आर्थिक दृष्टि से, पर्यटन से ही मांग की उत्पत्ति होती है। यह विभिन्न उद्योगों हेतु बाजार निर्मित करता है। अतएव वे सभी क्रियाएं जो पर्यटन से संबंधित हैं, पर्यटन उद्योग के अन्तर्गत आती हैं, राष्ट्रीय दृष्टिकोण से पर्यटकों से संबंधित हैं। पर्यटन उद्योग से अभिप्राय है, वे समस्त माध्यम जो पर्यटकों को सेवाएं व सुविधाएं प्रदान करते हैं। यह राष्ट्रीय दृष्टिकोण से पर्यटकों द्वारा किया गया सम्पूर्ण व्यय 'राष्ट्रीय पर्यटन' या 'पर्यटन उद्योग' कहा जा सकता है। अन्य शब्दों में, पर्यटन उद्योग से अभिप्राय है वे समस्त माध्यम जो पर्यटक सेवाएं व सुविधाएं प्रदान करते हैं। यह राष्ट्रीय व्यवस्था का एक हिस्सा है एवं पर्यटकों की इच्छाओं की पूर्ति करते हैं। यह एक भिन्न समूह है जिसके तहत वे समस्त व्यापार व उद्योग आते हैं जो यात्रियों की आवश्यकता पूर्ति करना अपना एक मुख्य उद्देश्य समझते हैं। विभिन्न प्रकार के यात्रियों जैसे—व्यापारी, सैलानी, वाणिज्य यात्री, विद्यार्थी, संबंधी एवं मित्रगण इत्यादि की जरूरतों की पूर्ति उद्योग द्वारा की जाती है। इस प्रकार हम देखते हैं कि यह उद्योग बहुत से व्यापारों का संगठन होते हुए स्वयं में एक अतिविशाल आर्थिक संगठन है।

प्रत्यक्ष रूप में असंख्य प्राथमिक एवं सहायक व्यापार यात्रियों पर ही पूर्णरूपेण निर्भर करते हैं। इन प्राथमिक व्यापारों में होटल व खानपान उद्योग व यात्रा अभिकरण इकाइयां आती हैं। सहायक व्यापारों के तहत वे व्यापार शामिल किए जा सकते हैं जो यात्रियों को सेवाएं उपलब्ध कराते हैं, उदाहरणार्थ—फुटकर दुकानें, स्मारिकाएं, पुरातत्वकालीन समान, उपहार विक्रेता इत्यादि, मनोविनोद के साधन (नाट्यशाला, चलचित्र गृह, विशेष पर्व व त्योहार, संगठित क्रीड़ाएँ), बैंक, केस शृंगारगृह, लान्ड्रीज, होटलों को सामान एवं सेवाएं पूर्ति करने वाले व्यापारी, खानपान इकाइयां, परिवहन से जुड़े उद्योग, सार्वजनिक सेवाएं, भवन-निर्माण संस्थाएं (विशेषकर पर्यटन क्षेत्रों में), बीमा कम्पनियां इत्यादि। इस प्रकार अनेक उद्योग पर्यटकों के लिए वस्तुओं एवं

सेवाओं की पूर्ति करते हैं। पर्यटन उद्योग इन सभी उद्योगों का एक सामूहिक मिश्रण है, जिससे पर्यटन में समानता आती है। परन्तु साथ ही पर्यटकों की विभिन्न प्रकार की मांगों से यह क्रियाएं आर्थिक अध्ययन की दृष्टि से कृषि उत्पादन एवं अवतरण उद्योगों की अपेक्षा अत्यधिक कठिन हैं।

पर्यटन के आधार

(1) पर्यटकों का रहन-सहन—उच्च रहन-सहन के कारण आय का अधिकतम भाग पर्यटन के कार्यों के लिए प्राप्त हो रहा है। साथ ही कार्य करने के घंटों में कमी हो रही है तथा वेतन सहित अवकाश भी उपलब्ध हो जाता है। इन सभी कारणों से पर्यटन की जरूरतें पूर्ण हो रही हैं—धन अवकाशकाल व्यतीत करने के लिए तथा समय जिसमें अवकाश लिया जा सके।

(2) अलीशान पर्यटन के विक्रय की व्यावसायिक क्षमता—ज्यादातर लोग अपने व्यवसाय में समर्थ हो चुके हैं तथा एक सामान्य आदमी को अपना धन व समय विदेशों में खर्च करके भ्रमण व यात्रा करने के लिए प्रोत्साहित करते हैं।

(3) पर्यटन का संबंध—ये यात्रा की परम्परा को कायम व जीवित रखने में मददगार होते हैं। निजी क्षेत्र व सरकारी पर्यटक संगठनों द्वारा की गयी प्रोत्साहन क्रियाएं राष्ट्र के पर्यटन के विकास में सहायक होती हैं। समाचार-पत्र, पत्रिकाएं, चलचित्र, रेडियो, टेलीविजन के माध्यम से विलासितापूर्ण क्रियाओं का उद्योग यात्रा व अवकाश इत्यादि से संबंधित समाचारों का विस्तार करके आराम व विलासिता के महत्व को समझता है।

(4) मूल्य वृद्धि एवं यातायात की गति—ये दोनों कारण या तो पर्यटन का उद्देश्य सीमित करते हैं या पर्यटन के विकास में मददगार होते हैं।

(5) भ्रमण में स्वतन्त्रता—पर्यटन उद्योग हेतु स्वतंन्त्रता का होना अत्यन्त आवश्यक है, अतएव इसको हमें संरक्षित एवं विकसित करना है। बहुत से देशों ने इस विचार को अत्यधिक सोच-समझकर ही अपने देशवासियों को आराम से भ्रमण व यात्रा करने की सुविधाएं प्रदान की हैं।

(6) पर्यटन क्षेत्र का सुन्दर जलवायु—सुन्दर जलवायु जिसमें गर्मी एवं सूर्य का प्रकाश हो, पर्यटन क्षेत्र का मुख्य आकर्षण है। कोई भी क्षेत्र जिसमें शिशिरकालीन आकर्षण हो, पर्यटन का आकर्षण बन जाता है। बहुत से क्षेत्र आनन्द एवं स्वास्थ्यपूर्ण जलवायु के कारण शक्तिशाली पर्यटन क्षेत्र बन जाते हैं।

(7) शिक्षित—अधिकांश यात्री पढ़े-लिखे व शिक्षित होते हैं। दूसरे विश्व युद्ध के बाद शिक्षा सुविधाओं में एक विस्फोटक उन्नति हुई। वस्तुतः बहुत ही समर्थशील व्यक्ति पर्यटक की श्रेणी में आने लगे हैं।

(8) पर्यटकों में प्रति-व्यक्ति आय—आस्ट्रेलिया, फ्रांस, कनाडा, अमेरिका, जर्मनी, इंग्लैण्ड, जापान इत्यादि राष्ट्रों में प्रति व्यक्ति आय में काफी वृद्धि हुई है, यही देश ज्यादातर पर्यटकों के स्रोत हैं।

(9) पर्यटकों में जिज्ञासा—विश्व के संचार के साधन जैसे—दूरदर्शन, आकाशवाणी, पत्रिकाएं, प्रकाशन तथा शिक्षा की नई-नई क्रियाओं के परिणामस्वरूप भावी पर्यटकों में जिज्ञासा विकसित हुई, अतएव इनकी विदेश भ्रमण की उत्कट इच्छा हुई ताकि वे स्वयं बाहरी घटनाओं की जानकारी प्राप्त कर सकें।

पर्यटन विकास से संबंधित कुछ मुख्य अंग जिन पर यह सम्पूर्ण उद्योग निर्भर है, निम्न प्रकार हैं—

सुन्दरता का होना—किसी भी पर्यटन स्थल का सुंदर होना अति आवश्यक है। इसी कारण पर्यटक आकर्षित होते हैं। सुन्दर पर्वतीय दृश्य व समुद्रतटीय सौन्दर्यता एक बहुत ही बड़ा आकर्षण है, जो पर्यटक इन स्थलों पर जाते हैं वे इनकी प्राकृतिक बनावट व सजावट अर्थात् सुन्दरता से बहुत ही प्रभावित हो जाते हैं।

पर्यटन की सुविधाएं— ये दो प्रकार की होती हैं :

प्राकृतिक सुविधाएं—नैसर्गिक सुविधाओं के अन्तर्गत समुद्र तट, सामुद्रिक स्थान, मछली पकड़ना, पर्वतारोहण व दृश्यावलोकन इत्यादि शामिल हैं।

मानवीय सुविधाएं—अनैसर्गिक सुविधाओं के अन्तर्गत विहारार्थ भ्रमणस्थल, स्नान के तालाब, हरियाली, नाट्यशालाएं, चलचित्र व अजनबी स्थल, झूलानुमा कुर्सी, सुसज्जित एवं प्रकाशयुक्त सज्जा इत्यादि शामिल हैं।

ऐतिहासिक एवं सांस्कृतिक कारण—समस्त पर्यटन-प्रिय राष्ट्र अपने इतिहास को पर्यटक आकर्षणों का एक माध्यम मानते हैं। इतिहास के विशिष्टांग व संस्कृति के आकर्षण पर्यटकों को अपनी ओर आकर्षित करते हैं।

विश्राम—कई पर्यटक किसी विशिष्ट स्थान पर इसलिए जाते हैं, क्योंकि वहां पर अच्छा खान-पान, आवास व अन्य सुविधाएं प्रदान करने वाले होटल मौजूद होते हैं।

पहुंच या नजदीकी—नजदीकी व पहुंच पर्यटन के प्रधान अंग हैं। विमानों के आविर्भाव से यात्रा के क्षेत्र में एक क्रान्ति आ गयी है। जो स्थान कुछ समय पूर्व तक पहुंच से बाहर थे वे आज बहुत ही समीप हो गये हैं।

अन्य कारण—किसी भी देश के नागरिकों का आगन्तुकों के प्रति मैत्री व आदरपूर्ण व्यवहार, पर्यटकों को आराम प्रदान करता है तथा अपने अवकाश को व्यतीत करने में आनन्द प्रदान करता है। सूचना-केन्द्रों के जरिए पर्यटकों को, जो देश के बावत अनजान हैं व क्षेत्रीय भाषाएं नहीं बोल सकते हैं और न ही समझ सकते हैं, उन्हें पर्यटन स्थलों, खेल की सुविधाओं, रात्रि की खेल क्रियाओं, खरीददारी संबंधी क्रियाओं आदि की बावत पूर्ण सूचना उपलब्ध होनी चाहिए, प्रशिक्षित कौशलपूर्ण (विशेषज्ञ) हर प्रकार की स्थानीय भाषा बोलने वाले मार्ग दर्शकों की जरूरत है। जहां तक हो सके पर्यटकों के ऊपर पारपत्र से संबंधित प्रवेश पत्र आदि का जितना भी दबाव है उसे पूरी तरह खत्म या कम करना चाहिए। सीमा शुल्क, मुद्रा नियन्त्रण व अन्य शिष्टाचारों को, जहां तक हो सके, कम करने की जरूरत है।

पर्यटन का विस्तार

पर्यटन को लेकर लोगों का बिल्कुल विभिन्न विचार है कुछ को यात्रा पंसद नहीं है इसके विपरीत दूसरे लोगों को अत्यधिक पंसद है। इस क्षेत्र में बहुत ही मनोवैज्ञानिक अध्ययन किए गए हैं। इन अध्ययनों से पता चलता है कि यात्रा की सीमाएं या अवरोधकों को निम्न भागों में बांटा जा सकता है।

(1) समय की कमी—यह प्रतिबन्ध मुख्यतः उन लोगों के लिए है जो अपना व्यवसाय या कार्य कुछ अवधि तक के लिए नहीं छोड़ सकते हैं।

(2) खर्च—उपभोक्ता धन की कमी की वजह से अपनी क्रियाएं नियंत्रित करते हैं। अन्य आकर्षित होने वाले व्ययों के साथ-साथ यात्रा के व्यय भी प्रतिद्वन्द्वी होते हैं। एक अध्ययन के अनुसार 62 प्रतिशत प्रतिपादकों के यात्रा न करने का यह कारण उनके उस उद्देश्य के लिए धन का अभाव ही प्रमुख था। यदि यह कहा जाए कि यात्रा अति खर्चीली है तो परोक्ष रूप में इसका आशय है कि यात्रा करना जरूरी नहीं है। अतएव घर से बाहर न निकलने का मुख्य कारण खर्च है।

(3) पारिवारिक परिस्थितियां—छोटे बच्चों के माता-पिता पारिवारिक जिम्मेदारियों एवं यात्रा की असुविधाओं की वजह से यात्रा करने के अभिलाषी नहीं होते हैं।

(4) भौतिक सीमाएं—अस्वस्थता व अन्य शारीरिक प्रतिबन्ध मुख्यतः वृद्धावस्था में लोगों को घरों में ठहरने के लिए बाध्य करते हैं।

2

पर्यटन की कला रचना

पर्यटन की सुविधाओं, यात्रा एजेन्सियों, पर्यटन खेलों, मनोरंजन उद्योग, यात्रा संचालन, कला, दस्तकारी और स्मारक सम्बन्धी उद्योगों का विकास करना पर्यटन की कला रचना का प्रमुख उद्देश्य है। इसके अलावा भू-गर्भ एवं ऊपरी निर्माण संबंधी गतिविधियाँ शामिल होती हैं जिनमें पर्यटकों के ठहरने, वैश्विक स्तर पर संबंध बनाने एवं मानव की व्यापक गतिविधियों हेतु आवश्यक होता है। अधिकतर प्रायः अर्थशास्त्री पर्यटन आधारित संरचना पर अधिक बल देते हैं। क्योंकि वह आर्थिक विकास की गति को तेज करने की एक अग्रिम शर्त है, इसीलिए प्रायः यह देखा गया है कि आधारित संरचनाओं के निर्माण में उच्च कोटि के पूंजी विनियोजन का होना विकासशील राष्ट्रों की एक प्रमुख विशेषता रही है। भारत में वर्तमान आधारित संरचना की स्थिति पर्यटक अर्थव्यवस्थाओं को पूर्ण नहीं कर सकती है। इस सम्बन्ध में, पर्यटन को बढ़ावा देने के लिए होटल व अन्य अनुपूरक आवास संबंधी सुविधाओं का एक प्रमुख स्थान है। विकासशील अर्थव्यवस्था व उसकी वृद्धि पर्यटन की आधारित संरचना की एक विशेषता है।

पर्यटकों द्वारा इन क्षेत्रों का आर्थिक दृष्टि से इस्तेमाल वहाँ के विस्तृत एवं अति दुर्गम प्राकृतिक व आर्थिक स्रोतों पर प्रयोग करता है जहां पर उपभोग की वस्तुओं

व निपुण कामगारों की पूर्ति एवं आधारित संरचनाओं को जोड़ने की कठिनाइयां जुड़ी रहती हैं। यात्रियों के काफी संख्या में आगमन के फलस्वरूप ही पर्यटन आधारित संरचना की आवश्यकता पड़ती है। नये पर्यटन स्थलों की खोज व विकास से व्यक्तियों, सामान व सेवाओं का अर्द्धविकसित क्षेत्रों में विचलन होता है।

जिस प्रकार की संरचना होती है उसी अनुसार होटलों, सामग्रियों के निर्माण और सेवा के उत्पादनों को निर्धारित किया जाता है। वर्तमान आधारित संरचना एवं अर्थव्यवस्था की बढ़ती हुई आवश्यकताओं के मध्य की अनियमितताओं के फलस्वरूप ही पर्यटन की संरचना के निर्माण की आवश्यकता महसूस हुई।

पर्यटन के माध्यम से बेरोजगारी को खत्म करने और विदेशी मुद्रा को अर्जित करने में मदद मिलती है। पर्यटन क्षेत्रीय विकास में भी महत्वपूर्ण भूमिका निभाता है। यदि रोजगार को ग्रामीण क्षेत्रों में वास्तविक रूप से नियोजित न किए जाय तथा अवस्थानुसार न लिया जाए तो पर्यटन विकास से सम्बन्धित क्रियाओं पर किए गये निजी व सार्वजनिक विनियोजन के नतीजों का कोई वास्तविक एवं अच्छा परिणाम नहीं निकलेगा। इसीलिए इन क्षेत्रों में पर्यटन की आधारभूत संरचना की योजनाओं को नियोजित करते वक्त पूरी सावधानी रखनी चाहिए।

विकासशील अर्थव्यवस्था को यातायात सेवायें सामाजिक और आर्थिक रूप से अधिक प्रभावित्त करती हैं। यह अनुभव किया गया है कि विकासशील अर्थव्यवस्था की एक मुख्य विशेषता यह है कि इसके उपरांत पर्यटन क्षेत्रों के मध्य परिवहन व्यवस्था की बहुत ही कमी होती है। नतीजन अन्य असंख्य आर्थिक असुविधाएं उत्पन्न होती हैं। विदेशी व स्वदेशी पर्यटकों के लिए महत्त्वपूर्ण नये पर्यटन स्थल खोजने तथा परिवहन साधनों के निर्माण की आवश्यकता है। पूंजी की कमी तथा आर्थिक पिछड़ेपन में योजनाओं की परेशानियों के कारण पर्यटन विकास हेतु एक उचित आधारित संरचना के निर्माण के मार्ग में अवरोधक रही है। यह भी देखा गया है कि वर्तमान आधारित संरचना की सामर्थ्य का पूर्ण उपभोग नहीं हो पाता क्योंकि सम्बन्धित आर्थिक क्षेत्र तत्काल फलीभूत नहीं होते हैं। कृषि प्रधानता के कारण हमारी अर्थव्यवस्था में आधारभूत संरचना पर बहुत ही असर पड़ता है।

पर्यटन विकास के परिणामस्वरूप समस्त संरचना व सेवाओं की आवश्यकता होती है जिसके कारण उत्पादकता में सुधार हुआ है। पर्यटन की आधारित संरचना के निर्माण का एक प्रमुख उद्देश्य अन्तर्क्षेत्रीय आर्थिक सम्बन्धों को विस्तृत करना है। क्योंकि इसके माध्यम से स्वदेशी बाजार के निर्माण का उत्थान व विकास होता है। परिवहन व्यवस्था के विस्तार से ग्रामीण क्षेत्रों व पर्यटन स्थलों के आपसी सम्बन्ध में

वृद्धि होती है एवं आर्थिक, प्राकृतिक सम्पदा को वितरण करने में सहायता प्राप्त होती है।

पर्यटन की व्यवस्था में प्रशिक्षित व योग्य व्यक्तियों की आवश्यकता में निरन्तर वृद्धि होती जा रही है। पर्यटन शिक्षा की तरफ विशेष ध्यान नहीं दिया गया है। पर्यटन उद्योग के सही एवं श्रम संचालन हेतु इस क्षेत्र में हुई आधुनिक विकसित तकनीकियों की जानकारी बहुत ही आवश्यक है। योजना के तहत क्षेत्रीय विकास के मिश्रित आर्थिक विकास के लिए यह जरूरी है कि क्षेत्र विशेष के लिए उचित आधारित संरचना को उस क्षेत्र की भूमि, जल स्रोतों के साथ-साथ प्राथमिकता मिलनी चाहिए साथ ही एक पर्यटन उद्योग के विविध पक्षों में अनुसंधान की अत्यधिक आवश्यकता है, जिससे भविष्य के विकास हेतु एक उचित आधारशिला रखी जा सके।

परिवहन सुविधाओं की मांग मौसमी उतार-चढ़ावों से प्रभावित होती है। बड़े-बड़े होटल व परिवहन व्यवस्थाएं मात्र अल्प अवधि के लिए ही उपयोग की जाती हैं तथा शेष अवधि में वे सिर्फ आंशिक रूप में ही पैदा की जाती हैं। अतएव उपलब्ध आधारित संरचनाओं के उचित प्रयोग का प्रश्न उठ खड़ा होता है। किसी भी नई संरचना के उच्चतम क्षमता उपयोग करने की बहुत ही जरूरत है, क्योंकि सम्पूर्ण संरचना को सुरक्षित/मरम्मत हेतु उष्ण जलवायु, मानसूनी वातावरण एवं परिवर्तित तापमान की अवस्था में बहुत ही धन व्यय करना पड़ता है। यह भी ज्ञात हुआ है कि कृषि विकास में भूमि व जल का उपयोग समस्त राष्ट्रीय कृषि विकास के लिए महत्त्वपूर्ण आधार है।

उद्योगों के विकास से विद्युत शक्ति की मांग में भी बढ़ोत्तरी होती है। पर्यटन विकास में पर्यटन क्षेत्रों के उपभोग व मूल्यांकन हेतु वित्तीय साधनों तथा तकनीकी विशेषज्ञों की आवश्यकता होती है। इससे आधुनिक पर्यटन की संभावना निश्चित होती है। प्रौद्योगिकी विशेषज्ञों का अभाव इस विशिष्ट उद्योग के संचालन में सबसे बड़ी बाधक रही है। दिल्ली, मुम्बई, कोलकाता व चेन्नई में चार महत्त्वपूर्ण क्षेत्रीय प्रबन्ध तकनीकी संस्थान स्थापित हैं एवं अन्य बहुत से संस्थान व खाद्यकला केन्द्र स्थापित हैं जो युवकों तथा युवतियों को मुख्यतः पर्यटन एवं आवास क्षेत्र से सम्बन्धित शिक्षा प्रदान करते हैं।

पर्यटन क्षेत्रों के आर्थिक विकास में पर्यटन सम्पदा की महत्वपूर्ण भूमिका होती है। भारत में पर्यटन आधारित संरचना का आर्थिक उन्नति में प्रमुख योग है, जिसके निर्माण में बहुत-सी बाधाएं हैं। सरकार को सम्पूर्ण अर्थव्यवस्था पर जबरदस्त दबाव डालना चाहिए जिससे अच्छे परिवहन साधनों, होटलों, बिजली व शक्ति साधनों तथा

इस उद्देश्य को प्राप्त करने से सम्बधित अन्य सुविधाओं का निर्माण हो सके। विद्युत शक्ति में वृद्धि होने से अर्थव्यवस्था की मूल आवश्यकताएं पूरी होती हैं एवं औद्योगिक विकास को प्रोत्साहन मिलता है।

भ्रमण के विभिन्न अंग

पर्यटन का निर्माण अनेक तत्वों के मिश्रण से होता है क्योंकि इसमें अधिक समय भी लग सकता है इसके लिए स्वास्थ्य सेवाएं, रहने की व्यवस्था, खान-पान आदि तत्वों का होना आवश्यक है। इसमें बहुत दिनों, हफ्तों व महीनों तक ठहरना भी शामिल है, जिसके लिए आवास, मनोरंजन हृदयावलोकन व मनोविनोद के लिए विभिन्न प्रकार की सुविधाओं एवं सेवाओं की आवश्यकता पड़ती है। पर्यटन की सफलता इन्हीं सभी तत्वों पर आश्रित होती है। यह एक बहुत ही महत्त्वपूर्ण कार्य है और जब तक पर्यटन व्यापार के विविध अंगों में सामंजस्य नहीं लाया जाता तब तक पर्यटन व्यापार में बढ़ोत्तरी नहीं हो सकती है।

अध्ययन करने की दृष्टि से इन सुविधाओं को निम्नांकित भागों में विभाजित किया जा सकता है—

1. परिवहन व परिवहन सम्बन्धी यंत्र।
2. प्राकृतिक संसाधन।
3. आधारभूत संरचना सम्बन्धी सुविधाएं।
4. उच्च संरचना सम्बन्धी सुविधाएं।
5. सरकार सम्बन्धी संसाधन एवं सुविधाएं।
6. पर्यटन के आकर्षण एवं मनोरंजन सम्बन्धी आवश्यकताएं।

पर्यटन के विभिन्न अंग

पर्यटन के निम्नलिखित विभिन्न अंग हैं—

1. परिवहन एवं परिवहन यंत्र,
2. यात्रा मार्गदर्शक,
3. प्राकृतिक स्रोत,
4. सैलानी संबंधी आधारभूत ढांचा,
5. यात्री यातायात टर्मिनल,
6. उच्च संरचना,
7. क्रीड़ा व खेलकूद,

8. प्रकाशन माध्यम,
9. हवाई अड्डों की इमारतें,
10. वित्तीय संस्थाएं,
11. खरीददारी के केन्द्र,
12. आधारित संरचना,
13. शिष्टाचार स्रोत तथा
14. पुलिस, सीमा शुल्क आदि प्रमुख अंग हैं।

जल शक्ति, सड़क, वायु, परिवहन, पर्यटन के आकर्षण आदि सभी सुविधाओं का एक पैकेज उद्देश्य प्राप्ति के लिए बहुत ही आवश्यक है। किसी राष्ट्र में पर्यटक यातायात की बढ़ोत्तरी इस बात पर काफी कुछ आश्रित होती है कि पर्यटकों को कितनी न्यूनतम सुविधाएं उपलब्ध करायी जा रही हैं। विकासशील अर्थव्यवस्था में पर्यटन उद्योग के विकास के लिए सामान्य राष्ट्रीय योजनाओं व सम्बद्ध कार्यक्रमों की काफी आवश्यकता है।

राष्ट्रीय योजना के तहत पर्यटन की आधारभूत संरचना और सामान्य सेवाओं में हुए खर्च की भिन्नता को ध्यान में रखते हुए, बाजार अन्वेषणों, सर्वेक्षणों एवं उत्पादन के मूल्यांकनों के द्वारा विनियोजन की दर व आय प्राप्ति को स्पष्ट किया जाता है। पर्यटन एक व्यापार ही नहीं है वरन् इसके अंतर्गत बहुत से उद्योग व जटिल क्रियाएं भी आती हैं। एक उचित प्रयोगात्मक कार्यक्रम बनाने तथा विनियोजकों, स्थानीय लोगों, राज्य एवं पर्यटकों को इसके क्रियान्वित करने के फायदे प्रदान करने के लिए इससे सम्बन्धित विभिन्न अंगों व तत्वों का मूल्यांकन करना अनिवार्य हो जाता है। राष्ट्रीय योजना में पर्यटन का स्थान इसको प्रदान की गयी प्राथमिकता पर निर्भर है। कई देश इसको विलासितापूर्ण उद्योग समझकर अपनी राष्ट्रीय योजनाओं में कम महत्व देते हैं।

राष्ट्रीय योजनाओं की प्राथमिकता वहाँ की आधारित संरचना, प्राप्त होने वाले कच्चे माल एवं सामाजिक एवं आर्थिक विकास के संतुलन पर निर्भर करती है। पर्यटन उद्योग की सफल स्थापना मूलरूप में देश की अर्थव्यवस्था की आधारित संरचना के उद्गम व विकास पर आधारित होती है।

आधारित सरंचना की आवश्यकताओं को प्रमुख भागों में बांटा जा सकता है जो निम्नलिखित हैं—

1. वे संरचनाएं जो कि बाहरी संसार से पहुंच व संचार सम्पर्क प्रदान करने में सहायक हैं।

2. वे संरचनाएं जो कि किसी भी उद्देशित स्थान पर लोगों के विचलन में सहायक होती हैं।
3. वे संरचनाएं जो कि विकास के लिए बिजली, शक्ति, जल परनाला व मल निष्कासन की जरूरी सेवाएं प्रदान करती हैं।

किसी भी पर्यटन स्थान पर पहुंचने के लिए पर्यटक सड़क, रेल, जलमार्ग, वायुमार्ग आदि यातायातों को अपनाता है। एक जरूरतमन्द आधारित संरचना के लिए सड़क, पुल, सेतु, बन्दरगाह, हवाई अड्डे और रेल की पटरी इत्यादि के निर्माण की आवश्यकता होती है। ये सभी साधन पर्यटन के भार को पर्यटन क्षेत्रों में अथवा उनसे बाहर ले जाने के लिए अनिवार्य हैं। साथ ही संचार सम्बन्धी साधनों की भी जरूरतें होती हैं। पहुंच से सम्बन्धित क्रियाएं भी जरूरी हैं ताकि संचार के साधनों की सुविधाएं प्राप्त हों तथा इस क्षेत्र में विचलन भी हो सके। इसका आवश्यकतानुसार सम्पूर्ण विकास क्षेत्र में सार्वजनिक सेवाओं का विस्तार करना होगा ताकि सार्वजनिक सुविधाएं जैसे–बिजली सम्बन्धी सुविधाएं पर्यटकों को प्राप्त हो सकें। एक सफल पर्यटन हेतु भूमि की आन्तरिक एवं संस्थापन सेवाएं बहुत ही महत्त्वपूर्ण हैं। इनके निर्माण में काफी समय लगता है। शुरुआत में ही उचित तथा चौड़ी सड़कों का निर्माण होना चाहिए अन्यथा छोटी व असुविधाजनक सड़कों को बाद में तोड़ना पड़ सकता है व परिवर्तन करके अच्छी व सुविधाजनक सड़कों का निर्माण करना पड़ता है। किसी भी सड़क को बनाने में अनेक प्रकार की सावधानियों पर ध्यान देना पड़ता है। जैसे–सड़कें ठोस एवं टिकाऊ होनी चाहिए, हर मौसम के अनुकूल हों तथा इनका उचित तौर पर निर्माण हो ताकि पानी व बरसात का प्रभाव न पड़े वे सुरक्षित होनी चाहिये एवं अन्तर्राष्ट्रीय मान्यताओं के मुताबिक उनका निर्माण होना चाहिए। सार्वजनिक संस्थानों जैसे सुविधाजनक सड़क सम्बन्धी सुविधाएँ, सड़कों के साथ-साथ पार्क एवं वनभोज की सुविधाएं, आराम के पार्क जिनमें जन सुविधाएं, दर्शनीय कक्ष, आटो मरम्मत व अन्य प्रकार की दूसरी सुविधाएं होनी चाहिये। हवाई अड्डों की पट्टियों का भी अन्तर्राष्ट्रीय मान्यताओं के अनुसार निर्माण होना चाहिए क्योंकि नये विमानों के आगमन प्रस्थान के लिए इनका उपयोग होता है। अतः भविष्य में इन पर आधुनिकीकरण का काम न करना पड़े।

सड़कों के किनारे के पार्क, पिकनिक स्थल, आराम-स्थल, दर्शनीय-स्थान इत्यादि इसी तरह की पार्क सुविधाओं का सतत् ध्यान रखना चाहिए ताकि उनकी साफ-सुथरे एवं उचित प्रकार से हिफाजत की जा सके।

पर्यटन से संबंधित आधारभूत सुविधाएं

पर्यटन से संबंधित निम्नलिखित आधारभूत सुविधाएं इस प्रकार हैं–

(1) पर्यटक संगठन : शिक्षा एवं प्रशिक्षण, श्रमिक संगठन, व्यापारिक संगठन, व्यक्तिगत/व्यावसायिक/कम्पनी संगठन।

(2) होटल : अन्तर्राष्ट्रीय होटल, विहाथल होटल, निवास युक्त होटल, वाणिज्य होटल, सराय, काफिला स्थल, क्लब, यात्री आराम स्थल, अवकाशकालीन पड़ाव, स्वास्थ्यवर्धक जल के स्थल, होस्टल, अल्पकालीन स्थिति के होटल, भुगतानशुदा मेहमान आवास, मध्यस्तरीय आवास, अनुपूरक आवास।

(3) खान-पान : रेस्टोरेन्ट, भुनने के कक्ष, मुख्य सेवाएँ भोजनालय, हस्पताल, दूध एवं अन्य भोजन कक्ष, रेलवे खान-पान व्यवस्था, वायु खान-पान व्यवस्था, सामुद्रिक खान-पान व्यवस्था, ड्राइभ इन्ज, वृताकार नृत्यकक्ष, क्लब, रात्रि क्लब, मधुशाला, चाय कक्ष, कॉफी कक्ष।

(4) ऐतिहासिक स्मारक : धार्मिक भवन, सैनिक एवं नागरिक भवन, कला/शिल्पकला, पुरातत्वकालीन अवशेष, रणभूमि, संग्रहालय, स्मारक, स्मारिकाएँ, पुस्तकालय।

(5) मनोरंजन स्थल : संगीत नृत्य, नाटक विशेष पर्व, ध्वनि एवं प्रकाश प्रदर्शन, प्रदर्शनी, संग्रहालय, लोकगीत एवं संगीत, गढ़ एवं किले, विध्वंस स्थल, मठ या आश्रम।

(6) क्रीड़ा एवं खेल कूद : जल-क्रीड़ा, नाव, लम्बी नाव डोंगी जल पर चलने की क्रीड़ा, चढ़ने की क्रियाएं, तरण-ताल क्रिया बन्दरगाह/ताल/नालिका विहार सुविधाएं, चमड़े से गोता मारना, अलांकन, टहलना, आखेट, पर्वतारोहण लहरों पर तैरना, कैलिपोत में यात्रा, समुद्र के ऊपरी भाग की यात्रा, नाविक दौड़, शीतकालीन क्रीड़ायें, बर्फ में चढ़ने का यंत्र, उठने की कुर्सी, टेलीफर, रेलवे, स्केटिग रिंक, बर्फ पर बिना पहिये की गाड़ी से दौड़, बर्फ की कक्षाएं, भूतल की क्रीड़ा व खेलकूद, गोल्फ,

		फुटबाल, वॉलीबाल, बैडमिन्टन, टेनिस, हैन्डबॉल, पिन्ग पोंग कक्ष, गोल्फ कक्ष, उबलते जल के या गर्म उद्यान पथ, सवारी करना, पशु सवारी, विद्यालयों के सवारी के पथ।
(7) प्रेस माध्यम	:	प्रेस, आकाशवाणी दूरदर्शन, यात्रा मार्गदर्शक, प्रकाशन आदि।
(8) वित्तीय संस्थाएं	:	बैंक, डाकघर, डाक, दूरभाष, टेलेक्स, बीमा कम्पनियां तथा अन्य वित्तीय संस्थाएं।
(9) खरीददारी केंद्र	:	कर युक्त विक्रय स्थल, भारतीय हस्तकला, वाणिज्य केन्द्र, खुले बाजार।

अन्य सुविधाएँ— हवाई अड्डे की इमारतें, यात्रियों के यातायात टर्मिनल, सीमा एवं देशान्तरवास।

पर्यटन के लिए सुविधाओं और सेवाओं की पूरी व्यवस्था करनी चाहिए। इन कक्षों पर विद्यमान कर्मचारी दयालु, मैत्रीपूर्ण व मृदुभाषी होने चाहिए। उन्हें निकटवर्ती पर्यटन सुविधाओं का पूरा ज्ञान होना चाहिये। साथ ही सैलानियों की आवास सुविधाओं, खरीददारी, मनोविनोद आदि की सलाह भी उपलब्ध करनी चाहिए।

पर्यटन की मूलभूत कला रचनाएं तथा आवश्यकताएं

पर्यटन पर आधारित संरचना में व्यक्तिगत से संबंधित जैसे यात्रियों के यातायात के टर्मिनल, स्वास्थ्य सेवाएं, होटल आदि इसमें सम्मिलित किये जाते हैं। ये सभी सुविधाएं काफी लागत पूर्ण हैं क्योंकि ये भवन निर्माण से सम्बन्धित हैं। इनमें से अधिकतर सुविधाएं निजी स्तर पर विकास करने वाले अथवा संचालकों द्वारा उपलब्ध करायी जाती हैं। या तो ये निजी स्तर पर निर्मित की जाती हैं अथवा समस्त योजना का एक अंग होती हैं। यात्री यातायात टर्मिनल आदि को आधारित संरचना से सम्बन्धित किया जा सकता है। किसी भी स्थानीय आवास सुविधा का अनुमान वहाँ पर ठहरने वाले पर्यटकों की संख्या से लगाया जा सकता है, अतएव पूर्ण आवास व्यवस्था का प्रबन्ध होना चाहिए।

संचालकों के लिए आर्थिक दृष्टि से व सम्बन्धित क्षेत्र की आर्थिक खुशहाली हेतु अन्य सुविधाएं जैसे—खान-पान, मनोविनोद, खरीददारी की विशेष योजना सम्बन्धी असुविधा नहीं होनी चाहिए, प्रायः ये व्यापारिक उद्यमियों द्वारा सरलता से की जाती हैं। इन सुविधाओं की प्रकृति एवं किस्म का स्तर बहुत ही मायने रखता

है, एवं पर्यटन स्थल मुहैया की आकृति के मूल अवस्था तक प्रवेश करती है। लगभग सम्पूर्ण पर्यटन योजनाओं में सभी सुविधाएं नियमित होती हैं। भवन निर्माण में वास्तु कलाकृति नमूनों की ओर विशेष ध्यान देना चाहिए। होटल की इमारतों की संरचना भी विशेष रूप में स्थानीय पर्यावरण के अनुसार होनी चाहिए। पर्यटक अपने दैनिक जीवन के अलावा स्वयं एक विभिन्न पर्यावरण में समाविष्ट होना चाहते हैं। वातानुकूलित तापमान, केन्द्रित ताप व जल पूर्ति साधन व अन्य आधुनिक सुविधाएं भी होटल की इमारत का निर्माण करते वक्त ध्यान में रखनी चाहिए। ये सुविधाएँ निम्न हैं–

क. पर्यटन विषयक उच्च आधारित संरचना

1. केटरिंग
2. यात्री यातायात
3. अन्य खरीददारी
4. आवास रेस्टोरेन्ट एवं मनोरंजन

ख. पर्यटकनिक ढांचा

1. प्राकृतिक एवं कृत्रिम गैस लाइन्स
2. मल निकासी की व्यवस्था
3. निकासी की व्यवस्था
4. सड़कें
5. रेलवे लाइन्स
6. विद्युत जोड़ने की व्यवस्था
7. राष्ट्रीय मार्ग
8. हवाई अड्डे पर सुविधाएं
9. जलपूर्ति की व्यवस्था
10. बस एवं रेलवे स्टेशन की सुविधाएं
11. पार्किंग के वाहन
12. वाहन
13. उद्यान
14. रात्रि प्रकाश
15. हवाई अड्डों की पटरियाँ

अन्य पर्यटन की संस्थापन व्यवस्था

पर्यटन को यदि प्रभावी बनाना है तो आवासीय से संबंधित सुविधाओं पर जोर डालना आवश्यक है जो पर्यटन का मुख्य स्त्रोत है। होटल की सुविधाओं में मध्यम वर्गीय के लिए भी आवास की सुविधाएं होनी चाहिए। वे सेवाएं विविध स्थानों पर प्रदान करनी चाहिए जो यात्रियों की इच्छाएं व आवश्यकताएं पूरी कर सकें। पर्यटन आवास की माँग पर्यटकों के सामाजिक स्तर, मूल्य स्तर एवं उनकी व्यय करने की सामर्थ्य व अन्य बातों पर निर्भर है, अतएव समस्त बाजार में उचित आवास व्यवस्था का प्रबन्ध होना चाहिए। पर्यटन आवास का नियोजन करते वक्त पर्यावरण की स्थिति, सम्भाव्य बाजार, परिवहन के साधनों, अन्य सम्बन्धित क्रियाएं : जिनमें लोग उद्‌देशित स्थान में कार्यरत हों, उद्‌देशित स्थान आदि का स्मरण रखना भी आवश्यक है। किसी भी होटल तथा इसी तरह की अन्य आवास सुविधाओं में विनियोजन करने से पूर्व सम्बन्धित अथितियों की यात्रा व आवास सम्बन्धी प्रवृत्तियों का ध्यान रखना भी आवश्यक है। अलीशान होटलों का निर्माण व अवकाशकालीन स्थलों के विकास को विस्तृत रूप में लेना चाहिए।

भारत में पर्यटन का विकास करने में आवासीय कठिनाइयां रुकावट उत्पन्न करती रही हैं। पर्यटन आवास की समस्या प्रमुख रूप से बड़े-बड़े या पर्वतीय स्थानों पर अधिक हैं। बहुत से स्थल सैलानियों के स्थल हैं परन्तु वहाँ पर कोई भी आवास सुविधाएं अनुपलब्ध हैं। भारत में हमें प्राथमिक एवं सहायक दोनों प्रकार की आवास सुविधाओं की जरूरत है।

परिवहन की सुविधा व प्रमुख साधन

आधारित संरचना के सम्बन्ध में यातायात का विकास बहुत ही महत्वपूर्ण है। परिवहन की आवश्यकताओं की पूर्ति के दौरान इसकी लागत, गति और आरामदेयता को ध्यान में रखना चाहिए। इसके अंतर्गत बसें, टैक्सी, ऑटोमोबाइल, समुद्री जहाज, रेल, हवाई जहाज, सामान वाहक व्यवस्था आदि सम्मिलित हैं। समस्त परिवहन सेवाएं उचित एवं कम व्यय वाली होनी चाहिए। मोटर कारों की बनावट अच्छी तरह होनी चाहिए। विस्तृत खिड़कियाँ, वातानुकूलित आरामप्रद सीटें तथा आराम करने के लिए सुविधाजनक कक्ष से भरपूर होनी चाहिए।

द्विभाषिए व पर्यटन प्रदर्शक उचित तौर पर शिक्षित एवं प्रशिक्षित होने चाहिए। उन्हें सम्बन्धित क्षेत्र के इतिहास, वास्तुकला, नृवंशविद्या, संस्कृति व आर्थिक

प्रणाली (अर्थव्यवस्था) का अच्छा ज्ञान होना चाहिए, साथ ही विदेशी भाषाओं का ज्ञान भी जरूरी है। सड़क परिवहन का ग्रामीण क्षेत्रों में आनन्द लेने के लिए अधिक प्रयोग होने के फलस्वरूप सड़कों के विकास का महत्व बढ़ता जा रहा है। वायु परिवहन के क्षेत्र में भी विकास की आवश्यकता है। हालांकि स्वदेशी हवाई निगम अपने मौजूद बेड़े के द्वारा वर्तमान जरूरतों को पूर्ण करने के हरसंभव कोशिश कर रहे हैं परन्तु साथ ही भविष्य में पर्यटकों की संख्या में वृद्धि को ध्यान में रखते हुए कुछ योजनाएं बनाना जरूरी हैं।

पर्यटन से सम्बन्धित मुख्य परिवहन के साधन

(1) **रेलवे**—ट्रेन्स, कोग्रेलवेज, एरियल, ट्रामवेज, मोनोरेल्स, होमर ट्रेन्स।
(2) **सड़कें**—मानव, जानवर, मोटर वाहन, कोचेज, कैब्ज, बसें, आरामदायक कारें, ऑटोमोबाइल्स।
(3) **सामुद्रिक**—जहाज, सामुद्रिक जहाज।
(4) **उड्डयन**—वायुयान, हेलीकाप्टर, नावें, जहाज, हाइड्रोफ्वोइल्ज, हूमर क्राफ्ट्स।

मेहमान सेवा संसाधन एवं सुविधाएं

प्रत्येक क्षेत्र की सांस्कृतिक सुविधाएं पर्यटकों के स्वागत में सहायक होती हैं। पर्यटकों का स्वागत न होने या इनको सन्तुष्ट न करने से उपलब्ध भौतिक सुविधाएं किसी भी उपयोग की नहीं हैं। सार्वजनिक सूचना केन्द्रों एवं दूरदर्शन तथा आकाशवाणी के प्रसारण द्वारा आगन्तुकों के प्रति सद्भावना का सृजन किया जाना चाहिए। निकटवर्ती लोगों को पर्यटन के महत्व से जाग्रत करना जरूरी है। होटल कर्मचारी, सुविधा केन्द्रों के कर्मचारी, प्रदर्शक आदि सीधे तौर पर पर्यटकों के सम्पर्क में आते हैं; उन्हें विशेष प्रशिक्षण की आवश्यकता होती है। उन्हें आगन्तुकों के स्वागतार्थ सूचनाएं प्रदान करने, मैत्री, सद्भावनापूर्ण सेवा की भावना, आकर्षक व्यक्तित्व आदि से सम्बन्धित अच्छे निर्देश देने चाहिए।

पर्यटनिक शिष्टाचार के प्रमुख आधार

पर्यटनिक शिष्टाचार के प्रमुख आधार सांस्कृतिक स्रोत हैं जिनमें इतिहास, ललित कला, शिल्प कला, साहित्य, संगीत, नाटक, कला, नृत्य, खरीददारी, रंगमंचशाला, चलचित्र, पुस्तकालय, ललित कला कक्ष, वनस्पति एवं जीव-जन्तु उद्यान आदि प्रमुख हैं। इसके अलावा सम्मेलन, संगीतशाला, विश्व प्रदर्शनियां,

ओलम्पिक खेलकूद, धार्मिक पर्व, तीर्थयात्रा, योग, भोजन एवं पेय पदार्थ, हस्तकारिता, साहित्य, धर्म, वाणिज्य उद्योग, व्यापार, कृषि भी इस स्रोत के अंतर्गत शामिल किये जाते हैं।

यात्रा व व्यवसाय से सम्बन्धित पर्यटन

यात्रा-अभिकर्ता, फुटकर यात्रा संचालक, यात्रा मार्गदर्शक सूचनाएँ, स्थानीय प्रतिनिधि, स्थानीय यात्राएँ अन्य।

सभी प्रकार के पर्यटन में खरीददारी एक मुख्य आधार है, जिससे पर्यटक के लक्षित क्षेत्र की यात्रा की सार्थकता प्रभावित होती है। यह आगन्तुकों की एक क्रिया है, जो विश्व की पूरी अर्थव्यवस्था को प्रभावित करती है। हवाई अड्डों के वातानुकूलन की व्यवस्था केन्द्रित होनी चाहिए।

रेल की सुरक्षा सुविधा तथा आरामदायक वातानुकूलित कारें व कोचों से मनोरम दृश्यों के अवलोकन के कारण बहुत से यात्री रेल यात्रा को अधिक पसन्द करते हैं। रेलों की तत्पर समय सारिणी तथा द्रुतगति के रेलों के आगमन के फलस्वरूप रेल यात्रा बहुत लोकप्रिय होती जा रही है। रेलवे स्टेशन से होटलों तक टैक्सी व बसों का समुचित प्रबन्ध होना चाहिए ताकि यात्री अपने उद्देशित स्थानों तक शीघ्र पहुंच सकें। रेलवे की क्षमता में सुधार अथवा वैकल्पिक रूप में रेलवे में विशेष कोच को संलग्न किया जाना चाहिए। टैक्सी एवं आरामदायक कारों की सेवाएं पर्यटन के आवश्यक घटक हैं। चालकों का व्यवहार, भाषा व अन्य गुण बहुत ही महत्वपूर्ण स्थान रखते हैं।

आरामदायक एवं आकर्षक ठहरने के स्थानों का प्रबन्ध होना चाहिए। उड़ान से सम्बन्धित सूचनाएं जैसे—आवागमन व उड्डयन के वितरण, समय तालिकाओं एवं आवश्यक सूचनाओं की उपलब्धि होनी चाहिए। सामुद्रिक यात्रा का भूमि व हवाई यात्रा की प्रगति में एक महत्त्वपूर्ण अंशदान रहा है। कुछ समय से सामुद्रिक यात्रा काफी प्रचालित हो गई है।

होटल के व्यवस्थापकों की सन्तुष्टि से सम्बन्धित स्पष्ट हिदायत देने के साथ-साथ ही पर्यटक कारों और कोचों की संख्या में भी साथ-साथ वृद्धि होनी चाहिए। आन्तरिक परिवहन सुविधाओं के सुधार के लिए न केवल नई सड़कों का निर्माण होना चाहिए वरन् साथ ही, वर्तमान सड़कों की भी अच्छी मरम्मत व देखरेख होनी चाहिए। कोचें, मनोरंजन व मनोविनोद की क्रिया हृदयावलोकन का एक प्रमुख हिस्सा है। विकासशील देशों में पर्यटकों की सुविधाओं की ओर विशेष ध्यान देने की

आवश्यकता है। विशिष्ट समारोह, विश्व प्रदर्शन संग्रहालय, क्रीड़ा आदि पर्यटकों के मुख्य आकर्षण हैं।

पर्यटन आधारित संरचना के लिए सुविधाएं एक अन्य मौलिक आधार हैं, इसके अंतर्गत न सिर्फ पर्यटकों के प्रोत्साहन व प्रतिबन्ध से सम्बन्धित क्रियाएं आती हैं अपितु साथ ही आवागमन के ये नियम भी आते हैं, जो पर्यटकों से सम्बन्धित हैं। स्वास्थ्य, सीमाकर, मुद्रा नियंत्रण इत्यादि के नियम न केवल सुविधाजनक होने चाहिए वरन् उनका प्रयोग सद्भावना के तौर पर होना चाहिए।

पर्यटन के क्षेत्र में मनोरंजन का महत्व तथा आवश्यकता

पर्यटन के क्षेत्र में कृत्रिम आकर्षण जैसे क्लब, जुआ व अन्य विकसित कार्यों आदि को प्रोत्साहन देने की कोई विशेष जरूरत नहीं है। अवकाश प्राप्त यात्रियों की संख्या में अत्यधिक आवश्यकता को देखते हुए वायु, समुद्रतटीय स्थल, पर्वतीय स्थल व दर्शनीय सुन्दरता को प्रोत्साहित किया जाना चाहिए एवं साथ ही इन स्थलों में बुनियादी आधारित संरचना सम्बन्धी सुविधाओं की व्यवस्था भी करनी चाहिए।

पर्यटन विनियोजन में मनोविनोद, पत्रकारिता व माध्यम, खान-पान केन्द्र, खरीददारी से सम्बन्धित सुविधाओं की संख्या तथा विशेषताओं का विकास हेतु एक अंश है। रेस्टोरेंट व होटलों की सफाई एवं स्वच्छता को अन्तर्राष्ट्रीय स्तर का होना चाहिए। प्रत्येक पर्यटन क्षेत्र के लिए एक अच्छे एवं लम्बे अर्से की दीर्घकालीन नियोजन व बाजार सर्वेक्षण एवं अनुप्रेषण की आवश्यकता है, ताकि पर्यटन उद्योग की जरूरतों का मूल्यांकन किया जा सके।

इसी तरह सांख्यकीय मापदण्ड प्रणाली से विभिन्न क्षेत्रों की कमियों को दूर किया जा सकता है एवं उचित समय में उद्देशित पर्यटकों को आकर्षित एवं सम्मिलित करने के लिए पर्यटन की जरूरी आधारित संरचना का निर्माण किया जा सकता है। पर्यटन की आधारित संरचना में समुचित विकास के लिए प्रशिक्षित व्यक्तियों की काफी आवश्यकता होती है ताकि पर्यटन का समुचित विकास हो सके।

पर्यटन से संबंधित जो कर्मचारी हैं उनको अपने कार्यों में कुशल होना चाहिए, जैसे—कैटरिंग से लेकर पर्यटक प्रदर्शक तक आदि। होटल उद्योग में तो काफी विकास की सम्भावना है। साथ ही अच्छे, उचित एवं प्रशिक्षित व्यक्तियों की जरूरत है यही परिस्थिति अन्य क्षेत्रों में भी लागू होती है। अन्तर्राष्ट्रीय स्तरों को कायम रखने अथवा उचित तौर पर संचालित करने के लिए प्रशिक्षण की उचित व्यवस्था की

आवश्यकता है ताकि पर्यटन व्यापार के सभी क्षेत्रों को अच्छी तरह कार्यबद्ध किया जा सके। इस तरह एक नियोजित एवं सुव्यवस्थित पर्यटन कार्यक्रम की काफी जरूरत है। अतः एक दीर्घकालीन नियोजन, जिसमें पर्यटक व्यापार के सभी अंश संयोजित व सम्पर्क रूप में शामिल हों, पर्यटन आधारित संरचना की अत्यन्त आवश्यकीय बुनियाद हैं।

पर्यटन की कला रचना में सरकार का योगदान

भारत के राज्य या केन्द्रशासित प्रदेशों में जहाँ-जहाँ पर्यटन स्थल हैं वहाँ की सरकार उन स्थलों की उचित देखभाल कर रही है जिसके कारण निजी क्षेत्रों को प्रोत्साहन व बढ़ावा मिलता है एवं नई-नई योजनाओं का उद्गम होता है। संसार के समस्त देशों की अर्थव्यवस्था में पर्यटन का महत्व बढ़ता जा रहा है। यह एक विकसित होने वाला उद्योग है। पर्यटन का विस्तार नई-नई रोजगार व्यवस्थाओं को जन्म देता है। नई इमारतों का निर्माण परिभ्रमण क्रियाओं का एक अवयव है, जिसमें पूँजी के विनियोजन की आवश्यकता पड़ती है तथा बहुत से लोगों के लिए रोजगारों का सृजन करता है।

देश में उत्पादित कुछ वस्तुओं हेतु राष्ट्र के भीतर ही बाजार उपलब्ध होता है तथा गुणक व संचित प्रभावों के फलस्वरूप समाज के समस्त वर्ग लाभान्वित होते हैं। इस प्रकार हम देखते हैं कि पर्यटन एक सार्वजनिक आधारित संरचना सुविधा है। सार्वजनिक अधिकारियों की पर्यटन की आधारित संरचना के निर्माण करने में अहम् भूमिका है। पर्यटन की आधारभूत संरचना में वे समस्त इकाइयां आ जाती हैं, जो कि पर्यटन की जरूरी संरचना में आवश्यक हैं। पर्यटकों की मांग के कारण परिवहन व संचार व्यवस्था की सुविधाओं, शहरीकरण, सिंचाई आदि का सृजन होता है जिसके फलस्वरूप आम अर्थव्यवस्था में समस्त निवासित व्यक्तियों को फायदे सुलभ होते हैं। सार्वजनिक अधिकारियों द्वारा राष्ट्रीय पर्यटन के प्रोत्साहन एवं विकास हेतु उठाए गए कदमों में पिछले कुछ वर्षों से एक नई झलक मिलती है। पर्यटन के क्षेत्र में सरकारों ने अनेक प्रकार के कार्य किये हैं, जैसे—पर्यटन की गतिविधि पर नियन्त्रण रखने हेतु संस्थानों का निर्माण, इन क्षेत्रों में संचालित निजी क्षेत्रों को आर्थिक सहायता प्रदान करना, विदेशों में विज्ञापन एजेन्सियों की स्थापना, पर्यटन अध्ययन एवं शोध के सर्वेक्षण केन्द्रों की स्थापना ताकि आन्तरिक एवं बाह्य बाजार की अच्छी जानकारी प्राप्त हो सके।

पर्यटन की आर्थिक संरचना के अंतर्गत प्रायः निम्न बातें आती हैं :

मूल-सार्वजनिक सुविधाएं एवं सेवाएं–इसके अंतर्गत धार्मिक स्थानों का प्रशासन, स्वास्थ्य शिक्षा, संस्कृति, वाणिज्य, पुलिस आदि आते हैं।

मुख्य सम्बद्ध सुविधाओं की पूर्ति–प्रायः इसके अंतर्गत जल, बिजली, गैस, संचार व्यवस्था, मूल-मार्ग, कूड़ा-करकट आदि स्रोत हैं।

वर्तमान आधारित संरचना का प्रमुख प्रयोजन लोगों की महत्वपूर्ण जरूरतों, जैसे–सुरक्षा प्रबन्ध, स्वास्थ्य, संचार पूर्ति आदि को व्यवस्थित ढंग से संचालित करना है। पर्यटकों की जरूरतों की पूर्ति का अभिप्राय केवल यह नहीं है कि पर्यटक अपने उद्देशित स्थान पर आराम से पहुँचें, शीघ्र पहुंचे, ठीक स्थान पर ठहरें व अच्छा भोजन करें, अपितु वे किसी विशिष्ट उद्देश्य से यात्रा करते हैं। इसलिए पर्यटकों को उसी तरह की सुविधाएं प्रदान करनी चाहिए।

परिवहन व आवास के अतिरिक्त पर्यटक स्रोतों के लिए एक बहुत विस्तृत अनुपूरक इकाइयों की भी आवश्यकता है। कुछ समारोह जैसे, अन्तर्राष्ट्रीय खेल, प्रदर्शनियां, सम्मेलन आदि के लिए स्थायी ढाँचों की जरूरत होती है। इन समारोहों के खत्म होने के उपरांत इमारतें शेष रह जाती हैं एवं इन्हें कुछ अन्य उद्देश्यों के लिए इस्तेमाल किया जा सकता है। पर्यटकों को आमंत्रित करने के इच्छुक राष्ट्र को उनका अति उत्तम स्वागत करने के लिए सेवाओं का उचित प्रबन्ध करना चाहिए।

यात्रा हेतु सड़कों, कार पार्क, समुद्रतट, स्वास्थ्य आदि की भी बहुत आवश्यकता होती है। किसी भी स्थल पर पर्यटक इन सभी सेवाओं की आवश्यकता अनुभव कर सकता है। ये सुविधाएं इस प्रकार उपलब्ध होनी चाहिए ताकि पर्यटकों को असुविधा न हो तथा गुणत्व का स्तर कम से कम इतना हो जो कि पर्यटक अपने देश में आवश्यकता अनुभव करते हैं। कुछ आधारित संरचनाएं जो पर्यटन में सहायक हैं पर्यटन के लिए ही निर्माण नहीं की जाती हैं परन्तु उनका सही संचालन पर्यटन के क्षेत्र में राष्ट्र की प्रसिद्धि की वृद्धि में मददगार होता है।

जल, गैस व बिजली सेवाएं, डाक सम्बन्धी सुविधाएं तथा स्वास्थ्य की सुविधाएं, मौजूद एवं उपलब्ध होनी चाहिए। उदाहरणार्थ, शीतकालीन स्थलों में स्वास्थ्य सेवाएं, बच्चों की नर्सरी एवं क्रीड़ा स्थलों का निर्माण होना चाहिए। एक उचित आधारित संरचना से राष्ट्र लाभान्वित होते हैं, जो विकास के मार्ग पर हैं। इसके अतिरिक्त सरकार द्वारा पर्यटन स्थलों पर विशेष पर्यटन के साज-सामान भी मुहैया कराये जाते हैं, जैसे–पर्यटक के बर्फ पर चलने के स्थल पर छोटे आनन्ददायक समुद्री

बन्दरगाह एवं हवाई अड्डों में साज-सामान। यह साज-सामान निजी स्वामित्व के द्वारा प्रदान करना सम्भव नहीं है क्योंकि इन पर किया गया पूंजी विनियोजन इनसे प्राप्त आय के अनुपात में बहुत ही ज्यादा है। एक संग्रहालय, रेस्टोरेन्ट व कलाकृति का निर्माण तथा साथ ही ग्रामीण क्षेत्र को सुरक्षित करना केवल सार्वजनिक वित्त द्वारा ही सम्भव हो सकता है। इस प्रकार हम देखते हैं कि सरकारें बहुत-सी क्रियाएं करती हैं, जो राष्ट्र के पर्यटन हेतु हैं, जिनमें कुछ क्रियाएं सामाजिक हैं एवं कुछ निजी क्षेत्र की सीमा के बाहर हैं।

सभी सरकारों की पर्यटन सीमा उसके देश की स्थिति और दशा पर आधारित होती है। विश्व के सभी देशों की विकास दर में चाहे वह सामाजिक हो, आर्थिक हो, राजनीतिक हो या संवैधानिक भिन्नता पाई जाती है। सामान्यतया समस्त केन्द्रीयकृत नियोजित अर्थव्यवस्था में सरकार ही पर्यटन कार्यक्रमों में एक आरम्भकर्ता, कार्यकारक और प्रशासनिक होती है।

पर्यटकों की सेवाएं

पर्यटकों की मुख्य सेवाएं निम्नलिखित हैं—

(1) सड़क परिवहन—उद्देशित स्थलों में तथा लघु एवं मध्यम दूरी के उद्देशित स्थलों में बसें एवं कोचें प्रत्यक्ष रूप में सतत् सेवाएं प्रदान करती हैं। कितने ही मार्गों में रेल परिवहन की सुविधाओं के आगमन से तो सड़क परिवहन की महत्ता और भी बढ़ गयी है।

(2) यात्री परिवहन—यह एक प्राथमिक पर्यटक सेवा है जो पर्यटकों द्वारा अपने उद्देशित स्थान तक पहुंचने के लिए विविध प्रकार की होती है। यह उद्देशित स्थान में भ्रमण का एक साधन है। यात्री परिवहन का वर्गीकरण विभिन्न प्रकार से किया जा सकता है, जैसे—निजी व सार्वजनिक, स्वदेशी व अन्तर्राष्ट्रीय, आदि-आदि।

(3) वायु परिवहन—सुदूर यात्रा करने का यह एक बहुत तीव्रगामी एवं आरामदायक साधन है तथा बहुत से मार्गों में परिवहन का एक मात्र साधन है।

(4) रेल परिवहन—यह देश के भीतर तथा दो देशों के मध्य की लघु या मध्यम दूरी के परिवहन का अति उत्तम साधन है। यह उद्देशित स्थान पर पहुंचने एवं उसके अंतर्गत यात्रा करने में सहायक होता है।

(5) सामुद्रिक परिवहन—छोटे सामुद्रिक मार्ग, जल रास्ते व सामुद्रिक यात्रा के लिए यह महत्त्वपूर्ण स्थान रखता है।

(6) ट्रेवेल एजेन्ट एवं टूर आपरेटर्स—परिवहन एवं आवास सुविधा मालिकों, संचालकों एवं ट्रेवेल एजेन्सियों के मध्य एक मध्यस्थता की भूमिका को निभाते हैं; टूर आपरेटर एक संगठनकर्ता का कार्य करते हैं, आवास का सम्मिश्रण करके एक उत्पादित वस्तु के रूप में परिणित करते हैं। इस निर्मित पदार्थ को उनके नाम व जोखिम पर बाजार में भुनाया जाता है।

(7) आवास एवं खान-पान—उद्देशित स्थल में यह एक ओर जरूरी पर्यटक सेवा है। पर्यटकों द्वारा आवास सुविधाओं की माँग की जाती, है, जैसे—होटल, अनुपूरक, आवास, मित्रों के सम्बन्धियों के साथ ठहराना, स्वयं अथवा निजी आवास सुविधा, जैसे—पड़ाव, काफिला इत्यादि।

(8) सूचनाएँ एवं दस्तावेज—पर्यटकों को निम्नांकित दस्तावेज एवं सूचनाओं की बाबत बहुत ही आवश्यक होती हैं—मुद्रा, दस्तावेज, सूचनाएँ, स्थलदर्शन, खरीददारी आदि।

(9) मनोविनोद—पर्यटकों द्वारा पर्यटन स्थलों में मनोरंजन तथा मनोविनोद की आवश्यकता अनुभव की जाती है।

इस तरह पर्यटक उपरोक्त सेवाओं को प्राप्त करना चाहते हैं एवं पर्यटक सेवाएँ ही बाजार की पूर्ति कही जा सकती हैं। पर्यटक सेवाओं के अंतर्गत वे सभी आर्थिक क्रियाएँ आ जाती हैं जिनका मुख्य उद्देश्य पर्यटकों की जरूरतों को पूरा करना है क्योंकि इन समस्त पूर्तियों की विशेष प्रकृति पर्यटकों की दैनिक क्रिया की जटिलता की निरन्तर पूर्ति के फलस्वरूप इन्हें पर्यटन की आधारित संरचना के नाम से पुकारा जाता है।

प्रमुख पर्यटन स्थल

उद्देशित स्थल का अर्थ उस स्थल से लगाया जाता है जिस स्थल से पर्यटक या यात्री यात्रा प्रारंभ करता है। परिवहन सेवाओं की तीव्रगति से प्रगति होने के फलस्वरूप व्यापार, वाणिज्य के केन्द्रों जहाँ पर ऐतिहासिक व सांस्कृतिक संगठन, सुसंचार व्यवस्था, मनोरंजन के साधन व खरीददारी की सुविधाएँ सघन हैं, स्वदेशीय एवं अन्तर्राष्ट्रीय पर्यटकों के लिए बहुत ही सम्पन्न स्थल बन चुके हैं। एक पर्यटन स्थल में निम्न विशेषताएँ होनी चाहिए :

आकर्षण—सभी प्रकार के पर्यटन स्थानों का महत्व उसके आकर्षण पर निर्भर होता है। आकर्षण दो तरह के होते हैं—

(1) समारोह के आकर्षण—इसके अंतर्गत कॉन्फ्रेंस, प्रदर्शिनयाँ व क्रीड़ा समारोह, व्यापारिक मेले, ओलम्पिक खेल इत्यादि सम्मिलित हैं।

स्थिति एवं समारोह आकर्षक दोनों ही पर्यटकों को आकर्षित करते हैं। अतः ये दोनों ही एक-दूसरे के अनुपूरक एवं सहपूरक हैं।

(2) स्थिति आकर्षण—इसके अंतर्गत जलवायु, प्राकृतिक सौन्दर्य एवं ऐतिहासिक आकर्षण आते हैं। हिमालय, ताजमहल, आल्पस, एटन्स, मेडिटेरियन, चैनल, द्वीप, लेक, डिस्ट्रिक पर्यटन स्थल जैसे—इडिम्बर्ग, स्ट्राटफोर्ट आन एमोन व कई मरुस्थल व सघन औद्योगिक केन्द्र आदि आकर्षण के उदाहरण हैं।

अभिगम्यता/पहुंचने योग्य—पर्यटन स्थलों की एक विशेषता यह होनी चाहिए कि पर्यटक वहां तक सरलता से पहुंच सकें। इसके अंतर्गत परिवहन द्वारा पहुँचना शामिल है। इसी अभिगम्यता की वजह से ब्रिटेन एवं पश्चिमी यूरोप पर्यटन स्थलों के लिए समुन्नत एवं महत्त्वपूर्ण क्षेत्र हैं।

सुविधाएं—सभी प्रकार के पर्यटन स्थानों पर सुविधाओं का योगदान महत्वपूर्ण होता है। इसके अंतर्गत निम्नलिखित सुविधाएं आती हैं :

1. आन्तरिक परिवहन व संचार सुविधाएँ
2. खान-पान सुविधाएँ
3. मनोविनोद सुविधाएँ
4. आवास सुविधाएँ

इन सुविधाओं की उपलब्धि में ही पर्यटक इन स्थलों में मनोरंजन तथा मनोविनोद के लिए इधर-उधर भ्रमण तथा विचरण कर सकते हैं।

परिभ्रमण के भेद

पर्यटन उद्योग एक आर्थिक प्रक्रिया है जो लोगों की आवश्यकता को पूर्ण करने में सहायता करता है इसलिए इसे बाजार की मांग और पूर्ति के उतार-चढ़ाव के अनुरूप ही निर्धारित किया जाता है। पर्यटक की अधिकतर माँगों का उद्देश्य व्यक्तियों को पर्यटन क्षेत्रों हेतु लालायित करना है। इस प्रकार पर्यटन के प्रलोभन भिन्न हैं तथा किसी भी यात्रा के निर्धारण में महत्त्वपूर्ण असर डालते हैं। इस प्रकार विभिन्न पर्यटक प्रलोभनों के परिणामस्वरूप अलग-अलग प्रकार के पर्यटनों का उद्गम होता है, जिन्हें निम्नांकित श्रेणियों में बांटा जा सकता है।

पर्यटन विस्तार के आधार पर—पर्यटन को उसके विस्तार की सीमानुसार निम्न प्रकार वर्गीकृत किया जा सकता है :-

(1) अन्तर्राष्ट्रीय पर्यटन या विदेशी पर्यटन—इसके अंतर्गत वह पर्यटक शामिल किया जाता है जिसमें नागरिक अपने देश के बजाय बाहरी देशों का भ्रमण करते हैं। यह विदेशी देश एक भिन्न इकाई होता है जिसका स्वयं का आर्थिक व राजनीतिक ढांचा होता है। भिन्न-भिन्न विदेशी मुद्राओं के कारण इस प्रकार के अन्तर्राष्ट्रीय पर्यटन का भुगतान के सन्तुलन पर प्रतिकूल प्रभाव पड़ता है : क्योंकि किसी भी राष्ट्र को विश्व के अन्य राष्ट्रों से अपने लेन-देन को संतुलित करना होता है। देश की सीमाओं को पार करने के लिए प्रशासनिकों द्वारा कुछ प्रतिबन्ध लगाये जाते हैं। परिणामस्वरूप कुछ विशिष्ट दस्तावेजों पासपोर्ट, वीजा व आवागमन एवं भ्रमण सम्बन्धी शर्तों की आवश्यकता पूर्ण करनी होती है। इन सब कदमों से पर्यटन का बहाव स्वतः प्रतिबन्धित हो जाता है।

स्वदेशीय एवं अन्तर्राष्ट्रीय पर्यटन में किसी विशेष तत्व के बजाय एक स्थिति/स्तर/दशा का ही भेद रह जाता है। वर्तमान समय में, भाषा की कुशलता, मुद्रा एवं सीमा-शुल्क में सुधार तथा विकसित राष्ट्रों के अंतर्गत नागरिकों के स्वतन्त्र भ्रमण के फलस्वरूप स्वदेशीय व अन्तर्राष्ट्रीय पर्यटन के अन्तर के मध्य दिन-प्रतिदिन कमी आती जा रही है।

(2) आन्तरिक पर्यटन या अन्तर्देशीय पर्यटन—ज्यादातर लोग अपने रहने के स्थान के बाहर अपने ही राष्ट्र के अन्य स्थानों की यात्रा करते हैं। देश के अन्दर यात्रा करना उनके लिए सरल होता है, क्योंकि भाषा, मुद्रा एवं दस्तावेज सम्बन्धी बाधाओं का सामना नहीं करना पड़ता है। उनकी स्वयं की भाषा परस्पर विचार-विनिमय के माध्यम की सुविधा मुहैया करती है। जो मुद्रा दिन-प्रतिदिन के जीवन में उपयोग की जाती है, वही विनिमय की जरिया होती है। पर्यटन सम्बन्धी विशेष दस्तावेज सम्बन्धी जरूरतों की पूर्ति करने की कोई आवश्यकता नहीं होती है। इसके अलावा स्वदेशी पर्यटन के अन्तर्गत भुगतान सन्तुलन से सम्बन्धित किसी भी प्रकार की कोई परेशानी नहीं होती है। इस तरह स्वदेशी पर्यटन विदेशी पर्यटन का विकल्प है जिससे राष्ट्र के नागरिकों को विदेशी मुद्रा के बचत से फायदे प्राप्त होते हैं।

विचलन की प्रकृति के आधार पर—सैलानी या तो किसी विशेष स्थान पर ठहरते हैं अथवा यात्रा करते हैं या विभिन्न क्रीड़ा-स्थलों पर थोड़ा-थोड़ा समय व्यतीत करते हैं। पर्यटकों द्वारा व्यतीत किये गये समय की प्रकृति के आधार पर पर्यटन को निम्नांकित तरीके से विभक्त किया जा सकता है—

(1) घुमाकार पर्यटन—कुछ लोग ऐसे होते हैं, जो अपनी छुटटी के पूरे समय को भ्रमण या घूमने-फिरने में ही गुजार देते हैं। वे एक स्थान से दूसरे स्थान की यात्रा

करते रहते हैं तथा पाँच रात्रि से ज्यादा का समय एक स्थल पर नहीं बिताते हैं, ये लगातार भ्रमण करते रहते हैं। इनकी यात्रा का उद्देश्य आवासी व मौसमी पर्यटकों से पृथक् है। ये बसन्त अथवा शरद ऋतु में अधिक तौर पर बिचरित रहते हैं।

(2) मौसमी पर्यटन—इस श्रेणी में वे पर्यटक आते हैं जिनकी अवकाशकालीन समय की अवधि एक महीने से कम होती है। ये पर्यटक एक माह से अधिक किसी विशेष स्थान पर नहीं ठहरते हैं।

इस प्रकार हम देखते हैं कि आवासीय पर्यटन एवं मौसमी पर्यटन में समय एक मूल कारक है।

(3) आवासीय पर्यटन—इसके तहत वे पर्यटक आते हैं जिनके पास एक माह से अधिक का समय है तथा किसी पर्यटन स्थल में अपना वक्त व्यतीत करते हैं। ये पर्यटक मध्यम जलवायु के शान्त एवं स्तब्ध स्थानों की खोज में रहते हैं। मुख्य रूप से आवासीय पर्यटन उच्च मौसम अर्थात् जब माँग उच्च स्तर पर होती है, में ही घटित होता है।

उद्‌गम एवं होनहारिता के अनुसार—इस आधार पर पर्यटन को निम्नलिखित भागों में बांट सकते हैं :

(1) बाह्यगमन पर्यटन—प्रत्येक राष्ट्र से बहुत से लोग दूसरे देशों का भ्रमण करते हैं। बाह्यगमन पर्यटन की श्रेणी में वही पर्यटक आते हैं, जो साधारणतया स्वदेश में निवास करते हैं परन्तु विदेशी भ्रमण करते हैं।

(2) आन्तरिक पर्यटन—किसी भी राष्ट्र के नागरिक अक्सर स्वदेश की सीमाओं के अंतर्गत ही एक-दूसरे स्थानों की यात्रा करते हैं, इसे ही आन्तरिक पर्यटक कहा जा सकता है।

(3) प्रवेश पर्यटन—इस श्रेणी के अंतर्गत वे पर्यटक आते हैं जो कि किसी भी राष्ट्र के अन्दर दूसरे राष्ट्र से प्रवेश करते हैं क्योंकि बाद वाले देश में पर्यटकों का उद्‌भव होता है।

प्रोत्साहन के आधार पर—विभिन्न प्रकार के प्रोत्साहनों पर पर्यटकों के बहाव की प्रकृति का उद्‌गम होता है। अतः पर्यटन को निम्नांकित श्रेणियों में विभाजित किया जा सकता है :

(1) खेल से सम्बन्धित पर्यटन—खेल से सम्बन्धित पर्यटन आज की सदी की एक साधारण प्रवृत्ति है। खेल की तरह यह प्राकृतिक दृष्टि से शक्तिशाली है तथा न इसके कुछ गुण हैं और न ही दोष हैं। परन्तु एक निरन्तर स्थायी योजना व मार्गदर्शन द्वारा इसके प्रभाव सकारात्मक हो सकते हैं। खेलों की अनवरत तैयारी व

प्रशिक्षण की इच्छा करना ही इस पर्यटन का प्रमुख प्रलोभन है। जल क्रीड़ा व शिशिरकालीन क्रीड़ा, आखेट एवं मछली पकड़ने की बहुत ही माँग है। इस क्षेत्र में खेलों की बड़ी-बड़ी मींटिग, जैसे—ओलंपिक खेल, विश्व व महाद्वीप स्तरीय बर्फ पर चलने की क्रियाएँ, फुटबाल, हॉकी, मुक्केबाजी, चैम्पियनशिप इत्यादि होते हैं, जो न केवल खिलाड़ियों को वरन् सभी खेल प्रेमियों व अनुयायियों को आकर्षित करते हैं।

इसी प्रकार ईश्वर भक्त लोगों के भी क्रीड़ा सम्बन्धी पर्यटन चलते रहते हैं। शिशिरकालीन खेल, प्राकृतिक खेल, पर्वतारोहण, ट्रैकिंग, हाइकिंग, जहाज की तैराकी, गोल्फ, आखेट, मछली पकड़ना, नृत्य, समुद्रतटीय क्रीड़ाएँ, लघु गोल्फ, हाइड्रोसाइकलिंग, नौका विहार, जल पर चलने की क्रीड़ा, स्केटिंग एवं दूसरे आन्तरिक मनोविनोद मनोरंजन प्रदान करते हैं एवं दिल बहलाते हैं। वे बहुत से लोग जहाँ क्रीड़ा व खेल के विकास को महत्व दिया जा रहा है, वहाँ पर्यटन की प्रवृत्ति के बहाव में बढ़ोत्तरी का बहुत अच्छा प्रभाव है, वे खेल व क्रीड़ा पर्यटन से लाभान्वित हो सकते हैं, उच्चस्तरीय खिलाड़ियों के लिए पर्यटन एक प्रमुख प्रलोभन शक्ति है, जो बहुत ही उत्तम है। परन्तु इसका आनन्द बरमिंघम व बेलग्रेड की तुलना में टोकियो में ज्यादा अच्छा लिया जा सकता है।

ऐसा माना जाता है कि किसी व्यक्ति को संसार की यात्रा करने के योग्य बनाने में उत्तम दर्शन अत्यधिक सहायक होता है। उच्चस्तरीय खिलाड़ी पर्यटन का एक प्रमुख प्रलोभन है। वर्तमान समय के खेलों हेतु भारत एक बहुत ही उचित स्थान है। भारत के विविध शहरों के क्लबों में पोलो खेला जाता है।

हिमालय में बर्फ पर चलने की क्रीड़ा, गोल्फ, क्रीड़ा के कक्ष, याच क्लब, ट्रैकिंग, सूर्य की किरणों में स्नान, जल संबंधी क्रीड़ा इत्यादि बहुत से खेल हैं, जो खेल जगत में विशिष्ट स्थान रखते हैं। गोवा तथा कोवलम के समुद्रतट तो आराम पसन्द पर्यटकों के लिए एक विशेषता है।

(2) आरामदायक पर्यटन अथवा अलीशान पर्यटन—इस प्रकार के पर्यटन के तहत वो सैलानी आते हैं, जो छुट्टी के समय में नये-नये स्थलों के दर्शन कर नैसर्गिक सौन्दर्यता, अनभिज्ञ, लोकगीत, शान्त व आधुनिक व्यक्त पर्यटन स्थल का आनन्द लेने के लिए एक जिज्ञासा की भावना से आराम से यात्रा करते हैं। इस श्रेणी के कतिपय पर्यटक यात्रा व भ्रमण के दौरान सतत् परिवर्तनशील स्थानों एवं वातावरण का आनन्द लेते हैं। औद्योगीकरण व यंत्रीकरण के विस्तार के कारण अनभिज्ञता व व्यस्त जीवन तथा कठिन जीवन क्रिया की दिनचर्या में आराम के अभाव के कारण आराम की क्रियाओं की आवश्यकता होती है तथा वे जिन्दगी के बढ़ते हुए दवाबों

व तनावों के कारण यात्रा हेतु विवश हो जाते हैं। ये पर्यटक अपने स्थलों में शान्त वातावरण देखना चाहते हैं। इन सैलानियों को निवासित पर्यटन अथवा अधिकतर मौसमी पर्यटन, जिसमें विचलन के कम गुण हैं, की श्रेणी में शामिल किया जा सकता है। इनको पुनः स्वास्थ्य लाभोन्मुख पर्यटन भी कहा जा सकता है। समुद्र तट व झरनों के दर्शन करना इस तरह के पर्यटन के उत्तम उदाहरण हैं।

(3) व्यापारिक या तकनीकी पर्यटन—कुछ समाजशास्त्री, अर्थशास्त्री और सिद्धान्तवादियों का मानना है कि व्यापार से सम्बन्धित यात्रा को पर्यटन यात्रा समझना गलत है। क्योंकि इसके अन्तर्गत ऐच्छिक कारकों का अभाव होता है। उनके अनुसार व्यापारिक यात्राएं व्यावसायिक यात्राएं हैं तथा सम्बन्धित व्यक्ति को यात्रा के स्थान व समय को छाँटने की आजादी नहीं होती है। अतः निजी आजादी जो कि पर्यटन का एक मूल तत्व है, इस प्रकार की यात्रा में विद्यमान नहीं है।

दूसरी ओर पर्यटन की समस्त परिभाषाओं की व्यापारिक यात्राओं को भी पर्यटन के अन्तर्गत शामिल करते हैं। व्यापारिक पर्यटन के अंतर्गत न केवल उद्योगपतियों की व्यावसायिक यात्राएं अथवा तकनीकी उद्देश्य प्राप्त करने के लिए उन व्यक्तियों द्वारा की जाती हैं जो उस व्यवसाय से सम्बन्धित नहीं हैं। परन्तु इससे आम लोगों को बहुत काम होता है। व्यापारिक पर्यटन को पर्यटन के अंतर्गत शामिल करने का एक मुख्य कारण यह भी है कि व्यापारी न केवल उपभोक्ता के तौर व्यवहार करते हैं वरन् प्रायः अपनी स्वतन्त्र समझ में एक वास्तविक सैलानी की तरह उद्देशित राष्ट्र के आकर्षणों का भी फायदा उठाते हैं।

पर्यटन के मौसमी गुणों के आधार पर—किसी विशेष मौसम में अगणित पर्यटकों के विचलन के बहुत से कारण हो सकते हैं। कुछ क्रीड़ाएं अथवा खेलकूद यात्रा से सम्बन्धित या प्रोत्साहित होते हैं एवं कुछ प्राकृतिक दशाओं एवं मौसम पर निर्भर करते हैं, जैसे आखेट, मछली पकड़ना, शिशिरकालीन खेल आदि। इसके अलावा धार्मिक पर्व भी हैं जैसे—मेले, कला एवं संगीत के उत्सव, व्यापारिक मेले, प्रदर्शनियाँ जो पर्यटकों को किसी विशेष मौसम में आकर्षित करते हैं। इस श्रेणी में निम्नलिखित तरह के पर्यटक आते हैं—

(1) ***घटनाकृत पर्यटन***—यह एक स्थानीय पर्यटन है क्योंकि यह धार्मिक कला, खेल, लोकप्रवृत्ति के रीति-रिवाजों के उत्सवों पर आधारित है।

(2) ***शरदकालीन पर्यटन***—अधिकतर यूरोपीय राष्ट्रों में इस तरह की पर्यटन की क्रियाएँ दिसम्बर माह से मार्च माह तक होती रहती हैं। शरदकालीन पर्यटन के दो महत्त्वपूर्ण प्रोत्साहन बर्फ एवं सूर्य की रोशनी की खोज हैं।

(3) ***ग्रीष्मकालीन पर्यटन***–ग्रीष्मकालीन पर्यटन ग्रीष्म ऋतु में ही होता है। इस प्रकार के पर्यटन के मुख्य प्रोत्साहन सूर्य की रोशनी व सूर्य स्नान विशेष रूप से समुद्र में सूर्य स्नान लेना है। यूरोपीय राष्ट्रों में यह विविध प्रकार का जनसाधारण पर्यटन है। अन्य मौसमों की अपेक्षा औसतन प्रवृत्ति दर बहुत ही निम्न है।

भ्रमण के प्रयोजन के आधार पर–यात्रा के उद्देश्य के अनुसार पर्यटन के निम्नांकित रूप मुख्य हैं–

(1) सामूहिक लाभ सम्बन्धी पर्यटन–इसके अंतर्गत वे सभी पर्यटक आते हैं जिनकी यात्रा का कुछ विशेष उद्देश्य होता है। इन्हें निम्नांकित श्रेणियों में विभाजित किया जा सकता है :

धार्मिक पर्यटन

अन्य उद्देशीय पर्यटन

मित्रों व सम्बन्धी पर्यटन

अध्ययन सम्बन्धी पर्यटन

(2) अवकाशकालीन पर्यटन–आधुनिक पर्यटन उद्योग में राष्ट्रीय और अन्तर्राष्ट्रीय क्षेत्रों के तहत छुटियों के समय के पर्यटन की खास मांग हो रही है। कुछ प्रोत्साहक क्रियाओं में जैसे–पर्वतारोहण एवं बर्फ पर चलने की क्रीड़ा में तो जोखिम का तत्व बहुत ही ज्यादा है। प्राचीन पर्यटन स्थलों का मुख्य आकर्षण मुख्यतः अवकाश व द्यूतक्रीड़ा पर आधारित था, जबकि नये पर्यटन स्थल अवकाश व खेलों पर आधारित हैं। अवकाशकालीन शिविरों पर बहुत से व्यक्ति शारीरिक मनोरंजन पर ध्यान देते हैं एवं एक अच्छे आचरण व खेलों में भाग लेने के दोनों पहलुओं में एक अच्छा संबंध है। यह भी देखने में आया है कि ऐच्छिक अवकाश की योजनाओं से शारीरिक क्रियाओं की ओर अवकाश के समय एवं स्वदेश लौटते वक्त, दोनों ही दशाओं में एक अच्छे व्यवहार में सुधार आया है, एक अच्छी धारणा बनी है, लोगों के प्रिय व्यापार तथा उत्तेजित अवकाश की ओर प्रबल इच्छाएं निर्मित हुई हैं। पारिवारिक यात्रा एवं सामूहिक छुट्टियों के साथ-साथ कैम्पिंग एवं दूसरे बहुत से पारिवारिक पर्यटक के मनोविनोद की क्रियाएं बहुत ही महत्त्वपूर्ण हैं।

माँग के सामाजिक एवं आर्थिक पहलू के अनुसार–आर्थिक माँग के अनुसार निम्नलिखित तरह से पर्यटन को विभाजित किया जा सकता है :

(1) योवन पर्यटन–योवन पर्यटन का विभाजन सैलानियों द्वारा दी गई सूचना और उनके शिक्षण के प्रयोजन पर निर्भर होता है।

(2) सामाजिक पर्यटन—यह पर्यटन जनसंख्या के उस वर्ग में दिखाई देता है जिनकी आय माँग को पूरा करने में सक्षम नहीं है। सामाजिक पर्यटन को सरकार एवं कुछ निजी उद्यमियों द्वारा प्रदान की गयी सुविधाओं के द्वारा सम्भव बनाया जा सकता है। यह देखने में आया है कि सामाजिक पर्यटन का फायदा देश की जनता तक पहुंचता है, क्योंकि अधिकतर पर्यटन का महत्व बाह्य पर्यटन से सम्बन्धित श्रेणी को जाता है तथा बाहर जाने वाले पर्यटक को बहुत कम श्रेय जाता है।

(3) स्ववित्तीय पर्यटन—इसके अंतर्गत वे लोग आते हैं, जो अपनी माँग को पूर्ण करने हेतु स्वयं अपने साधनों को प्रयोग में लाते हैं। समस्त प्रवेश करने व बाह्यगमन व अन्तर्देशीय पर्यटक इस श्रेणी में आते हैं। निजी सैलानी उद्योग व राष्ट्रीय अर्थव्यवस्था दोनों के लिए स्ववित्तीय पर्यटन फायदेमंद है।

सामाजिक कारकों के आधार पर—प्रारंभ में ही पर्यटन के लिए कुछ विशेष स्वतन्त्र शब्दों का उपयोग किया गया है, जैसे—जन पर्यटन, लोकप्रिय पर्यटन, सामाजिक पर्यटन, आदि। अतः यह उचित होगा कि इन शब्दों का आशय समझ लिया जाए।

(1) जन पर्यटन—जिस पर्यटन में अनेक लोग शामिल होते हैं, उसे जन पर्यटन कहते हैं। यह पर्यटकों की संख्या की भावना का अहसास कराता है तथा पर्यटन में हिस्सा लेने वाली जनसंख्या के अंश अथवा पर्यटन क्रियाओं के आकार पर आधारित होता है।

(2) लोकप्रिय पर्यटन—यह गुण सम्बन्धी भावना का आभास कराता है एवं जन पर्यटन को जन्म देता है। इसका अभिप्राय उन क्रियाओं से है, जो अपने आकर्षण एवं उपलब्धता की वजह से लोगों को सर्वमान्य होती है। यह लोकप्रिय इसलिए है क्योंकि यह उसकी इच्छा के अनुसार है तथा कम मूल्यों में उपलब्ध है।

(3) सामाजिक पर्यटन—कम साधनों वाले लोगों को पर्यटन में शामिल करने अथवा उन्हें पर्यटन में भाग लेने के लिये प्रेरित करने की प्रक्रिया को सफल बनाने के लिये की गई कोशिश को सामाजिक पर्यटन कहते हैं। सामाजिक पर्यटन का सिद्धान्त मुख्यतः आर्थिक मदद एवं विशेष सुविधाओं व तौर-तरीकों पर आधारित है। ये सरकार एवं अन्य तीसरे पक्ष द्वारा अथवा सहकारिता के आधार पर प्रदान किए जाते हैं।

डॉक्टर डब्ल्यू. हुन्जीकर ने सामाजिक पर्यटन की व्याख्या इस प्रकार की है। "पर्यटन का वह रूप जो निम्न वर्गीय लोगों द्वारा प्रयोग किया जाता है, सामाजिक पर्यटन है तथा यह एक बिल्कुल अलग व आसानी से मान्यताप्राप्त

सेवाओं द्वारा ही सुलभ है।" श्री एम. अन्धेरी पेलेमोण्ट के मतानुसार "सामाजिक पर्यटन, पर्यटन का वह रूप है, जो उन व्यक्तियों द्वारा निर्मित किया गया है, जो सामाजिक दखलंदाजी के बिना इसकी लागत के भार को वहन नहीं कर सकते हैं।" उदाहरणार्थ बिना किसी संगठन की सहायता के जिसके वे स्वयं सदस्य हैं।

सामाजिक पर्यटन के लक्षण

सामाजिक पर्यटन के निम्न लक्षण हैं :

(1) आर्थिक सहायता उपलब्ध पर्यटन—ये वे सामाजिक पर्यटन हैं जिनमें सरकार और श्रमिकों की सदस्यता है, सरकार द्वारा मदद दी जाती है।

(2) बाह्य पर्यटन—सामाजिक पर्यटन में वे यात्री शामिल किए जाते हैं, जो कि अपने सामान्य निवासित स्थान के बाहर किसी विशेष वातावरण की यात्रा करते हैं।

(3) सीमित साधन—सामाजिक पर्यटन कामगारों तथा उन व्यक्तियों तक सीमित है, जो यात्रा व आवास के भुगतान के लिए उचित धनराशि का संचय नहीं कर सकते हैं क्योंकि उनकी आय काफी कम होती है तथा उनके फायदे बहुत ही ज्यादा होते हैं।

सामाजिक पर्यटन के विकास में बाधाएँ—विश्व में ज्यादातर कर्मचारियों को छुट्टियों के साथ वेतन भी प्रदान किया जाता है जिसके कारण वह घर पर ही रहना पंसद करते हैं जो सामाजिक पर्यटक के विकास में मुख्य रूप से बाधक हैं। जिसका मुख्य कारण यह है कि उनके पास पर्यटन सम्बन्धी व्यय एवं सूचनाओं की कमी होती है। परिवहन की कठिनाइयाँ व उचित प्रकार की आवास सुविधाओं की कमी होती है। इस तरह की समस्त समस्याएँ संगठित एवं सामाजिक पर्यटन द्वारा पूरी की जा सकती हैं।

आर्थिक समस्याओं का समाधान आर्थिक सहायता व बचत योजनाओं द्वारा संरचनाओं की पूर्ति, अनुभव व सम्पर्क के द्वारा परिवहन समस्याओं को सामान ले जाने वालों के साथ पैकेज योजनाओं द्वारा तथा आवास की समस्याओं का हल पर्यटन स्थल के अनुबंध द्वारा किया जा सकता है। इस प्रकार संगठित सामाजिक पर्यटन उन सभी लोगों तक पहुँच सकता है, जो यात्रा करने योग्य नहीं हैं। वित्त संबंधी समस्याओं को बचत योजनाओं द्वारा पूरा किया जा सकता है। प्रत्येक संगठन बचत योजनाओं में सक्रिय भाग ले सकते हैं। इन संगठनों के सदस्य मिलकर योगदान कर

सकते हैं तथा आवास के समय उन्हें उनकी जमा बचत से लाभकारी धनराशि निकालने की अनुमति दी जा सकती है।

सेवाकाल के समय के अनुपात के अनुसार ही धनराशि को निर्धारित किया जाता है। बेल्जियम में तो कानूनी तौर पर सामान्य वेतन से दोगुना अवकाशकालीन वेतन के रूप में भुगतान किया जाता है तथा यदा-कदा तो मालिक यात्रा हेतु नगदी के रूप में अनुदान भी देते हैं। रूस में, मजदूर संघ आराम-गृहों के सम्पूर्ण रहन-सहन के व्यय का ज्यादातर अंश भुगतान करते हैं। पोलैण्ड में सरकार यात्रा एवं आवास के कुछ अंश का स्वयं भुगतान करती है। न्यूजीलैंड में होटलों के किराये में कमी करके सरकारी कर्मचारियों को लाभान्वित किया जाता है। सरकारों द्वारा उपलब्ध कराई गई अन्य वित्तीय सुविधाएँ हैं–अवकाशकालीन आवास सुविधाओं का प्रबन्ध सरकार द्वारा आर्थिक मदद व करों में रियायतें आदि। रूस में तो सरकार द्वारा अवकाशकालीन केन्द्रों में प्रत्यक्ष विनियोजन किया जाता है।

बचत से सम्बन्धित योजनाओं द्वारा फ्रांस, आस्ट्रेलिया, बेल्जियम आदि देशों में अत्यधिक लाभ हुये हैं। इसी कामगार अवकाश के फण्ड संगठनों या मालिकों द्वारा परिवहन व होटल उद्योग, सरकार व स्थानीय संस्थाओं के सामूहिक सहयोग द्वारा नियोजित किए जा सकते है। फ्रांस में लाभकारी समितियाँ यात्रा के लिए वित्तीय मदद प्रदान करती हैं तथा साथ ही श्रमिकों व बच्चों के लिए अवकाशकालीन आवासीय गृहों को संचालित करती हैं। साथ ही इनको आंशिक रूप में मालिकों व कामगारों के योगदान द्वारा आर्थिक सहायता प्रदान की जाती है। नीदरलैण्ड में तो कुछ मालिक अवकाशीय वेतन के अलावा पूरे वर्ष के वेतन का 2 प्रतिशत बोनस के रूप में कामगार को योगदान के रूप में देते हैं। यूनाइटेड किंगडम में कामगारों को अवकाशकालीन बोनस दिये जाते हैं।

अमेरिका में कुछ सामान्य प्रेक्षक पहले यात्रा फिर भुगतान जैसी उदार योजनाओं को संचालित करते हैं। फेडरल रिपब्लिक जर्मनी में रेलवे यात्रा करने के लिए 12 या अधिक संख्या के समूहों को 33 प्रतिशत छूट, 25 अथवा इससे ऊपर के समूहों को 50 प्रतिशत छूट एवं विद्यार्थी, स्कूल के बच्चों या युवा संगठनों को इससे भी अधिक छूट दी जाती है। बेल्जियम में कामगारों एवं उनके परिवार के सदस्यों को अवकाशकालीन समय में सामाजिक सुरक्षा कार्ड मालिकों का सर्टीफिकेट प्रस्तुत करने पर रेलवे के सस्ते दाम के टिकट निर्गमित किए जाते हैं।

फ्रांस के अन्तर्गत सामान्य शिक्षा, खेल, पर्यटन, संस्कृति से सम्बन्धित फायदा प्राप्त करने वाली जितनी भी संस्थायें हैं, वो सभी बिक्री व मनोरंजन कर

से पूर्णरूप से मुक्त हैं। सरकार कैम्पों व युवा संगठनों को यंत्रों के रूप में सहायता प्रदान करती है। पूर्वी जर्मनी में समस्त सेवाओं की वित्तीय जरूरतें सरकार द्वारा पूर्ण की जाती हैं। नार्वे में अवकाशकालीन केन्द्रों के विकास के लिए ब्याजमुक्त ऋण अथवा कम ब्याज पर ऋण दिए जाते हैं। डेनमार्क सरकार ने इन केन्द्रों के विकास हेतु 50 प्रतिशत कोष/धनराशि का सृजन किया है। परिवहन की असुविधाओं के कारण परिवहन कम्पनियाँ निम्न भाड़े को विशेष दरों की योजनाओं में मदद प्रदान करती हैं। चार्टेड फ्लाइड निजी यात्रा खर्च को कम करने का एक उत्तम साधन है।

अमेरिका के लगभग सभी बड़े शहरों में सूचना प्रदान करने के केन्द्रों को स्थापित किया गया है। ये कार्यालय कामगारों को अवकाश व्यतीत करने सम्बन्धी सलाह देने के लिए अपनी प्रकाशन पत्रिकाएँ निर्गमित करते हैं। कनाडा में कामगार संघ के कार्यालय व अन्य संगठनों द्वारा बुलेटिनें प्रकाशित की जाती हैं। हमारे देश की विकसित होती हुई अर्थव्यवस्था में सामाजिक पर्यटन के विचार की महत्ता से हम काफी प्रभावित्त हुए हैं। सामाजिक पर्यटक का कार्य सिर्फ यह नहीं है कि निम्न स्तर के लोगों को जो कि सामान्य तौर पर अवकाश का आनन्द लेने में अक्षम हैं, उन्हें अवकाश दिया जाए, उन्हें नई विकसित योजनाओं में श्रमिकों को वेतन सहित तथा यवन यात्राओं में मदद देकर देश के सभी क्षेत्रों में यात्रा सम्बन्धी समस्त योजनाओं में सम्मिलित किया जाए। हमारा उद्देश्य एक कल्याणकारी देश का निर्माण करना है।

लेनिन में एक ऐसा स्वास्थ्य पर्यटन स्थान है, जो अत्यन्त ही मशहूर है। परन्तु नवीन औषधियों में प्रगति होने के कारण ऐसा लग रहा है कि यह स्थान स्वयं लुप्त हो जाएगा, परन्तु कुछ संगठनों की सहायता से यह स्थल नये पर्यटकों को आकर्षित करने लगा है। छोटे-छोटे गोल्फ कक्ष, छोटे तरण-ताल, टेनिस कोर्ट्स व अन्य योजनाएँ बनाकर एवं स्वास्थ्य केन्द्र व होटलों के सैलानियों की माँग के अनुसार परिवर्तित कर दिया गया है। काफिलों के पड़ाव व होटल युवा पर्यटकों को बहुत ही पसन्द हैं। कैम्पिंग तो अवकाश समय बिताने का बहुत ही कम व्ययशील साधन है। फ्रांस व अन्य कुछ राष्ट्रों में कैम्पों के लिए आर्थिक मदद प्रदान की जाती है।

हंगरी में वेतन अवकाश प्राप्त लोगों को 50 प्रतिशत की छूट दी जाती है। चेकोस्लोवाकिया में हरेक श्रमिक के अवकाश के फार्म के साथ एक रेल का टिकट निःशुल्क दिया जाता है। सामाजिक पर्यटन के विकास में आवास की सुविधा एक मुख्य समस्या है। होटल तो निम्न आय वर्ग की पहुँच के बाहर हैं। अतः इस पर्यटन

में अनुपूरक आवास का एक महत्त्वपूर्ण स्थान है। कई श्रमिक संगठनों व मालिकों ने श्रमिकों की सुविधा के लिए अवकाश केन्द्रों का निर्माण किया है। अर्जेंटीना में अति सौन्दर्य-स्थलों, समुद्र तटों पर विशेष होटल, बंगले व चैलेट की सुविधाएँ हैं। अवकाश केन्द्र खोले गये हैं। इनमें तरण-ताल, मधुशाला, खेल के मैदान, पार्क एवं अन्य मनोरंजन सम्बन्धी सुविधाएं मुहैया करायी जाती हैं। स्विट्जरलैंड में तो सामाजिक पर्यटन के लिए अनेक अवकाशकालीन केन्द्र हैं।

ग्रीक में बहुत बड़ी औद्योगिक इकाइयाँ अपने श्रमिकों के लाभ के लिए विविध कैम्पों का संचालित करती हैं। अधिकांश देशों में ये कैम्प कैम्पिंग क्लब व युवाओं के अभियान द्वारा संचालित होते हैं। सूचनाओं की कमी के कारण यह कहा जा सकता है कि सार्वजनिक प्राधिकरणों, श्रमिक संघों, राष्ट्रीय पर्यटन संगठनों एवं अन्य सम्बन्धित संस्थाओं को विशिष्ट प्रकार के पर्यटन स्थलों की सूचना प्रदान करने की तरफ विशेष ध्यान देना चाहिए।

श्रमिक वर्ग व कृषक इस देश की 98% की जनसंख्या में हैं इनकी सहायता पर ही एक जनकल्याणकारी देश का अस्तित्व निर्भर करता है। एक जनकल्याणकारी सरकार को इन फायदों की प्राप्ति हेतु कामगारों को कार्य करना पड़ता है, अन्यथा इसका कोई अर्थ नहीं रहता है। जनकल्याणकारी देश का मुख्य लक्ष्य यह है कि कामगारों को वेतन सहित अवकाश व राष्ट्र के युवा वर्ग को यात्रा सम्बन्धी धनराशि की मदद प्रदान की जाए, यह एक महत्त्वपूर्ण घटक है। राष्ट्र के युवा वर्गों हेतु यह एक शिक्षा है ताकि वे स्वयं राष्ट्र के दर्शन कर सकें, सांस्कृतिक प्रिय लोगों से मिल सकें एवं औद्योगिक क्षेत्रों को देख सकें। इसी के द्वारा देश में रहने वाले एक-दूसरे व विविध प्रकार के लोगों की आवश्यकताओं को समझा जा सकता है तथा पूरा किया जा सकता है। इसके अलावा इस प्रकार का सामाजिक पर्यटन आजकल की औद्योगिक क्रांति हेतु एक गहराई व अगाध रुचि एवं उत्तेजना की भावना जाग्रत कर सकता है।

नियोजित अर्थव्यवस्था को बहुसंख्य व्यक्तियों के घरों तक तब तक नहीं पहुंचाया जा सकता, जब तक उन्हें स्वयं की आधुनिक योजनाओं को चालू करने का मौका प्रदान नहीं किया जाएगा। भाखड़ा नंगल, सिंदरी स्टील प्लांट, चित्तरंजन, बंगलौर व अन्य केन्द्रों हेतु रियायती यात्राओं के प्रबन्ध करने चाहिए, उन व्यक्तियों के लिए जो इनका दर्शन करना चाहते हैं और औद्योगिक क्रांति की भावना से इन जगहों की स्वयं जानकारी प्राप्त करना चाहते हैं।

होटल स्तर एवं दर आकार कमेटी 1958 ने निम्नांकित सुझाव दिए थे :

प्रत्येक होटल उपक्रमों में कामगारों की वेतन सहित अवकाश की योजना शुरू होनी चाहिए एवं अन्य इकाइयों में भी इसका विस्तार किया जाए। यह एक जरूरी प्रावधान होना चाहिए।

रेलवे द्वारा विद्यार्थी समुदाय एवं युवा वर्ग के लिए आर्थिक मदद सहित यात्राओं का वृहद पैमाने पर प्रबन्ध करना चाहिए।

धार्मिक स्थानों की यात्रा करने वाले यात्रियों के लिए बड़े पैमाने पर उनके घरों से सम्पूर्ण भारत के धार्मिक स्थलों तक धन सम्बन्धी मदद की यात्राओं का प्रबन्ध होना चाहिए। कुछ देशों के सामाजिक पर्यटन सम्बन्धी यात्राओं के लिए एक बहुत बड़ी धनराशि अलग से रखी जाती है।

संस्कृति से परिचय कराता पर्यटन

पर्यटन के द्वारा लोगों को देश-विदेश की संस्कृति का पता चलता है जिसके द्वारा हम उनके रहन-सहन, पहनावे आदि के बारे में जानकारी प्राप्त करते हैं। इस प्रकार यह एक राष्ट्रीय तथा अंतर्राष्ट्रीय संपर्क का प्रमुख साधन है। सांस्कृतिक सम्बन्धों तथा अन्तर्राष्ट्रीय सहयोग के विकास में पर्यटन एक महत्त्वपूर्ण स्थान रखता है। किसी भी राष्ट्र के भीतर सांस्कृतिक पक्षों के विकास को पर्यटन सम्बन्धी स्रोतों में वृद्धि के रूप में समझा जा सकता है। एक देश द्वारा स्वयं को पर्यटकों के सामने प्रस्तुत करने के तरीकों को ही सांस्कृतिक पहलू समझा जा सकता है। खान-पान, अतिथि-सत्कार, आचार-व्यवहार, कला की वस्तुएं, देश के अन्य पहलू, यात्रियों अथवा पर्यटकों के आकर्षण हैं।

पर्यटन की सफलता सिर्फ परिवहन व होटलों की अच्छी व्यवस्था पर ही निर्भर नहीं करती बल्कि इसके साथ ही राष्ट्रीयता की भी झलक होनी चाहिए तथा सेवाओं को उचित मूल्य प्रदान करना चाहिए। सांस्कृतिक पर्यटन सेवा-भाव से प्रेरित होना चाहिए। जैसे कि एक विख्यात रहन-सहन व शिक्षण केन्द्र में अध्ययन व ज्ञान प्राप्त करने की इच्छा, विदेशी लोगों के रहन-सहन के संस्थान एवं आकार का अध्ययन, ऐतिहासिक महत्व की इमारतें, भूतपूर्व सभ्यता के अवशेषों, आधुनिक प्राप्तियां, ललितकला कक्ष, प्रसिद्ध धार्मिक स्थानों के दर्शन एवं कला, पर्वों व समारोहों में शामिल होना, गायन, नृत्य, थियेटर, लोकगीत, मेलों आदि में सक्रिय भाग लेना आदि। भारत एक अतिविशाल एवं लोकप्रिय देश है, जिसका स्वयं का अपना पृथक् एक इतिहास है व अपनी महान् परम्परा है।

संग्रहालय व स्मारक तो हर प्रकार के पर्यटन यात्रावृत में एक आवश्यक स्थान रखते हैं। हमारी योजनाओं में सांस्कृतिक पर्यटन एक प्रमुख लक्षण है। यह भी प्रस्ताव है कि मास्टर प्लान के अंतर्गत कुछ निर्धारित पुरातात्विक केन्द्रों के चारों ओर के क्षेत्रों के लिए विशेष योजनाएँ बनाई जाएँ ताकि इन केन्द्रों के पर्यावरण व प्राकृतिक व्यवस्था को सुरक्षित रखा जा सके। यह भी सुझाव है कि यूनेस्को व यू. एन.डी. पी. के अंतर्गत सांस्कृतिक पर्यटन के विशेषज्ञों की सेवाएँ प्राप्त करके इन योजनाओं को बनाने में सहायता प्राप्त की जाए। इस प्रस्ताव के अन्तर्गत कुछ स्मारकों को सम्मिलित किया जा सकता है, जो निम्नलिखित हैं–

1. बीजापुर, ऐहोल, बदामी, पट्टादरखाल एवं हम्पी।
2. सारनाथ, गया, नालन्दा, राजगीर, कुशीनगर, श्रावस्ती व सांची के बौद्ध केन्द्र।
3. खजुराहो, भुवनेश्वर, पुरी व कोणार्क।
4. एलीफेंटा, अजन्ता एवं एलोरा के गुफा मन्दिर।

यह पिछले चार सालों से एक सभ्य जीवन के रूप में गौरवान्वित है। जिस कारण इसे संसार की एक महान् सभ्यता के रूप में माना जाता है। यह एक जीती-जागती सभ्यता है, जो कि स्पष्ट तौर पर प्राचीन सभ्यता से विकसित हुई है। इन्हीं कारणों से किसी भी विदेशी की भारत यात्रा पर एक सांस्कृतिक प्रभाव अवश्य पड़ता है। अतएव सर्वप्रथम एवं विस्तृत अर्थ में भारत में हर दृष्टि से पर्यटन का सांस्कृतिक सम्पर्क होता है एवं इस पर्यटन को सांस्कृतिक पर्यटन के नाम से भी जाना जाता है। रंगशाला, पुस्तकालय, संग्रहालय व अन्य राष्ट्रीय स्तरीय संस्थान, यात्रियों को आकर्षित करने के लिए बहुत ही बड़ी सम्पत्ति हैं।

भारत की कलाकृतियाँ

कलाकृतियाँ प्रत्येक देश का प्रमुख विषय होती हैं जिसमें भारत भी अग्रणीय है। भारत में मुख्य रूप से जैसे–चित्रकारी, मूर्तिकला, भवन निर्माण कला, चित्रकला, पानी के रंग, रेखाचित्र के वृत्त, प्राकृतिक शिल्पकला आदि यात्रा करने के प्रमुख आकर्षण हैं। भारत देश में आने वाले यात्रियों के लिए कला के आकर्षणों की संतुष्टि बहुत ही प्रमुख है, इसीलिए इसे सांस्कृतिक खरीददारी कहा जा सकता है।

विभिन्न पर्यटन स्थानों से खरीदी गई सामग्रियों को पर्यटक स्मारिकाओं के रूप में अपने देश ले जाते हैं। अतः इन वस्तुओं की बिक्री एक प्रमुख स्थान रखती है। कला एवं कृति सांस्कृतिक पर्यटन के महत्वपूर्ण अवयव हैं क्योंकि कुछ यात्री पारंपरिक कला और कृतियों के केन्द्रों को देखना चाहते हैं। उदाहरणार्थ–नवम्बर 1976 में 89 होटल

उद्योगपति केटरिंग टाइम्स ऑफ लंदन द्वारा प्रायोजित करने पर भारत की यात्रा पर आये, वे इतने प्रभावित हुए कि उसमें से कुछ भारतीय आभूषणों का कुछ और अध्ययन करने हेतु भारत में रुक गये। कला, धातु के काम, नक्काशी, आभूषणों एवं कढ़ाई के क्षेत्र में भारतीय कलाकारों में एक उच्चकोटि के विकसित रंग व नमूनों की प्रवृत्ति है।

मथुरा में मिट्टी के बर्तन तो अपनी बनावट एवं सुन्दरता के लिए काफी प्रसिद्ध हैं। ग्वालियर में निर्मित मिट्टी के बर्तनों का प्रचलन यूरोप में बहुत अधिक है। भारत अपनी कला एवं कृतियों के लिए विश्वविख्यात है।

कुम्हार, बढ़ई, लोहार, जुलाहे (बुनकर) आदि में अधिकांश कार्य-कुशलता अपनी पारिवारिक परंपरा पर आधारित व्यवसाय पर निर्भर होने के कारण है। पंजाब एवं सिंध की चमकीली मिट्टी के बर्तन प्रसिद्ध हैं। भारत के कई स्थानों पर उच्चकोटि की काली मिट्टी के बर्तन निर्मित किये जाते हैं।

मुर्शिदाबाद एवं बनारस की जरीदारी, जयुपर व दिल्ली के आभूषण, हैदराबाद व मुरादाबाद की आन्तरिक तरीके से लगाई हुए धातु की कृतियाँ, सेनेगर जयपुर में रंगने के गृह, औरंगाबाद के 'हेमरू' जुलाहे, वाराणसी एवं काँचीपुरम के रेशमी जुलाहे, स्वामीलाई के तांबे धातु की खोज करने वाले, महाबलीपुरम् के पत्थर की इमारत बनाने वाले शिल्पी, मैसूर, दिल्ली एवं राजस्थान के हाथीदाँत की नक्काशी करने वाले, आगरा के संगमरमर के कारीगर, मैसूर व बंगलौर के चन्दन की लकड़ी को खोदकर मूर्ति बनाने वाले/नक्काशी करने वालों के हस्त से कढ़े हुए दुशाले, फिरोजाबाद की कांच की चूड़ियाँ हमारे देश में काफी विख्यात हैं। जिन स्थानों पर यह वस्तुएं बनाई जाती हैं, उन कलाकृतियों के दर्शन उनके मनोरंजन के लिए आकर्षण के केन्द्र हैं। यदि यात्री स्वयं इन केन्द्रों में कलाकृतियों को निर्मित होते हुए देखें तथा वहीं से अथवा पास के विक्रय केन्द्रों से इनकी खरीददारी करें, तो यह विक्रय की महत्त्वपूर्ण तकनीक बन जाती है। होटलों अथवा इनके निकटवर्ती स्थानों में भी स्थानीय कला-कृतियाँ व हस्तकारिक वस्तुएँ प्रदर्शित की जा सकती हैं ताकि पर्यटक स्थानीय लोगों द्वारा निर्मित कलाकृतियों से परिचित हो सकें, ललित कला एवं कलाकृतियां कला वस्तुओं संगीत पारम्परिक सामरिक सेना सम्बन्धी सामग्रियां सांस्कृतिक आकर्षणों की कला प्रदर्शनियाँ आयोजित कर सांस्कृतिक अनुभवों को प्रदर्शित किया जा सकता है।

उद्योग-धन्धे

प्रत्येक क्षेत्र का व्यावसायिक विकास पर्यटन हेतु अत्यन्त ही महत्त्वपूर्ण आकषर्ण है। असंख्य अन्तर्राष्ट्रीय पर्यटक क्षेत्रीय अर्थव्यवस्था में रुचि दिखाते हैं। ये

यात्राकृत देश के उद्योग, वाणिज्य एवं उत्पादित वस्तुओं को देखना चाहते हैं। विदेशी पर्यटकों द्वारा उद्योगों का दर्शन करने से स्थानीय संस्कृति में रुचि उत्पन्न होती है एवं उन्हें स्थानीय वस्तुएँ एक स्मारिका के रूप में खरीददारी के लिए लालायित करती हैं। वे स्वदेश लौटने के उपरांत भी उन वस्तुओं का क्रय कर सकते हैं।

संगीत विद्या

संगीत भी किसी राष्ट्र की पारम्परिक चीज है। राष्ट्र के संगीत की अभिव्यक्ति एवं स्रोत आकर्षण एवं आनन्द प्राप्ति के पहलू हैं। एक पर्यटक किसी भी राष्ट्र की यात्रा के दौरान उस राष्ट्र के बाबत अधिक से अधिक जानकारी भी करना चाहता है। भारत की संस्कृति में संगीत को महत्वपूर्ण स्थान प्राप्त है। पर्यटकों के लिए भारतीय संस्कृति विशेषतः शाम के समय एक सांस्कृतिक एवं सामान्य मनोविनोद का साधन है। होटल में मेहमानों के लिए स्थानीय संगीत को मनोविनोद का साधन बनाया जा सकता है। सायंकालीन मनोविनोद के कार्यक्रम, संगीतशाला, प्रमाण संरक्षित किले एवं ध्वनि निर्माण इत्यादि सभी ये किसी भी राष्ट्र की इस विशेष कला में योगदान देते हैं। परेड व स्वागत समारोह, संगीत समूहों के साथ स्थानीय संगीतों के स्रोतों से मनोविनोद में सहायता मिलती हैं। इसी तरह टेपरिकार्डिंग के फोटोग्राफ

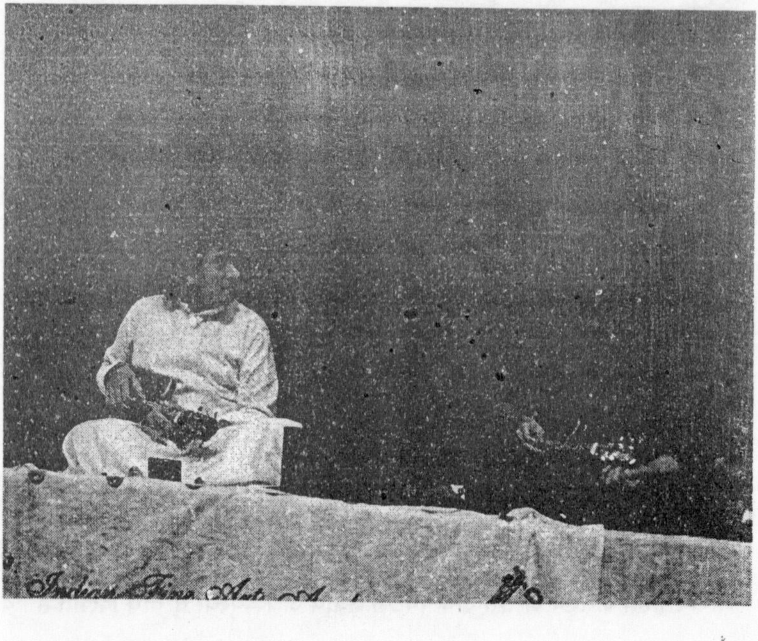

जो पर्यटक प्रयोग के लिए स्वदेश में खरीद सकते है, वे भी सांस्कृतिक स्मृतियों हेतु व उस राष्ट्र की संगीत की सुन्दरता व खुशी को पुनः याद करने के साधन हैं।

नृत्य-विद्या

प्रत्येक देश में नृत्य उसके सांस्कृतिक आकर्षण और उत्तेजना के महत्वपूर्ण पहलू हैं। नृत्य अपनी प्रकृति एवं सौन्दर्यता व आकार को प्रस्तुत करता है एवं पर्यटकों को प्रेरित व प्रोत्साहित करता है। रंग, वेशभूषा, संगीत ढांचा व कार्यान्वियन की कुशलता व प्रदर्शन का ढंग आकर्षण को प्रभावित करते हैं। भारतीय प्रवृत्ति एवं नृत्य मौलिक हैं। नृत्य एवं मौलिक संगीत भारतीय संस्कृति के अभिन्न अंग हैं एवं पर्यटकों के मनोरंजन के साधनों में सम्मिलित किए जाने चाहिए। नृत्य एवं मनोविनोद होटल कार्यक्रमों का एक हिस्सा होना चाहिए।

रूसी रंगमंच नृत्य, पूर्वी यूरोपियन राष्ट्रों के लोक नृत्य, मैक्सिको के लोक नृत्य, थाई, नृत्य, जापान के कबूकी नृत्य, अफ्रीका के बहुत से राष्ट्रों के नृत्य, उन देशों की संस्कृति के द्योतक हैं।

भारतीय नृत्य व संगीत हालांकि ये स्वयं प्राथमिक आकर्षण नहीं हैं किन्तु ये एक द्वितीयक मनोविनोद के रूप में उपयोग किए जाने चाहिए। संगीतकार, गायक व नृत्यकारों को पर्यटन उद्योग द्वारा प्रोत्साहन देने से एक उपयोगी उद्देश्य प्राप्त किया जा सकता है।

बिजनेस

खरीददारी पर्यटन उद्योग का एक महत्त्वपूर्ण पहलू है। यदि इसे उचित तरीके से संगठित किया जाए, तो व्यापार व खरीददारी हमारे आगन्तुकों का एक महत्त्वपूर्ण अनुभव हो सकता है।

खेती और किसान

देश के प्रत्येक क्षेत्र की खेती से उस जगह की संस्कृति की एक रोमांचक व आकर्षक पक्ष का परिचय प्राप्त होता है। काफी पर्यटक खेती, पालतू जानवर, मुर्गीपालन, दुग्ध उद्योग, फसलें, ताजे फल व सब्जियाँ, शहद, पेय इत्यादि में रुचि रखते हैं एवं खेती में उत्पादित अन्य वस्तुओं का शीघ्र आनन्द लेते हैं। इस तरह कृषि एवं सम्बन्धित सेवाओं का विकास स्थानीय यात्राओं का एक रोमांचकारी पहलू है।

भोजन शैली

भारतीय पारम्परिक शराब, मद्यपान जीयर, स्प्रिट आदि का सही तरीके से उपयोग करने की बहुत संभावना है। हमारे पास विभिन्न क्षेत्रीय खान-पान एवं स्वास्थ्यवर्द्धक व प्राकृतिक पेय पदार्थ हैं, जो कि पर्यटकों के लिए महत्त्वपूर्ण सिद्ध हो सकते हैं। पर्यटक स्थानीय खान-पान की वस्तुओं जो कि स्थानीय/मौलिक प्रकार

की हैं को पसन्द करते हैं। खान-पान उद्योग द्वारा पर्यटकों पर एक अमिट छाप छोड़ी जा सकती है। संरक्षित खान एवं पेय पदार्थ शराब व बीयर, उपभोग हेतु ताजा खाना इत्यादि खान व पेय पदार्थों में सांस्कृतिक आकर्षण के परिचायक हैं।

वन्य जीव-जन्तु

भारत के वन, सरोवर व नदियाँ बहुत से जीव-जन्तुओं के निकास स्थल हैं जैसे—शेर, बाघ, एक सींग वाला रेण्डियर, बर्फीला बाघ, जंगली गधा व अनेक प्रकार के विदेशी पक्षियों को यहाँ बहुत से वन्य प्राणी शरण स्थलों में सुरक्षित किया जा रहा है, जिनमें प्रकृति प्रेमी पर्यटक पशु एवं पक्षियों के प्राकृतिक कार्यकलापों व आदतों को देखने के अवसर हासिल कर सकते हैं।

विश्व प्रसिद्ध स्मारक

इसके अंतर्गत भारत की कला एवं भवन-निर्माण कला आती है। यूरोप एवं अमेरिका के ज्यादा पर्यटक इसके दर्शनाभिलाषी हैं। विरासती-स्मारकों को प्रमुख रूप से चार भागों में विभाजित किया जा सकता है, जो इस प्रकार हैं—

1. बौद्धकालीन स्मारक।
2. हिन्दूकालीन स्मारक।
3. यूरोप व ब्रिटिश एसोसिएशन से सम्बन्धित स्मारक।
4. भारतीय मुस्लिम स्मारक।

इन समस्त स्मारकों का उपयोग सांस्कृतिक पर्यटन हेतु किया जा सकता है। इस प्रकार के पर्यटन की बहुत सम्भावनाएँ हैं तथा इसके वर्तमान प्रयोग की अपेक्षा भविष्य में अधिक प्रयोग करने के बहुत अच्छे आसार हैं।

प्राकृतिक सम्पदाएं

लगभग 50 प्रतिशत पर्यटक यह अनुभव करते हैं कि उनकी भारत यात्रा की इच्छा का महत्त्वपूर्ण कारण यहाँ की सुन्दरता एवं प्राकृतिक सौन्दर्यता है।

सामूहिक पर्यटन

वर्तमान समय में पर्यटन सम्मेलन को महत्वपूर्ण स्थान प्राप्त है। अनेक अन्तर्राष्ट्रीय सम्मेलन, महासम्मेलन, अन्तर्राष्ट्रीय संगठनों की वैधानिक संस्थाओं

की बैठकें व अन्य असंख्य सम्मेलन, सभाएं, राष्ट्रीय सम्मेलन इत्यादि हैं जो प्रत्येक वर्ष भिन्न-भिन्न देशों में आयोजित होते हैं। स्थानीय, राज्यस्तरीय, राष्ट्रीय व अन्तर्राष्ट्रीय सम्मेलन निरंतर आयोजित किये जाते हैं। देश संस्था संघों, संगठनों, मैत्रिक संघों, सोसायटी, शिक्षण संस्थाओं, व्यावसायिक समूहों, शिक्षण विज्ञापकों व अन्य संस्थाओं की वार्षिक सभाएँ होती रहती हैं। इन सम्मेलनों एवं सभाओं में सहस्त्रों की संख्या में लोग भाग लेने हेतु शामिल होते हैं। ये हिस्सा लेने वाले लोग बहुत दिनों तक उन राष्ट्रों के शहरों में निवास करते हैं जहाँ सम्मेलन होते हैं।

विश्व के कई देश इस प्रकार के सम्मेलनों के गम्भीर आर्थिक पहुलुओं से अच्छी तरह परिचित हैं। उन्हें उचित इमारतें निर्मित करनी होती है, यन्त्रों से सुसज्जित कक्षों का निर्माण उनके संचालन के लिए करना होता है। भारत में पर्यटक विभिन्न उद्देश्यों से आते हैं, उनकी भारत की यात्रा के मुख्य उद्देश्य देश का गहन अध्ययन, पारम्परिक खाद्य-पदार्थों का स्वाद लेना, पेय पदार्थों का उपभोग व जितना हो सके स्थानीय लोगों के निकट रहना है। बहुत से पर्यटक परिचित वातावरण से अनजान स्थानों का सर्वेक्षण करना चाहते हैं। भारत में समस्त प्रकार के पर्यटकों के लिए आराम व सुविधाएं उपलब्ध हैं। यह एक रंगीला एवं सौन्दर्य सम्पन्न राष्ट्र है, इसके स्मारक, इमारतें, रीति-रिवाज, पर्व, धर्म, दर्शनशास्त्र, कला आदि इसकी प्राचीन व सम्पन्नता के परिचायक हैं। यहाँ के वन, पर्वत व घाटियाँ, समस्थल व मरुस्थलीय दीवारें आदि प्रकृति प्रेमी व जीव वैज्ञानिकों के लिए बहुत ही रोमांचक विषय हैं।

तीर्थयात्राएँ और पर्यटन

भारत को तीर्थ स्थानों की नगरी कहा जाता है तथा पर्यटन में उसका उच्च स्थान है। कुछ तीर्थ स्थल एवं कलाकृतियां प्राचीन समय से ही आकर्षण का केन्द्र रही हैं। हिमालय पर्वत, पवित्र गंगा के किनारे स्थित हरिद्वार (स्वर्गद्वार), मथुरा और वाराणसी के मंदिर, भारत के ऐतिहासिक स्थान, खुजराहो के मूर्ति निर्माण, अजन्ता और एलोरा की गुफाएँ आदि प्रमुख हैं। काफी संख्या में हिन्दू, मुस्लिम, ईसाई, बौद्ध इत्यादि धार्मिक ग्रंथों में वर्णित विविध स्थलों की धार्मिक यात्रा पर जाते हैं, जिनमें अधिकांश हिन्दू धर्म के लोग होते हैं। हिन्दू धर्म के द्वारा तीन और धर्मों की उत्पत्ति हुई है जिनमें पहला सिक्ख, दूसरा जैन और तीसरा बौद्ध धर्म है। इन धर्मों के अलावा भी मुस्लिम, ईसाई, जोरोस्थर्स, सलिज भी अपना अस्तित्व रखते हैं। किसी भी विदेशी को भारत यात्रा से एक सांस्कृतिक प्रभाव का असर अनुभव होता है।

विस्तृत अर्थों में भारत में पर्यटन से सांस्कृतिक सम्पर्क के किसी न किसी पहलू पर विद्यमान होता है। अतः यह एक सम्भावित सांस्कृतिक पर्यटन है, जो विदेशी भारतीय संस्कृति की जानकारी करना चाहते हैं, उन्हें भारतीय जीवन को देखना चाहिए। भारत की सांस्कृतिक यात्रा में, मन्दिर स्थलों का दर्शन, जहाँ राष्ट्रीय नृत्य का प्रदर्शन होता है, वस्तुओं और आभूषण निर्माण के स्थान, जैसे—बनारस की काल्पनिक जरीदारी जहाँ पर विदेशी भारत की सांस्कृतिक सम्पदा का ज्ञान प्राप्त कर सकें तथा साथ ही आधुनिकता के विकास की जानकारी कर सकें, को शामिल किया जा सकता है। 1969 में पर्यटन मन्त्रालय द्वारा लिए गए कई व्यक्तिगत साक्षात्कारों में यह प्रतीत हुआ कि 54 प्रतिशत पर्यटक भारत में यात्रा करना सिर्फ इसलिए पसन्द करते हैं, क्योंकि यहाँ इतनी सुन्दर मानवकृत इमारतें, मन्दिर व गिरजाघर हैं।

भारत की पंचवर्षीय योजनओं के तहत सांस्कृतिक पर्यटन को महत्वपूर्ण स्थान प्रदान किया गया है।

यह सुझाव है कि कुछ चुने हुए पुरातत्व केन्द्र के चारों तरफ कुछ इस प्रकार के मास्टर प्लान बनाये जायें, जिससे उनके पर्यावरण व प्राकृतिक सौन्दर्य की बनावट को सुरक्षित रखा जा सके। कुछ साल पहले भारत ने यूनेस्को के अंतर्गत सांस्कृतिक पर्यटन के क्षेत्र में डॉ. एफ. आर. अलचिन की मदद इन योजनाओं को तैयार करने के लिए प्राप्त की। एलीफेन्टा, अजन्ता व एलोरा की गुफाएं, बौद्ध धर्म के प्रमुख

केन्द्र, सारनाथ, गया, नालन्दा, राजगिरी, कुशीनगर, श्रावस्ती, साँची तथा बीजापुर, एहोल, बदामी, पटाड़रवाल, हम्पी, खजुराहो, भुवनेश्वर, पुरी व कोणार्क को योजनाओं के अंतर्गत लिया गया है।

मानव के द्वारा उन स्थानों पर जहां उन्हें ईश्वर के वास का अनुभव हुआ है, प्राचीन काल में ही यात्राएं की गई थीं जिस कारण धार्मिक यात्रायें प्रारम्भ हुईं। जो सदियों से स्पष्ट थीं तथा आज भी स्थानीय परिस्थितियों के अनुरूप परिवर्तित होती रही हैं एवं समयानुसार परिवर्तित होती रहेंगी।

भारत में एक प्राथमिक सिद्धान्त यह रहा है कि देवता किसी विशेष स्थान में ही निवास करते हैं अथवा यहाँ से अपनी शक्ति का प्रदर्शन करते हैं। मिस्र, ज्यूज व ग्रीक के लोगों में तो धार्मिक स्थलों की यात्रा करने की बहुत ही गहन धारणा रही है। इस्लाम धर्म के अंतर्गत यह बहुत ही चरम सीमा पर पहुंच गया है। घोषित देवदूतों में मुस्लिम धर्म के तहत प्रत्येक मुसलमान के लिए अपनी जिन्दगी में एक बार मुहम्मद के जन्म स्थान मक्का की यात्रा करना बहुत ही जरूरी माना गया है।

ईसाइयों में तीर्थयात्रा का उद्भव ईसामसीह की स्मृति की भक्ति के फलस्वरूप हुआ है।

ईसाई धर्म के अनुयायी इस बात का अनुसरण करते हैं कि उन धार्मिक स्थानों की तीर्थयात्रा की जाए जो उनके ईश्वर की सांसारिक जिन्दगी के स्मरणों से सम्बद्ध रहे हैं। वे मुख्य स्थल जिनकी ईसाई धर्म के अनुयायी तीर्थयात्रा करते हैं वे हैं—वेथालियम व येरुशलम। तीर्थयात्रियों की संख्या में वृद्धि होने के साथ-साथ उनके द्वारा किये गए तीर्थस्थानों की संख्या में भी वृद्धि होती है। पूर्व दिशा में जबकि तीर्थस्थानों की संख्या घट रही थी उसी समय पश्चिम में उनकी संख्या में बढ़ोत्तरी होती रही है।

मध्यकाल में असुरक्षित जीवन एवं उत्तेजित लुटेरों तथा नबाबों के युद्ध के कारण तीर्थयात्री बड़े-बड़े झुण्डों में हथियार से युक्त सुरक्षाकर्मियों सहित भ्रमण करते थे। वे आक्रमणकारियों तथा विदेशी शासकों के विरुद्ध भी धर्मयुद्ध के लिए काम करते थे। भारत प्राचीनकाल से ही तीर्थयात्राओं के लिए प्रसिद्ध स्थान रहा है। धार्मिक ज्ञान की खोज व आध्यात्मिक शान्ति प्राप्त करने के लिए व्यक्ति हर समय भारत की यात्रा करने के लिए आते हैं।

पश्चिम में यह यात्रा धार्मिक शहरों में श्रद्धांजलि व उनके समाधि स्थानों हेतु शुरू की गई थी। धार्मिक यात्री मुनियों और ऋषियों के समाधि स्थलों की तरफ अधिक आकृष्ट होते थे, क्योंकि इनका विश्वास उनकी दैविक सार्थकता और वरदान

में निहित होता था। चमत्कारिक शक्तियों को करने वाले सुप्रसिद्ध व्यक्तियों की तरफ भी दूर-दूर से भक्तजन व श्रद्धालु आकर्षित होते थे। अफ्रीका में सर्वाधिक शहीदों के मकबरे हैं। रोम को तो विश्व के महत्त्वपूर्ण पवित्र स्थल होने का गौरव प्राप्त है क्योंकि यहां गौरव प्राप्त सेन्ट पाल व सेन्ट पीटर के मकबरे हैं। तीर्थयात्रा का एक अन्य उद्देश्य यह है कि पवित्र स्थानों से धार्मिक अवशेषों को प्राप्त किया जाए।

कुछ लोगों की मान्यता है कि धार्मिक यात्रा करने से उनके द्वारा किये गये पाप मिट जाते हैं। इस प्रकार की जाने वाली यात्रा को खोजी यात्रा भी कहा जा सकता है। इसी प्रकार मनोकामना पूरी होने पर भी बहुत से भक्त देवी-देवताओं को धन्यवाद देने हेतु तीर्थयात्रा करते हैं। इस प्रकार हम देखते हैं कि धार्मिक उद्देश्य से की गई यात्रा के पीछे ईश्वर की इच्छानुसार भक्त की आज्ञा पालन करने की भावना होती है।

बहुत से स्थानों में यह सही भी माना जाता है कि कुछ वस्तुयें जिसका मृत-आत्मा से सम्बन्ध है उसे प्राप्त किया जाए। धर्म प्रेमी पवित्र कागज से गिरी हुई मोम को लेते है। कैथोलिक भक्त गोवा में एक सन्त की ममी युक्त देह की उंगली को मोड़ते हैं। कुछ भक्त संन्यासी एवं शहीदों के मकबरे से धूल को ले जाते हैं। कुछ तीर्थयात्री धार्मिक मन्दिरों से 'प्रसाद' ले जाते हैं।

ऐतिहासिक तथ्यों के आधार पर यह प्रमाण मिलते हैं कि बहुत से व्यक्ति अपने घर व निवास स्थान धार्मिक व राजनीतिक कारणों से छोड़ते हैं। तीर्थयात्रा का बहाना बनाकर वे दूसरे स्थानों पर प्रवास के लिए चले गए हैं। उदाहरणार्थ—अग्रजपूरिन्टेस, जिन्हें तीर्थयात्रा का जन्मदाता कहा जाता है, ने मैसायूसेटस में प्लेमाउथ कालोनी को बसाया। 1620 में उन्होंने पलावर में एक साथ यात्रा की।

धर्म प्रेमी पर्यटकों को भारत की स्मारक सम्पदा में ज्यादा रुचि है। हमारे देश में कोई भी ऐसा स्थान नहीं है जो किसी-न-किसी कारण पवित्र स्थान न माना जाता हो, यहाँ की नदियाँ एवं ताल, पर्वत, समुद्रतट, कस्बे व मन्दिर समस्त स्वदेशी व विदेशी पर्यटकों के आकर्षण रहे हैं। भारत बहुत से धर्मों का जन्म स्थल रहा है हिन्दू, बौद्ध, जैन व आज का इस्लाम धर्म, ईसाई, सिक्ख, जोरेस्टर सब हमारे धर्मनिरपेक्ष राष्ट्र में एक साथ रहकर शान्तिपूर्वक पूजा करते हैं।

इस वर्ग में तीर्थ स्थानों की पर्याप्त संख्या है जिन्हें अग्रलिखित हिस्सों में बांटा गया है—

1. सतत् यात्रा करने वाले तथा भारत में निर्वासित व्यक्ति व पर्यटक विश्व के विभिन्न भागों में हैं। ये व्यक्ति अपनी प्राचीन पीढ़ियों के धर्मस्थानों में सांस्कृतिक क्षेत्रों की खोज करने की जिज्ञासा रखते हैं।

2. जापान, थाईलैण्ड, लंका इत्यादि से बढ़ते हुए असंख्य पर्यटक जो कि बौद्ध धर्म से सम्बन्धित केन्द्रों की यात्रा करना चाहते हैं।
3. अमेरिकन व यूरोपियन हिन्दू, बौद्ध, ईसाई व सिक्ख धर्म के कुछ पहलुओं में रुचि रखते हैं।
4. हिन्दू, मुस्लिम, सिक्ख, ईसाई, जैन, बौद्ध व दूसरे सम्प्रदाय व धर्मों के अनुयायियों को पवित्र स्थलों की अपनी जिन्दगी में एक बार जरूर यात्रा करनी होती है।

उपर्युक्त विवरण से यह ज्ञात होता है कि धार्मिक यात्रा पर्यटन उद्योग का अत्यन्त ही विशाल हिस्सा है। अब यह आम बात हो गई है कि यात्री अपनी व्यावसयिक यात्रा में पर्यटन स्थल व पवित्र स्थानों की यात्रा भी सम्मिलित करते हैं—जिसकी वजह से अन्य राष्ट्रों से भारत में यात्री सतत् आते रहते हैं। कुछ वर्ष पहले भारत सरकार ने धार्मिक पर्यटन के विकास पर बल दिया था। आज हम देखते हैं कि बहुत से पवित्र स्थानों, जो कि पर्यटन के आकर्षण हैं, की बिल्कुल ही उपेक्षा की गयी है। दक्षिण-पूर्व एशिया के राष्ट्रों से पर्यटकों को बौद्ध रुचि स्थलों के विविध पैकेज प्रदान करने से, पर्यटकों को प्रोत्साहित करने का बड़ा स्वर्णिम मौका है। इन यात्राओं में बौद्ध गया, जहां भगवान बुद्ध को आत्मज्ञान प्राप्त हुआ था, सारनाथ, जहाँ शान्त हिरन वाटिका में बुद्ध भगवान ने अपना प्रथम उपदेश दिया था, साँची जहाँ बहुत बड़ा स्तूप है व बिहार में स्थित राजगीर को सम्मिलित किया जा सकता है। प्रसिद्ध बौद्ध स्थल जैसे अजन्ता व एलोरा व प्राचीन नालन्दा विश्वविद्यालय पर्यटकों में एक विशेष रुचि उत्पन्न करते हैं।

भारत में पर्यटकों द्वारा यहाँ के धार्मिक स्थलों की विस्तृत यात्रा की जा सकती है जिसमें दिल्ली का कमल मंदिर, अक्षरधाम मंदिर, उत्तराखण्ड का हरिद्वार तथा अन्य धार्मिक स्थल, उड़ीसा का ईसाई गिरजाघर और दक्षिण भारत के कलाओं से सुशोभित आदि मंदिर शामिल हैं। धार्मिक यात्राओं की सफलता के लिए यात्रा निर्देशक का अच्छा चुनाव होना चाहिए, उन्हें विविध धार्मिक स्थलों का ज्ञान हो तथा ये अच्छी बुद्धिमत्ता से पर्यटकों के प्रश्नों का जवाब दे सकें।

बौद्ध धर्म से सम्बन्धित यात्रा का स्थान

यह धर्म पहले तीर्थ यात्रा के रूप में प्रसिद्ध नहीं था लेकिन अब तीर्थ यात्राओं में बौद्ध धर्म का प्रमुख स्थान है। वे स्थल जहाँ बुद्ध ने जन्म लिया, परम ज्ञान प्राप्त किया, प्रवचन दिया, निवारण निर्वाण प्राप्त किया, उन दर्शनाभिलाषी लोगों के लिए

निर्वाण की प्राप्ति का वचन पूर्ण करते हैं। अपनी पूर्ण निवारण प्राप्ति के पहले भगवान बुद्ध ने चार स्थानों को बताया जिनकी हर पवित्र अनुयायी को दर्शन करने चाहिए। ये स्थल हैं—लुम्बनी वाटिका, जहाँ भगवान का जन्म हुआ, बौद्ध गया, जहाँ उन्हें परम ज्ञान प्राप्त हुआ, सारनाथ की हिरन वाटिका, जहाँ उन्होंने कानून के चक्र को चलाया एवं कुशीनगर जहाँ पर उन्हें निवारण प्राप्त हुआ था।

बौद्ध धर्म के पवित्र ज्ञान के चार अन्य स्थान हैं—श्रावस्ती, अन्यावस्था राजगीर व वैशाली अथवा अन्य चार के साथ मिलकर आठ पवित्र महत्व के स्थल बनते हैं। प्राचीन समय से ही तीर्थयात्री इन स्थानों की पवित्र भावना से यात्रा करते हैं। उनका मानना है कि भगवान बुद्ध आज भी यहां मौजूद हैं। सम्राट अशोक स्वयं एक बार इस तीर्थ यात्रा पर गये एवं उन्होंने इस तीर्थयात्रा को 'धर्मयात्रा' धर्मनिष्ठा की यात्रा का नाम दिया। वर्तमान तीर्थयात्री अपनी प्राचीन यात्राओं में बौद्ध धर्म के बड़े-बड़े केन्द्रों को शामिल करते हैं, जैसे—साँची, भरहुत, अमरावती, नागार्जुनकोन्डा, अजन्ता व एलोरा, कारला व भाजा। ये केन्द्र चित्रकारी, नक्काशी कला, वास्तुकला, भवन निर्माण कला के भण्डार हैं एवं यह भगवान की जीवनी व उपदेशों से ओतप्रोत हैं, इनकी सुन्दरता व वैभव ने स्वदेशी एवं विदेशी पर्यटकों को बिना किसी धर्म की भावना से आकर्षित किया है। ये स्थान बौद्ध धर्म की रचनात्मकता के स्मारक चिह्न हैं।

पर्यटन के आकार

पर्यटन के कई आकार हैं। ये विविध आकार, समय की अवधि, समूह की संख्या व पर्यटन के आर्थिक प्रभावों पर आधारित हैं। ये विविध आकार निम्नलिखित रूप में वर्गीकृत हैं—

यात्रियों की संख्या पर आधारित पर्यटन

यात्रियों की संख्या के आधार पर पर्यटन को निम्न भागों में बांटा गया है—

(1) सामूहिक पर्यटन—सामूहिक पर्यटन का अभिप्राय उस पर्यटन से है जिसके अन्तर्गत पर्यटक केवल एक यात्रा का क्रय करता है। क्योंकि वह पूर्व भुगतान की गयी राशि में भाड़ा व अन्य मदों की कीमतों को होटल के किराये से अलग नहीं कर सकता है। इसे हम एकमुश्त दौरा भी कह सकते हैं। पर्यटक एक व्यक्ति विशेष सैलानी समूह के सदस्य के रूप में पूर्व निर्धारित प्रबन्ध व्यवस्थानुसार विचलन करता है। ट्रेवेल एजेन्सी, इस आकार के पर्यटन में एक पूर्व नियत व्यवस्था

के तहत किसी निहित समस्त समूह को एक विशेष उद्देशित स्थान अथवा स्थलों के लिए पूर्व भुगतान के बदले यात्रा में सम्मिलित होने व बिक्री का प्रस्ताव देती है, जिसमें कि सभी समूहों के लिए या एक पूर्व निश्चित समय सूची होती है। सदस्यों को अपने विषय में कुछ नहीं सोचना पड़ता है क्योंकि सभी सुविधाओं व सेवाओं का संगठनकर्ता स्वयं प्रबन्ध करते हैं।

(2) निजी पर्यटन–जब कोई मनुष्य या व्यक्तियों का समूह यात्रा पर निकलता है तथा वे स्वयं उद्देशित स्थान का चयन करते हैं व कार्यक्रम नियत करते हैं तथा इसमें किसी भी समय स्वतन्त्रतापूर्वक परिवर्तन कर सकते हैं, उसे निजी पर्यटन कहा जाता है। पर्यटक एक निजी या समूह के सदस्य के रूप में भ्रमण करते हैं तथा उन्हें यात्रा एवं ठहरने के स्थान के प्रबन्ध से कोई अर्थ नहीं रहता है। पर्यटक तो स्वयं सीधे मालवाहक, होटल या अन्य पूर्तिकर्ताओं से अथवा ट्रेवेल एजेन्ट्स के जरिए, परिवहन, आवास व अन्य सुविधाओं का पृथक् रूप से व्यवस्था कर सकते हैं। व्यक्ति विशेष अथवा परिवार स्वयं इन सुविधाओं का अपनी आवश्यकतानुसार प्रबन्ध करता है, इसीलिए इसे स्वतन्त्र यात्रा भी कहा जा सकता है।

प्रबन्ध अनुसार

इस प्रकार के पर्यटन में पर्यटक अपने पर्यटन स्थल के लिए भोजन, वाहन, निवास आदि की व्यवस्था स्वयं करते हैं–

1. समय पर आधारित

समय पर आधारित यात्रा को अधोलिखित भागों में बांटा जा सकता है–

(1) कम समय का पर्यटन–इस प्रकार के पर्यटन में कम समय की यात्रायें शामिल होती हैं, जो अधिकतम आठ या दस तक की होती हैं। इस प्रकार की यात्राओं की संख्या बढ़ने से देश के आर्थिक महत्व पर बड़ा अच्छा असर पड़ता है। सामाजिक तौर पर इस प्रकार का पर्यटन उन ग्राहकों के लिए उचित है जो कि दीर्घकालीन अवकाश प्राप्त नहीं कर सकते हैं।

(2) अधिक समय का पर्यटन–इस प्रकार के पर्यटन में अधिक समय वाली यात्रायें सम्मिलित होती हैं। जैसे–एक पर्यटक के लिए कई सप्ताह व महीनों की यात्रा, आधुनिक पर्यटन में विशेष तौर पर निजी मोटरों का इस्तेमाल होने के कारण यात्राएँ लचीली हो गयी हैं। परिणामस्वरूप एक दीर्घकालीन यात्रा कई राष्ट्रों में की जा सकती है। परन्तु साथ ही कई पर्यटन स्थलों की छोटी यात्रा भी की जा सकती

है। यात्रा एवं ठहरने की अवधि किसी विशेष स्थल में कम-से-कम 24 घंटे की होनी चाहिए। इसे लघुयात्रा अथवा दर्शन भेंट यात्रा कहा जा सकता है।

(3) मनोरंजक साहसिक पर्यटन—इस प्रकार की यात्रा के तहत रात्रि के समय रुकने की व्यवस्था नहीं होती है। प्रायः इसे आन्तरिक अन्तर्राष्ट्रीय दिन की यात्रा या दिन की भेंट यात्रा के रूप में वर्णित किया जाता है। मनोरंजन के लिए भ्रमण को आई. ओ. टी. ओ. की पर्यटन परिभाषा में भी शामिल किया गया है। यह यात्रा 24 घंटे से अधिक नहीं होती है एवं इसमें आवास का प्रबन्ध नहीं होता है। ये यात्राएँ उन स्थानों में अधिक विख्यात हैं जहाँ अनुकूल विनिमय दरें होने से यात्रा के समय खरीददारी की सुविधाएँ प्राप्त होती हैं।

2. भुगतान के सन्तुलन पर प्रभाव के आधार पर

इस प्रकार के पर्यटन को दो भागों में विभाजित किया गया है—(1) आगवानी पर्यटन तथा (2) निष्क्रिय पर्यटन।

(1) आगवानी पर्यटन—इसका सम्बन्ध विदेशी पर्यटकों के आगमन से है जिसके परिणामस्वरूप राष्ट्र के अंतर्गत विदेशी मुद्रा के पदार्पण से भुगतान के सन्तुलन पर अनुकूल असर पड़ता है।

(2) निष्क्रिय पर्यटन—इसका आशय स्वदेश के नागरिकों द्वारा विदेशी भ्रमण से है। यह पर्यटन स्वदेश में अर्जित विदेशी मुद्रा को लेकर भुगतान के सन्तुलन पर प्रतिकूल प्रभाव डालते हैं।

3. प्रयुक्त होने वाले पर्यटन साधनों पर आधारित

इस प्रकार के पर्यटन को साधनों के आधार पर बांटा गया है जो मुख्य रूप से इस प्रकार हैं—

1. पैदल पथिक पर्यटक
2. सड़क पर्यटक
3. रेल पर्यटक
4. मोटर पर्यटक मोटरकार
5. वायु पर्यटक

सांख्यकीय पर्यटन-आम पर्यटन

सांख्यिकीय पर्यटन या आम पर्यटन को दो भागों में विभाजित किया गया है—

1. गुणत्वशील पर्यटन

इस पर्यटन को भी दो भागों में विभाजित किया गया है–

1. सामाजिक पर्यटन,
2. जनप्रिय पर्यटन।

2. पर्यटन संगठन

वैश्विक पर्यटन में तीव्र गति से होने वाले विकास के परिणामस्वरूप विगत कुछ सालों में कई प्रमुख संगठन पर्यटन उद्योग से सम्बद्ध हैं।

3

पर्यटन की क्रमिक प्रगति

प्राचीन समय से ही मनुष्य ने भ्रमण करना प्रारंभ कर दिया था। इसका मुख्य उद्देश्य जीवन निर्वाह करना था। वह अपनी भूख शांत करने के लिए जंगलों में कन्दमूल, फल, जानवरों का माँस आदि वस्तुओं को प्राप्त करने के लिए जंगलों में भटकता फिरता था। जैसे-जैसे मनुष्य का मानसिक विकास हुआ उसे खाने-पीने, सोने आदि के लिए सुरक्षित स्थानों की आवश्यकता महसूस हुई। इस प्रकार मनुष्य का भ्रमण जारी रहा है। इसलिए कह सकते हैं कि मनुष्य ने शुरू-शुरू में अपनी जीविका तथा सुरक्षा कारणों से भ्रमण करना आरंभ किया।

मनुष्य को एकाकी जीवन की तुलना में समूह में रहना अच्छा लगा, अतएव उसने नये लोगों से जानकारी, परिचय एवं दोस्ती करने के उद्देश्य से यात्रा एवं भ्रमण प्रारंभ किया।

मनुष्य का सांस्कृतिक विकास होने के साथ-साथ उसकी जिज्ञासाएं भी बढ़ती रहीं। उसे जिज्ञासा हुई कि नये क्षेत्र देखे जायें जहां भोजन आसानी से उपलब्ध हो सके, रहने के स्थान भी सुरक्षित हों, पीने का जल भी निकटवर्ती क्षेत्र में ही उपलब्ध रहे एवं अन्य मूलभूत आवश्यकता की वस्तुएँ भी एकत्र की जा सकें। अतः मानव

ने दूर-दराज के क्षेत्रों को देखने की इच्छा से भ्रमण किया एवं जो व्यक्ति अधिक भ्रमणशील रहा उसकी जानकारी अन्य लोगों से काफी बढ़ गई। अन्य लोगों ने भी देखा-देखी नई जानकारियों अर्थात् ज्ञान की खोज में यात्राएं प्रारंभ कीं। नये क्षेत्रों की जानकारी तथा ज्ञान की खोज में निकलने वालों की संख्या में सतत् वृद्धि होती गयी। लोग समूहों/काफिलों में भ्रमण पर निकलने लगे। दूसरे क्षेत्रों में जाने पर अदला-बदली की क्रिया प्रारंभ हुई जिसने कालान्तर में व्यापार का रूप ग्रहण कर लिया तथा लोग व्यवसाय के उद्देश्य से भ्रमण पर निकलने लगे।

समय के साथ-साथ जैसे-जैसे मनुष्य का धार्मिक विकास हुआ तथा उनमें धर्म प्रचार की भावना उत्पन्न हुई इसके परिणामस्वरूप धार्मिक यात्राएं प्रारंभ हुईं जिसमें यूरोप में ईसाई, यूरेशिया में इस्लाम, पूर्वी एशिया में बौद्ध एवं भारत में हिन्दू धर्म के प्रचार-प्रसार के लिए काफी यात्राएं की गयीं।

इस प्रकार कह सकते हैं कि मनुष्य अपने धार्मिक, सामाजिक, आर्थिक, राजनैतिक एवं सांस्कृतिक कार्यों से अपने शहर, प्रान्त एवं राष्ट्र से बाहर जाते रहे हैं।

पर्यटन का वर्गीकरण

पर्यटन का वर्गीकरण निम्न प्रकार से है–

(1) आनन्द और मनोरंजन हेतु यात्रा–अत्यंत प्राचीनकाल में यात्रा तथा भ्रमण का प्रयोजन आनन्द एवं मनोरंजन का नहीं था। रोमन काल में यात्रा रोमन साम्राज्य के विकास के साथ शुरू हुई। इन्हें आनन्ददायक यात्रा कहा गया। रोमन नागरिकों ने लगभग 150 से 50 ईसा वर्ष पूर्व के पिरामिड्स मन्दिर, धर्म एवं धार्मिक स्थानों के दर्शन के लिए अवकाश के वक्त में यात्रा करना प्रारंभ की। इस समय के मुख्य यात्री डायोडोरस, पोलिवियस, मेला, स्ट्रैवो थे।

रोमन साम्राज्य का अन्त हो जाने के पश्चात् अन्धयुग का प्रादुर्भाव हुआ। इस युग में आनन्ददायक यात्रायें बहुत प्रभावित हुईं एवं साहसी यात्री ही यात्रा पर निकले। इस काल के मुख्य यात्रियों में व्यापारी सुलेमान 851 ई. अलमसूदी अलबरूनी थे। दसवीं शताब्दी में यूरोप के व्यापारियों ने यूरोप की हडसन खाड़ी की यात्राएं कीं।

(2) व्यापार और व्यवसाय हेतु यात्रा–ईसा के पैदा होने से कई वर्ष पूर्व व्यापारी और दुकानदार नये देशों और जनजातियों से व्यावसायिक सम्पर्क बनाने के लिये दूर-दूर तक यात्रायें करते थे। ईसा से 3000 वर्ष पूर्व मुद्रा की खोज ने यात्रा की और योगदान दिया।

(३) ज्ञानार्जन के लिए यात्रा--प्राचीन काल में नवीन जानकारी एवं नये स्थानों की जानकारी की भावना से यात्राएं की गयीं। होमर के महाकाव्य में यूनान को रोमन के ज्ञानार्जन के लिए भ्रमण की भूख का वर्णन है।

पर्यटन का प्रारम्भ

पर्यटन का प्रारंभ ईसा से 1200 सदी के समय से माना जाता है। इनमें नई यात्राओं की खोज की गयी जिस कारण इसे पुनर्जागरण काल भी कहा जाता है। इस युग में मार्कोपोलो (1254 से 1324 ई.) ने मध्य एशिया होते हुए चीन तक अर्थात् भूमध्य सागर से प्रशान्त महासागर तक यात्रा की एवं मध्य युग का महान् यात्री माना गया। अरब यात्री इब्नबूता सन् 1304 से 1337 ए. डी. में हुआ।

सामुद्रिक अभियान

मध्यकाल के दौरान राजकुमार हैनरी (1482 ई.), कोलम्बस (1492 ई.), वास्कोडिगामा (1498 ई.), मैगलन (1520 ई.) आदि पुर्तगालियों ने महासागर की यात्राएं कीं। उसके अलावा 1534 में फ्रांस से जैकिक्स, कार्टियर ने सेंटलारेंस नदी क्षेत्र की ओर तथा सन् 1660 ई. में इंग्लैड के अंग्रेजों ने वर्जीनिया क्षेत्र की यात्रा की। 17वीं सदी में विकास का कुछ कार्य वारेनियस द्वारा किया गया।

धार्मिक यात्राएं : धर्म प्रचार के उद्देश्य से विभिन्न देशों में भी यात्राएं की हैं जिनमें प्रमुख हैं–पूर्वी एशिया में बौद्ध धर्म के प्रचार के लिए, भारत में हिन्दू धर्म के प्रचार के लिए, यूरोप में ईसाईयत के प्रचार के लिए तथा यूरेशिया में इस्लाम धर्म के प्रचार के लिए।

उल्लेखनीय है कि 18वीं शताब्दी यात्रा के विकास का स्वर्णिम समय था, जिसमें विशेषकर सन् 1763 से 1793 के मध्य 30 वर्ष तक यात्राएं की गयीं एवं ये यात्राएं अधिकतर इटली में की गयीं। इटली उस समय प्रबुद्ध व्यक्तियों की राजधानी थी। अठारहवीं शताब्दी में न सिर्फ धनाढ्य वर्ग के लोग इटली की यात्रा पर आये वरन् अनेक कवि, लेखक तथा बुद्धिजीवियों ने भी इटली की यात्रा की।

18वीं शताब्दी के दौरान यूरोप के अंतर्गत वार्षिक यात्रा का शुभारंभ किया गया जो कि एक ऐसी महत्वपूर्ण घटना बनी जिसने व्यक्तियों को यात्रा करने के लिए मौके एवं समय प्रदान किया तथा इसी की वजह से बीसवीं सदी में पर्यटन का असाधारण विकास हुआ। वर्तमान समय में छुट्टी के दिनों का अर्थ उन दिनों से

समझा जाता है जो प्रतिदिन के कार्यों से हटकर फुरसत और मनोरंजन के लिए होता है एवं पुनः नैत्यिक कार्यों के लिए स्फूर्ति व ताजगी लाता है।

19वीं सदी में यात्रा का विकास औद्योगिक विकास, यातायात साधन एवं संचार साधनों का विकास, नए-नए आविष्कार जैस–छपाई मशीन, दूरबीन, कैमरा इत्यादि के कारण हुआ। कारखानों में अधिक मेहनत करने वाले श्रमिकों ने मानसिक और शारीरिक आराम हेतु सालाना अवकाश दिये जाने की मांग की। साथ ही साथ कई कर्मचारी मिलकर मनोरंजन हेतु पर्यटन पर जाने लगे एवं अवकाश कैम्प लगाये। होली डे होम्स का शुभारंभ हुआ, परन्तु कम ही व्यक्ति अभी तक इसका फायदा उठा पा रहे थे। 19वीं सदी के अन्तिम चौथाई भाग तक कर्मचारियों के अवैतनिक अवकाश पर विचार किया गया एवं ध्यान दिया गया, तब पश्चिम यूरोप के कुछ कारखानों ने अपने कामगारों को सचेतन अवकाश देना शुरू किया।

20वीं सदी के प्रारंभ होने तक आधुनिक पर्यटन की सभी मुख्य विशेषतायें भ्रमण के तौर पर स्पष्ट होने लगी थीं। मानव के स्वभाव में आनन्द की खोज, शिक्षा, ज्ञान के लिए यात्रा, भौतिक सम्पदा एवं साधनों की वृद्धि, यात्रा को सामाजिक प्रतिष्ठा का मापदण्ड, मानव के दैनिक कार्यों में आराम करने की बढ़ती आवश्यकता, यात्रियों की परिवहन व्यवस्था में सुधार इत्यादि ने यात्रा के लिए एक उपजाऊ धरातल तैयार किया। इस शताब्दी में मनोरंजन यात्राओं का काफी विस्तार हुआ। सदी के प्रथम 25 सालों तक आनन्दमय यात्रा एवं विलासिता की वस्तु जो समाज के उन्हीं विशेष वर्गों तक सीमित थी, जिनके पास पर्याप्त खाली समय व धन था। इनके द्वारा आम व्यक्तियों तक पर्यटन एवं आनन्दमय यात्रा कम पैसे में कर, लाभ उठाने का लालच दिया गया। यह कार्य यात्रा पर्यटन से सम्बद्ध संस्थाओं द्वारा भी किया जाने लगा।

प्रथम विश्व युद्ध की वजह से की गई ये कोशिशें बहुत ही प्रभावित हुईं। क्योंकि पर्यटन का विकास शान्ति में ही निहित था। प्रथम विश्व युद्ध के उपरांत के काल में पर्यटन पुनः विश्व युद्ध से पूर्व के स्तर को प्राप्त हो गया एवं लोगों ने वैश्विक पर्यटन की प्रकृति में बदलाव एवं क्षेत्र में भी वृद्धि की। युद्धोतर काल में यूरोप एवं अमेरिका में श्रमिक एवं मध्यम वर्ग के लोगों का रहन-सहन का स्तर उठने लगा तथा कई ऐसे क्षेत्रों में पर्यटन प्रारंभ हुआ जहां अभी तक लोग इसे जानते ही नहीं थे। सन् 1929 व 1930 में 15 लाख पर्यटक, स्विट्जरलैंड, 10 लाख के लगभग इटली तथा 20 लाख पर्यटक आस्ट्रेलिया गये। फ्रांस, स्पेन, ग्रेट ब्रिटेन से भी काफी मात्रा में पर्यटक गये। प्रथम विश्व युद्ध में मोटरकार

का विकास हो जाने पर अमेरिका में पर्यटन को बढ़ावा मिला। इसी के साथ-साथ सार्वजनिक सड़क परिवहन व सड़कों के विकास ने पर्यटन में तेजी से विकास किया। नौकरी करने वाले कर्मियों को कुछ छुट्टियां दी जानी शुरू हुईं, परन्तु उन्हें इस छुट्टी का वेतन नहीं दिया जाता था। तत्पश्चात् सन् 1938 में सालाना सवैतनिक अवकाश मिलना प्रारंभ हुआ। श्रमिकों एवं कर्मचारियों में सामूहिक पर्यटन की प्रवृत्ति विकसित हुई। इसी समय श्रम संघ प्रतिनिधियों ने प्रथम कन्वेंसन की। बाद में कई दूसरे देशों ने सवैतनिक अवकाश देना स्वीकार किया। बाद में कई अन्य राष्ट्रों ने भी इसे अपने यहां क्रियान्वित किया। बाद में यह अवकाश संसार के सभी देशों में क्रियान्वित किया गया जो पर्यटन के लिए एक वरदान सिद्ध हुआ।

दूसरे विश्व युद्ध के दौरान पर्यटन पर प्रतिकूल प्रभाव पड़ा जिससे अर्थव्यवस्था का संतुलन अनुकूल नहीं रहा लेकिन अर्थव्यवस्था में धीरे-धीरे सुधार होते ही पर्यटन एक उद्योग के रूप में सामने आया। वर्तमान समय में विश्व के कई देशों में पर्यटन को एक उद्योग के रूप में मान्यता दे दी गई है।

स्वातंत्र्योत्तर भारत में पर्यटन की प्रगति

विश्व में पर्यटन की दृष्टि से भारत का प्रमुख स्थान है। भारत के असीमित भण्डार में पर्यावरण सम्बन्धी मनमोहक सम्पदा, सुन्दरता तथा अटूट सांस्कृतिक परम्पराओं की बहुत ही विविधता व निरन्तरता है। यहां की परिवर्तित होने वाली ऋतुएं, जलवायु की विविधता, हिमाच्छादित पर्वतमालाएं, रमणीय समुद्री तट, सदाबहार हरा-भरा केरल, रणबांकुरों एवं स्वाभिमानी राजाओं की जन्मस्थली राजस्थान, चार पवित्र धामों की भूमि, अलग-अलग धर्मों एवं धार्मिक स्थलों का देश भारत पर्यटन की दृष्टि से दुनियाभर के पर्यटकों को सदैव से ही आकर्षित करता रहा है। मोहनजोदड़ो एवं हड़प्पा के अवशेष, मौर्यकालीन कौशाम्बी, वैशाली और उज्जैन, अशोककालीन पाटलिपुत्र और नालंदा, महाराष्ट्र में अजन्ता, एलोरा एवं एलीफैण्टा की गुफायें, मध्य प्रदेश की खजुराहो की मूर्तियाँ यहां के प्रमुख पर्यटन स्थल हैं। कुतुबमीनार, ताजमहल, फतेहपुर सीकरी, लालकिला, सारनाथ, अजमेर शरीफ, बौद्ध गया, इलाहाबाद का संगम, उत्तराखंड में गंगोत्री, यमुनोत्री, बद्रीनाथ, केदारनाथ, आन्ध्र प्रदेश में भगवान तिरुपति जी लाखों पर्यटकों को अपनी ओर खींच लेते हैं।

प्राचीन समय से ही भारतीय पर्यटन ने विश्व के प्रसिद्ध लोगों को अपनी तरफ खींचा जिसमें फाह्यान ह्वेनसांग, वास्कोडिगामा, कैप्टन कुक आदि शामिल हैं।

भारतीय पर्यटन स्थल : वर्तमान परिदृश्य

स्वतंत्रता के बाद भारत ने पर्यटन के क्षेत्र में सराहनीय उन्नति की है। वर्तमान में पर्यटन एक उद्योग का रूप ले चुका है। पूर्व में पर्यटन को शिक्षा का एक जरिया समझा जाता था। पर्यटन उद्योग एक ऐसी ताकत के तौर पर हमारे समक्ष आ रहा है जो विश्व के सांस्कृतिक एकीकरण में एक मील का पत्थर सिद्ध होगा। पर्यटन के महत्व को देखते हुए स्वतंत्रता के उपरांत सरकार ने सन् 1952 में पर्यटन हेतु प्रथम समुद्र पार कार्यालय न्यूयार्क में खोला। सन् 1957 में फ्रेंकफर्ट में भी दूसरा पर्यटन कार्यालय स्थापित किया गया। टूरिज्म ट्रैफिक ब्रांच के स्थान पर परिवहन एवं संचार मंत्रालय के अधीन पर्यटन विभाग की स्थापना की गयी।

सन् 1961 में भारत आने वाले पर्यटकों की संख्या 1,39,804 थी जो भारत-चीन युद्ध के कारण सन् 1962 में कम होकर 1,34,036 रह गयी। सन् 1963 में तत्कालीन वित्त सचिव एल. के. झा की अध्यक्षता में सरकार ने एक तदर्थ समिति का गठन किया। तत्कालीन पर्यटन महानिदेशक एस. एन. चैव को इस समिति में नियुक्त किया गया। इस समिति ने पर्यटन को बढ़ावा देने के लिए सुझाव दिये।

सन् 1965 में 'झा' समिति की सिफारिशों को क्रियान्वित करने के लिए एक उच्च अधिकार सम्पन्न समिति की स्थापना की गयी जिसके आधार पर निम्नलिखित तीन निगमों की स्थापना का प्रावधान किया गया।

1. इण्डिया टूरिज्म कार्पोरेशन लि.।
2. होटल कार्पोरेशन ऑफ इण्डिया लि.।
3. इण्डिया टूरिज्म ट्रान्सपोर्ट अण्डरटेकिंग लि.।

उपर्युक्त सभी निगम पर्यटकों हेतु आवास-सुविधा, खान-पान, आवागमन के लिए ट्रेवल एजेंसियों की सुविधाएं एवं अन्य बहुत-सी सुविधाओं को उपलब्ध कराने में जुट गये, परन्तु इस निगमों के मध्य उचित तालमेल न हो पाने के कारण इससे कोई वांछित परिणाम नहीं प्राप्त हो सका। इस कमी को दूर करने के लिए एक नये निगम भारतीय पर्यटन विकास निगम की स्थापना की गयी।

तत्कालीन प्रधानमंत्री श्रीमती इन्दिरा गांधी ने पर्यटन के महत्व को स्वीकारते हुए इसकी वृद्धि एवं विकास के लिए एक पृथक् मंत्रालय पर्यटन और नागरिक उड्डयन मंत्रालय की स्थापना 13 मार्च, 1967 को की तथा प्रथम बार डॉ. कर्ण सिंह को इसका केन्द्रीय मंत्री बनाया गया। भारत के पर्यटन विकास में यह एक प्रमुख कदम था। सन् 1985 में तत्कालीन प्रधानमंत्री श्री राजीव गांधी ने एक राज्यमंत्री की सहायता से इस

मंत्रालय का कार्यभार स्वयं संभाला। पर्यटन विभाग का महानिदेशक इस विभाग का सर्वोत्तम अधिकारी होता है जो सीधे उपर्युक्त मंत्रालय के अधीन होता है। पर्यटन विकास के लिए योजना, प्रकाशन, प्रचार, होटल, विपणन, सर्वेक्षण, वन्य जीवन-संरक्षण और आवास इत्यादि प्रभाग इसके अधीनस्थ के तौर पर कार्यरत हैं।

जनवरी 1999 से अगस्त 1999 के दौरान भारत में 1.55 मिलियन पर्यटकों ने भारत की यात्रा की जिससे भारत को लगभग 1.83 बिलियन डॉलर लगभग 7850 करोड़ रुपये की विदेशी मुद्रा अर्जित हुई।

सन् 1978 में चीन और भारत दोनों ही देशों में लगभग 7 लाख विदेशी पर्यटक आये थे। वर्ष 1998 में चीन में यह संख्या 2 करोड़ 30 लाख हो गयी जबकि भारत में यह संख्या केवल 23.7 लाख ही हुई। इस तरह भारत की तुलना में चीन में पर्यटकों की संख्या में 10 गुनी वृद्धि हुई। भारत में आने वाले विदेशी पर्यटकों की अन्तर्राष्ट्रीय स्तर पर संख्या 0.4 फीसदी है।

भारत में आने वाले विदेशी पर्यटकों की संख्या

वर्ष	*विदेशी पर्यटकों की संख्या (लाखों में)*
1950	0.53
1960	1.23
1970	2.80
1978	7.20
1994	18.20
1995	23.00
1996	24.00
1997	25.00
1998	23.70
1999	24.20
2000	24.50
2001	25.00

विदेशी पर्यटक प्रायः भारत में गोवा को पर्यटन के रूप में सबसे अधिक पंसद करते थे लेकिन अब राजस्थान, महाराष्ट्र, उत्तर प्रदेश, तमिलनाडु तथा दिल्ली आदि उनके लिए आकर्षण के केन्द्र बन गए हैं।

भारतीय राज्यों में विदेशी पर्यटकों की संख्या

राज्य	*विदेशी पर्यटकों की संख्या (लाखों में)*
राजस्थान	6.07
उत्तराखंड	2.00
महाराष्ट्र	9.80
बिहार	2.13
दिल्ली	11.62
उत्तर प्रदेश	7.19

जिन राष्ट्रों से भारत में सर्वाधिक पर्यटकों आते हैं, उनमें क्रमशः इंग्लैण्ड, संयुक्त राज्य अमेरिका, श्रीलंका, म्यांमार, चीन, कनाडा और थाइलैण्ड उल्लेखनीय हैं।

क्षेत्रीय पर्यटक

पूरे भारत की यात्रा करने वाले क्षेत्रीय पर्यटकों की संख्या लगभग 1.67 करोड़ है। इसके अतिरिक्त 4 लाख स्वदेशी पर्यटक हर साल विदेशों में अवकाश मनाने जाते हैं।

भारत में पर्यटन से सम्बन्धित खर्च

पर्यटन को प्रोत्साहित करने में ज्यादा खर्च करने वाले देशों में भारत भी शामिल है। केन्द्र सरकार ने वर्ष 1995 में मात्र 1.54 करोड़ डालर व्यय किए, जो आस्ट्रेलिया द्वारा इसी मद पर किये गये व्यय का पांचवां भाग एवं सिंगापुर, थाइलैण्ड का एक-तिहाई एवं दक्षिण कोरिया का आधा है।

पर्यटक प्रोत्साहन पर खर्च की गयी राशि—1998

देश	*प्रोत्साहन व्यय (करोड़ डालर में)*
दक्षिण कोरिया	3.05
भारत	1.54
आस्ट्रेलिया	7.58
थाइलैण्ड	5.10

देश	प्रोत्साहन व्यय (करोड़ डालर में)
सिंगापुर	4.47
मलेशिया	2.55

भारत के प्रमुख पर्यटन स्थल

भारत के प्रमुख पर्यटन स्थल—दिल्ली, गोवा, गुजरात, महाराष्ट्र, केरल, तमिलनाडु, उत्तर प्रदेश, उत्तराखंड, मध्य प्रदेश, आन्ध्र प्रदेश एवं उड़ीसा हैं। कश्मीर जो धरती का स्वर्ग कहा जाता है, आतंकवाद की चपेट में है। पूर्वोत्तर भारत के भविष्य पर्यटन स्थल हैं। कन्याकुमारी का सूर्योदय तथा सूर्यास्त, दार्जिलिंग की स्वर्णिम कंचनजंगा चोटी, असम का काजीरंगा अभयारण्य, बिहार में बौद्ध परिपथ, उड़ीसा का कोणार्क का सूर्य मंदिर, डल एवं वूलर झीलें, गोवा का सतरंगी समुद्रतटीय भू-दृश्य विश्व पर्यटकों के मुख्य आकर्षण केन्द्र हैं।

भारत में पर्यटन-आवास सुविधा की वर्तमान स्थिति

31 मार्च, 1998 तक पर्यटन विभाग द्वारा मान्यताप्राप्त 1164 होटल भारत में थे जिनमें कुल 65,000 कमरे पर्यटकों के लिए उपलब्ध थे। इनमें से 43 होटल पांच सितारा और डीलक्स, 56 पांच सितारा, 74 चार सितारा, 46 हैरिटेज, 289 तीन सितारा, 314 दो सितारा, 142 एक सितारा थे। लगभग 200 होटल ऐसे थे जिनकी सितारा होटल के रूप में मान्यता नहीं थी।

पर्यटन उद्योग में संभावनाएं

सन् 1998 में भारत ने विदेशी सैलानियों के माध्यम से लगभग 11,745 करोड़ रुपये की विदेशी मुद्रा अर्जित की। इस प्रकार वस्त्र उद्योग एवं हीरे जवाहरात उद्योग के उपरांत विदेशी मुद्रा प्राप्त करने वाला पर्यटन सबसे बड़ा उद्योग है।

पर्यटन से प्राप्त विदेशी मुद्रा

वर्ष	पर्यटन से प्राप्त विदेशी मुद्रा
1996	10,000
1997	11,051
1998	11,745
1999	12,124

भारत में पर्यटन के विकास से सम्बन्धित किये गये प्रमुख कार्य

पर्यटन उद्योग को आर्थिक विकास के एक साधन के रूप में प्रोत्साहित करने हेतु सन् 1982 में भारतीय सरकार ने एक व्यापक पर्यटन नीति की घोषणा की थी।

इस पर्यटन नीति में पर्यटन के लिए भारत के कई आकर्षणों की सही योजना, सही सीमांकन एवं पूरी तरह एकीकृत राष्ट्रीय कार्यक्रम के द्वारा पर्यटन की वृद्धि एवं विकास को वास्तविकता में परिवर्तन करने का संकल्प लिया गया।

वर्ष 1986 में पर्यटन को एक उद्योग की मान्यता दी गयी इसके लिए आर्थिक मदद, राज्य वित्तीय संस्थानों द्वारा ऋणों की मंजूरी में प्राथमिकता, बिजली पानी कनेक्शनों की संख्या में वरीयता देने की घोषणा की गई। जुलाई 1991 में पर्यटन को विदेशी निवेश हेतु प्राथमिकता के क्षेत्र के तौर पर भी घोषित कर दिया गया एवं पर्यटन को प्रोत्साहन देने के लिए वर्ष 1991 को पर्यटन वर्ष एवं वर्ष 1999-2000 को 'भारत यात्रा वर्ष' के रूप में घोषित किया गया।

पर्यटन विकास में आने वाली रुकावटें व उनका निवारण

1. प्रशिक्षित गाइड एवं टूर आपरेटर्स का न होना।
2. रेलगाड़ियों, वायुयानों आदि का समय से न चलना।
3. महंगे होटल।
4. प्रचार-प्रसार विज्ञापन आदि का अभाव।
5. आंतकवाद का न रुक पाना।
6. उचित परिवहन सुविधाओं का अभाव/घरेलू वायु सेवाओं का अभाव।
7. होटल मालिकों द्वारा ठगी।
8. आवासीय सुविधाओं की कमी।
9. कानून-व्यवस्था की खस्ता हालत।
10. करों की दर अधिक।
11. आधारभूत सुविधाओं का विकास।
12. कानून और शान्ति व्यवस्था।
13. टैक्सी एवं आटो रिक्शा चालकों की धोखाधड़ी।
14. यात्रा-परिपथ गन्तव्य स्थानों आदि का सम्पूर्ण विवरण।
15. **अतिथि देवो भव की मान्यता का आचरण।**
16. **बाधाओं के निदान के लिए सुझाव।**

17. वाहनों के संचालन सम्बन्धी समय की पाबन्दी।
18. सस्ती एवं उच्च स्तरीय आवासीय व खान-पान सेवाएं।
19. पर्यटन को भारत जैसे कृषि प्रधान राष्ट्र के लिए सच्चे मन से रोजगार-अवसर सर्जक उद्योग के रूप में मान्यता दिए जाने की राजकीय पहल।
20. सरकारी उदार नीतियां।

भारत में पर्यटन विकास की संभावनाएं

भारत प्राकृतिक सौंदर्य से परिपूर्ण है। भारत का हर क्षेत्र पर्यटन विकास के नजरिये से बहुत ही महत्वपूर्ण है। इसके उत्तर में 50,000 वर्ग कि. मी. क्षेत्र में 2500 कि. मी. लम्बी हिमालय की पर्वत शृंखलायें हैं। इसी पर्वत शृंखला में संसार की सबसे ऊंची पर्वत चोटी एवरेस्ट अवस्थित है। यह पर्वत अपने सौन्दर्य के अतिरिक्त उपयोगिता के कारण पर्वतारोहियों एवं पर्यटकों के स्वर्ग हैं। हिमालय के अतिरिक्त राष्ट्र के पूर्वोतर भाग में पूर्वांचल की पर्वत शृंखलायें, उत्तर-पश्चिम में अरावली की पहाड़ियां, देश के मध्यवर्ती भाग में विन्ध्याचल तथा सतपुड़ा की पर्वतमालायें, दक्षिण में नीलगिरी के पर्वत शिखर तथा राष्ट्र के पूर्वी व पश्चिमी तट पर पूर्वी घाट व पश्चिमी घाट की पर्वत मालायें हैं। इन पर्वत शिखरों से निकलने वाली गंगा, यमुना, सिंधु, नर्मदा, ताप्ती, गोदावरी, महानदी, कृष्णा, कावेरी इत्यादि नदियां हैं जिनके किनारे-किनारे कई पर्यटन नगरों एवं केन्द्रों का विकास हुआ है। लगभग 75,165 कि.मी. लम्बे समुद्र तट में भी कई स्थलों पर मीलों लंबी सुरंगे एवं मनोहरी पर्यटन केन्द्र हैं। अजन्ता-एलोरा की गुफाओं सहित राष्ट्र के दक्षिणी भाग में कई कलात्मक एवं उत्कृष्ट मन्दिर हैं। यहां के स्मारक, नगर, पुरानी इमारतों के अवशेष, मध्ययुगीन किले, राजप्रासाद, मकबरे सभी पर्यटकों को आकृष्ट करने की सामर्थ्य रखते हैं।

यदि देखा जाए तो भारत जनसंख्या और क्षेत्रफल के नजरिए से इतना बड़ा है कि कई यूरोपीय देश इसमें समा सकते हैं, परन्तु पर्यटकों की संख्या की दृष्टि से भारत यूरोप के कई छोटे राष्ट्रों से भी पिछड़ा हुआ है। विश्व पर्यटन में भारत का अंश सिर्फ 0.36 फीसदी है। विश्व पर्यटन में भारत की भागीदारी को बढ़ाने के लिए पर्यटन के सुनियोजित विकास की आवश्यकता है। यदि इस हेतु गंभीर प्रयत्न किये जायें तो देश में आने वाले विदेशी पर्यटकों की संख्या में 25 से 30 प्रतिशत प्रतिवर्ष की वृद्धि की जा सकती है। सन् 1981 में हमारे राष्ट्र में लगभग 12 लाख 80 हजार पर्यटक आये। 1991 में यह संख्या बढ़कर लगभग 16 लाख 78 हजार हो गयी।

1996 में लगभग 22 लाख 87 हजार 860 पर्यटक देश में आये और इस तरह 1996 में पर्यटकों की संख्या में 8 प्रतिशत की वृद्धि हुई।

इस प्रकार भारतीय अर्थव्यवस्था में भी पर्यटन की महत्वपूर्ण भूमिका रहेगी क्योंकि आज भी पर्यटन रत्न-आभूषण तथा सिले-सिलाए वस्त्रों के उपरांत देश के लिए सर्वाधिक विदेशी मुद्रा अर्जित करने वाला उद्योग बन गया है। इसके साथ ही पर्यटन व्यवसाय ने 80 लाख व्यक्तियों को रोजगार उपलब्ध कराये हैं।

पर्यटन उद्योग के इस महत्व को दृष्टिगत रखते हुए ही पृथक्-पृथक् पंचवर्षीय योजनाओं के अन्तर्गत पर्यटन को विकसित करने के लिए अधिक ध्यान दिया गया। प्रथम पंचवर्षीय योजना में पर्यटन संबंधी स्पष्ट नीति का अभाव था। द्वितीय पंचवर्षीय योजना में विदेशी एवं घरेलू दोनों प्रकार के पर्यटन विकास पर जोर दिया गया। तृतीय पंचवर्षीय योजना मुख्य पर्यटकों के आवास एवं आवागमन की सुविधाओं पर केन्द्रित थी। चतुर्थ पंचवर्षीय योजना में पर्यटन संबंधी मुख्य प्रयोजन विदेशी मुद्रा अर्जित करने के साथ-साथ रोजगार के लिए मौके उपलब्ध कराना रहा। पांचवीं पंचवर्षीय योजना में पर्यटन विकास के लिए संसाधनों की कमी को देखते हुए इस दिशा में कोई विशेष उन्नति नहीं हुई।

छठी पंचवर्षीय योजना पर्यटन विकास के क्षेत्र में महत्वपूर्ण रही। इसमें योजना बनाने की अपेक्षा कार्यप्रणाली पर अधिक ध्यान दिया गया तथा विदेशी मुद्रा अर्जित करने के अतिरिक्त वृहत् आर्थिक विकास के पहलुओं पर ध्यान दिया गया।

7वीं पंचवर्षीय योजना में पर्यटन को उद्योग का दर्जा दिया गया साथ ही निजी क्षेत्रों से मिलने वाली सहायता को भी प्रोत्साहित किया गया तथा गैर-परम्परागत क्षेत्रों जैसे ट्रेकिंग एवं शीतखेलों के विकास पर बल दिया गया।

आठवीं पंचवर्षीय योजना में भी निजी भागीदारी बढ़ाने पर बल दिया गया तथा सभी राज्यों में पर्यटन सम्बन्धी आधारभूत सुविधाओं के विकास के लिए मास्टर प्लान बनाने पर ध्यान केंद्रित किया।

नौवीं पंचवर्षीय योजना में भी जोर पर्यटन को बढ़ावा देने के लिए आधारभूत सुविधाओं के विकास पर बल देने के साथ पृथक्-पृथक् राज्यों की पर्यटन नीति में पारस्परिक सामंजस्य का उद्देश्य सम्मिलित किया गया। योजना आयोग ने इस पंचवर्षीय योजना के समापन वर्ष 31 मार्च, 2002 तक पर्यटन विकास हेतु बजट में लगभग 511 करोड़ रुपये के व्यय का प्रावधान किया गया है।

पर्यटन के बढ़ते महत्व को देखते हुए ही वर्ष 1982 में सरकार ने राष्ट्रीय पर्यटन नीति घोषित की। आर्थिक उदारीकरण को दृष्टिगत रखते हुये राष्ट्रीय पर्यटन

नीति में सन् 1991 में कुछ बदलाव किए गए। इन बदलावों के प्रमुख उद्देश्य इस प्रकार थे–

1. आर्थिक विकास में पर्यटन को आगत के रूप में उपयोग किया जाये क्योंकि इससे रोजगार उपलब्ध कराने एवं संसाधन जुटाने में सहायता मिलती है।
2. पिछड़े क्षेत्रों के सामाजिक-आर्थिक विकास में पर्यटन की भूमिका को स्वीकार किया जाये।
3. इन उद्देश्यों की प्राप्ति के लिए पर्यटन का विकास इस प्रकार किया जाये जिससे कि वातावरण एवं परितन्त्र में संतुलन बनाये रखा जा सके।
4. इनके साथ ही राष्ट्रीय एवं अन्तर्राष्ट्रीय सद्भावना में वृद्धि हेतु पर्यटन की भूमिका को स्वीकार किया जाये।

इस प्रकार पर्यटन विकास के लिए राष्ट्रीय पर्यटन नीति के इन उद्देश्यों को ध्यान में रखने के साथ-साथ कुछ सुविधाओं जैसे–आवास की व्यवस्था, हवाई अड्डों, सड़कों एवं रेल व्यवस्था में सुधार तथा पर्यटन केन्द्रों पर स्वच्छता बनाये रखने तथा अल्प बजट वाले पर्यटकों को अतिरिक्त सुख-सुविधाएं एवं रियायतें प्रदान करने की व्यवस्था की जानी चाहिए। आवासीय परेशानी से निपटने हेतु पर्यटकों को पेइंग गेस्ट के तौर पर लोगों को घरों में रखने के लिए तैयार किया जा सकता है। कुछ पर्यटकों पर यह योजना प्रारंभ भी की जा चुकी है।

यदि नई शताब्दी में भारतीय अर्थव्यवस्था का विकास और सुधार होता रहा तो घरेलू एवं विश्वस्तर पर पर्यटकों की अतिशय वृद्धि को देखते हुए इन सुविधाओं के और अधिक विकास की आवश्यकता होगी। अतः इन सुविधाओं के विकास के साथ हमें अपनी पर्यटन विकास नीति इस तरह बनानी होगी कि हमारी पारिस्थिति में किसी तरह का असंतुलन उत्पन्न न होने पाये।

भौगोलिकता से परिचय कराता पर्यटन

1. क्षेत्र
2. वातावरण–सुहावना दृश्य, बादल, तापक्रम की दिशा एवं बर्फ का गिरना इत्यादि।
3. स्थल पर पहुंचने की सुगमता एवं स्थिति
4. दृश्य

   (i) जल-नदियां, झीलें, जल-प्रपात, हिमखंड, गर्म पानी के झरने, समुद्र इत्यादि।

(ii) वनस्पति वन, रेगिस्तान घास का प्रदेश, बंजर प्रदेश, स्थान।

(iii) भूमि की बनावट : यथा पर्वत, सुरंग, समुद्री किनारा इत्यादि।

5. वन्य जीवन

(क) वन्य जीवन—पक्षी विहार, चिड़ियाघर इत्यादि।

(ख) शिकार, मछली मारना इत्यादि।

6. स्थायी आकर्षण

(i) शहर, नगर, ग्राम

(ii) ऐतिहासिक विरासत एवं खंडहर

(iii) पुरातन खंडहर

7. संस्कृति—जीवन के पहलू, परम्पराएं, लोक कथाएं, कला एवं हस्तकला।

(Elements of Tourism) तीन तत्वों (Transport, Local and Accommodation) के अतिरिक्त पर्यटन के तत्व और भी हैं जिन पर संक्षिप्त प्रकाश निम्न शीर्षकों में डाला जा रहा है।

(1) दृश्यों का आकर्षण—सुहाने मौसम की तरह यदि पर्यटन स्थान का दृश्य अच्छा होता है तो पर्यटक उसके प्रति अधिक आकर्षित होते हैं। पर्वत, झील, हिमखण्ड, झरने, वन, रेगिस्तान इत्यादि के दृश्य पर्यटक आकर्षित करने में अपना खास महत्व रखते हैं। आस्ट्रेलिया और स्विट्जरलैंड में आल्पस पर्वत की उत्तरी शृंखला इटली में दक्षिणी शृंखला, भारत एवं नेपाल में हिमालय पर्वतों की शृंखला पर्यटकों को प्रथम दृष्टि में ही आकर्षित करती हैं। महान् प्राकृतिक आश्चर्य जैसे—अमेरिका में विशाल सुंरग, उत्तरी आयरलैंड की विशाल ऊंची सड़कें, नाइजीरिया के झरने, आइसलैंड के गर्म जल स्रोत, आल्पस पर्वत के हिमखंड, अफ्रीका के जंगल, विशाल नदियां, झीलें एवं रेगिस्तान पर्यटकों को आकर्षित करने और पर्यटन उद्योग को विकसित करने में अपना विशिष्ट महत्व रखते हैं।

(2) सुविधाएं—पर्यटक स्थलों पर सुविधाओं का होना अति आवश्यक है। समुद्र के किनारे तैरने, नाव चलाने, नाव में घूमने तथा दूसरी सुविधाएं जैसे—नाचने, मनोरंजन व मन बहलाने की होनी आवश्यक हैं। सुविधा दो प्रकार की हो सकती है—एक प्राकृतिक यथा—समुद्र का किनारा, समुद्र में नहाना, मछली पकड़ने की सम्भावनाएं, चढ़ाई के अवसर, दृश्यों का अवलोकन इत्यादि। दूसरी मानव द्वारा निर्मित यथा—विभिन्न प्रकार के मनोरंजन के साधन और सुविधाएं, ताड़ एवं नारियल के वृक्षों के नीचे शरण स्थल तथा नहाने की सुन्दर व्यवस्था आदि पर्यटकों को अपनी तरफ सहज ही आकर्षित कर लेते हैं।

(3) सुहाना मौसम—जिस पर्यटन स्थल का मौसम सुहाना होता है वहां पर्यटक अधिक आकर्षित होते हैं। अवकाश के लिए अच्छा मौसम तथा एक महत्वपूर्ण स्थान का होना जरूरी है, जहां पर्यटक अपने खट्टे-मीठे अनुभव सहेजता है। करोड़ों पर्यटक अपने देशों से समुद्र के किनारे सुहावने मौसम एवं सूर्य की किरणों का आनन्द लेने आते हैं। समुद्र के किनारे की सूर्य की किरणें एवं शीतल मंद समीर पर्यटकों को एक दीर्घ अवधि से आकर्षित करती चली आ रही हैं।

वास्तविक रूप से जो पर्यटन स्थान समुद्र के किनारे पर स्थिति हैं उनके विकास में मौसम और सूर्य की किरणों का महत्वपूर्ण योगदान होता है। स्पेन, इटली और ग्रीस जैसे यूरोपीय देशों में सुन्दर (Beach Resorts) का विकास किया गया है। इटली ने अपने दक्षिणी भाग केलेब्रीया, सिसली एवं सरडिनिया आदि नगरों में भूमध्यसागरीय किनारों पर चमकती हुई सूर्य की किरणों का आनंद लेने के लिए (Resorts) बनाए हैं। संयुक्त राज्य अमेरिका में मैक्सिकन (Resort), पेसिफिक एवं कैलिफोर्निया, फ्लोरिडा एवं हवाई द्वीप का उदाहरण लिया जा सकता है, जहां पर्यटकों को आकर्षित करने के लिए रिजार्ट्स (Resorts) बनाए गए हैं। भारत, श्रीलंका, थाइलैण्ड, इन्डोनेशिया, आस्ट्रेलिया एवं दूसरे राष्ट्रों में भी सुहावने मौसम के लिए कुछ कार्य किये गये हैं। वास्तव में ये देश अब अच्छे मौसम से सम्बन्धित पर्यटन स्थल विकसित करने में अपना विशिष्ट स्थान रखते हैं।

शरद ऋतु का वातावरण, गर्म हवाएं और सूर्य की किरणें, पर्यटकों को आकर्षित करने में प्रमुख भूमिका निभाती हैं। बहुत से क्षेत्र शरद ऋतु में पर्यटकों को आकर्षित करते हैं। इसका उदाहरण भारत में शिमला का है।

(4) ऐतिहासिक एवं सांस्कृतिक महत्व—पर्यटकों को आकर्षित करने में ऐतिहासिक व सांस्कृतिक पर्यटन स्थानों का भी महत्वपूर्ण योगदान होता है। हमारे देश में आगरा अपने ताजमहल, इटली में पीसा अपनी झुकी मीनार की वजह से ही पर्यटकों को आकर्षित करते हैं। सहस्त्रों अमेरिकन-कनेडियन यूरोप की पुरानी ऐतिहासिक धरोहर एवं मूल देश से सम्बद्ध भावुकताओं के कारण यूरोप आते हैं। इंग्लैंड आने वाला कोई भी विदेशी पर्यटक लंदन को राजधानी एवं राष्ट्र के बड़े शहर होने की वजह से ही नहीं बल्कि उसके ऐतिहासिक महत्व, परम्पराएं और बहुत से सांस्कृतिक आकर्षणों के कारण देखने आते हैं। इसी तरह बहुत से राष्ट्र अपने पर्यटन उद्योग को विकसित करने के लिए अपनी ऐतिहासिक धरोहरों का प्रयोग कर रहे हैं।

(5) सुगमता—सुगमता, जैसा कि नाम से स्पष्ट है कि पर्यटन स्थल तक सैलानी का सरलता से पहुंचना पर्यटकों को आकर्षित करने में अपना विशिष्ट महत्व

रखता है। जिस स्थान पर यातायात की सुविधा नहीं है उसे पर्यटक विशेष महत्व नहीं देते हैं। जिस स्थान तक पहुंचने के लिए यातायात का मार्ग सुलभ है तथा यह सही मार्ग पर स्थित है तो यह स्थान पर्यटक को बहुत बड़ी संख्या में अपनी ओर आकर्षित कर सकता है। दूरी का कारण भी पर्यटकों की अभिलाषा को प्रभावित करता है। अधिक दूरी की अपेक्षा पास का पर्यटक स्थल यात्रियों को अधिक प्रभावित करता है। विश्व के पर्यटक बाजार जैसे–संयुक्त राज्य अमेरिका, यूरोप, कनाडा, जापान व आस्ट्रेलिया से दूर पर्यटन स्थल होने पर सैलानियों की संख्या प्रभावित होती है। इन राष्ट्रों से भारत में आने पर काफी राशि व्यय करनी पड़ेगी। सुगमता के ही कारण यूरोप और उत्तरी अमेरिका से लगातार पर्यटक विश्व के देशों में आते हैं तथा सर्वाधिक पर्यटक इन राष्ट्रों से देखने आते हैं। एक ही क्षेत्र के देशों के पर्यटकों की संख्या भी इन देशों की ज्यादा है। यूरोप एवं उतरी अमेरिका में आने वाले लोगों में 80 प्रतिशत पर्यटक हैं। अतः पर्यटक स्थल पर पहुंचने की सुगमता पर्यटन की वृद्धि और विकास में अपना विशेष महत्व रखती है।

(6) आवास–सभी प्रकार के पर्यटन में आवास की प्रमुख आवश्यकता होती है। दूर-दराज के क्षेत्रों में जाने वाले पर्यटकों हेतु आवास की सभी सुविधाओं का होना अति आवश्यक है। आवास में सिर्फ खाने की जरूरत तथा निवास की गणना की जाती है। पिछले 25 सालों से इसमें कुछ सुधार किया गया है। कुछ बोर्डिंग हाउस एवं छोटे निजी होटलों के उपयोग में गिरावट आई है। बड़े होटल प्रसिद्ध पर्यटक स्थलों पर ही बनाए गए हैं। हाल के वर्षों में इन आवास स्थलों में सुधार में साथ कुछ अनौपचारिक सुविधाएं भी उपलब्ध कराई जाने लगी हैं। नए प्रकार के आवास विशेषतः (holiday villages, apartment house) तथा पर्यटक झोपड़ी (tourist cottage) आदि बहुत लोकप्रिय हो रहे हैं। पर्यटक के लिए आवास स्वयं में एक महत्वपूर्ण आकर्षण के केन्द्र हैं।

वास्तविक रूप से ज्यादातर पर्यटक किसी पर्यटन स्थान पर इसलिये जाना चाहते हैं, क्योंकि वहां सभी सुविधाओं से युक्त अति उत्तम प्रकार के होटल हैं। कुछ राष्ट्र यथा–स्विट्जरलैंड, हॉलैंड, आस्ट्रेलिया एवं नीदरलैंड अपने होटलों में उत्तम सुविधाएं, आरामदायक एवं स्वच्छता प्रदान करने के कारण ही अपना स्थान बनाए हुए हैं। दूसरे राष्ट्रों में भी बनाए गए रिजार्ट्स (Resorts) में खाने-पीने एवं स्वच्छता की अच्छी व्यवस्था है।

उपर्युक्त तथ्यों के अतिरिक्त कुछ दूसरे उत्तम घटक भी हैं जो पर्यटकों को प्रभावित करते हैं। इनमें अतिथि सत्कार मुख्य है। आने वाले देशवासियों से दोस्ती

एवं प्रेम का व्यवहार रखने पर ही दूसरे पर्यटकों को आकर्षित किया जा सकता है। ये संतुष्ट पर्यटक ही अपने देश जाकर स्थिति का वर्णन करेंगे। फ्रेंच सरकार ने अपने देशवासियों को पर्यटकों से होने वाले फायदे से अवगत कराने के लिए एक अभियान चलाया था। भारत सरकार ने भी एक योजना तैयार की है जिसमें पर्यटकों के स्वागत का महत्व बताया और विविध तरह के नारे जैसे—सफाई रखिए, दर्शकों का स्वागत करिये। इस प्रकार पर्यटकों से मित्रतापूर्ण व्यवहार रखने से पर्यटन को बढ़ावा मिलता है।

जलवायु-पर्यावरण

भारत की कृषि जलवायु पर निर्भर है तथा उसी के अनुसार भिन्न-भिन्न प्रकार की फसलें उगाई जाती हैं जिनको काटने का समय भी अलग-अलग होता है। यह सब कुछ जलवायु और मौसम पर निर्भर करता है। खेती के उपरांत आराम का समय आता है, जिसे हम भारतीय उल्लास और आनन्द में बदल देते हैं। इस प्रकार भारतीय सामाजिक जीवन में कई पर्वों और मेलों का उदय हुआ। तमिलनाडु का पोंगल, शान्ति-निकेतन का पौष मेला, मेघालय का नोंमकेम तथा केरल का ओणम पर्व ऐसे ही कुछ पर्व हैं। इसके अलावा राष्ट्र के विविध भागों में ऐतिहासिक और धार्मिक महत्व के कई पर्व भी मनाए जाते हैं। वस्तुतः भारत मेलों तथा पर्वों का राष्ट्र है जहां श्रद्धा, उत्साह और आनन्द के साथ प्रतिवर्ष सैकड़ों पर्व मनाए जाते हैं। यह पर्व अपने आप में अद्वितीय हैं। उनमें रंग है, रूप है और सौन्दर्य का अपार भण्डार है।

पिछले कुछ सालों में भारतीय मेले भी पर्यटकों का प्रमुख आकर्षण रहे हैं। अतः इनमें से कई पर्व नियोजित तरीके से मनाए जाते हैं एवं पर्यटकों को आकृष्ट करने के लिए प्रचार भी किया जाता है। इनमें से कुछ पर्व पूरे अथवा देश के ज्यादातर हिस्से में और कुछ देश के खास हिस्सों में मनाए जाते हैं। उदाहरणार्थ होली देश के विभिन्न भागों में मनाई जाती है पर मथुरा और वृंदावन की होली का अपना रंग है और पर्यटक भी इसी होली की ओर आकृष्ट होते हैं। इसी तरह मैसूर या कुल्लू का दशहरा विश्वविख्यात है।

भारत में पर्यटन काल तथा स्थान

भारत में कुछ पर्यटन स्थल ऐसे हैं जिसमें पर्यटक एक निश्चित मौसम में आते हैं। प्रायः अनेक लोग यह कहते रहते हैं कि यह मौसम पर्यटन का मौसम है। इस

बात का क्या मतलब है? वास्तविक रूप से सम्पूर्ण संसार में पर्यटक बिखरे पड़े हैं और तदनुसार उनकी गतिविधियां और सेवाएं चलती रहती हैं। आपने किसी होटल के मालिक को यह कहते सुना होगा कि इस बार का मौसम खराब था। यहां वह जलवायु के बारे में नहीं, अपितु अपने व्यवसाय के संदर्भ में बात कर रहा है कि एक निर्धारित समय में उसे उम्मीद के मुताबिक फायदा नहीं हुआ। इस इकाई में हमने पर्यटन के विद्यार्थियों को यह बताने की कोशिश की है कि पर्यटन में मौसम का क्या अभिप्राय है? प्रारंभ में जलवायु और मौसम सम्बन्धी परिस्थितियों का जिक्र करते हुए बताया गया है कि इससे लोगों का एक स्थान से दूसरी स्थान पर जाना किस सीमा तक प्रभावित होता है। इसके उपरांत इसे विभिन्न प्रकार की मौसमीयता, इसके माप और स्थल पर पड़ने वाले असर की चर्चा की गई है। पर्यटन स्थल की कुछ विशेषताओं का भी उल्लेख किया गया है। इस प्रकार की सूचनाओं से युक्त पर्यटन पेशेवर अपने ग्राहकों की शंकाओं का निवारण कर सकते हैं। आम जागरूकता कुछ करने में सहायता कर सकते हैं।

भारत की जलवायुविक स्थितियां तथा उसका परिचय

उत्तरी अक्षांश में मौसम के अनुसार एक साल को प्रमुख रूप से चार खण्डों में विभाजित किया जा सकता है–

1. बसन्त (21 मार्च से 20 जून)।
2. ग्रीष्म (21 जून से 22 सितम्बर)।
3. शरद (23 सितम्बर से 21 दिसम्बर)।
4. शीत (22 दिसम्बर से 20 मार्च)।

मौसम में बदलाव जलवायु परिवर्तन के कारण होता है जिसका प्रभाव मानवीय जीवन के हर क्षेत्र पर पड़ता है। मनुष्य भी जलवायु में होने वाले परिवर्तनों से अछूता नहीं रह पाता। किसी क्षेत्र में पिछले सालों से जो मौसम बना होता है, वही उस क्षेत्र की जलवायु होती है। मौसम के कुछ प्रमुख जलवायु निम्नलिखित हैं–

1. हवा का तापमान।
2. बरसात का प्रकार और आकार।
3. वायुमंडलीय दबाव।
4. घटा का प्रकार और आकार।
5. हवा की गति और दिशा।
6. हवा की आर्द्रता।

उपर्युक्त तथ्यों का मापन करने से एक जगह पर किसी भी समय में मौसम को निर्धारित किया जा सकता है। कालखंड एवं स्थान के परिवर्तन से मौसम के निर्धारण में बदलाव होता रहता है। परन्तु सामान्य रूप में एक स्थान पर वर्षों तक समान मौसम बना रहता है। पिछले वर्षों के औसत मौसम को ही उस स्थल विशेष की जलवायु के रूप में माना जाता है।

पृथ्वी अपनी धुरी पर घूमने के साथ-साथ सूर्य का भी चक्कर लगाती है और 23.27 डिग्री पर झुकी है। भूमध्य रेखा का झुकाव भी इस ओर है। इन सब कारणों के अतिरिक्त मौसम अन्य कारकों से भी प्रभावित होता है।

पर्यटन पर जलवायु परिवर्तन के प्रभाव

भारत के अधिकांश पर्यटन स्थल जलवायु पर निर्भर करते हैं। गर्मियों के मौसम में लोग समुद्रतटी इलाकों की तरफ आते-जाते हैं एवं जाड़े के दिनों में गर्म स्थलों की ओर रवाना होते हैं। इसी वजह से भारत में हिल स्टेशनों का विकास हुआ। गर्मी के दिनों में जब भारत के मैदानी क्षेत्रों में तापमान बढ़ जाता है तब पहाड़ियों पर ठंड होती है एवं आनन्ददायक मौसम होता है। ब्रिटिश शासन के दौरान शासकों ने अपने आराम के लिए हिल स्टेशनों का निर्माण और विकास किया। गर्मी से बचने के लिए वे हिल स्टेशनों की शरण लेते थे। भारत के मुख्य हिल स्टेशनों में श्रीनगर, गुलमर्ग, पहलगाम, शिमला, कुल्लू, मनाली, मसूरी, शिलांग, मांउट आबू, कोडइकनाल और ऊटी शामिल हैं। गर्मी के मौसम में लाखों अंतर्राष्ट्रीय एवं घरेलू पर्यटक इन पहाड़ियों एवं झीलों पर जाते हैं। वस्तुतः यहां गर्मी से राहत मिलती है इसमें से कुछ स्थान जाड़े के दिनों में शीतकालीन खेल खेलने के लिए भी प्रसिद्ध हैं।

भारत में 5,700 कि. मी. लम्बा समुद्रतट है जो गर्मी के दिनों में तुलनात्मक दृष्टि से ठंडा और जाड़े के दिनों में अपेक्षाकृत गर्म रहता है। गोवा, कोवलम, मसल्लपुरम, पुरी, कोणार्क, गोपालपुर, समुद्रतट आदि बीच अंतर्राष्ट्रीय स्तर के हैं और इन बीचों के विकास से सामुद्रिक परिभ्रमण को अत्यधिक प्रोत्साहन मिला है। भारत की जलवायु में विभिन्नता होने के कारण यहां समुद्रतटीय और पहाड़ पर्यटन के विकास में काफी मददगार हैं।

सम्पूर्ण विश्व में पर्यटकों के भ्रमण पर जलवायु का प्रमुख प्रभाव पड़ता है। प्रत्येक सैलानी उचित मौसम में ही घूमना चाहते हैं। मसलन पर्यटन की दृष्टि से गर्मी का मौसम पहाड़ी क्षेत्रों के लिए सर्वश्रेष्ठ मौसम होता है। ग्रीष्म ऋतु में लाखों पर्यटक यूरोप में भूमध्यसागर पर बने तटों एवं संयुक्त राज्य के पश्चिमी तटों की

तरफ भागते हैं। शीतकाल में पर्यटक गर्म प्रदेशों की तरफ जाते हैं। इनमें भारत भी शामिल है। भारत में गर्मी भी ज्यादा पड़ती है और ठंड भी जबकि केरल में ऊष्णकटिबंधीय समुद्री जलवायु है, जहां गर्मी होती है। जलवायु में इस विविधता के कारण लाखों लोग पृथक्-पृथक् मौसम में एक जगह से दूसरी जगह पर आते-जाते हैं। साभान्यतया गर्मी के दिनों में उत्तर से दक्षिण की ओर तथा उत्तर-पूर्व मानसून में दक्षिण से उत्तर की ओर पर्यटकों का प्रस्थान होता है।

पर्यटकों को आकर्षित करता भारतीय मौसम

भारत में नये साल की शुरुआत विभिन्न उत्सवों से होती है। नए साल के पहले दिन को पूरे राष्ट्र में अत्यन्त धूमधाम एवं उत्साह से मनाया जाता है। इसी महीने में तमिलनाडु, कर्नाटक और आंध्र प्रदेश में सफल कटाई का त्योहार पोंगल मनाया जाता है। इस दिन नये धान का भात बनाया जाता है और इसे सूर्य भगवान को चढ़ाया जाता है। शान्तिनिकेतन में पशुओं को सजाकर पशु मेले में ले जाया जाता है। उत्तर प्रदेश में इस दिन मकर संक्राति मनाई जाती है। इस दिन लाखों लोग इलाहाबाद में गंगा स्नान करते हैं।

उत्तराखंड प्रकृति का अद्वितीय उपहार है, उस पर मेले, पर्व त्योहारों का आयोजन इसे स्वर्ग बना देता है। जहां तक पर्वों की बात है उत्तराखंड में सभी त्योहार विधि-विधानपूर्वक मनाए जाते हैं, जिन त्योहारों में व्रत रखा जाता है वे भी उतनी ही निष्ठा से मनाए जाते हैं जितनी निष्ठा से देश के अन्य भागों में मनाए जाते हैं। अतः यह कहना उचित होगा कि उत्तराखण्ड के त्योहार और मेले से यहां की प्राचीन जीवन शैली की आत्मा की झलक दिखाई देती है। ये मेले हमारी सभ्यता, संस्कृति तथा धार्मिक भावनाओं के प्रेरणास्रोत हैं। साथ ही इनसे हमारा समाज एकसूत्र में बंधता है और भाईचारे को बढ़ावा मिलता है। 26 जनवरी के दिन हमारे देश में गणतंत्र दिवस मनाया जाता है। यह एक राष्ट्रीय पर्व है और सन् 1950 में भारत के गणतंत्रात्मक देश घोषित होने के उपलक्ष्य में यह त्योहार मनाया जाता है। इस दिन प्रत्येक राज्य की राजधानियों में राष्ट्रीय ध्वज फहराया जाता है और परेड आयोजित की जाती है।

गुजरात के अहमदाबाद की पतंगबाजी बहुत ही ज्यादा प्रसिद्ध है। इस दिन पूरा आकाश रंग-बिरंगे पतंगों से भर जाता है। इस त्योहार के दौरान पतंगबाजी होती है, सांस्कृतिक कार्यक्रम आयोजित किए जाते हैं, भू-जनोत्सव होता है एवं खूबसूरत कलाकृतियों की प्रदर्शनी लगाई जाती है। पर्व, त्योहार और मेले, किसी भी समाज

के लोकजीवन के अभिन्न भाग होते हैं। इनमें लोक-कलाओं का सहज प्रदर्शन दिखाई देता है। बहुत से लोकनृत्यों का आयोजन विशेष पर्वों तथा मेलों में होता है। उत्तराखंड मेलों, पर्वों और त्योहारों की धरती है। यहां के समाज को त्योहार प्रिय हैं। वे बड़ी आतुरता से ऐसे आयोजनों की प्रतीक्षा करते हैं। जीवन की समस्याओं और चुनौतियों को वे हँसकर झेलते हैं।

26 जनवरी को दिल्ली में व्यापक समारोह का आयोजन होता है जिसमें सभी प्रदेशों के प्रतिनिधि अपनी-अपनी कला को प्रदर्शित करते हैं। जनवरी के महीने में हाथियों का जुलूस पर्यटकों के आकर्षण का मुख्य केन्द्र है। इस दिन त्रिचुर में 100 से भी अधिक हाथियों को सजाकर उनकी दौड़ आयोजित की जाती है। जनवरी के अन्त एवं फरवरी के शुरू में पश्चिम बंगाल में गंगा के मुहाने पर गंगा सागर नामक मेला लगता है। जहां लाखों तीर्थयात्री स्नान करने पहुंचते हैं। फरवरी का महीना भी पर्वों तथा त्योहारों के रंग से रंगा हुआ है।

प्रत्येक वर्ष फरवरी-मार्च के महीने में थार मरुस्थल में स्थित नागौर में एक अद्वितीय पशु मेला लगता है। इस मेले में हजारों घोड़े, ऊंट एवं अन्य जानवर लाये और खरीदे-बेचे जाते हैं। अतिथियों के मनोरंजन के लिए इस मेले में लोकनृत्य, संगीत और सांस्कृतिक कार्यक्रमों का आयोजन किया जाता है। इसी महीने में जैसलमेर का मरुभूमि मेला और कोटा का हादोती उत्सव मेला लगता है। जहां गांव के माहौल में देश से आए हस्तशिल्पों की प्रदर्शनी लगाई जाती है। यहां ग्रामीण भारत की झलक दिखाई देती है और इसी से सम्बद्ध सांस्कृतिक कार्यक्रम आयोजित किए जाते हैं तथा विभिन्न प्रकार के खाद्य पदार्थों की दुकानें लगाई जाती हैं। यहां सूरजकुण्ड में ग्रामीण भारत की छटा दिखाई पड़ती है और वहीं खजुराहो के मंदिरों में सप्ताह भर का नृत्य समारोह आयोजित किया जाता है। यहां देश भर के शास्त्रीय नृत्य का आयोजन होता है। इस उत्सव के समय शास्त्रीय नृत्य एवं मंदिर के गहरे सम्बन्ध को दिखाया जाता है।

भारत के दक्षिणी क्षेत्र मदुरई में हर साल देवी मीनाक्षी और उनके पति सुन्दरेश्वर का ब्याह रचाया जाता है जो चित्रई त्योहार के नाम से प्रसिद्ध है। इस पर्व के अन्त में रथों पर देवी-देवताओं की एक शोभायात्रा निकाली जाती है। इस तरह केरल के त्रिचूर में पुरम नाम से एक ऐसा ही मंदिर उत्सव मनाया जाता है और हाथियों पर देवी-देवताओं की शोभा यात्रा निकाली जाती है और मंदिर में ढ़ोल बजाए जाते हैं।

भारत के दक्षिण प्रदेशों के अलावा पूरे देश में होली के पर्व को अत्यन्त ही धूमधाम से मनाया जाता है। प्रत्येक उम्र के लोग रंग के इस त्योहार में भाग

लेते हैं और एक-दूसरे पर रंग, अबीर उड़ेलते हैं। इस पर्व के ठीक पूर्व बसंत पंचमी का त्योहार आता है जिस दिन विद्या की देवी सरस्वती की पूजा अर्चना की जाती है। मार्च का महीना भी कम रंगीला नहीं है। शिवरात्रि के दिन परमपिता शिव की पूजा की जाती है। इस दिन मंदिर में भक्तों की भारी भीड़ लग जाती है एवं पूरे दिन और रात लोग शिवरात्रि का त्योहार मनाते हैं। अप्रैल के महीने में बैशाखी का त्योहार आता है। पूरे भारत में हिन्दुओं के नववर्ष का आरम्भ इसी दिन से माना जाता है और इसी उपलक्ष्य में यह पर्व मनाया जाता है। पंजाब में यह पर्व पूरे उत्साह के साथ मनाया जाता है और हर ग्राम में लोग भांगड़ा नृत्य करते हैं। जयपुर में महीने भर का संतोत्सव मनाया जाता है। इसे गंभीर पर्व के नाम से जाना जाता है। उत्सव के अन्तिम दिन गौरी देवी की पूजा अर्चना की जाती है एवं रथ यात्रा निकाली जाती है। इसमें नृत्य, संगीत के साथ-साथ सजे हुए ऊटों, घोड़ों तथा हथियारों पर गौरी देवी की मूर्ति रखकर शोभायात्रा निकाली जाती है।

मई के महीने में सम्पूर्ण भारत में मुसलमान ईद-उल-फितर का त्योहार मनाते हैं। रमजान, उपवास के महीने के अन्तिम दिन नमाज अदा की जाती है, परिवार के लोग एक-दूसरे से मिलते हैं और तरह-तरह के पकवान बनाये जाते हैं। सूफी संत ख्वाजा मोइनुदीन चिश्ती के ईश्वर से प्रतीकात्मक मिलन के उपलक्ष्य में राजस्थान में अजमेर की दरगाह पर उसे मनाया जाता है।

इस महीने में गंगटोक में आयोजित किया जाने वाला अंतर्राष्ट्रीय फूल मेला अपने आप में अद्भुत है। इसमें 500 से अधिक फूलों की किस्मों की प्रदर्शनी लगाई जाती है। जून के महीने में भारत के मैदानी क्षेत्रों में जब गर्म हवा चल रही होती है उस समय लेह में हेमिस पर्व महोत्सव मनाया जा रहा होता है। इस दिन हेमिस मठ के आंगन में मुखौटा नाच पेश किया जाता है। जून-जुलाई में पुरी में आयोजित होने वाली रथयात्रा एक प्रमुख मंदिर उत्सव है। इस दिन 14 मीटर ऊँचे प्राचीन एवं विशाल रथ पर भगवान की शोभा यात्रा निकाली जाती है।

इसके अलावा इस शोभा यात्रा में और भी अनेक रथों को सम्मिलित किया जाता है। इन सभी रथों को भक्तगण खीचते हैं और भगवान जगन्नाथ के प्रति अपनी श्रद्धा एवं लगाव प्रकट करते हैं। जुलाई के महीने में पूरे भारत में मुसलमान ईद उल जुहा अथवा बकरीद का पर्व मनाते हैं। इसी दिन अल्लाह के आदेश का पालन करते हुए पैगम्बर इब्राहिम नें अपने पुत्र की बलि चढ़ाई थी। इसी की याद में यह पर्व मनाया जाता है।

सम्पूर्ण भारत में हर साल 15 अगस्त के दिन को स्वतंत्रता दिवस के रूप में मनाया जाता है। 15 अगस्त, 1947 को भारत अंग्रेजों की गुलामी से आजाद हुआ था, इसकी स्मृति में हम इसे याद करते हैं। ध्वजारोहण, परेड, सांस्कृतिक कार्यक्रम इस दिन के मुख्य आकर्षण होते हैं। इस दिन प्रधानमंत्री राष्ट्रीय ध्वज तिरंगे को फहराते हैं तथा लालकिले की प्राचीर से राष्ट्र को संबोधित करते हैं।

अगस्त और सितम्बर के महीने में प्रभु श्रीकृष्ण के जन्म की खुशियां होती हैं जिसे पूरे देश में जन्माष्टमी पर्व के रूप में मनाया जाता है। इस पर्व में कृष्ण लीलाएं होती हैं एवं मंदिरों में पूजा अर्चना की जाती है।

मणिपुर में 'रास-लीला' नृत्य का आयोजन किया जाता है। मथुरा और वृंदावन, जहां कृष्ण ने अपना बालपन और जवानी व्यतीत की थी, में यह उत्सव अत्यधिक धूमधाम के साथ मनाया जाता है। केरल में फसल कटाई का त्योहार ओणम उत्सव एक उमंग से भरा पर्व है। इस छोटे से राज्य के सभी लोग आयु, लिंग, जाति और धर्म का भेदभाव भूलकर प्रिय राजा महाबलि की वापसी का स्वागत करते हैं तथा इस उपलक्ष्य में अपने घरों को साफ-सुथरा कर खूब सजाते हैं।

एल्लेच्पे और अरनमुला में आयोजित होने वाली नेहरू नौका दौड़ इस उत्सव का प्रमुख आकर्षण होती है। महाराष्ट्र, तमिलनाडु, आंध्र प्रदेश एवं कर्नाटक में सितम्बर माह में गणेश चतुर्थी मनाई जाती है। सात दिनों तक गणेश भगवान की पूजा अर्चना की जाती है और फिर इसकी शोभायात्रा निकालकर समुद्र, नदी अथवा झील में इन्हें विसर्जित किया जाता है। शोभायात्रा के दौरान गीत गाते एवं झूमते भक्तगण नजर आते हैं। 2 अक्तूबर को राष्ट्रपिता महात्मा गांधी का जन्म दिन है। इसे गांधी जयंती के रूप में मनाया जाता है। इस दिन पूरे राष्ट्र में प्रार्थना सभाएं आयोजित की जाती हैं। दीपावली भारत का एक प्रमुख त्योहार है। इस दिन सम्पूर्ण राष्ट्र दीपों से जगमगा उठता है। लोग एक-दूसरे को मिठाइयां खिलाते हैं। क्रिसमस साल का आखिरी पर्व होता है जिसे 25 दिसम्बर को मनाया जाता है। इस दिन गिरजाघरों में पूजा-अर्चना करने के साथ-साथ उपहार वितरित किए जाते हैं।

संचार माध्यमों का पर्यटन विकास में महत्व

संचार का अर्थ है आपस में बांटना। संचार को अंग्रेजी भाषा में कम्यूनिकेशन कहा जाता है। इस शब्द की उत्पत्ति लैटिन भाषा के "कम्यूनिको" शब्द से हुई है।

जब हम किसी व्यक्ति को जानकारी प्रदान करते हैं या किसी से जानकारी ग्रहण करते हैं तो हम उस व्यक्ति के साथ सूचना, विचारों और भावों का आदान-प्रदान करते हैं। इस प्रकार के ही विचारों, सूचनाओं इत्यादि के आदान-प्रदान की क्रिया को संचार क्रिया की संज्ञा दी गई है।

संचार एक ऐसी प्रक्रिया है जिसके द्वारा दो से अधिक व्यक्ति अपने कथ्यों, विचारों, अनुभवों का इस प्रकार आदान-प्रदान करते हैं कि प्रत्येक व्यक्ति सन्देश के आशय, अभिरुचि तथा उपयोग की सामान्य जानकारी प्राप्त कर लेता है। संचार का मतलब लोगों के पास इस प्रकार ज्ञान पहुँचाना है कि वे कुछ उपयोगी परिणामों को प्राप्त करने के लिए उस ज्ञान के अनुसार कार्य करें। संचार का आशय एक व्यक्ति से दूसरे व्यक्ति अथवा समुदाय तक सूचना, विचार, मनोवृत्ति या मनोभावों का संचारण है।

संचार श्रोताओं एवं ग्राहकों के बीच एक सम्बन्ध स्थापित करने की प्रक्रिया है जिसके आधार पर चिह्न, संकेत, संदेश ऐसे तात्पर्य देते हैं जिन्हें जनसम्पर्क या अन्तर्वैयक्तिक के जरिये से बांटा जाता है। "स्पीड" (SPEED) संचार का महत्वपूर्ण लक्षण है। यहाँ S से अभिप्राय (Simple) अर्थात् सरल, P से Practical अर्थात् व्यावहारिक, E से Economical सस्ता अथवा कम खर्च वाला, E से Effective प्रभावशील और D से Duplicate अर्थात् दोहराने योग्य है। उपर्युक्त विवेचना संचार की व्यावहारिक परिभाषा प्रदर्शित करती है। यह वह परिभाषा है जो व्यावहारिक संचार माध्यम की विशेषता को प्रस्तुत करती है।

संचार की परिभाषा

संचार की मान्यताप्राप्त प्रमुख परिभाषाओं का विवरण निम्न प्रकार है–

जे पाल लिंगन के अनुसार–संचार एक ऐसी पद्धति है जिसमें दो या दो से अधिक व्यक्ति आपस में विचारों व तथ्यों का आदान-प्रदान करते हैं।

टी. एस. मैथ्यूस के मतानुसार–संचार एक जैसी जटिल एवं सामान्य प्रक्रिया है जिसे हम सामान्य शब्दों में व्यक्त नहीं कर सकते।

पैटर लिटिल ने अपनी किताब कम्यूनिकेशन इन बिजनिस में संचार को निम्नांकित तरीके से व्याख्यायित किया है–संचार वह प्रक्रिया है जिसमें किसी सूचना को एक व्यक्ति या संगठन के मध्य इस प्रकार पहुंचाया जाता है कि हरेक व्यक्ति उस विषय के आशय को तथा उसके प्रयोग को अच्छी तरह समझता है।

कोलेमेन एवं मार्श के मतानुसार—वे तमाम शैक्षिक प्रयोगात्मक कार्यक्रम जो पर्यटन एवं परिवहन प्रबन्धन के क्षेत्र में चलाये जायें, संचार कहे जाते हैं।

उपरोक्त परिभाषाओं के माध्यम से संचार और उसके साधन संकीर्ण दायरे में बंध जाते हैं। इस दायरे की सीमाएं पर्यटन एवं परिवहन तक सिमट कर रह जाती हैं। वास्तव में संचार अथाह सागर की भांति विस्तृत है एवं सागर में विद्यमान अनन्त जीवों की तरह संचार के तार इतने विस्तृत हैं कि सम्पूर्ण विश्व को समेट कर रख दिया है। संचार का उपयोग मात्र पर्यटन के क्षेत्र में ही नहीं वरन्, चिकित्सा, वैज्ञानिक खोज एवं आविष्कारों इत्यादि में भी किया जाता है।

वैरन वीवर के मतानुसार—वे समस्त विधियाँ जिनके द्वारा एक व्यक्ति का मस्तिष्क दूसरे व्यक्तियों के दिमागों को प्रभावित करता है, संचार कहलाते हैं।

संचार के प्रमुख कारक

संचार के कारक विभिन्न हैं उनमें से प्रमुख कारकों का वर्णन निम्नलिखित है—

संप्रेषण

शिक्षा साधन विहीन प्रणाली नहीं है। इसके अन्तर्गत विषय अथवा सामग्री की जरूरत होती है। प्रसार शिक्षा व संचार प्रक्रिया में श्रोताओं को देने के लिए मुख्यतया विषय सामग्री पर्यटकों में मनोवैज्ञानिक कथ्यों का विवरण, व्यक्तियों और समूहों द्वारा किए गए निर्णयों का विवरण इत्यादि अन्य संदेश हो जाते हैं।

संदेश या विषय सामग्री वह सूचना है जिसको सूचना वाहक अपने श्रोताओं को उस ढंग से पहुंचाता है कि वे उसे समझें, अपनायें तथा उस पर अमल करें।

सूचना वाहक

सूचना वाहक के माध्यम से संचार प्रक्रिया का शुभारम्भ होता है, जिसे संचार प्रक्रिया की प्रथम अवस्था माना जाता है। सूचना वाहक बोलने वाला, लिखने वाला, कोई शिक्षक, प्रबन्धक, टैक्सी ड्राइवर, छोटा फुटवियर विक्रेता, बड़ा होटल मालिक या होटल के प्रबंधक से लेकर छोटा कर्मचारी भी हो सकता है। सूचना उपयोगी विषय के सम्बन्ध में विचार प्रकट कर सूचना ग्राहक को आकर्षित करता है। सूचना वाहक उचित माध्यमों जैसे—रेडियो, टी. वी., फोटोग्राफ आदि के द्वारा सूचना का संचार करता है।

संचार प्रक्रिया को प्रभावित करने वाले कारकों में सूचना वाहक प्रमुख भूमिका निभाता है। उसके बोलने का ढंग, उसका सूचना के विषय में ज्ञान, उसकी छवि इत्यादि समस्त बातें सूचना के संचार को प्रभावी बनाती हैं।

संचार प्रक्रिया तभी सार्थक हो सकती है जब सूचना वाहक में अग्रांकित विशेषतायें विद्यमान होंगी–

1. सूचना वाहक अपने उद्देश्यों, श्रोताओं की विषय सामग्री, उपयुक्त संचार माध्यम के विषय में उत्तम ज्ञान रखता हो।
2. संचारकर्ता को संचार प्रक्रिया के दौरान शिक्षण योजना संचार सामग्री एवं मूल्यांकन विधि की पूर्ण जानकारी होनी चाहिए।
3. संचार कार्यकर्ता में अपनी योग्यता व कमियों के निष्पक्ष विश्लेषण की सामर्थ्य होनी चाहिए।

सूचना वाहक को अग्रांकित कार्यों को करना आवश्यक होता है–

(1) संचार माध्यम का ठीक प्रकार से उपयोग करने में।

(2) संचार सामर्थ्य को कैसे बढ़ाया जाए।

(3) सूचना वाहक, शिक्षक व शिक्षण सम्बन्धी सामग्री व उसके मूल्यांकन के लिए योजना तैयार करता है।

सूचना ग्राहक

किसी भी तरह की सूचना को प्राप्त करने के इच्छुक व्यक्ति को ही सूचना ग्राहक कहते हैं। यह सूचना किसी भी प्रकार की हो सकती है। देश-विदेश में हो रहे परिवर्तन, बाजार में आ रही नई वस्तुओं के विषय में जानकारी, बाजार भाव, पर्यटन के नए आयाम, पर्यटन सम्बन्धी अन्य प्रमुख जानकारियां जो यात्रा एवं पर्यटन के लिए हितकारी हैं।

संचार के स्रोत

संचार माध्यम एक ऐसी प्रक्रिया है जिसके माध्यम से सूचना देने वाला भिन्न-भिन्न साधनों का प्रयोग कर अपनी सूचना को सूचना प्राप्त करने वाले तक भेजता है। परिवर्तित परिपेक्ष्य में संचार माध्यम एक सरल प्रपत्र, पर्चे से लेकर जटिल कम्प्यूटर इंटरनेट भी हो सकता है।

संचार के विभिन्न रूप

हर व्यवसाय में दो प्रकार की संचार व्यवस्था देखने को मिलती है—बाह्य संचार तथा आन्तरिक संचार।

आन्तरिक संचार : आन्तरिक संचार के अंतर्गत संगठन के अन्दर ही संचार होता है। आन्तरिक संचार दो प्रकार का होता है। औपचारिक या आधिकारिक एवं अनौपचारिक संचार। अनौपचारिक संचार नीचे अथवा ऊपर की ओर किया जाता है।

संचार का ऊपरी भाग

यदि प्रबन्धक किसी सूचना को श्रमिक वर्ग तक संचारित करता है तो वह निम्न वर्ग अर्थात् कर्मचारी वर्ग से सूचना ग्रहण भी करता है। इस प्रकार का संचार जहाँ संदेश निम्न वर्ग से उच्च वर्ग की ओर संचारित होता है अर्थात् अधिकारियों या विभाग प्रमुख की तरफ संचारित होता है। इस तरह की संचार प्रक्रिया ऊपर की ओर संचार कहलाती है।

संचार का निचला भाग

इस प्रकार के संचार के अंतर्गत आज्ञा, व्यक्तिगत निर्देश, नीति विवरण पत्र, कार्य लेख इत्यादि शामिल रहते हैं। प्रबंध निदेशक विभागीय प्रमुखों को संचारित करता है। विभागीय प्रमुखों द्वारा कार्य निदेशकों को संचारित किया जाता है जो इस संचार प्रक्रिया को आगे बढ़ाते हुए कामगारों, निम्न वर्ग तक संदेश का संचार करता है।

संचार के स्रोत

संचार का निचला भाग मौखिक और लिखित दोनों ही तरह का होता है। किसी भी कार्य को प्रारम्भ करने हेतु प्रमुख निर्देश, पत्रों के द्वारा नीतियों एवं कार्य विधियों को परिपत्रों, नियमावली, पत्रिकाओं द्वारा घोषित किया जाता है, कार्य निदेशक के लिए अपने श्रमिकों को मौखिक निर्देश देना अधिक सहज हो जाता है। प्रबन्ध निदेशक अपने समस्त विभाग प्रमुखों की सभा बुलाकर स्वयं को अधिक सुविधायुक्त पाता है एवं एक लम्बे पत्र को लिखने के स्थान पर आमने-सामने का सम्पर्क प्रभावपूर्ण होता है।

संचार के सिद्धान्त

संचार को और ज्यादा प्रभावशाली बनाने में सहायक और लिखित संचार के लिये उपादेय सिद्धान्तों का विस्तृत वर्णन निम्न प्रकार किया गया है–

स्पष्टता–संदेश देने वाले का संचार चक्र मन से आरम्भ होता है। इस स्थिति में शुद्धता की बड़े पैमाने पर जरूरत होती है। यदि शुरू में ही वह स्पष्ट नहीं होगा तो सम्पूर्ण संचार प्रक्रिया अव्यवस्थित हो जाती है। सन्देश देने से पूर्व निम्नांकित तीन बातें स्पष्ट होनी चाहिए–

संचार का प्रधान उद्देश्य क्या है?

क्या संचारित होना है?

इस उपयोग के लिए कौन-सा माध्यम उपयोगी साबित होगा?

संक्षिप्तता

जो व्यक्ति सूचना को पढ़ता है उसका समय बहुत कीमती होता है इसलिए उसे यह महसूस नहीं होना चाहिए कि वह बेकार की सूचना पढ़कर अपना समय बर्बाद कर रहा है। जितना हो सके, संदेश संक्षिप्त लिखना चाहिए। विचारों में संक्षिप्तता पढ़ने वाले का ध्यान केन्द्रित करती है।

भावगत स्पष्टता

अधिकांश सूचनायें शब्दों की मदद से संचारित होती हैं। इसलिए सूचना वाहक को शब्दों के संगठन एवं शब्दों के विषय में सावधान रहना चाहिए।

संचार के माध्यम

संगठनों द्वारा संचार माध्यमों का उपयोग सूचना भेजने के लिये किया जाता है। ये संचार माध्यम सूचना देते हैं, ध्यानाकर्षण करते हैं, रुचि जगाते हैं, इच्छा जाग्रत करते हैं और यह अनुरोध करते हैं कि उसके उत्पाद सहयोग एवं संरक्षण के योग्य हैं। संचार माध्यमों के द्वारा इस प्रकार के संवाद प्रेषण को प्रचार के नाम से जाना जाता है तथा विभिन्न संचार माध्यमों के द्वारा प्रचार किया जाता है।

तथ्यात्मक सम्पूर्णता

व्यावसायिक संचार में सम्पूर्णता अत्यधिक महत्वपूर्ण है। अपूर्ण सन्देश को पढ़ कर व्यक्ति उलझन में पड़ जाता है एवं उस संदेश के अनुसार कोई कार्य किया जाता

है तो यह व्यावसायिक संगठन के लिए नुकसानदायक सिद्ध होता है। इसलिए हमें संदेश को इस प्रकार संगठित और व्यवस्थित करना चाहिए कि संदेश ग्रहण करने वाले को उसमें संदेह न रहे।

संचार माध्यम को निम्नांकित तौर पर विभक्त किया जा सकता है–

प्रसारण जनसंचार माध्यम

प्रसारण जनसंचार माध्यम के तहत टी.वी., रेडियो, केबल, ट्रांसमीशन, सेटेलाइट ट्रांसमीशन आदि को सम्मिलित किया जाता है। इन संचार माध्यमों के द्वारा श्रोता/दर्शक अपने स्थान पर ही संदेश ग्रहण कर लेता है।

मुद्रित माध्यम

इसके अन्तर्गत दैनिक समाचार-पत्र, साप्ताहिक समाचार पत्र, सामान्य अथवा विशिष्ट रुचि की पत्रिकाएं, महिला जगत, बाल पत्रिकाएं, साहित्यिक पत्रिकाएं, फैशन डिजाइनर, वन्य जीव-जन्तु इत्यादि से सम्बन्धित पत्रिकाएं, चिकित्सकों, वास्तुकारों, वकीलों, परामर्शकों इत्यादि के लिए पेशेवर पत्रिकाएं अथवा व्यापारिक पत्रिकाएं आती हैं। इनमें लेख, समाचार अथवा विज्ञापन छपते हैं। उदाहरणार्थ–एच एण्ड एफ एस रिव्यू जो कि मुम्बई से प्रकाशित होने वाला ऐसा मुद्रित माध्यम है जिसके द्वारा अभावगत उद्योग और इसके उत्पाद अपने ग्राहकों से संवाद स्थापित करते हैं। विभिन्न पर्यटन विभाग द्वारा प्रकाशित पुस्तकें जो कि हवाई जहाज और पर्यटन के लिए चलाई जा रही रेलगाड़ियों में पर्यटकों को दी जाती हैं।

प्रदर्शन संचार माध्यम

प्रदर्शन संचार माध्यम के अन्तर्गत निम्न प्रकार के माध्यम आ जाते हैं–

1. चमकदार बत्तियां एवं इनसे बने चिह्न जो रात में चमकते हैं।
2. होर्डिंग जिन्हें सार्वजनिक स्थलों जैसे व्यस्त सड़क या पार्क में लगाया जाता है।
3. विक्रेताओं द्वारा लगाए गए चिह्न एवं बैनर।
4. आकाश गुब्बारा और लेखन।
5. वाल पोस्टर, पर्चा इत्यादि।

संचार माध्यम को एक अन्य रूप में भी विभाजित किया जा सकता है–

1. मौखिक संचार।
2. लिखित संचार।
3. श्रव्य दृष्टिगत संचार।
4. दृष्टिगत संचार।

इलेक्ट्रॉनिक संचार माध्यम

इसमें आडियो एवं वीडियो टेप आते हैं। इन माध्यमों का उपयोग पर्यटन विभाग, टूर ऑपरेटर आदि प्रचारात्मक कार्यों के लिए करते हैं। अब संग्रहालय इत्यादि में बतौर गाइड इन आडियो टेपों को उपयोग में लाया जा सकता है।

मौखिक संचार

रेडियो, टेलीफोन, टेपरिकॉर्डर, मानवीय संसाधन जैसे—पर्यटक स्वयं पर्यटन स्थलों के सम्बद्ध में जानकारी आदान-प्रदान करते हैं, टैक्सी ड्राइवर, होटल मालिक, होटल में कार्य वाले कर्मचारी इत्यादि इसके अंतर्गत आते हैं।

मौखिक संचार का महत्व

मौखिक संचार का प्रमुख महत्त्व इस प्रकार है—

1. वाणी प्रोत्साहन तथा नियंत्रण का सर्वाधिक शक्तिशाली माध्यम है। इसलिए यह श्रोता/ग्राहक को प्रोत्साहित करने में ज्यादा प्रभावशाली है।
2. मौखिक संचार की स्थिति में संदेशवाहक अपने आप को संतुष्ट महसूस करता है, क्योंकि उस दशा में उसे स्पष्टीकरण एवं जवाबदेही का अवसर मिलता है।
3. मौखिक संचार में व्यय बहुत कम होता है।
4. मौखिक संचार प्रणाली के द्वारा तत्क्षण यह पता लगाया जा सकता है कि उसका संदेश ग्राहक को ठीक प्रकार से समझ में आ रहा है या नहीं। इसी आधार पर वह अपने संदेश को व्यवस्थित कर सकता है।
5. वक्ता अपनी वाणी की विधिवत तीव्रता में परिवर्तन कर अपने मत के अर्थ को अच्छी तरह से समझा सकता है।
6. अनेक विषयों में जब किसी सन्देश पर शीघ्रातिशीघ्र प्रतिक्रिया की आवश्यकता होती है तो उस स्थिति में मौखिक संचार अधिक प्रभावी होता है। समय की बचत के साथ यह लिखने में हो रहे परिश्रम की बचत भी करवाता है।

मौखिक संचार की प्रमुख कमियां

मौखिक संचार में प्रमुख दोष पाए जाते हैं जिनका विवरण इस प्रकार है–

1. मौखिक संदेश में यदि कोई त्रुटि हो जाती है तो उस स्थिति में यह पता लगाना कठिन हो जाता है कि गलती किसके द्वारा हुई।
2. इस स्थिति में जब सूचना देने वाला एवं ग्रहण करने वाला एक दूसरे से दूर होते हैं, उनके मध्य सम्पर्क का कोई साधन नहीं होता तो मौखिक संचार असम्भव हो जाता है।
3. अगर मौखिक संचार में स्पष्टीकरण के पूरे अवसर मिलते हैं तथापि ऐसा नहीं है कि गलतफहमी की संभावनाएं ही नहीं हैं।
4. मौखिक संचार के लिए लम्बे संदेश उपयुक्त नहीं होते।
5. मौखिक संदेश की तब तक कानूनी एवं विधिक मान्यता नहीं होती जब तक उसे स्थायी प्रतिवेदन का अंग न बना लिया जाये।

लिखित संचार

लिखित संचार के तहत निम्नलिखित माध्यमों को सम्मिलित किया जाता है–

प्रकाशित सामग्री, टेलीग्राम, न्यूज लेटर, रूपपत्र, विवरण पुस्तिका, प्रतिवेदन, पत्रिकाएं, प्रश्नावली, प्रेस विज्ञप्ति, प्रायोजित लेख, सार संग्रह इत्यादि।

लिखित संचार की प्रमुख विशेषताएं

लिखित संचार की मुख्य विशेषताएं इस प्रकार हैं–

1. यह जिम्मेदारियों के नियम पालन करने में सुगमता प्रदान करता है।
2. यह एक स्थायी लेख होता है।
3. यह अनेक बार निर्देशित किया जा सकता है।
4. यह सही एवं सुतथ्य होता है।

लिखित संचार की प्रमुख कमियां

लिखित संचार की प्रमुख कमियां निम्न हैं–

1. अधिक समय व्यय होता है।
2. तीव्र स्पष्टीकरण सम्भव नहीं है।

दृश्य-श्रव्य संचार

इस संचार में निम्न प्रकार के साधन आते हैं–

1. वीडियो टेप सिस्टम,
2. केबिल ट्रान्समीशन,
3. सैटेलाइट ट्रान्समीशन एवं
4. टेलीविजन।

आज के समय में उपरोक्त सभी साधनों ने अपना विस्तृत विकास कर लिया है और दर्शकों के ऊपर अपनी स्थिति को भी अत्यधिक मजबूत कर लिया है। केबिल प्रसारण का जाल जटिल होने के साथ प्रभावी भी है। इन सभी में वीडियो टेप सिस्टम की अवधारणा बिल्कुल नवीन है। वीडियो टेप प्रणाली का आविष्कार ब्रिटिश टेलकाम द्वारा बिना संगठन के कम्प्यूटरीकृत जानकारी प्रेषित करने के लिए किया जाता है। वीडियो टेप सिस्टम हेतु एक दूरभाष लाइन अर्थात् टेलीफोन लाइन और रंगीन टी. वी. की आवश्यकता है। रंगीन टेलीविजन रिसीवर एवं एडाप्टर की सहायता से की-बोर्ड और डीकोडर से सम्बद्ध होता है। वीडियो टेप सिस्टम कम खर्च में ट्रेवेल एजेन्ट को आरक्षण सुविधा के साथ-साथ अपने ग्राहक एवं पूर्ण पर्यटन व्यवसाय से सीधे सम्पर्क करने में सहायक सिद्ध होता है। यह संचार जनसंघ शिक्षा, जनसंघ प्रचार, जनसंघ प्रचार समिति आदि क्षेत्रों में महत्वपूर्ण स्थान रखता है। विशाल व्यवसाय समुदाय मुख्यतः अपने मजदूरों की शिक्षा के लिए अपने उत्पादों के प्रचार के लिए पूंजीपति इस तकनीक का प्रयोग करते हैं। इस तरह की तकनीक को अधिक प्रभावी बनाने के लिए छोटी, प्रभावपूर्ण, आकर्षक एवं रुचिकर फिल्मों का निर्माण अनिवार्य हो जाता है। समाचार पत्र, सिनेमा, रेडियो एवं टी. वी. जैसे जन संचार के साधन जिन्हें विज्ञापन नहीं चलाता मीडिया के नाम से जाने जाते हैं। प्रत्यक्ष डाक सेवा, प्रदर्शन, पोस्टर बुलेटिन आदि को Below the Line मीडिया कहा जाता है। इस प्रकार का विभाजन लागत के अनुसार किया जाता है। लागत निश्चित करते वक्त एजेंट एक विभाजन रेखा देता है। इस रेखा के ऊपर प्रथम प्रकार के तथा द्वितीय प्रकार के संचार माध्यम रखे जाते हैं।

आमने सामने का संचार

ऐसा माना जाता है कि आमने-सामने का संचार और मौखिक संचार दोनों ही एक समान हैं, परन्तु विभिन्न परिस्थितियों में इन दोनों के मध्य कुछ अन्तर

दिखाई देते हैं। उदाहरण—दो व्यक्तियों के दूरभाष पर सम्पर्क को हम आमने-सामने का सम्पर्क नहीं कहेंगे। दो राष्ट्रों के प्रधानमंत्री आपस में हाथ मिलाते हैं एवं हर्षोल्लास के साथ सम्मानित करते हैं तो इस स्थिति को हम आमने-सामने का संचार कह सकते हैं, लेकिन इस संचार को मौखिक संचार नहीं कहा जा सकता है।

आमने-सामने संचार की प्रमुख विशेषताएं

आमने-सामने संचार की प्रमुख विशेषताओं को निम्नलिखित प्रकार से बताया गया है—

1. आमने-सामने का संचार वाद-विवाद के लिए उपयुक्त है। इसमें सम्मिलित होने वाले जरूरी व्यक्ति के भावों तथा स्फुरण से तुरन्त उत्तर मिल जाता है। यह आवश्यक नहीं कि शब्दों द्वारा जबाव दिया जाये। इस प्रकार के संचार में वक्ता को सामंजस्य हेतु उपयुक्त अवसर मिलता है।
2. चेहरे का भाव तथा आकृति संचार को बेहतर बनाने में उपयोगी सिद्ध होता है। मौखिक संचार को कई कारणों से लिखित संचार से श्रेष्ठ सिद्ध किया गया है। इसमें वक्ता की आवाज तथा तीव्रता वाणी की तीव्रता की सहायता से सन्देश का संचार किया जाता है। इसकी उपादेयता यह भी है कि जो संदेश संचारित किया जाता है, उसके संबंध में चेहरे की भाव भंगिमा का भी ज्ञान होता है।

आमने-सामने के संचार की प्रमुख कमियां

इस संचार की प्रमुख कमियां इस प्रकार से हैं—

1. विशाल जनसमूह में यह प्रभावी नहीं है।
2. आधुनिक तथा विकसित संगठनों में इसका उपयोग काफी कठिन है क्योंकि उनके कार्यालय तथा विभाग अलग-अलग होते हैं।
3. यदि ध्यानपूर्वक नहीं सुना जाए तो यह निष्क्रिय अथवा निष्प्रभावी हो जाता है।

देशाटन और मानव सम्पदा

जब मानव संसाधनों को पर्यटन की दृष्टि से देखा जाता है तो इसका अभिप्राय पर्यटन के विभिन्न क्षेत्रों जैसे—होटल, पोषाहार, परिवहन सुविधाओं, गाइड सेवाओं,

प्रकाशनों इत्यादि में मनुष्य के महारत हासिल करने से है। वास्तव में हम पर्यटन का वास्तविक विकास तभी कर सकते हैं जब हमें पर्यटन के विभिन्न आयामों का बौद्धिक एवं शारीरिक प्रशिक्षण प्राप्त हो। उदाहरणार्थ, हम होटल व्यवसाय को पर्यटन के क्षेत्र में तभी लाभकारी बना सकते हैं जब हमारे पास होटल प्रबन्धकों की पर्याप्त कुशलता हो।

मानव संसाधन के विकास पर लोकतान्त्रिक और कल्याणकारी सरकारों द्वारा ध्यान देने के पश्चात् व्यक्ति को अन्य दूसरे आर्थिक संसाधनों के बराबर पूंजी मान लिया गया है एवं उसे एक संसाधन के रूप में विकसित किया जाने लगा है। सर्वेक्षणों द्वारा स्पष्ट भी हो चुका है कि जिस क्षेत्र में भी मानव संसाधन पर्याप्त तौर पर विकसित है, उन क्षेत्रों का विकास तुलनात्मक तेजी के साथ सम्पन्न हुआ है। मानव संसाधनों के अभिप्राय को स्पष्ट करते हुए सर आर्थर लुईस ने इस प्रकार वर्णित किया है :–

मानव संसाधन के रूप में तभी परिवर्तन हो सकता है, जब उसे किसी विशेष क्षेत्र में ज्ञान, बुद्धि प्रशिक्षण इत्यादि आधारों पर विशेषज्ञता प्राप्त हो चुकी हो। बौद्धिक परिपक्वता के साथ-साथ शारीरिक एवं मानसिक स्वास्थ्य को भी मानव संसाधन के अन्तर्गत रखा गया है।

भारत सरकार ने पर्यटन के क्षेत्र में मानव संसाधन का और ज्यादा विकास करने हेतु कई प्रकार के कार्यक्रम और योजनाओं को तैयार किया है। इन कार्यक्रमों और योजनाओं का अध्ययन अग्रांकित तरीके से किया जा सकता है।

(क) गाइड प्रशिक्षण संस्थान

गाइड प्रशिक्षण संस्थान राष्ट्रीय और अन्तर्राष्ट्रीय दोनों ही स्तर के सैलानियों हेतु आधारभूत संरचना की जरूरत के महत्वपूर्ण कारक होते हैं। देश में विदेशी पर्यटकों के आवागमन में तेजी से वृद्धि हुई है और यह संख्या 3.2 मिलियन के आँकड़े को पार कर गयी है। फलस्वरूप गाइडों की मांग में भी बेतहाशा वृद्धि हुई है और गाइडों की बढ़ती हुई मांग को पूरा करने के लिए 1996 से 3 टियर गाइड प्रणाली प्रारंभ की गयी है जिसके अंतर्गत क्षेत्रीय स्तर, राज्य स्तर तथा स्थानीय स्तर पर गाइड प्रशिक्षण कार्यक्रम चलाये जा रहे हैं।

इन प्रशिक्षण कार्यक्रमों में अन्तर्राष्ट्रीय भाषाओं–जर्मन, फ्रेंच, अंग्रेजी भाषाओं के पाठ्यक्रम तैयार किये जाते हैं।

(ख) भारतीय पर्यटन एवं यात्रा प्रबन्ध संस्थान

मानव संसाधन विकास क्रियाओं में प्रशिक्षण कार्यक्रमों, पर्यटन तथा यात्रा उद्योग के हित से सम्बन्धी अनुसंधान कार्यक्रमों को प्रारंभ करने के लिए पर्यटन मंत्रालय के अंतर्गत एवं निबंधित सोसायटी के रूप में 1983 में भारतीय पर्यटन एवं यात्रा प्रबन्धक संस्थान की स्थापना की गयी।

भारतीय पर्यटन एवं यात्रा प्रबन्ध संस्थान के वित्तीय पोषण की जिम्मेदारी परिभ्रमण विभाग एवं भारत सरकार की होती है। इसकी स्थापना उच्च शिक्षा की आवश्यकताओं को पूर्ण करने, यात्रा एवं पर्यटन उद्योग से सम्बन्धित प्रशिक्षण एवं इस क्षेत्र में उत्कृष्टता के एक केन्द्र के रूप में की गयी है। 1992 में इसे ग्वालियर में विपणन कर दिया गया।

इस संस्थान द्वारा पर्यटन प्रबन्ध में डिप्लोमा, गन्तव्य स्थल प्रबन्धन में डिप्लोमा, यात्रा प्रचालन सेवा प्रबंधन, पर्यटन विपणन भाषा पाठ्यक्रम, वन्य जीव पर्यटन प्रबन्धन, इत्यादि पाठ्यक्रम संचालित किये जाते हैं। पाठ्यक्रम के संचालन का कार्य विश्वविद्यालयों द्वारा भी किया जाता है। इस संस्थान द्वारा प्रबन्ध विकास कार्यक्रमों में निम्न पाठ्यक्रमों को शामिल किया गया है–

(1) पर्यटन एवं यात्रा में प्रबन्ध समस्याएं

(2) प्रभावी पर्यटन विकास

(3) पर्यटन में कम्प्यूटर तकनीकी

(4) यात्री सेवाओं का प्रबन्ध

(5) पर्यटन विकास का पर्यावरणीय क्षेत्र

(ग) होटल प्रबन्ध कैटरिंग प्रौद्योगिकी

यद्यपि सरकार ने होटल प्रबन्ध कैटरिंग तथा पोषाहार के क्षेत्र में प्रशिक्षण कार्यक्रम 1962 से ही शुरु कर दिया था, परन्तु व्यावसायिक आधार एवं पर्यटन की दृष्टि से इसका विधिवत् श्रीगणेश 1984 से हुआ जब पर्यटन मंत्रालय ने राष्ट्रीय होटल प्रबन्ध एवं कैटरिंग प्रौद्योगिकी परिषद् की स्थापना की। इस परिषद् के मुख्य उद्देश्य निम्नलिखित हैं–

1. राष्ट्रीय स्तर पर प्रमाणन तथा सत्यापन करना।
2. पाठ्यक्रम एवं सुविधाओं की आवश्यकताओं का मानकीकरण करना।

3. सम्बद्ध संस्थाओं में स्टाफ के सदस्यों के लिए शैक्षिक एवं अन्य अहर्ताएं निर्धारित करना।
4. उचित प्रशिक्षण कार्यक्रम को तैयार करने हेतु एक तकनीकी भण्डार घर के रूप में कार्य करना।
5. होटल प्रबन्ध कैटरिंग प्रौद्योगिकी से संबंधित विकास पर सरकार को राय-मशविरा देना।
6. इस क्षेत्र में अन्तर्राष्ट्रीय स्तर की नवीनतम क्रियाओं का लेखा-जोखा रखना।

इस प्रकार होटल प्रबन्धन एवं कैटरिंग के क्षेत्र में हुए अत्याधुनिक परिवर्तनों के परिणामस्वरूप इन संस्थानों में छात्रों की रुचि एवं दृष्टिकोण में व्यापक बदलाव हुए हैं। वर्तमान समय में देश में 18 होटल प्रबन्धन संस्थान तथा 15 भोजन कला संस्थान स्थापित किये जा चुके हैं। इसके अतिरिक्त पूर्वोतर राज्यों में डिप्लोमा स्तर के पाठ्क्रमों को प्रारंभ किया गया है।

उपरोक्त संस्थानों के वित्त पोषण का कार्य केन्द्र सरकार द्वारा अनुदान के माध्यम से होता है।

(घ) भारतीय स्कीइंग एवं पर्वतारोहण संस्थान

भारतीय स्कीइंग और पर्वतारोहण संस्थान को सन् 1968 में स्थापित किया गया था, जिसका प्रमुख प्रयोजन पर्यटकों को निम्न क्रीड़ाओं के स्थान के विषय में जानकारी प्रदान करना है--

1. गुलमर्ग हिम क्रीड़ा परियोजना
2. पैराग्लाइडिंग
3. रिबर रनिंग स्कीइंग
4. जल क्रीड़ा
5. पर्वतारोहण एवं चट्टारोहण
6. स्कीइंग

वर्तमान वित्तीय वर्ष में इस संस्थान ने औली एवं पटनीटाप में स्कीइंग, हिमाचल में ट्रेकिंग, ऋषिकेश में रनिंग इत्यादि में प्रशिक्षण उपलब्ध कराया है।

(ङ) जल क्रीड़ा के लिये मशहूर संस्थान

सम्पूर्ण देश में जल खेलों का सम्वर्धन करने के उद्देश्य से पर्यटन विभाग ने सन् 1990 में गोवा में एक राष्ट्रीय जल क्रीड़ा संस्थान की स्थापना की है। इस संस्थान के सम्बन्ध में प्रमुख बातें निम्न हैं--

1. वर्तमान में इसे भारतीय पर्यटन तथा यात्रा प्रबन्धन संस्थान को हस्तान्तरित किया जा चुका है।
2. यह संस्थान सेलिंग, विंड सर्फिंग, जल स्कीइंग, आउटबोर्ड मोटर रख-रखाव, जीव रक्षा तकनीकों एवं जल क्रीड़ा के सम्बन्ध में प्रशिक्षण प्रदान करता है।
3. गोवा सरकार ने इस संस्थान के लिए 18 एकड़ अतिरेक भूमि एवं 60 प्रतिशत उपकरण प्राप्त कराये हैं।

संस्थान ने अपने प्रशिक्षण को गोवा से बाहर 11 राज्यों तक पहुँचाया है।

4

पर्यटन की महत्ता

विश्व के किसी भी देश की अर्थव्यवस्था में पर्यटन उद्योग का महत्त्वपूर्ण योगदान होता है। किसी भी देश की आर्थिक क्रियाएं इससे प्रभावित होती हैं। कुछ वस्तुओं का वितरण समान रूप से नहीं होता है वह अपनी जरूरतों के अनुसार हासिल करता है। उसी प्रकार मनुष्य अपनी छुट्टियां बिताने देश-विदेश के विभिन्न प्रकार के जलवायु, प्राकृतिक स्थलों, आदि स्थलों का भ्रमण करता है। इस तरह पर्यटक पर्यटन स्थल के क्षेत्रों का विकास करने में सहयोग करते हैं। वर्तमान समय में विकसित देश के पर्यटक विकासशील देशों का आनंद लेते हैं।

देशाटन की भौगोलिक विशेषताएं

किसी भी देश की भौतिक विशिष्टताएं यथा—भू-आकृतियां (पर्वत, पठार, चट्टान, हिमनद, वन इत्यादि) मैदानी क्षेत्र, जलस्रोत (झील, नदी, समुद्र, जल-प्रपात इत्यादि) समुद्र तट एवं प्राकृतिक स्थल, द्वीप घाटियां शताब्दियों से मनुष्य को अपनी ओर आकर्षित करती रही हैं। इसलिए यह कहा जा सकता है कि इसी सौन्दर्य को देखने के लिए पर्यटक इन भौगोलिक क्षेत्रों की तरफ आनन्दमयी जीवन को बिताने

के लिए आकर्षित होते हैं। अतः पर्यटन क्षेत्रों में भौगोलिक क्षेत्र का अपना विशेष महत्व है।

पर्यटन का महत्व : पर्यटन एक अनूठा उद्योग है क्योंकि इसमें बिना धुएं के उद्योग, बिना कक्षा की शिक्षा, बिना कानून निर्माण की एकता एवं यथाविधि आधार के राजनीतिक कौशल प्राप्त होते हैं। पर्यटन का महत्व उस समय समक्ष आया जब यूनाइटेड नेशन की आम सभा में 1967 को अंतर्राष्ट्रीय पर्यटन वर्ष घोषित किया गया। यह माना गया कि पर्यटन एक आधारित एवं ऐच्छिक मानवीय गतिविधि है जिसे सभी लोगों एवं सरकारों द्वारा प्रशंसा एवं बढ़ावा देना चाहिए।

बहुचर्चित मनीला घोषणा से इस विचार को ताकत मिलती है कि पर्यटन एक ऐसी आवश्यक गतिविधि है जो देश के विकास के लिए जरूरी है क्योंकि इसका सीधा असर समाज के भौतिक, सामाजिक, सांस्कृतिक, शैक्षणिक तथा आर्थिक व्यवस्था पर पड़ता है।

पर्यटन का भौतिक महत्त्व

पर्यटन भूगोल के अंतर्गत पर्यटन का भौतिक महत्व, प्रमुख तत्वों का स्थान प्राकृतिक घटक है। विश्व के पर्यटक प्राकृतिक दर्शनीय स्थलों को देखने के लिए आते हैं। इस प्रकार जलवायु, मौसम, धरातलीय स्थलाकृतियां, वन्य जीव-जन्तु, जलराशियाँ, समुद्री बीच, जलसंसाधन, नई-नई स्थलाकृतियों की खोज इत्यादि प्राकृतिक दृश्य मनुष्य को अपनी ओर आकर्षित करते हैं। पर्यटन के भौतिक महत्व के मुख्य घटक निम्नांकित हैं–

(1) वनस्पति तथा जीव-जन्तु–मानव वनस्पति तथा जीव-जंतुओं की विविधताओं को देखकर आनन्द की अनुभूति करता है। भारत में सुन्दर वन, दक्षिण एवं उत्तर भारत में विभिन्न प्रकार के वनस्पति एवं जीव-जंतु, भूमध्यरेखीय सदाबहार कांगो एवं अमेजन के वन, कोणधारी वन तथा कई तरह के जीव-जन्तु मानव को अपनी ओर आकर्षित करते हैं।

इसकी विभिन्न प्रकार की किस्मों को देखने के लिए एक पर्यटक की दिलचस्पी बढ़ती है, जिसके कारण वे इनकी तरफ आकर्षित होकर पर्यटन करने के लिए बाध्य हो जाते हैं।

(2) हिमनद तथा जलराशियां– हिमनद, जलप्रपात, झील, झरने इत्यादि को देखना मानव को हमेशा रोमांचित करता है। अनेक हिमचद जैसे–गंगोत्री, यमनोत्री हिस्पर, पिण्डार तथा यूरोप एवं अमेरिका के कई हिमनद तथा जलराशियां मानव के पर्यटन स्थल हैं।

(3) धरातलीय विविधतायें—पर्यटक को धरातलीय विभिन्नता भी अपनी ओर आकर्षित करती है। इसमें पर्वत, पठार, नदी, मैदान, घाटी इत्यादि स्थलाकृति आते हैं। उच्चतम पर्वत श्रेणियां, हिमाच्छादित पर्वत चोटियाँ, सुन्दर हिमनद, बुग्याल सभी मनुष्य को आकर्षित करते हैं। पठारी एवं चट्टानी भू-दृश्यावली पर्यटक को लुभाती हैं। कई सुरम्य घाटियां एवं स्थल पठारों तथा पहाड़ों पर स्थित होते हैं, जो पर्यटक को आकर्षित करते हैं। इसी प्रकार कई पठार खनिज तत्वों के भण्डार होते हैं जो मनुष्य की आर्थिक क्रियाओं के उद्गम के स्रोत बन जाते हैं। ये भी पर्यटक के लिए आकर्षण के केन्द्र होते हैं।

(4) जलवायु एवं मौसम—मानव जिस वातावरण में रहता है, परिवर्तन के लिए उसके प्रतिकूल वातावरण को पसंद करता है, जैसे—भीषण गर्मी सहन करने वाले को हिमालय की वादियाँ पसंद आती हैं एवं घोर ठण्डे प्रदेशों में रहने वाले थार मरुस्थल के गर्म स्थानों को पसंद करते हैं। गर्मियों के मौसम में व्यक्ति समुद्र तट और पहाड़ी क्षेत्रों जैसे ठण्डे इलाकों की ओर जाते हैं और जाड़े के दिनों में गर्म स्थलों की ओर जाते हैं। इसी वजह से भारत में हिल-स्टेशनों का विकास हुआ। भारत के प्रमुख हिल-स्टेशनों में प्रसिद्ध हैं, श्रीनगर, शिमला, कुल्लू-मनाली, मसूरी, नैनीताल, शिलांग, दार्जिलिंग, माउण्ट आबू, कोडइकनाल, ऊटी आदि। भारत में 5700 कि. मी. लंबा समुद्रतट है जो गर्मी के मौसम में अपेक्षाकृत ठण्डा एवं जाड़े के मौसम में अपेक्षाकृत गर्म रहता है।

गोवा, कोवलम, कन्याकुमारी, ममलपुरम, पुरी, समुद्र तट आदि 'बीच' अंतर्राष्ट्रीय स्तर के हैं, जिनका विकास होने के कारण यहाँ समुद्रतटीय एवं पहाड़ी पर्यटन के विकास में काफी मदद मिली है। पूरे संसार में पर्यटकों का आना-जाना वस्तुतः जलवायु के कारणों से प्रभावित होता है। पर्यटक इस मौसम में घूमना पसंद करते हैं। भारत में मौसम की इतनी ज्यादा विविधता है कि लोग हर मौसम में एक स्थान से दूसरे स्थान आते-जाते रहते हैं। उदाहरणार्थ—उत्तर में पंजाब में महादेशीय मौसम है जहां गर्मी भी ज्यादा पड़ती है और ठण्ड भी, जबकि केरल में उष्णकटिबंधीय समुद्री जलवायु है, जहां गर्मी होती है एवं हवा में नमी। इसी प्रकार असम में उमस होती है, तो राजस्थान में सूखापन। जलवायु में इस विविधता की वजह से लाखों पर्यटक भिन्न-भिन्न मौसमों में एक स्थान से दूसरी स्थान पर आते-जाते रहते हैं। साधारणतया गर्मी के दिनों में उत्तर से दक्षिण की तरफ और उत्तर-पूर्व मानसून में दक्षिण से उत्तर की ओर पर्यटकों का प्रस्थान होता है।

उपर्युक्त वर्णित देशाटन की भौगोलिक विशेषताओं के अलावा इसकी कुछ अन्य विशेषताएं भी हैं जिसे निम्न शीर्षकों के अंतर्गत स्पष्ट किया गया है–

अन्य विशेषताएं

(1) जलसंसाधनों का जल-क्रीड़ा में प्रयोग–भारत में जल संसाधन की पर्याप्त मात्रा में उपलब्धता के कारण जैसे–नदियां, हिमनद, झील, जलप्रपात इत्यादि अनेक प्रकार के जल-विद्युत प्लांट, बाँध आदि बनाये गये हैं। उदाहरणार्थ–भाखड़ा नांगल बांध, नागार्जुन सागर आदि इन सबके अलावा मसूरी, डल झील आदि नदियों तथा बांधों में पर्यटक खेल-क्रीड़ा का आनन्द लेते हैं। ये सभी अपने वर्तमान प्रौद्योगिकी एवं मनोरम दृश्य तथा अत्याधुनिक सुविधा के कारण देश-विदेश के पर्यटकों को अपनी ओर आकर्षित करते हैं।

(2) नयी स्थलाकृतियों की खोज–वर्तमान काल में एक विशेष स्थान के कारण वहाँ पर जनसंख्या घनत्व बढ़ जाता है, पर्यावरण प्रदूषित होने लगता है तथा मेहमान एवं मेजबान में संघर्ष बढ़ जाता है। परिणामस्वरूप सरकार एवं निजी संगठनों द्वारा दूसरी जगहों पर नयी-नयी स्थलाकृतियों की खोज की जाती है जिससे पर्यटकों का ध्यान उस ओर आकर्षित होता रहे। इस प्रक्रिया से नयी जगह का वन अधिसंरचनात्मक विकास होता है एवं वहाँ की जनता आर्थिक लाभ प्राप्त करती है।

(3) पर्यटकों का स्थलाकृतियों भू-दृश्यों का अधिक ज्ञान होना–आजकल संसार के प्रत्येक क्षेत्र में विकास होने के कारण पर्यटकों का भी आनन्दमय जीवन खुशहाल बनाने हेतु तरह-तरह के मानसिक ज्ञान-विज्ञान का विकास हुआ है।

(4) एक दूसरे की संस्कृति को समझना–सैर के माध्यम से दो क्षेत्रों अथवा दो देशों के मध्य संस्कृति का आदान-प्रदान होता है। उदाहरणार्थ जब यूरोपियन पर्यटक भारत में आता है तो यहाँ की ग्रामीण एवं आध्यात्मिक संस्कृति से प्रभावित होता है। वह अपनी भौतिक संस्कृति से ऊब चुका होता है, जिसके कारण यहाँ की संस्कृति एक नई स्फूर्ति उत्पन्न करती है फलस्वरूप वह यथाशीघ्र उसे अपना लेता है। इधर मेजबान भी सैलानी की भौतिक संस्कृति को उतना ही तीव्र स्वीकार करता है चूंकि वह उसके लिए एक नवीन बाढ़ होती है।

(5) जातिवाद एवं धार्मिकवाद रूढ़िवादिता का कम होना–पर्यटकों द्वारा बार-बार देश-विदेशों में सैर करने से एक-दूसरे के जाति एवं धर्म के विषय में जानकारी बढ़ती है। अतः एक दूसरे के कुरीतियों को दूर करने की शक्ति मिलती है।

(6) मेजमान तथा मेहमान की अच्छी समझ—परिभ्रमण के कारण मेजमान एवं मेहमान में पारस्परिक समझ का विकास होता है। तरह-तरह के अंधविश्वास और छुआछूत इत्यादि कुरीतियाँ का अन्त होता है। आपसी वार्तालाप से भाषा का विकास होता है एवं एक-दूसरे की भावनाओं एवं विचारों का आदान-प्रदान होता है। फलस्वरूप आपसी समझ में वृद्धि होती है।

(7) भाईचारे की भावना—जब सैलानी ऐसे स्थान पर जाता है जहां के समाज एवं संस्कृति की जानकारी कम ही पायी जाती है तो उसके भीतर वहाँ के विषय में अलग-अलग विचार उत्पन्न होते हैं। लेकिन कुछ समय व्यतीत करने के उपरांत वहाँ के मेजबान लोगों के आपसी संबंध के आदान-प्रदान से उसकी सोच इतनी परिवर्तित जाती है कि पर्यटक उस जगह को अपना समझने लगता है, जिससे भाईचारे का विकास होता है।

परिभ्रमण की आर्थिक जरूरत

परिभ्रमण की आर्थिक जरूरतों को निम्न प्रकार से स्पष्ट किया जा सकता है—

(1) आय—वर्तमान में पर्यटन से कई तरह की वस्तुओं और सेवाओं की माँग होती है, जैसे—आवास, खाद्य और पेय, लंबी दूरी के एवं स्थानीय परिवहन मनोरंजन, खरीददारी स्थल, दर्शन हेतु मार्गदर्शक आदि पर्यटकों द्वारा किए गए व्यय से इन सेवाओं को प्रत्यक्ष आय ही नहीं होती है बल्कि इससे होटलों, रेस्तरां, परिवहन, संचालक एवं श्रमिकों, ट्रेवेल एजेंटों एवं उनके कर्मचारियों, मनोरंजन गृहों के मालिकों तथा वहाँ के कर्मचारियों, कलाकारों, शिल्पियों और कला की वस्तुएं बनाने वाले लोगों को भी रोजगार उपलब्ध होता है। इन सभी सेवाओं से जुड़े रोजगार तथा इनसे होने वाली आय निम्नलिखित कारकों पर निर्भर करती है—

(क) पर्यटकों की संख्या

(ख) उनके खर्च करने का अवसर

(ग) पर्यटकों द्वारा किया गया व्यय

विभिन्न प्रकार के करों, चुंगी और प्रवेश शुल्क इत्यादि से केन्द्र एवं राज्य सरकार को अच्छी आय होती है।

(2) विदेशी मुद्रा—सैर से विदेशी मुद्रा की प्राप्ति होती है। विदेशी पर्यटन से अंतर्राष्ट्रीय व्यापार को बढ़ावा मिलता है। विदेशी पर्यटक आभूषण और कीमती पत्थर, कालीन, कपड़े, हस्तशिल्प एवं चमड़े का बना सामान खरीदते हैं, फलस्वरूप विदेशी मुद्रा देश में आती है।

ये विदेशी पर्यटक अपने-अपने राष्ट्रों में इन वस्तुओं के प्रचार के माध्यम भी बनते हैं। इससे विदेशों में इन चीजों की मांग बढ़ती है और इनके निर्यात में वृद्धि होती है। भारतीय हस्तशिल्प की श्रेष्ठ गुणवत्ता एवं पर्यटकों के द्वारा दुनियाभर में इनके प्रचार होने से इस क्षेत्र की निर्यात संभावनाएँ अत्यधिक बढ़ी हैं।

(3) लघु उद्योगों का विकास—भारत में जब-जब विदेशी पर्यटक भ्रमण करने के लिए आते हैं तो यहां के विभिन्न प्रकार के लघु उद्योगों के सामानों (हस्तशिल्प/हथकरघा/खादी/ग्रामोद्योग/खाद्य सामग्री एवं परंपरागत सामानों) की ओर अधिक ही आकर्षित होते हैं। वे इसे खरीदते हैं और निशानी तथा यादगार के तौर पर अपने राष्ट्र में ले जाते हैं, जिसकी वजह से वहां पर इन सामानों की मांग बढ़ती है। परिणामस्वरूप इन सामानों की निर्यात हेतु लघु उद्योगों का विकास होता है। लघु उद्योगों में हथकरघा को भारतीय चित्र पर दिखा सकते हैं, जो भारत में पर्यटन के विकास के साथ-साथ विकसित हुआ है।

(4) रोजगार—पर्यटन उद्योग में प्रत्यक्ष एवं अप्रत्यक्ष रोजगार उत्पन्न करने की अद्भुत सामर्थ्य है। होटल से लेकर विविध टूर-ऑपरेटरों, पर्यटन कार्यालयों, परिवहन संचालकों, टूरिस्ट गाइडों इत्यादि के तौर पर लोगों को इससे प्रत्यक्ष रोजगार हासिल होता है। होटलों में सब्जी, मांस, मछली, मुर्गी, अनाज आदि आपूर्ति करने वालों, बिजली मिस्त्री, प्लम्बर, बढ़ई और बिस्तर, पर्दे आदि देने वालों एवं अन्य सभी प्रकार की सहायक सामग्रियों के आपूर्तिकर्ताओं एवं टैक्सी ड्राइवरों इत्यादि को इससे रोजगार मिलता है।

परिभ्रमण की राजनीतिक जरूरत

परिभ्रमण की राजनीतिक जरूरत को निम्न प्रकार से बताया गया है—

(1) रियायत देना—पर्यटकों को आकर्षित करने के लिए वह क्षेत्र आवास, होटल, परिवहन एवं खान-पान की आपूर्ति, खरीददारी आदि पर भारी छूट देकर पर्यटकों को आकर्षित करते हैं इससे उस क्षेत्र के आर्थिक विकास में मदद मिलती है।

(2) अच्छी नीति बनाना—पर्यटकों को अपने देश अथवा क्षेत्र में आकर्षित करने के लिए उनके लिए सुरक्षा, नियम, कानून, स्वास्थ्य सुविधाएं तथा दूसरी आधारभूत सुविधाओं का विशेष ध्यान रखा जाता है जिससे पर्यटक सुरक्षित एवं आनन्दपूर्वक पर्यटन का मजा उठाते हुए भ्रमण करते हैं। जैसे-कश्मीर में पर्यटकों की कमी तथा उत्तरांचल में वृद्धि इसका प्रमाण है।

(3) वीजा प्रणाली का सरल होना—वर्तमान समय में हर राष्ट्र पर्यटन उद्योग की तरफ विशेष ध्यान दे रहा है, क्योंकि इससे वहाँ की आर्थिक उन्नति होती है। अतः वीजा प्रणाली जो एक पर्यटक के लिए कानूनी दस्तावेज होता है तथा एक क्षेत्र से दूसरे क्षेत्र तक आने-जाने का अधिकारिक प्रमाण-पत्र एवं लाइसेंस होता है, विकासशील देश विकसित देशों के पर्यटकों को लुभाने के लिए वीजा प्रणाली को अत्यधिक सरल कर रहे हैं क्योंकि ये सैलानी विदेशी मुद्रा के प्रमुख स्रोत हैं।

(4) दो राष्ट्रों के बीच अच्छे संबंध होना—सैर के कारण दो देशों में पारस्परिक संबंध स्थापित होता है, जैसे भारत एवं पाकिस्तान। जब विदेशी पर्यटक किसी दूसरे देश में आते-जाते हैं तो वहाँ की केंद्रीय एवं क्षेत्रीय सरकारें उनके मेहमान नवाजी पर विशेष ध्यान देती हैं। कारण स्पष्ट है कि इससे वहाँ पर पर्यटन उद्योग को बढ़ावा मिलता है, जिसके कारण आय, रोजगार व विदेशी मुद्रा में वृद्धि होती है। अतः एक दूसरे देशों का संबंध उत्तम होना स्वाभाविक है।

(5) सहयोग की भावना—जब दो देशों के बीच पर्टयन काफी विकसित हो जाता है तो उनके मध्य एक पारस्परिक सहयोग की भावना पैदा हो जाती है। जैसे—भारत एवं नेपाल, आस्ट्रेलिया एवं न्यूजीलैण्ड इत्यादि समय-समय पर प्राकृतिक आपदाओं तथा विश्व संकट के समय एक-दूसरे की मदद करते हैं।

उपर्युक्त तथ्यों के आधार पर यह निष्कर्ष निकलता है कि यह उद्योग किसी भी राष्ट्र की जीवनशैली को निर्धारित करता है। विश्व के पर्यटन उद्योग में पर्यटकों को लुभाने, अपने देश में विदेशी मुद्रा के भण्डार में वृद्धि करने की एक होड़-सी मची हुई है। कुछ देश तो पर्यटकों के लिए विशेष कार्यक्रम नियोजित करते हैं। सिंगापुर एक अच्छा उदाहरण है, जिसकी लगभग 80 प्रतिशत आर्थिक व्यवस्था पर्यटन उद्योग पर आधारित है। भारत में भी पर्यटन उद्योग बहुत ही तीव्र गति से बढ़ रहा है जो भविष्य का सबसे महत्त्वपूर्ण उद्योग बनने जा रहा है। अतएव भारत सरकार को पर्यटन उद्योग में विशेष रुचि लेकर अपने पड़ोसी राष्ट्रों के साथ सम्बन्ध अच्छे किए जा सकते हैं।

अतः हम कह सकते हैं कि वर्तमान समय में पर्यटन उद्योग किसी भी राष्ट्र की आर्थिक स्थिति को निश्चित करने में प्रमुख भूमिका निभाता है।

5

आर्थिक विकास में पर्यटन का महत्त्व

विश्व के अधिकतम विकासशील देशों में उच्च आर्थिक दर प्राप्ति में पर्यटन की महत्त्वपूर्ण भूमिका होती है। देश-विदेश के क्षेत्रों में सामान्य स्थान से बाहर जो धन अर्जित किया जाता है वह धन पर्यटक स्थलों पर खर्च होता है। पर्यटकों के खर्च से क्रय शक्ति में जो वृद्धि होती है उससे अर्थव्यवस्था को ऊपर उठने में सहायता मिलती है। सम्पूर्ण विश्व में पयर्टन में जो धन सामान्य निवास स्थान से बाहर अर्जित किया जाता है वह पर्यटन के स्थलों में व्यय होता है। आगन्तुकों के व्यय से प्राप्त होने वाले क्षेत्रों में जो क्रय शक्ति में वृद्धि होती है यही पर्यटन का सबसे आश्चर्यजनक आर्थिक प्रभाव है। पर्यटन व्यय के कारण जो धन के प्रभाव का सृजन होता है, वह गुणक के प्रभावों के माध्यम से समस्त अर्थव्यवस्था में अपना स्थान बना लेता है।

राष्ट्र के आर्थिक विकास में पर्यटन की भूमिका

राष्ट्र के आर्थिक विकास में पर्यटन उद्योग की महत्त्वपूर्ण भूमिका होती है क्योंकि वर्तमान समय में पर्यटन उद्योग का लगातार विकास हो रहा है जो राष्ट्रीय आय का प्रमुख स्त्रोत है। राष्ट्रीय और अन्तर्राष्ट्रीय दोनों ही स्तर पर होने वाली आय से राष्ट्रीय आय में वृद्धि होती है। हालांकि स्वदेशी पर्यटन के क्षेत्र में यह तथ्य सत्य नहीं उतरता है, क्योंकि जहाँ तक खान-पान के व्यय का प्रश्न है यह अर्थव्यवस्था में एक स्थान से दूसरे स्थान में परिवर्तित होता रहता है। पर्यटन के तहत सेवाओं व उपभोक्ताओं की वस्तुओं की मांग सम्मिलित है, अन्यथा इनका उत्पादन ही नहीं होता। इससे उत्पादन के क्षेत्रों को बढ़ावा मिलेगा, जिससे स्वयं राष्ट्रीय उपभोग में वृद्धि होगी। क्योंकि पर्यटन व्यय के आकार का प्रभाव उत्पादन की वृद्धि में परिलक्षित होता है, अतः विभिन्न क्षेत्रीय कच्चे माल के माध्यम से उस क्षेत्र व अन्य क्षेत्रों में आय वृद्धि की क्रिया से देखा जा सकता है। यह विकास उपभोग से बिल्कुल स्वतंत्र है एवं इससे समस्त अर्थव्यवस्था पर धारावाहिक प्रतिक्रिया के रूप में प्रभाव पड़ता है। आवास का निर्माण, पर्यटकों हेतु शहरीकरण, आधारिक संरचना की संस्थापना, बढ़ते हुए पर्यटन के परिवहन व यन्त्र आदि कुछ उदाहरण हैं, जिनसे जाहिर होता है कि पर्यटकों की माँग से किस तरह विनियोजन को बढ़ावा मिलता है। इस तरह यहाँ एक वेग वृद्धि सिद्धान्त लागू होता है। साधारण अर्थव्यवस्था में जब कभी उपभोग में संचित दर से वृद्धि होती है, तो यह सिद्धान्त लागू होता है परन्तु यदि प्रति समय यह कम दर से बढ़ रही हो तो यह बिल्कुल लागू नहीं होगा एवं समान रूप में पर्यटन में भी लागू होता है। पर्यटन के क्षेत्र में उन्नति का द्विमुखी विकास व संचित विकास कायम रहना चाहिए।

अतः पर्यटकों की माँग के विनियोजन का वेग वृद्धि को विभिन्न प्रकार से प्रभावित करता है जिसे निम्नलिखित पंक्तियों के अंतर्गत स्पष्ट किया गया है–

1. होटल व अन्य सहायक आवास का निर्माण।
2. संस्थापना तथा आधारित संरचना के कार्य, सड़कों का जाल, स्वछता, जल व बिजली की पूर्ति, हवाई अड्डे, शहरीकरण की योजनाएँ।
3. खाद्य सामग्री, मांस, पेय पदार्थ तथा अन्य कृषि सम्बन्धी वस्तुओं का उत्पादन।
4. परिवहन के साधन व सह-कल पुर्जों आदि का उत्पादन।
5. पर्यटकों के माध्यम से क्रय की जाने वाली वस्तुओं के उत्पादन में वृद्धि।
6. वित्तीय क्षेत्र, बैंकिंग, बीमा व अन्य व्यावसायिक सेवाओं का पुनः निर्माण।

7. नये व्यापारिक संस्थानों का निर्माण।

पहले दो का राष्ट्र की अर्थव्यवस्था पर बहुत ही आधिक प्रतिकूल प्रभाव पड़ता है।

ज्यादातर देशों में पर्यटन का उसकी वित्तीय व्यवस्था में महत्वपूर्ण स्थान होता है। देश वित्तीय व्यवस्था में पर्यटन के योगदान को ज्ञात करने के लिए आवश्यक हो जाता है कि स्वदेशी और विदेशी दोनों ही स्तर पर प्राप्त होने वाली आय की सही-सही जानकारी प्राप्त हो। अतएव अन्तर्राष्ट्रीय यात्रा से प्राप्ति व राष्ट्रीय आय की तुलना इस महत्ता की सूचक हो सकती है। इन आँकड़ों से प्रतीत होता है कि बारवडोज में राष्ट्रीय आय का एक बहुत ही अच्छा अंश पर्यटकों के माध्यम से योगदान किया जाता है। इसके अतिरक्त मैक्सिको, जोर्डन, पनामा व स्पेन में पर्यटन एकमात्र महत्वपूर्ण निर्यात रहा है एवं कनाडा में अखबारी कागज के पश्चात् द्वितीय स्थान रखता था। ब्रिटेन में यह बहुत ही काफी रकम स्टर्लिंग में अर्जित करता है। पर्यटन 24 से भी अधिक राष्ट्रों में निर्यात की जाने वाली तीन मुख्य मदों में से एक है। पर्यटकों से प्राप्त आय का अदृष्टिगत निर्यात के क्षेत्र में आस्ट्रेलिया, कनाडा, कॉस्टारीका, डोमिन रिपब्लिक, इक्वेडर, इल सलबडोर, हन्डूराज, आयरलैंड, इटली, मैक्सिको, पुर्तगाल, स्पेन व स्विटजरलैंड में प्रथम स्थान है।

बहुत से राष्ट्रों में राष्ट्रीय आय के साथ पर्यटन से प्राप्त आय का एक अच्छा अनुपात है। भारत में पर्यटन आय, राष्ट्रीय आय में बहुत ही कम योगदान करती है। 1971-72 के संयुक्त राष्ट्र संघ के विकास कार्यक्रमों व इण्डियन मार्केट रिसर्च ब्यूरो के सर्वेक्षणों के अनुसार, पर्यटन प्राप्ति के सरकारी आँकड़े असली पर्यटन अर्जन के प्रत्यक्ष व अप्रत्यक्ष स्रोतों से 50 प्रतिशत से अधिक नहीं है। पर्यटन की महत्ता तभी परिदर्शित की जा सकती है, जबकि अवकाश लेने वाले एवं व्यापार-यात्रा आदि करने वाले लोगों के कुल व्यय का अनुमान लगाया जाए एवं इस व्यय की तुलना राष्ट्रीय आय की मदों के आँकड़ों से की जाए। इस तरह एक व्यक्तिगत परिवार हेतु पूरे एक वर्ष हेतु कुल आय-व्यय के बजट की समस्या है। यही समस्त राष्ट्र पर लागू होती है। पारिवारिक खर्च के अतिरिक्त केन्द्रीय, राज्य व स्थानीय अधिकारी व व्यापारिक संगठनों का भी ऐसा होता है। इन सबके हिसाब रखने के सिद्धांत व अर्थशास्त्र के नियम एक समान हैं।

प्रति वर्ष राष्ट्र की कुल व्यय व उत्पादित माल व सेवाओं का मूल्य आँका जाता है। राष्ट्र में व्यक्तिगत व संघीय संस्थाओं की आय के जोड़ को ही राष्ट्रीय आय कहते हैं। यह एक वर्ष के अन्दर उत्पादित वस्तुओं तथा संस्थाओं का मूल्य है इसकी गणना

करने के ढंग को उत्पादन प्रणाली कहते हैं। इसकी आय प्रणाली जिसमें व्यक्तिगत व संघीय आय का जोड़ करते हैं, एवं व्यय प्रणाली जिसमें लोगों के क्रम के मूल्य, जिसमें जो क्रय व संचय किया गया, के योग से तुलना की जाती है। इस तरह ये तीन प्रणालियाँ, उत्पादन प्रणाली, आय प्रणाली व व्यय प्रणाली एक-दूसरे पर नियंत्रण का कार्य करती हैं। इन तीनों प्रणालियों के अनुमान का सार एक ही है और वह है कुल राष्ट्रीय उत्पादन। कुल राष्ट्रीय उत्पादन में से कुछ राशि पूंजी को अटल रखने अथवा इसके ह्रास हेतु रखा जाता है, ताकि पर्यटन के माध्यम से निर्मित राष्ट्रीय आय ज्ञात हो सके। यही एक वर्ष की न्यूनतम राशि है जो कि पर्यटन व्यय के माध्यम से उत्पादित होती है। ज्यों ही पर्यटक व्यय करना आरंभ करते हैं राष्ट्रीय आय में वृद्धि होने लगती है। वह होटल मालिकों, रेस्टोरेन्टों, यात्रा संचालकों आदि को धन भुगतान करते हैं। यह धन प्राप्त करने वालों की आय में जुड़ जाता है एवं राष्ट्रीय आय का अंश बन जाता है। इसी तरह होटल मालिक भी रेस्टोरेन्ट, टूर, आपरेटर आदि को धन भुगतान करते हैं। यह धन प्राप्तकर्ताओं की आय में जुड़ जाता है एवं राष्ट्रीय आय का अंश बन जाता है। ठीक इसी भांति रेस्टोरेन्ट मालिक व टूर संचालक इस धन को वेतन, मजदूरी, ब्याज पूर्ति व किराये के रूप में भुगतान करते हैं और इससे फिर राष्ट्रीय आय में वृद्धि हो जाती है। यह प्रक्रिया तब तक कार्यान्वित रहती है जब तक कि वित्त व्यवस्था में संचालित होती रहती है लेकिन इसकी औसत क्रमशः अल्प होती रहती है।

राष्ट्रीय आय में पर्यटन आय का स्थान

पर्यटन से होने वाली विदेशी आय का लाभ कुछ ही लोगों तक सीमित.है जो देश के वित्त प्रबन्ध की उन्नति व विकास में एक विशिष्ट विकट समस्या है। मध्यकाल में बचत को सोने, चांदी, जेवर अथवा मूल्यवान धातु के रूप में परिवर्तित किया जाता था। द्वितीय अवस्था में, बचत को भूमि व वन सम्पत्ति के रूप में किया जाता था। परिवर्तनशील पूंजी नियोजन कृषि हेतु उत्पादनहीन भूमि में किया जाता था। पर्यटक ने राष्ट्रीय आय को कुछ ही व्यक्तियों के हाथ में संचय होने से रोका है, इसने व्यय के अवसरों का सृजन किया। इससे बचत को यात्रा, सामुद्रिक यात्रा, ललित कला दर्शन, स्वास्थ्य लाभ, पर्यटन स्थल, अन्तर्राष्ट्रीय क्रीड़ा समारोह, महा सम्मेलन, प्रदर्शनियों, मेलों आदि में व्यय करने का अवसर मिलता है। इन सबसे धन का पुनः वितरण होता है, जिससे रोजगार के अवसरों में वृद्धि होती है, अन्य क्षेत्रों में बेरोजगारी घटती है। इन सबका यह प्रभाव पड़ता है कि अविकसित क्षेत्रों जिनकी

तरफ कोई ध्यान नहीं दिया गया था, में विकास होने लगता है। सामाजिक दृष्टि से आय के वितरण में सुधार की क्रिया से आय में वृद्धि अधिक प्रभावशाली होगी। आय का स्थिर होने से अथवा वितरण न होने से कोई विकास नहीं होता है।

विदेशी निर्यात तथा पर्यटन

व्यापार से जुड़े प्रत्येक व्यक्ति आयात और निर्यात के अन्तर्गत आते हैं, जिसे भौतिक व्यापार भी कह सकते हैं। परन्तु इससे महत्वपूर्ण निर्यात का एक अन्य तरीका सेवाएं हैं जो अदृष्टिगत निर्यात हैं। इसके तहत निम्न हैं–

(1) बीमा सेवायें,

(2) पर्यटन व्यापार,

(3) परिवहन सेवायें,

(4) बैंकिंग सेवायें,

पर्यटन व्यापार को अदृष्टिगत निर्यात कहते हैं, क्योंकि यह एक सेवाओं का व्यापार है हालांकि इसके तहत कुछ भौतिक मदें जैसे खान-पान आदि भी शामिल हैं।

यहाँ पर सवाल पैदा होता है कि पर्यटन का विदेशी या अन्तर्राष्ट्रीय व्यापार में क्या उपयोगिता है? सर्वप्रथम पर्यटन उद्योग के अन्तर्गत आता है जहाँ पर यह पर्यटन शब्द व्यापार से संबंधित है। इसकी उत्पादित वस्तु है–सेवायें प्रदान करना। जो कि एक अदृष्टिगत वस्तु है एवं इसको अन्य व्यापारिक वस्तुओं की भांति ही व्यापार में विनिमय व विपणन किया जाता है। अधिकतर राष्ट्रों में यह बहुत ही विशद है। भारत ने कुछ की समय पूर्व से कठिन परिश्रम करने के पश्चात् पर्यटन व्यापार को बहुत ही विकसित किया है। परिणामस्वरूप यह राष्ट्र की आय अर्जन का बहुत ही महान उद्योग बन चुका है। कभी-कभी पर्यटन के विपरीत प्रभावों के कारण अन्तर्राष्ट्रीय व्यापार में इसको समझना बड़ा ही मुश्किल हो जाता है।

जब किसी भी देश के व्यक्ति हमारे देश (भारत) में आते हैं तो उनका हमारे देश में उपस्थित रहने से हम उन्हें पर्यटन सेवा उपलब्ध करते हैं। परंतु यदि हम ब्रिटेन जायें एवं वहां पर अपना धन व्यय करें तो उन्हीं के देश में रहकर ब्रिटेन की पर्यटक सेवाओं का आयात करते हैं। इस तरह की अजनबी परिस्थिति इसलिए पैदा होती है क्योंकि हम व्यक्तियों अथवा आदमियों से सम्पर्क कर रहे हैं। अभौतिक वस्तुओं से तो उद्योगों में विशेषकर कर से सम्बन्धित समस्यायें पैदा होती हैं, परन्तु इसके कुछ फायदे भी हैं, इससे सर्वाधिक लाभ वस्तु को बाहर भेजने से है। क्योंकि

फुटकर खर्चे व सेवाओं के खर्च विदेशी एजेन्टों को देने के बजाय भारतीय कम्पनियों को दिए जाते हैं। इससे राष्ट्र को बहुत ही आय अर्जित होती है।

पर्यटन के अन्तर्गत कई प्रकार की सामग्रियां आती हैं। अपनी सेवाओं पर प्रत्येक देश का एक विशिष्ट अधिकार होता है। हर राष्ट्र अपने खान-पान हेतु प्रसिद्ध है, इसके अपने रीति-रिवाज, भव्य दृश्य, इतिहास, विशेष समारोह हैं, जिनमें यात्रियों की विशेष रुचि होती है। भारतीय पर्यटन की ब्रिटेन के पर्यटन से तुलना नहीं की जा सकती है, ये दोनों बिल्कुल भिन्न हैं। ब्रिटेन में यात्रा करने वाले पर्यटकों की तुलना में भारत में आने वाले पर्यटकों के कुछ भिन्न उद्देश्य होते हैं।

यह कथन अपने आप में एकदम सही स्पष्ट होता है कि व्यापार की संरचना जितनी विशाल होगी पर्यटन का स्वरूप उतना ही बड़ा होगा। क्योंकि यह एक दूसरे राष्ट्रों के मध्य सम्पर्क स्थापित करते हैं एवं इससे राष्ट्रीय लोगों का आपस में विचार विनिमय होता है जो कि अपने मित्रों को व्यापार हेतु किसी विशेष कारण हेतु तथा अन्त में शुद्ध आनन्दनीय मनोरंजन यात्रा हेतु लालायित करते हैं। यह मुख्य धारणा है कि यात्रा के विशद आकार का जन्म धनी राष्ट्रों में होता है। धनी राष्ट्र वही हैं जहाँ अन्तर्राष्ट्रीय पर्यटन व्यापार उचित रूप में विकसित है। क्योंकि राष्ट्रीय अस्तित्व का यह एक बहुत ही लाभकारी स्वरूप या आकार है।

अन्तर्राष्ट्रीय व्यापार के महत्व को एक परिवार के उदाहरण के माध्यम से भली-भांति समझा जा सकता है। अगर हर परिवार को स्वावलम्बी बनना है, तो उनके पास इतना वक्त नहीं होगा कि किसी व्यवसाय व व्यापार में दक्षता हासिल कर सकें। सेवाओं का विनिमय न होने से राष्ट्र गरीब हो जायेगा। जितना ज्यादा विशिष्टीकरण का स्तर होगा उतना ही व्यापार से विनिमय होगा, परिणामस्वरूप इससे व्यापार में भाग लेने वाले लोगों में धन की वृद्धि होगी।

यह कहा जाता है कि यात्रा करना जीवन हेतु अनिवार्य नहीं है। यात्रा केवल उन्हीं राष्ट्रों के माध्यम से की जा सकती है जहाँ कि निवासित लोगों के जीवन की न्यूनतम आवश्यकताओं को पूर्ण कर सकते हैं एवं इसके पश्चात् उनके पास इतनी आय हो कि वे यात्रा में भाग ले सकें। यात्रा एक तरह की वस्तु है जिसके उपभोग करने से और अधिक उपभोग करने की प्रबल इच्छा होती है। कुछ वस्तुएं हैं जैसे डबल रोटी। आय में वृद्धि होने से इसकी मांग में घटौती होती है। इसी तरह कुछ वस्तुएं हैं जिनकी मांग अपरिवर्तित रहती है। कुछ अन्य मामलों में आय वृद्धि से मांग में वृद्धि नहीं होती है। उदाहरणार्थ, स्याही। यात्रा के बारे में बिल्कुल विपरीत नियम लागू होता है, जितना ज्यादा आप धनी होंगे उतना ही आप विश्व यात्रा करना

चाहेंगे एवं लोगों से मिलना चाहेंगे। यात्रा के आनन्द कुछ लोगों हेतु समाप्त नहीं होते हैं। आमतौर पर कहा जा सकता है कि यात्रा करने से सबसे अच्छा भविष्य उन लोगों का है जो अधिकतर यात्रा करते रहते हैं। एक यात्री जो विदेशी यात्रा कर चुका है फिर सें भविष्य में अन्य राष्ट्रों की यात्रा करना चाहता है। यदि उसके पास अधिक धन हो तो वह अपना अधिक धन व समय अन्तर्राष्ट्रीय पर्यटन का आनन्द लेने में व्यय करना चाहेगा। अर्थशास्त्रियों का सुझाव है कि अन्तर्राष्ट्रीय पर्यटन पर किए गये व्यय की रकम का स्वतंत्र रूप में खर्च करने योग्य आय के साथ प्रत्यक्ष सम्बन्ध होता है। अर्थात् इस आय को स्वेच्छानुसार आय भी कह सकते हैं, जो कि कुल आय से कर घटाने एवं अन्य आवश्यकताओं को पूर्ण करने के पश्चात् शेष बचती है। यह एक सैद्धान्तिक विचार है जो कि व्यावहारिक जीवन में मान्य है। इसका अर्थ आप की उस रकम से है जो कि यात्रा जैसी अनावश्यकताओं पर व्यय करने हेतु जीवन की जरूरी चीजें जैसे खान-पान, रहन-सहन, कपड़ा इत्यादि को पूर्ण करने के पश्चात् उपलब्ध होती है। यह देश की आमदनी की रचना से जुड़ी है। आमदनी को खर्च करने में तमाम सामग्री व उपभोक्ता की चीजें जैसे—रेफ्रिजरेटर, अच्छा घर, कार, अति प्रिय क्रियायें, फोटोग्राफी आदि पर्यटन से सम्बन्धित हैं। परन्तु इन्हें बिल्कुल अलग रूप से क्रय किया जा सकता है, अक्सर ये वस्तुयें यात्रा उद्योगों की प्रतिस्पर्धी होती हैं।

तुलनात्मक तौर पर अन्तर्राष्ट्रीय व्यापार में पर्यटन का एक महत्वपूर्ण पहलू होना देखा जाए तो आधुनिक प्रगति का द्योतक है। विशद यात्रा आन्दोलन के पैमाने में आय का एक समान वितरण होने का मुख्य कारण उन्नीसवीं व बीसवीं सदी में व्यापार उद्योग व परिवहन का विशद होना एवं साथ ही आय में वृद्धि एवं सामाजिक क्रान्ति है। इसी वक्त अन्तर्राष्ट्रीय व्यापार में एक बहुत बड़ा फैलाव हो रहा था; जिससे न केवल माल व वस्तुओं के विचलन से विश्व भर में बहुत बड़े पैमाने पर प्रत्येक किस्म के अन्तर्राष्ट्रीय लेन-देनों में वृद्धि हुई अपितु साथ ही राष्ट्र की अर्थव्यवस्था का भी विकास हुआ है।

देश के निर्यात व्यापार और आयात यात्रियों के परिवहन की अपेक्षा की जाए तो यह मालूम होता है कि इन दोनों में एक व्यापक व घनिष्ठ संबंध है। बिना किसी शंका के कहा जा सकता है कि यात्रा से मार्केट का सृजन होता है; क्योंकि यात्रियों में अधिकतर व्यापारी होते हैं अतएव एक अच्छे फलते-फूलते पर्यटक आन्दोलन से प्रभावित गतिशील सम्पदा इन पर्यटकों की आगामी आर्थिक व्यवहार तथा बर्ताव पर अवश्य प्रभाव डालती है।

पर्यटन के माध्यम से भिन्न-भिन्न राष्ट्रों के मध्य पैदा होने वाले लेन-देन व धन सम्बन्धी विचलन चालू एक पूंजी दोनों ही प्रकृति के होते हैं ये निम्न हैं–

(1) विदेशी मुद्रा विनियोजन–विशेषकर आवास व्यवस्था में विदेशी पूंजी विनियोजन भी भुगतान संतुलन में पूंजी खाते के तहत सम्मिलित होती है।

(2) वित्त समस्या–विदेशों में पर्यटन उद्योगों में कार्यरत राष्ट्रीय नागरिकों के माध्यम से हस्तांतरित धन विदेशों में पर्यटन कार्यालय व विज्ञापन हेतु प्रदान किए गए वित्त का हस्तांतरित। यह मदें भी भुगतान सन्तुलन में अन्य अदृश्यगत मदों के तहत सम्मिलित की जाती हैं।

(3) यात्रा किए गए देश में पर्यटकों के माध्यम से किया गया व्यय–ये व्यय आवास, खान-पान, खरीददारी तथा स्थानीय परिवहन आदि पर किए जाते हैं। यह मद भुगतान के संतुलन के पर्यटन खाते में अलग रूप में परिलक्षित होते हैं।

(4) कर लाभांश–जिन राष्ट्रों में पूंजी का उद्गम हुआ उनमें पूंजी पर प्राप्त फायदे को प्रेरित करने के धन-सम्बन्धी हस्तांतरण के लेन-देन होते हैं। ये भी भुगतान संतुलन में अन्य अदृश्यगत मदों के तहत सम्मिलित किए जाते हैं।

(5) विदेशी, अन्तर्राष्ट्रीय, हवाई व सामुद्रिक कम्पनियों के माध्यम से किया गया भुगतान–इससे भी व्यापार सन्तुलन के अन्य उद्देश्य दृश्यगत मदों में दिखाया जाता है अर्थात् सेवाओं व धन सम्बन्धी लेन-देन के अन्य आयात व निर्यात।

(6) पर्यटकों के माध्यम से क्रय किया गया पूंजी सम्बन्धी सामान–मोटरकार, पुरानी कीमती वस्तुयें आदि व्यापार के सन्तुलन में आयात-निर्यात माल के तहत सम्मिलित की जाती हैं।

संतुलित भुगतान के धनात्मक प्रभाव

विदेशी यात्रियों द्वारा किये गये खर्च का देश के भुगतान पर अत्यधिक असर पड़ता है। अन्तर्राष्ट्रीय पर्यटकों के व्यय किये बिना बहुत से प्रशान्त व सुदूर पूर्व के राष्ट्रों में बहुत ही कमजोर भुगतान सन्तुलन रहता था। फलतः बहुत से राष्ट्रों को आयात में कमी करने को बाध्य होना पड़ता था। ब्रिटेन, स्पेन, इटली, आयरलैंड, न्यूजीलैंड, स्विट्जरलैंड व मैक्सिको में पर्यटन एक महत्वपूर्ण योगदान कर सकता है। एक व्यक्ति विशेष के बावत जो उसकी है, ठीक उसी के विपरीत एक राष्ट्र को बहुत धनी होना अनिवार्य नहीं है। निरंतर सकारात्मक सन्तुलन होने से भी राष्ट्र की अर्थव्यवस्था पर मुद्रास्फीति के प्रभाव पड़ते हैं। इसका उदाहरण जर्मन की 1968-69 की आर्थिक परिस्थितियाँ हैं। बहुत से जर्मन लोगों ने पर्यटन के रूप में सम्पूर्ण

यूरोपियन राष्ट्रों में यात्रा करके जर्मन मार्क्स को व्यय करना आरम्भ किया ताकि आन्तरिक मुद्रास्फीति को घटाया जा सके। दूसरी तरफ हम एक विकासशील राष्ट्र का विचार करते हैं। क्योंकि इस राष्ट्र को आधारित संरचना व मूल उद्योगों के निर्माण के वास्ते, बहुत से यन्त्र व अन्य सामान विदेशों से आयात करने होते हैं। अतः इसके भुगतान सन्तुलन के घाटे में होने की सम्भावना होती है। अतएव उनके लिए लाभदायक होगा कि नकारात्मक व्यापार सन्तुलन के क्षतिपूर्ति हेतु विदेशी पर्यटकों को आमन्त्रित करें, सैद्धान्तिक रूप में, पर्यटन के आर्थिक प्रभाव तभी अच्छे होते हैं जबकि पर्यटकों के सन्तुलन के सकारात्मक भुगतान सन्तुलन (बिना पर्यटन के) की अपेक्षा अच्छे हों, अतएव सकारात्मक भुगतान सन्तुलन (पर्यटन को सम्मिलित न करके) में यह उम्मीद होगी कि पर्यटक सन्तुलन नकारात्मक हो, एक राष्ट्र जो सकारात्मक पर्यटक सन्तुलन से फायदा प्राप्त करता है उसके लिए यह जरूरी होगा कि विदेशी पर्यटक जो उनके देश में यात्रा करते हैं उनकी अपेक्षा उसके स्वयं के विशेष श्रेणी नागरिक (धनी वर्ग) विदेशों में कम व्यय करें। इस अवस्था में, विनिमय नियंत्रण जरूरी हो जाता है।

किसी भी व्यक्तिगत राष्ट्र का व्यापार खाता उस राष्ट्र की अर्थव्यवस्था में पर्यटन के महत्व को मापने का तरीका है। विदेशी पर्यटकों के माध्यम से किया गया व्यय किसी भी देश के भुगतान संतुलन की स्थिति को सकारात्मक रूप में प्रभावित करता है। पर्यटन से अर्जित आय निम्न तथ्यों पर निर्भर करती है :

(1) उनके ठहरने का औसत समय।

(2) आने वाले पर्यटकों की संख्या।

(3) किसी विशेष समय में मूल्यों का स्तर।

पर्यटन के मुख्य रूप से तीन भागों में बांटा गया है–

(1) स्थानीय पर्यटन–स्थानीय पर्यटन हेतु निर्माण किए गए आवास में कुछ-न-कुछ आयातित पूर्ति व यन्त्र भी सम्मिलित होते हैं। बहुत से स्थानीय पर्यटन आयातित ईंधन के माध्यम से संचालित परिवहन के माध्यम से यात्रा करते हैं एवं आयातित खान व पेय पदार्थ का उपभोग करते हैं, दूसरी तरफ स्थानीय पर्यटक के माध्यम से निर्मित यात्रा यन्त्र विदेशी पर्यटकों से विदेशी मुद्रा अर्जित करने का साधन बन जाते हैं। इस तरह अन्तर्राष्ट्रीय पर्यटन से भुगतान संतुलन निकालने में बड़ी सुविधा मिलती है। यह स्पष्ट है कि पर्यटन के भिन्न-भिन्न आकार आपस में पूर्णतः सम्बन्धित हैं एवं इनका भुगतान संतुलन की स्थिति पर विभिन्न प्रभाव डालते हैं।

(2) आन्तरिक पर्यटन—वित्त नजरिए से आन्तरिक पर्यटन अत्यधिक लचकशील दिखाई पड़ता है। चूंकि इसके द्वारा दो प्रकार के पर्यटनों का विकास होता है। प्रगतिशील पर्यटन व्यापार एक प्रभावशाली पर्यटन पर ही निर्भर करता है। इस पर्यटन यन्त्र के माध्यम से ही स्थानीय पर्यटन व्यापार के सृजन के साधन, जैसे—होटल, परिवहन, दर्शनीय स्थल आदि निर्मित होते हैं। आन्तरिक पर्यटन से आय का सृजन होता है जिसके परिणामस्वरूप स्थानीय नागरिक स्वदेश में यात्रा कर सकते हैं इस तरह की आय से स्थानीय लोगों की आय में वृद्धि होगी जो कि विदेशी यात्रा करने की क्षमता रखते हैं एवं देश में असंख्य पर्यटकों के आगमन के प्रभाव से स्थानीय लोगों में विदेश भ्रमण करने की इच्छा पैदा होती है।

(3) बाह्य पर्यटन—बाह्य पर्यटन से किसी भी पर्यटन का निर्माण नहीं होता। बाह्य पर्यटन के द्वारा नये विचारों या ख्यालों व भावनाओं का विकास होता है। एक राष्ट्र हेतु जो बाह्य पर्यटन हैं वही दूसरे राष्ट्र हेतु आन्तरिक पर्यटन कहा जा सकता है।

कुछ अर्थशास्त्रियों का मत है कि अन्तर्राष्ट्रीय पर्यटकों के व्यय की महत्ता का अध्ययन स्थानीय नागरिकों के माध्यम से विदेशों में यात्रा किए गए व्यय को ध्यान में रखकर करनी चाहिए। इस तर्क का प्रयोग प्रायः पर्यटन की महत्ता को करने के लिए किया जाता है कि अन्तर्राष्ट्रीय पर्यटकों (अन्तर्राष्ट्रीय पर्यटन) के खर्च की तुलना विदेशों में यात्रा करने वाले नागरिकों (बाह्य पर्यटन) के माध्यम से किए गये खर्च से करनी चाहिए ताकि पर्यटन के भुगतान संतुलन की स्थिति पर हुए प्रभावों का सही ज्ञान हो सके। यह तर्क अक्सर पर्यटन की महत्ता को कम करने हेतु किया जाता है।

6

वित्तीय उद्देश्य तथा पर्यटन

विदेशी पर्यटन वित्तीय उद्देश्य का प्रमुख स्त्रोत है। इसके द्वारा अंतर्राष्ट्रीय पर्यटकों तथा अर्जित स्वदेशी लोगों द्वारा अंतर्राष्ट्रीय वित्त को विदेशों में खर्च करने के लिए उपयोगी बनाया जा सकता है। पर्यटन किसी भी देश के संतुलित आर्थिक विकास का लक्ष्य प्राप्त करने का माध्यम होते हैं। पर्यटन को बढ़ावा देने के लिए कुछ सामान्य आर्थिक उद्देश्य हैं जिनको निम्नलिखित तरीके से बताया जा सकता है–

(1) पर्यटन के माध्यम से श्रमिकों की बहुतायत वाले स्थानों में बेरोजगारी समाप्त करने में मदद मिलती है।

(2) पर्यटन से क्षेत्रीय विकास में सुविधा तथा वृद्धि होती है।

(3) एक निर्यात व्यापार के समान पर्यटन विदेशी मुद्रा अर्जित करता है जो कि व्यापार संतुलन को सुधारने में सहायक होती है।

आने वाले पर्यटकों के खर्च के अन्तर्गत सम्पूर्ण व्यय भी सम्मिलित हैं, उदाहरणस्वरूप बाहर से आने वाले पर्यटकों के द्वारा खुद के रुकने से किये गये व्यय का भुगतान इस प्रकार से होता है जैसे—आवास, खान-पान, स्थानीय परिवहन, दर्शनीय स्थल (दर्शक सेवायें सहित), मनोविनोद, खरीददारी (स्मारिकायें सहित) व स्थानीय यात्रा अभिकर्ताओं के माध्यम से अर्जित कमीशन।

प्रायः आने वाले पर्यटकों की संख्या अथवा पर्यटक राशि की संख्या को पर्यटक खर्चों के समानान्तर समझा जाता है। इसका अर्थ यह है कि पर्यटक संख्या में वृद्धि स्वतः ही पर्यटन से अर्जित आय में वृद्धि का सूचक है, जब कभी यह सम्बन्ध सही नहीं उतरता तो इसके मुख्य कारण निम्न हो सकते हैं–

मुद्रा के मूल्य ह्रास अथवा पुनर्मूल्यांकन के माध्यम से दखलंदाजी करने से।

सीमापार से यात्रियों का अचानक पदार्पण, जिसका विदेशी राष्ट्रों से प्राप्त आय पर थोड़ा-सा ही प्रभाव पड़ता है, पर्यटकों के व्यय का कोई अंश मापदंड के तहत सम्मिलित होने से वंचित होता है। अंतर्राष्ट्रीय पर्यटन के भुगतान पर पड़े प्रभाव निम्नलिखित हैं–

1. यह स्थल बाहर भी रोजगार के अवसर उपलब्ध कराने में सहायक सिद्ध होते हैं तथा इसका श्रेय पर्यटन के उद्योग को ही माना जाता है। पर्यटन के माध्यम से सृजन किए गए रोजगारों की सीमा अदक्ष कार्यों से लेकर प्रबन्ध कार्यों तक फैली हुई है। यह उद्योग अन्य उद्योगों की अपेक्षा लोगों को प्रत्यक्ष व अप्रत्यक्ष रूप से रोजगार प्रदान करता है। पर्यटन उद्योग से लाभान्वित वे व्यक्ति हैं जो कि होटलों, ट्रेवल एजेन्सी, हवाई कम्पनियों आदि में काम करते हैं अथवा टैक्सी चालक, पुष्प-विक्रेता, स्मारिका-विक्रेता, सब्जी-विक्रेता, गोस्त-विक्रेता, हाउसकीपर, भोजन विशेषज्ञ, प्रतीक्षक, मनोविनोदकर्ता आदि के साथ-साथ बहुत से दक्ष व अदक्ष श्रमिकों जैसे स्वागतकर्ता कर्मचारियों, चैम्बरमेड, पार्टर, किचन स्टाप, माली इत्यादि की भी जरूरत होती है।
2. रोजगार से जुड़ा पर्यटन एक उद्योग है। यात्रियों की सामग्री पूर्ति हेतु रोजगार के मौके उपलब्ध होते हैं। इस क्षेत्र में नये रोजगार सृजन करके एवं साथ ही अन्य क्षेत्रों जैसे निर्माण कार्य, परिवहन आदि में बढ़ावा देकर रोजगार के स्तर में वृद्धि करता है। अन्य क्षेत्रों, विशेषकर कृषि के क्षेत्रों में यह अप्रत्यक्ष रूप में उत्पादकता की वृद्धि करता है, क्योंकि यह मानव-शक्ति को रोजगार के लुप्त साधनों अथवा मौसमी अर्द्ध-रोजगारों को स्वयं की तरफ आकर्षित करता है। विशद रूप में, पर्यटन को श्रमिकों का उपयोग करने वाला क्षेत्र समझा जाता है एवं इसके रोजगार पैदा करने का अधिकतर प्रभाव उन क्षेत्रों में समझा जा सकता है, जहाँ पर अन्य आर्थिक क्रियाओं हेतु सीमित साधन विद्यमान हैं।
3. पर्यटन के माध्यम से रेस्टोरेन्ट व्यावसायिकों, यात्री स्थानों को यातायात एवं हस्त व्यवसायों के पथ-प्रदर्शक इत्यादि के रूप में प्रत्यक्ष कार्यों का विकास

होता है। पर्यटकों के माध्यम से क्रय की गयी वस्तुओं के अलावा, प्रायः हस्तकारिता वस्तुएं, कढ़ाई और चीनी मिट्टी व पीतल के बर्तन, जेवर एवं चित्रकारी सम्बन्धी वस्तुओं पर भी पर्यटक बहुत आकर्षित होते हैं। पर्यटन उद्योग में प्रत्यक्ष रूप में रोजगारित लोगों के माध्यम से किया गया व्यय होटल उद्योग में अप्रत्यक्ष पेशों को जन्म देता है। ये रोजगार पर्यटन उद्योग से सम्बन्धित हैं, एवं पर्यटन के विस्तार के परिणामस्वरूप वे निर्माण में रोजगार के अवसरों में वृद्धि करते हैं। जो अप्रत्यक्ष रूप में पर्यटन उद्योग में रोजगारित है, वे माल व सेवाओं की मांग के कारण अन्य क्षेत्रों में रोजगार के अवसरों का सृजन करते हैं। इस तरह की अतिरिक्त मांग को पूर्ण करने हेतु माल व वस्तुओं के उत्पादन हेतु अधिक व्यक्तियों की जरूरत होती है। आय व रोजगार वृद्धि का चक्र तब तक धीरे-धीरे चलता रहेगा जब तक यह पूर्णतः समाप्त नहीं हो जाती है, यह प्रवृत्ति रोजगार से गुणित के कारण विद्यमान रहती है।

4. रेस्टोरेन्ट और अन्य यात्री से जुड़ी सेवाओं के लिए मजदूरों की आवश्यकता पड़ती है। यह माननीय है कि विकासशील राष्ट्रों में विकसित राष्ट्रों की अपेक्षा होटल के कर्मचारियों एवं कमरे-बिस्तरों का अनुपात प्रायः अधिक होता है। अनुभव से ज्ञात होता है कि एक बिस्तर की व्यवस्था से एक व्यक्ति को रोजगार मिलता है। उदाहरणार्थ, 1000 बिस्तर वाली इकाई में अनुमानतः 500 व्यक्तियों को उस होटल में रोजगार मिलता है और अन्य 500 स्थान अन्य सहायक व्यवसायों जैसे—लान्ड्रीज, विक्रय केन्द्रों, स्मारिका निर्माताओं व व्यापारियों, मनोविनोद स्थलों, सामान-वाहक, वाहन-चालकों, उत्पादक व विक्रेताओं, मनोविनोद व्यापारियों, आदि क्षेत्रों में उपलब्ध हो जाते हैं। निर्माण उद्योग पर तो अत्यधिक प्रभाव पड़ता है क्योंकि पर्यटन में बहुत-सी इमारतें, होटल, अतिथि गृह, पर्यटक बंगले, घर चैलटे, युवा होस्टल, पर्यटक-लघु निवास स्थल, निवासित कक्ष, सामूहिक आवास आदि का निर्माण होता है। हजारों कक्षों के निर्माण के परिणामस्वरूप असंख्य हस्तकारिकों को रोजगार प्राप्त होते हैं।
5. होटल निर्माण हेतु आवश्यक विदेशी पूंजी पर प्रेक्षित ब्याज व लाभ होता है।
6. अधिकतर लागत लाभों के तर्कों पर ही केन्द्रित रही, क्योंकि विनियोजन की लागत को पर्यटन के लाभों की दृष्टि में रखकर आँका जाता है। श्री रिचर्ड के अनुरूप "पर्यटन व्यापार के प्रत्येक प्रत्यक्ष 100 कार्यों हेतु अन्य उद्योगों

में अनुमानतः 63% कार्य अप्रत्यक्ष रूप में पर्यटन व्यापार पर आश्रित होते हैं। साधारण आर्थिक शब्दों में यह रोजगार गुणित 1.6 है। ब्रिटेन में दस लाख पाउन्ड पर्यटन व्यय से राष्ट्र के अन्दर (स्वदेशी पर्यटन से भी रोजगार को जोर मिलता है) पर्यटन के व्यापार में सम्भवतः 860 कार्यों को प्रोत्साहन मिलता है अथवा आकर्षित किया जाता है एवं सशक्त अर्थव्यवस्था में लगभग कुल 1255 कार्यों को सहायता मिलती है।

7. विदेशी पर्यटकों के उपभोग हेतु खान-पान, पेय, ईंधन, विशेष पूर्ति सहित आयातित वस्तुओं की विदेशी मुद्रा लागत का भुगतान संतुलन पर प्रभाव का ज्ञान प्राप्त करना होगा।
8. विदेशी विज्ञापन व प्रमोशन सहित पर्यटन विकास के कार्यक्रमों के संचालन को चालू रखने हेतु विदेशी मुद्रा की लागत।
9. पर्यटन उस पूंजी को अर्जित करता है जिसको लोग विनियोजन करने हेतु निर्णय नहीं कर पाते हैं। अधिकांश राष्ट्रों में पर्यटन एक बहुत ही महान उद्योग है, जिसमें लोग बड़ी संख्या में आवास, खान-पान, परिवहन, मनोरंजन, सेवा उद्योग व स्मारिका व्यापार में तल्लीन हैं।
10. पर्यटन के माध्यम से विदेशी व सार्वजनिक कोष का बोझ हल्का हो जाता है।
11. पर्यटन एक श्रमिक-कृत उद्योग है और अन्य साधारण उत्पादक उद्योग की तुलना में कई गुना अधिक रोजगार प्रदान करता है, कितनी ही तरह की व्यापारिक कम्पनियां जैसे—होटल, रेस्तरां, ट्रेवल एजेन्ट्स, टूर आपरेटर्स, उपहार विक्रय केन्द्र, परिवहन आदि पर्यटन डालर का एक बहुत बड़ा अंश अर्जित करता है। इसके परिणामस्वरूप बहुत सारी अन्य सहायक अथवा उच्चस्तरीय व्यापारिक कम्पनियां, वाहन चालक, फोटोग्राफर, केश-शृंगारक, चिकित्सक, दन्त चिकित्सक आदि लाभान्वित होते हैं।
12. पर्यटन तथा रोजगार—पर्यटन बेरोजगारी समाप्त करता है।
13. पर्यटन बेरोजगारी समाप्त करने का एक बहुत बड़ा पहलू है, जिन क्षेत्रों में श्रमिकों का बाहुल्य है वहां बेरोजगारी है। उनके लिए पर्यटन एक समाधान है, क्योंकि पूंजी नियोजन की दृष्टि से श्रमिक शक्ति के इस्तेमाल का प्रतिशत अधिक है। बेरोजगारी को समाप्त करने में पर्यटन के महत्व को जानने हेतु निम्न बातों की तरफ ध्यान देना चाहिए :
    (i) सरकारी आँकड़ों में, पर्यटन सेवाएं और अन्य सेवाएं जो पर्यटन से संबंधित नहीं हैं व समानान्तर अन्य खान-पान क्रियाओं में रोजगारित

व्यक्तियों में कोई अन्तर नहीं रखा जाता है। इसी तरह पर्यटन के रोजगार को छोटे-छोटे क्षेत्रों जैसे ट्रेवल एजेन्सी आदि से सम्बन्ध होने पर भी अलग दिखाने के बजाय परिवहन के विभिन्न भागों के तहत प्रदर्शित किया जाता है।

(ii) पर्यटकों से सम्बन्धित अधिकतर क्रियाएं मौसमी होती हैं। इन पर रोजगारित लोगों की संख्या वर्ष में तेजी व मन्दी के समय के अनुरूप परिवर्तित होती रहती हैं।

(iii) यह भी देखा गया है कि रोजगार के आँकड़ों में केवल वेतनभोगी वर्ग को ही शामिल करते हैं एवं मालिक व स्वरोजगारित व्यक्तियों को इसके अन्दर शामिल नहीं किया जाता है। पर्यटन से सम्बन्धित सेवाओं की कई छोटी-छोटी इकाइयों के माध्यम से पूर्ति की जाती है जो कि पूर्ण मानव शक्ति का एक बहुत बड़े अंश का प्रतिनिधित्व करता है।

इस प्रकार उपर्युक्त आधारों की तुलना करने से विकासशील राष्ट्रों में पर्यटन के विस्तार के लाभों का अनुमान लगाने में मदद मिलती है। इससे पर्यटन रोजगार में वृद्धि के अवसरों का अनुमान लगाने में एक सही आधार प्राप्त होता है। साथ ही यह पर्यटन विकास हेतु सार्वजनिक वित्त प्राप्त करने का एक सुदृढ़ तर्क भी है। प्रशान्त तथा सुदूर पूर्व के राष्ट्रों में पर्यटन के एक सर्वेक्षण के अनुरूप पर्यटन खर्च का 64% वेतन व मजदूरी पर व्यय होता है।

यदि सभी पर्यटन उद्योग या व्यवसाय को पर्यटन बाजारीकरण के रूप में ग्रहण किया जाए तो इस व्यवसाय में रोजगार से संबंधित सभी लोगों का आंकड़ा जल्द ही पर्याप्त अनुमानों से ज्ञात किया जा सकता है। इन आँकड़ों से यह प्रतीत होता है कि अन्य उद्योग व सेवाओं में कुल रोजगारित व्यक्तियों के अनुमानतः 2% व्यक्ति होटल, रेस्तरां व लान्ड्रीज, फुटकर सेवाएं, ट्रेवल एजेन्सीज, स्मारिका व्यापार, पेय पदार्थ, दवाइयां, फोटोग्राफी का सामान आदि में कार्य करने वाले व्यक्तियों को शामिल नहीं किया गया है। पर्यटन के संबंध में **श्री मेडेलिक ने** कहा है कि "पर्यटन रोजगार का एक स्रोत है, विशेषतः उन क्षेत्रों हेतु जिनमें रोजगार के वैकल्पिक साधन सीमित हैं। जैसे कि उद्योग विहीन के वे क्षेत्र जिनमें दर्शनीय आकर्षक व जलवायु के अतिरिक्त प्राकृतिक साधनों की कमी है।"

श्री लिकोरिश के अनुमानों के अनुरूप ब्रिटेन में कुल कार्य करने वाले लोगों का कम-से-कम 7% पर्यटन उद्योग में रोजगारित है। स्विट्जरलैंड में किए गए एक अध्ययन के अनुसार यह उद्योग ग्रीष्मकाल में 140,000 लोगों को रोजगार प्रदान

करता है, जिसमें से 70,000 होटलों में कार्य करते हैं, पर्यटन उद्योग में कार्य करने वालों में से करीब आधा प्रतिशत आवास प्रदान करने वाली इकाइयों में कार्यरत है।

श्री रोउथ के अनुमानों का सुझाव है कि ब्रिटेन में कुल कार्य करने वाली जनसंख्या का 5% पर्यटन उद्योग के माध्यम से अपनी जीविका अर्जित करती है। ब्रिटेन में बहुत से क्षेत्र हैं जहाँ पर कुल कार्य करने वाली जनसंख्या के 10% से अधिक लोग मुख्य तौर पर पर्यटकों को सेवायें प्रदान करने में व्यस्त हैं। कुछ क्षेत्रों व स्थानों में यह अनुपात विशेषतः अधिक है एवं कुछ पर्यटन स्थलों में 20% से भी अधिक है।

7

पर्यटन वाणिज्य के लाभ

पर्यटन वर्तमान समय में उद्योग का रूप धारण कर चुका है। यह किसी भी देश की अर्थव्यवस्था में महत्वपूर्ण भूमिका निभाता है। पर्यटन के माध्यम से विदेशी मुद्राएं प्राप्त होती हैं। इन सबके अतिरिक्त देश के लिए राष्ट्रों के बीच आपसी संबंध मजूबत होता है।

किसी भी अर्थव्यवस्था में पर्यटन के आर्थिक मूल्यों का मूल्यांकन इसके राष्ट्रीय आय में किए गए योगदान, विदेशी मुद्रा, विनिमय अर्जन, सरकारी राजस्व में योगदान एवं सृजित किए गए रोजगारों के अवसरों से की जा सकती है। पर्यटकों के माध्यम से प्रत्यक्ष व अप्रत्यक्ष रूप में सृजन किए गए समस्त रोजगार का राष्ट्रीय श्रम-शक्ति में वही अनुपात है जो कि पर्यटन के योगदान से। पर्यटन से सरकार को जो राजस्व अर्जित होता है उसको अर्जन अथवा पर्यटन आय से भिन्न समझना चाहिए। पर्यटन आय या अर्जन को इस प्रकार भी समझ सकते हैं कि वह अंतर्राष्ट्रीय लेन-देन जो पर्यटक प्रक्रिया के उपरांत उत्पन्न होता है तथा भुगतान व्यवस्था को समृद्ध करता है। पर्यटन आय से आशय आरम्भ में पर्यटन व्यय के माध्यम से राष्ट्रीय आय में गुणित के प्रभावों के योगदान से होता है—पर्यटकों के माध्यम से क्रय अथवा प्रयोग

किए जाने वाले सामान व सेवाओं पर करारोपण। पर्यटन राजस्व मालूम करते वक्त पर्यटन उद्योग के माध्यम से क्रय की गयी वस्तुओं व सेवाओं पर लगे करों तथा फायदों के विभिन्न कारणों को भी ध्यान में रखना चाहिए।

राष्ट्र के उत्पादन या वित्त प्रबन्ध के कुछ व्यापार में दूसरे देश के पर्यटन से उत्पन्न किये गये प्रत्यक्ष अथवा अप्रत्यक्ष व्यापार का प्रतिशत अनुमान विदेशी पर्यटन के योगदान से लगाया जा सकता है। यह अनुमान है कि पर्यटन आय के गुणित प्रभाव, पर्यटन रोजगार के गुणित प्रभावों के बराबर होते हैं। यदि हम अर्थव्यवस्था की रोजगार की प्रबलता में सम्भावित मूल्य और पर्यटन रोजगार के सामान्य अनुपात की दर को माप सकें तो हम औसत रोजगार का अनुमान लगा कर यह प्रदर्शित कर सकते हैं कि किसी विशेष अवधि में प्रति इकाई मूल्य वृद्धि अथवा प्रत्येक अतिरिक्त राष्ट्रीय वस्तु में कितने अतिरिक्त रोजगारों का सृजन किया जा सकता है। इस तरह रोजगार-आय अनुपात के सिद्धान्त के माध्यम से रोजगार उपलब्धि में पर्यटन के योगदान का अनुमान लगाया जा सकता है। क्योकि हमारे पास अनुमान लगाने व पर्यटन क्षेत्र के रोजगार व आय के विशद आँकड़े एवं पर्यटन क्रियाओं के उत्पादन के कार्यों के प्रयोग के आंकड़े उपलब्ध नहीं हैं।

पर्यटन के वित्त प्रबन्ध में व्यवसाय का रोजगार उद्भव करने के सम्पूर्ण असर का आंकड़ा उस आमदनी से है जो कि उपभोगी वस्तुओं के उद्योगों में पर्यटन रुपये के प्रथम व आगामी व्यय के चक्र द्वारा पैदा होती है। यह एक आम धारणा है कि पर्यटन व्यय द्वारा पैदा वेतन व मजदूरी उस व्यय का 50% होती है।

पर्यटक के एक वर्ष हेतु किये गये व्यय के परिणामस्वरूप हुई आर्थिक क्रिया से वेतन व भजदूरी का भुगतान किया जा सकेगा। 1000 रुपये के पर्यटन व्यय पर गुणित के प्रभाव का अध्ययन करने पर यह प्रतीत होता है कि इससे एक ऐसी अर्थव्यवस्था पैदा होती है, जिसमें 540 रु. (कुल पर्यटन व्यय का 54%) वेतन व मजदूरी के रूप में भुगतान किये जा सकते हैं। यह सम्भव है कि विदेशी यात्रियों को प्रदान की जाने वाली सेवाओं में सम्बन्धित उद्यमों में कर्मचारियों की संख्या का अनुमान लगाया जा सकता है। भारत में कर्मचारियों तथा होटल कक्षों का अनुपात 2.5:1 स्वीकार किया गया है। यूरोप में यह अनुपात 1:1 व उत्तरी अमेरिका में 008:1 अनुमानित किया गया है। यात्रा उद्योग के तहत इतने सारे हिस्से आते हैं कि यह अनुमान लगाना बड़ा ही कठिन है कि पर्यटन के विभिन्न उद्यमों जैसे रेस्तरां, अतिथि गृह, विक्रय केन्द्र, परिवहन सेवायें, संग्रहालय आदि जो पर्यटकों की आवश्यकताएं अप्रत्यक्ष रूप में पूर्ण करते हैं, में कितने रोजगार आरोपणीय हैं। इसके अतिरिक्त,

अनेक प्रोत्साहित रोजगार हैं जो कि पर्यटन स्थलों के विकास, सड़क निर्माण, निर्माण-उद्योग आदि के परिणामस्वरूप होते हैं। भारत में पर्यटन रोजगार के प्रभावों का अभी तक कोई अनुमान नहीं लगाया गया है। कुछ अर्थशास्त्रियों का विचार है कि पर्यटन के माध्यम से सृजित प्रत्यक्ष व अप्रत्यक्ष रोजगारों का राष्ट्र की कुल रोजगार व्यवस्था में लगभग उतना ही अनुपात है जितना कि पर्यटन का राष्ट्रीय आय के योगदान में है।

भारत जैसे उन्नतिशील राष्ट्र में पर्यटन को संवर्धन से वित्त उद्भव तथा व्यवसाय में अत्यधिक बल मिलेगा। 1968 में एँगलिया में किये गये अध्ययन के अनुरूप यह सुझाव दिया गया कि प्रति 2,400 के पर्यटन खर्च के लिये एक नया रोजगार सृजन किया गया है।

श्री जी. एच. ब्लाके व आर. एल. लोलेस के मतानुसार, "प्रत्येक छह नये होटल शयनों हेतु प्रत्यक्ष रूप में एक एवं अप्रत्यक्ष रूप में बहुत से अन्य रोजगारों का सृजन होगा। राष्ट्र के महत्वहीन भागों हेतु यह बहुत ही महत्वपूर्ण तथ्य है। उदाहरणार्थ—दक्षिणी क्षेत्र, जहाँ रोजगार के स्रोत बहुत ही सीमित हैं। इस तरह क्षेत्रीय विकास हेतु विशेषतः कार्य करने के साधन सृजन की दृष्टि से पर्यटन एक महत्वपूर्ण पहलू है।"

उपर्युक्त परिभाषा से स्पष्ट होता है कि पर्यटन व्यवसाय में दक्षता देने की वित्त की तीव्र गति व उच्च कोटि के अनुभवी विदेशी कार्मिकों की आवश्यकता होती है। परन्तु साथ-ही साथ यह असंख्य स्थानीय लोगों को रोजगार प्रदान करती है। पर्यटन सेवाओं से कम दक्ष कर्मचारियों का अनुपात बहुत ही उच्च होता है, भारत जैसे राष्ट्र में जहाँ कि अदक्ष मजदूरों की अधिकता होने से, यह विशेषतया बहुत ही अनमोल है। यात्रा उद्योग में प्रारम्भिक अवस्था में ये सेवायें उच्चस्तरीय निपुण/दक्ष नहीं लगती, परन्तु इन सेवाओं में कुछ दक्षता के स्तर की जरूरत होती है। यात्रा-पर्यटक उद्योग में प्राचीन समय से ही कार्य के दौरान प्रशिक्षण आदि की व्यवस्था रही है, ताकि श्रमिकों को पर्यटन से सम्बन्धित दक्षता के विकास हेतु स्थानीय शैक्षणिक व व्यावसायिक प्रशिक्षण स्रोतों पर निर्भर न रहना पड़े। स्थानीय व्यक्तियों को प्रशिक्षण देकर जटिल तथा उच्च कोटि के दक्षतापूर्ण स्थानों में नियुक्त किया जा सकता है। "रोजगार सृजन में विदेशी पर्यटन के प्रत्यक्ष योगदान का अनुमान अर्थात् प्रारम्भ में खर्च की गयी पर्यटक खर्च की किस्त, सृजन किए गये रोजगार का अनुमान लगाना बड़ा ही मुश्किल है एवं समान मापदण्ड भी नहीं लगाया जा सकता है, क्योंकि भारतीय अर्थव्यवस्था में विशद आँकड़े अथवा पर्यटन

व्यय व रोजगार का पर्यटन क्रियाओं में प्रति व्यक्ति आय के वर्गीकरण का बड़ा अभाव है।"

पर्यटन एवं यातायात

पर्यटन स्थल पर यातायात की समुचित व्यवस्था होनी चाहिए क्योंकि पर्यटकों के आने से यातायात के साधनों को अधिक लाभ होता है। बस व मोटरों पर किये गये व्यय का सर्वेक्षण करके उसकी मान्यता को जांचने की प्रणाली एक संतोषजनक पद्धति प्रतीत नहीं होती है, क्योंकि पर्यटकों की यात्रा का यह एक बहुत ही महत्वपूर्ण साधन है। अतएव, एक बड़ी राशि इस साधन अथवा मद पर खर्च की जाती है। इसके अतिरिक्त बहुत-सी टैक्सियों, स्वचालित वाहनों, भाड़े की कारें व निजी कारें हैं, जो कि पर्यटन यातायात से प्रभावित होती हैं। पर्यटकों की समुद्री व वायु यातायात के माध्यम से सैर आन्तरिक भू-परिवहन हेतु भी लाभकारी है।

अनूठी वस्तुओं हेतु नवीन विक्रय स्थल

स्वदेशी उद्योगों, नवीन बाजारों आदि का निर्माण पर्यटन से होता है। हाल ही में यूरोप में किये गये पर्यटन के सर्वेक्षण से ज्ञात होता है कि पर्यटक अपने कुल धन का 33 प्रतिशत देशीय वस्तुओं के तत्कालीन क्रय हेतु संचित रखते हैं। पर्यटकों के माध्यम से क्रय किया गया सामान निर्यात व्यापार का अंश है। इसके निर्यातक को न तो कोई परिवहन का खर्च देना पड़ता है न ही आन्तरिक व्यापार के जैसा इसमें कोई जोखिम उठाना पड़ता है। कुछ वस्तुएं तो अन्तर्राष्ट्रीय स्तर पर प्रसिद्ध होने से पर्यटकों के माध्यम से मांग में रहती हैं। उदाहरणार्थ–स्विट्ज़रलैंड की घड़ियाँ, स्काटलैंड के ऊनी वस्त्र, इटली का रेशम व लाइलॉन, फ्रांस की शराब व फैशन, स्केन्डिनेबिया के घरेलू सामान के गैडगेट व पूर्वी देशों के गलीचे व ऊनी सामान।

पर्यटन की संनिर्माण प्रक्रिया की मांग

पर्यटन को बढ़ावा देने के लिए एक से निर्माण आधारित प्रक्रिया की जरूरत होती है। प्राकृतिक और कृत्रिम दोनों प्रकार की आधारित संनिर्माण की उन्नति से प्रगति की प्रक्रिया पर थोड़ा-बहुत असर पड़ता है। पर्यटन मांगों के कारण परिवहन व संचार सुविधाओं, शहरीकरण, मरीना और समुद्रतट के विकास व उद्यमों के निर्माण की जरूरत पड़ती है। इन स्पष्ट लाभों के अतिरिक्त जन-सूचनाओं में सुधार होने से स्थानीय लोगों को भी लाभकारी प्रभाव प्राप्त होते हैं। परिणामस्वरूप समस्त निवासियों को आर्थिक सुधार से फायदे प्राप्त होते हैं।

विकास की दर का पर्यटन पर प्रभाव

विकास की दर पर पर्यटन का अधिक प्रभाव पड़ता है इसका मतलब है कि जो वित्त सामग्रियों की पैदावार पर निर्भर है, इससे पूर्ति में वृद्धि होगी व मांग में वेग वृद्धि होगी। परिणामस्वरूप बहुत ही तेज गुणक के प्रभाव होंगे। राष्ट्र की अर्थव्यवस्था में धन को पुनः नियोजित किए जाने के परिणामस्वरूप वेग वृद्धि प्रभाव के अतिरिक्त गुणक के प्रभाव भी होते हैं।

प्रशासनिक आय के रूप में पर्यटन

पर्यटन प्रशासनिक आय का प्रमुख स्त्रोत है। सरकार को सम्पूर्ण यात्रा उद्योग से कर के रूप में प्राप्त राशि, खान-पान सेवाओं पर सीमा तथा उत्पादन कर जो कि

पर्यटकों हेतु क्रय व प्रयोग की जाती हैं, राजस्व आय का एक मुख्य स्रोत है। केन्द्रीय, राज्य व स्थानीय सरकारों के माध्यम से विभिन्न तरह के कर प्राप्त किये जाते हैं। पर्यटकों के खर्च न करने पर पर्यटन पर निर्भर उद्योग पनप नहीं सकते हैं, परिणामस्वरूप केन्द्रीय, राज्य व स्थानीय सरकार को कर से बहुत ही कम राजस्व प्राप्त होगा।

भारत जैसे विकासशील देश में पर्यटन के व्यय का अनुमान अथवा गुणांक करीब 302 है एवं इस गुणित के परिणामस्वरूप पर्यटकों के माध्यम से किये गये व्यय से राष्ट्रीय आय निर्मित होती है। गुणित के प्रभाव का वर्णन करते वक्त यह स्पष्ट किया गया है कि कुछ रकम कर के रूप में चली जाती है। प्रत्यक्ष व अप्रत्यक्ष करों से केन्द्रीय व राज्य सरकारों को प्राप्त होने वाले कर के अंश का राष्ट्रीय आय में अनुमान लगाने हेतु मूल लागत का एक पुराना तथा प्राचीन तरीका अपनाया जाता है। विदेशी पर्यटन के कारण औसत राष्ट्रीय आय जो उच्च अथवा निम्न गुणित से पैदा होती है, उससे इस अनुपात से गुणा किया जाता है ताकि विदेशी पर्यटन से अर्थव्यवस्था में प्राप्त कर राजस्व आय मालूम हो सके।

राष्ट्र को पर्यटन के माध्यम से प्राप्त कुल कर राजस्व आय का अनुमान राष्ट्र की कुल अर्जित आय व कुल अनुमानित राष्ट्रीय आय के परस्पर सम्बन्ध से मालूम की जा सकती है। यह अनुमान लगाया गया है कि यह सम्बन्ध प्रायः 10% है, अतएव कुल राजस्व आय इस तरह होगी—

कुल राजस्व आय=पर्यटकों के माध्यम से किया गया वार्षिक खर्चा×गुणक×व्यापार द्वारा पैदा किये गये कर का प्रतिशत। पर्यटन उद्योग के प्राप्त करों से सरकार को मुख्य-मुख्य परियोजनाओं को चलाने में वित्तीय मदद मिलती है। ये कर हवाई अड्डा कर, शराब व तम्बाकू कर, ईंधन कर एवं देश में आयातित वस्तुओं पर कर आदि के रूप में प्राप्त होता है। एक बहुत बड़ी राशि जो कि पर्यटन के बिना अर्जित नहीं हो पाती, सरकार को सड़क, स्कूल, चिकित्सालय आदि के निर्माण एवं अन्य जरूरी सामाजिक सेवाओं की पूर्ति हेतु उपलब्ध हो जाती है।

पर्यटन व्यय, कर-दर, कर राजस्व आदि का विभिन्न पर्यटक क्रियाओं हेतु पूर्ण वर्गीकरण न होने से यह सम्भव नहीं कि कर राजस्व में विदेशी पर्यटन के प्रत्यक्ष योगदान का सही अनुमान लगाया जा सके। पर्यटन से आय की गणना करने हेतु पर्यटन उद्योग में कार्य करने वाले लोगों के माध्यम से द्वितीय व पुनः होने वाले चक्र में क्रय किये गये माल, सामान व सेवाओं के कर को भी ध्यान में रखना होगा। पर्यटन के स्वभाव को देखते हुए भारत जैसे राष्ट्र में यह बहुत ही असंभव है कि सरकारी

आय का अनुमान लगाया जाये, केवल सरकार के माध्यम से अर्जित राजस्व कर, जो पर्यटकों पर लगाया जाता है, का ही अनुमान लगाया जा सकता है।

उदाहरणस्वरूप कई राष्ट्रों में पर्यटकों के माध्यम से प्रस्थान के वक्त भुगतान किया गया प्रति व्यक्ति कर होटल बिलों पर अतिरिक्त चार्ज व हवाई, रेल व सड़क यात्रा-पत्रों पर विक्रय कर। प्रत्यक्ष कर जो सरकार के माध्यम से प्राप्त किया जाता है वह तो अर्जित राजस्व का एक छोटा-सा अंश होता है, कुछ अर्थशास्त्रियों का विचार है कि पर्यटन से प्राप्त सरकारी राजस्व का अनुमान उस अनुपात से लगाया जा सकता है जितना कि करों का राष्ट्रीय आय में स्वयं का योगदान है।

पर्यटन एक अन्तर्राष्ट्रीय वित्त व मुद्रा को एकत्रित का प्रमुख साधन है। यह निर्यात उद्योग पर्यटन को विस्तृत मानने वाले पर्यटन के द्वारा एकत्रित विदेशी धन की ओर संकेत करते हैं। अथवा वे कुछ स्तर पर प्रत्यक्ष रूप में अथवा गुणित के संचालन से अर्जित मुद्रा का कुछ देश की वस्तुओं पर हुए प्रभावों को वर्णित करते हैं। हर राष्ट्र का यह उद्देश्य है कि पर्यटन उद्योग को एक निर्यात की भांति विकसित किया जाए, ताकि औद्योगिक यंत्र प्राप्त हो सकें व स्वावलम्बी विक्रय के लिये साधारण वातावरण निर्मित किया जा सके। व्यापार के विशद आकार, वस्तुओं की विभिन्नता, अत्यधिक संख्या में होने एवं साथ ही वित्त लाभों के कारण पर्यटन एक बहुत बड़ी आर्थिक क्रिया एवं विदेशी पूंजी के अर्जन का एक महत्वपूर्ण स्रोत बन चुका है।

यह कथन अमान्य अथवा अस्वीकार नहीं कर सकते कि अत्यधिक उन्नतिशील देशों में पर्यटन वित्त विकास के लिए और इसके विपरीत बिना पर्यटन की स्थिति में सही करने का एक माध्यम है। साथ ही साथ यह भी सत्य है कि जिन राष्ट्रों में रहन-सहन का स्तर अधिक ऊँचा है उनमें पर्यटन विदेशी मुद्रा के असंतुलन को ठीक करने में बड़ा योगदान करता है। कुछ वर्षों से पर्यटक प्राप्त करने वाले राष्ट्रों में विदेशी मुद्रा की प्राप्ति में बहुत तीव्र गति से वृद्धि हुई है। पर्यटन के माध्यम से बहुत से राष्ट्रों जैसे—जापान, हांगकांग, स्पेन, इटली, मौरिसस, यूगोस्लाविया, फ्रांस आदि की अर्थव्यवस्था में सुधार आया है। यहाँ तक कि पूर्वी यूरोप के राष्ट्र जो आरम्भ में शर्मीले व इच्छुक न थे—वे भी पर्यटन के आधारभूत सिद्धांतों को मानने लगे हैं एवं पर्यटन यातायात को विदेशों में युगोस्लाविया प्रवाहित होने हेतु कानून व नियमों में ढील दे रहे हैं। युगोस्लोवाकिया प्रथम राष्ट्र था जिसने प्रवेश-पमर बीजों को समाप्त किया एवं रूस व भारत ने होटलों व अनुपूरक आवास विधाओं का निर्माण करने हेतु बड़े-बड़े कार्यक्रमों का संचालन किया है।

पर्यटन ही एक ऐसा निर्यात व्यापार है, जो बिना राष्ट्रीय स्रोतों का ह्रास किये हुए और वास्तविक तौर पर माल का निर्यात किये बिना एक बहुत बड़ी विदेशी मुद्रा अर्जित करता है। अतएव पर्यटन एक निर्यात व्यापार के रूप में अन्य निर्यात उद्योगों की अपेक्षा अपना एक महत्वपूर्ण स्थान रखता है एवं स्वयं अद्वितीय स्थान स्थापित किए हुए है। विदेशी पर्यटन की सदैव से ही प्रशंसा की गयी है, क्योंकि इससे अन्तर्राष्ट्रीय अर्थव्यवस्था में विदेशी मुद्रा प्रवाहित होती है। प्राप्त करने वाले राष्ट्र को वरदान स्वरूप मिली हुई क्रय शक्ति उस राष्ट्र की विदेशी सम्पत्तियों में पूर्णतः घुल-मिल जाती है। इससे विदेशों से क्रय करने के वरदान सम्पूर्ण राष्ट्र में बिना किसी भेदभाव के फैल जाते हैं। क्योंकि इन सम्पत्तियों से परिवहन के यन्त्र व उद्योगों हेतु माल का आयात प्रत्यक्ष रूप में उन वर्गों की उत्पादन शक्ति वृद्धि करने में इस्तेमाल किया जाता है, जिनका पर्यटन से कोई सम्बन्ध नहीं होता है। पर्यटन के उत्तेजित प्रभाव से आय में फायदे होते हैं, क्योंकि इससे बाह्य आय के स्रोतों का विकेन्द्रीकरण होता है, जिससे विशेष गुण वाली निर्यातित वस्तु के भाग्यहीन प्रभाव को जीत लिया जा सकता है। उदाहरणस्वरूप एक रेस्टोरेन्ट में निर्यात आमदनी उन सामग्रियों से उपलब्ध होती है जो आम तौर पर निर्यात नहीं की जा सकती हैं जैसे—दूध, अण्डे, मक्खन व अन्य खान-पान की वस्तुएं।

पर्यटन एक अदृश्य निर्यात है। वाणिज्य की दृष्टि से यह व्यापार व उद्योगों से भिन्न है। एक वाणिज्य के आयात व निर्यात व्यापार में सदैव देश से वस्तुओं का अन्दर आने एवं बाहर जाने का क्रम बना रहता है। हर निर्यात करने वाले राष्ट्र भुगतान सन्तुलन की स्थिति पर नजर रखते हुए भाव व वस्तुओं का आन्तरिक उपयोग न करके विदेशों को निर्यात करते हैं। निर्यात का यह उद्देश्य नहीं है कि निर्यातक राष्ट्रों में इन वस्तुओं की अधिकता है, अपितु विदेशी मुद्रा के अभाव के एकमात्र कारण एवं भुगतान सन्तुलन को अनुकूल बनाने हेतु निर्यात किया जाता है। निर्माणकर्ताओं की स्वयं की असुविधाओं के बल पर माल का निर्यात किया है। पर्यटन के क्षेत्र में ग्राहक, स्वागतकर्ता की तरफ अग्रसर होता है। यह एक बहुत ही लाभदायक स्थिति है। पर्यटन उद्योग में पारस्परिक स्तर पर माल का निर्यात किये बिना एक बहुत बड़ी विदेशी मुद्रा अर्जित होती है। यही एक व्यापार है जिसमें निर्यातक देश बिना किसी बलिदान के फायदा प्राप्त करता है। पर्यटन में न तो किसी माल को जहाज के माध्यम से विश्व भर में भेजा जाता है और न ही किसी किस्म के भाड़े व कर का भुगतान करना पड़ता है। इसका मार्केट स्वदेश में ही होता है। खान-पान की वस्तुएं, स्मारिकाओं व अन्य विलासितापूर्ण वस्तुओं जैसे गलीचे, जेवरात

आदि वस्तुओं हेतु बहुत से नये क्रेता मिलते हैं जो कि अन्यथा कभी भी देश में नहीं आने देते। फुटकर व्यापारियों, हस्तकारों व कृषकों हेतु बहुत धन आ जाता है, हर खर्चीले पर्यटक हेतु योजनाएं बनाई जाती हैं, ताकि उच्च किस्म की विलासितापूर्ण वस्तुओं की बिक्री से विदेशी मुद्रा अर्जित की जा सके। बढ़ती हुई अत्यधिक जनसंख्या के साथ-साथ हम स्वयं किन्हीं भी वस्तुओं की बचत बलिदान के रूप में करके देश के बाहर भेजने के परिश्रम के बिना, विदेशी मुद्रा अर्जित कर सकते हैं। विदेशों से आये पर्यटक ठीक हमारे दरवाजे पर ही विदेशी मार्किट का प्रतिनिधित्व करते हैं। ऋण व मदद से भविष्य की पीढ़ी पर बहुत बड़ा बोझ पड़ता है क्योंकि मूलधन व ब्याज के भुगतानों के अतिरक्त राष्ट्र के स्वाभिमान व स्वालम्बन की महत्ता पर विपरीत पर्यटक विकसित राष्ट्रों की तरफ से विकासशील व अर्द्ध-विकसित राष्ट्रों हेतु सहायता का स्रोत है एवं उनकी राष्ट्रीय अर्थव्यवस्था का एक बहुत बड़ा साधन है।

पर्यटन एक उद्योग है जो निश्चित क्षेत्र का निर्माण करता है। पर्यटन कुछ पृथक्-पृथक् प्रकार से अपने संचालन इलाके में परिवर्तन करता है। कुछ परिस्थितियों में और व्यवसायों की तुलना में यह तथ्य इस व्यवसाय का उद्योग में खरा साबित होता है, सम्बन्धित क्षेत्र की जीवनचर्या, रहन-सहन व संस्कृति ही इस उद्योग के कच्चे माल व निर्मित माल के प्रतीक हैं। बहुत से ऐतिहासिक महत्व के शहर व स्थान एवं इमारतें व स्मारक हैं जो कि पर्यटन के स्रोत हैं, जिनका और कोई विकल्प नहीं है। ये अपनी साधारण विशेषताओं के कस्बे हैं, जिन्होंने अपनी वास्तुकला की श्रेष्ठता एवं आगन्तुकों को आकर्षित करने की विशेषता को आज भी सुरक्षित रखा है एवं जिनकी वजह से पर्यटक के माध्यम से उनकी पुरानी सम्पदा का नया प्रयोग हुआ है। किले, बड़े-बड़े गिरिजाघर, स्मारक, संग्रहालय, चित्रशाला, पुराने विश्वविद्यालयों के जन्म स्थल, राजमार्ग, चौकोर स्थल आदि इन स्थलों को पर्यटक उद्देशित स्थान बना देते हैं। मोटर वाहनों के कारण व्यक्तिगत गतिशीलता में वृद्धि के फलस्वरूप आगन्तुकों हेतु इन कस्बों के आकर्षण में बढ़ोतरी हुई। फलस्वरूप इनकी विशेषताओं को संरक्षित करने का एक उत्तरदायित्व भी हुआ। एक शहर में पर्यटक विशेष रूप से एक पैदल अनुवेषक (पथिक) है। वह शहर उसके स्वागत हेतु लालायित है ताकि उसके लिए अनुवेक्षण करना सम्भव हो सके एवं वह अपना वाहन खड़ा कर सके। प्राचीन इमारतें, जो कि अपने बुनियादी उद्देश्यों को भूल गयी हैं, का नया उपयोग करके संरक्षित किया जा सकता है।

पुराने वातावरण में प्राचीन इमारतों के नमूनों को पुरानी सजावट में एकान्त तथा शान्ति प्रदान की जानी चाहिए। ये विचार **श्री माइकल मिडिलटोन**, डाइरेक्टर, सिविल टूरिस्ट के माध्यम से ब्रिटेन में 1971 में पर्यटन तथा पर्यावरण के सम्मेलन में भाषण देते वक्त व्यक्त किये गये थे। इनके अनुसार आक्सफोर्ड, स्ट्रेट फोर्ड, बिन्दसार व इडिंगबर्ग के अतिरक्त अन्य बहुत से शहर पहले से अपनाये गये छोटे उद्योगों की भांति पर्यटन को भी अपना सकता है। स्ट्रेटफोर्ड-आन-अमोन इंग्लैंड में स्थिति एक छोटा कस्बा-बाजार है, जो अधिक महत्वपूर्ण नहीं है। विश्वभर के पर्यटकों के माध्यम से इस स्थान की बहुत ही यात्रा की जाती है। प्रति वर्ष हजारों पर्यटक अपने व्यस्त जीवन के बावजूद सम्पूर्ण विश्व से स्ट्रेटफोर्ड विश्वविख्यात साहित्यकार विलियम शेक्सपियर के ख्याल को श्रद्धांजलि पूर्व मृत व्यक्ति की कहानी पर विकसित हो रहे हैं। जहाँ व्यापारी वर्ग फलते-फूलते हैं, वहीं व्यावसायिक व अवकाश प्राप्त व्यक्ति इसे एक तमासे के रूप में देखते हैं। बजाय कि उसे एक तुच्छ व खिलवाड़ आलिंगन को निरर्थक स्थल के रूप में एवं इसे स्वस्थ चिड़चिड़ेपन का आधार मानते हैं।

पर्यटन और विविध उद्योग

पर्यटन उद्योग को उद्योगों का समूह कहा जा सकता है क्योंकि इस उद्योग के साथ विविध प्रकार के उद्योग जुड़े हुए होते हैं। पर्यटन उद्योग अर्थव्यवस्था का एक भाग/हिस्सा है जो कि पर्यटकों की इच्छाओं की पूर्ति करता है एवं उन कम्पनियों तथा संस्थाओं को भी इसमें शामिल किया जाता है, जिनका समान उद्देश्य पर्यटकों की आवश्यकताओं की पूर्ति करना होता है। कुछ मूल पर्यटन-उद्योग व कम्पनियां जैसे होटल व रेस्टोरेन्ट, परिवहन, यात्रा अभिकर्ता, यात्रा संचालन आदि पूर्णतः व मुख्यतः अपने व्यापार हेतु पर्यटकों पर ही आश्रित हैं। इन उद्योगों को प्रत्यक्ष व अप्रत्यक्ष की श्रेणी में विभक्त किया जा सकता है, प्रत्यक्ष श्रेणी में उन उद्योगों को शामिल किया जा सकता है जो कि पर्यटकों से सीधी आय प्राप्त करते हैं, जैसे होटल मालिक, परिवहन मालिक, फुटकर व्यापारी आदि। अप्रत्यक्ष श्रेणी में उन उद्योगों शामिल किया जा सकता है जो कि पर्यटक व्यय के प्रारम्भिक प्राप्तकर्ताओं के माध्यम से वितरण करने से लाभान्वित होते हैं। बहुत-सी वस्तुओं व सेवाओं की पूर्ति करने वाले पूर्तिकर्ता जो पर्यटकों को प्रत्यक्ष रूप में पूर्ति नहीं करते परन्तु ये अप्रत्यक्ष रूप में पर्यटन स्थानों पर विद्यमान सार्वजनिक सेवाएं हैं।

पर्यटक समस्त मूल तथा सहायक उद्योगों का उचित मार्गदर्शन करता है। इन उद्योगों का विकास व उन्नति अन्य सहयोगी उद्योगों अथवा व्यापार, जैसे—इमारतें, कपड़ा उद्योग, शृंगार की वस्तुओं व कुटीर उद्योगों आदि पर्यटक से घनिष्ठ रूप से सम्बन्धित हैं। यदि किसी राष्ट्र में पर्यटकों के आगमन में बढ़ोतरी होती है तो इससे स्वयं ही अन्य उद्योगों के सम्पूर्ण समूह का विकास होता है। लघु निर्माणकर्ता, फर्मे, सेवापूरक व्यापार, मूल उद्योगों आदि की भी उन्नति होती है, जिससे प्राकृतिक साधनों का पूर्ण उपयोग होता है। इस व्यवसाय या उद्योग के उद्‌भव से पैदावार में सविस्तार एवं स्वराष्ट्रीय उद्योग धन्धों में पैदावार शक्ति की मांग का निर्माण व विकास होता है। पारम्परिक शिल्पकारिक वस्तुओं में कुछ विस्तार होता है। अन्य आयात के वैकल्पिक उद्योग भी विकास की आशा कर सकंते हैं ताकि विदेशी पर्यटकों की जरूरत की पूर्ति हो सके। इन नई क्रियाओं की सेवा जैसे उद्योगों में केन्द्रित होना संभव है जो कि यात्रा सेवाओं की पूर्ति में इस्तेमाल किये जाने वाले यंत्रों की मरम्मत हेतु आवश्यक हैं। इस तरह उद्योगों का विकास सम्पूर्ण औद्योगीकरण की क्रिया को क्रियान्वित करता है, जिससे रोजगार के अवसरों में वृद्धि होती है एवं उच्च रहन-सहन का अवसर प्राप्त होता है। इन समस्त क्रियाओं से लोगों को प्रबन्ध के क्षेत्रों की आधुनिक जानकारी में मदद मिलती है। विदेशियों के माध्यम से स्थानीय वस्तुओं की जानकारी से विदेशों में पारम्परिक वस्तुओं की मांग में वृद्धि होती है।

आगन्तुक देश के माध्यभों या स्रोतों एवं आर्थिक दशा का दूसरे देशों में जानकारी होने से राष्ट्र के अन्तर्गत विदेशी पूंजी विनिमय को बल मिल सकता है। इसलिए निर्यात उद्योग के लिए देश के प्रयोग न करने वाले माध्यमों को निर्मित किया जा सकता है अर्थव्यवस्था में पर्यटन के प्रभाव व योगदान हेतु कच्चा व निर्मित माल विश्लेषण प्रणाली सबसे उचित है। यह प्रणाली ब्रिटेन व आयरलैंड में सेर्र विश्वविद्यालय के श्री जी. रिचर्ड के माध्यम से विकसित की गयी एवं पर्यटन के क्षेत्र में लागू की गयी। उनके अनुसार अर्थव्यवस्था पर पर्यटन के प्रभाव असंख्य उद्योगों में फैल जाते हैं। विश्व के बहुत से राष्ट्रों में इस बात के उदाहरण हैं, जहाँ कि स्थानीय उद्योगों का विकास पर्यटकों के उस स्थान में आवागमन के परिणामस्वरूप हुआ है, इस बात का मुख्य कारण यह था कि पर्यटक या तो उत्पादक (निर्माणकर्ता) अथवा उद्योगपति या व्यापारी था और उससे उन स्थानों में अपने व्यापार के अवसर दिखाई दिये जहाँ उसने अपना अवकाश व्यतीत किया।

पर्यटक का आकर्षण नये उद्योगों के उद्गम का स्रोत है। आजकल उद्योगों को आकर्षित करने का सबसे उचित ढंग पर्यटकों को आकर्षित करना है। इसका मुख्य कारण यह है कि उद्योगपति यंत्रों व कल-पुर्जों का स्थान ढूंढने से पहले अपने अतिथियों हेतु आवास की सुविधाएं, अच्छे सार्वजनिक संस्थाओं व विभिन्न किस्म के मनोरंजक साधनों की तरफ अधिक ध्यान देते हैं। उद्योगों को बाहर से आकर्षित करने के साथ-साथ एक जनसमुदाय जिसने पर्यटन व्यापार को विकसित किया है, वह स्वयं के औद्योगिक विकास को प्रोत्साहित करेगा। साधारणतः एक समाज पर्यटन व्यापार को मदद देने हेतु व्यावसायिक दक्षता को विकसित करके पूंजी एवं प्रबन्ध दक्षता का संचय करता है। परिणामस्वरूप दूसरी ओर औद्योगिक विकास सम्भव हो सकता है, तथ्यों से प्रतीत होता है कि औद्योगिक प्रतिस्पर्द्धा न होकर आपस में सम्बन्धित है। यह भी सत्य है कि जिस समुदाय ने उद्योगों को आकर्षित किया गया है वहां पर्यटक स्वयं आकर्षित होते हैं।

पर्यटन और लघु उद्योग

पर्यटन एक ऐसी प्रक्रिया है जो उच्च स्तर पर खाने-पीने और रहन-सहन के तौर-तरीकों को बरकरार रखने में सहयोग प्रदान करती है। बहुत से फुटकर व्यापारी व बहुत विशद परिधि के जिन्दगी के व्यवसायी व व्यापारी जैसे वकील, लेखपाल आदि भी पर्यटक से फायदे प्राप्त करते हैं। पर्यटक विक्रय केन्द्रों में एक बहुत बड़ी रकम खर्च करते हैं। पर्यटन के विकास से बहुत उच्च कोटि के विक्रय केन्द्रों का निर्माण व उनकी देखभाल होती है, जिनकी मौजूदगी से क्षेत्र में निवास करने वाले लोगों को बहुतायत फायदे प्राप्त होते हैं। पर्यटक के फुटकर क्रय के मुख्य अंग निम्नलिखित हैं–

1. कैमरा, फिल्म, फोटोग्राफिक कार्य से सम्बन्धित विस्तारण के खर्चे।
2. कपड़े व जूतों का क्रय।
3. खाद्य-क्रय व मिष्ठान आदि पर व्यय।
4. अन्य खरीददारियां जैसे पोस्टकार्ड, डाक-टिकटें आदि।
5. श्रृंगार-सामान व इत्र पर व्यय।
6. जेवरातों तथा हस्त-थैलों आदि पर व्यय।
7. श्रृंगार-सामान एवं स्मारिकाएं आदि।
8. उपहार के रूप में क्रय-सामान जैसे खिलौने।

पर्यटक सुविधाओं में सुधार

वस्तुतः पर्यटन के समय में विभिन्न सामग्रियों की खपत अत्यधिक सीमा पर होती है जैसे—जल, विद्युत, गैस व अन्य सामग्रियों की मांग में वृद्धि हो जाती है। पर्यटकों की मांगें सार्वजनिक सेवाओं में पूंजी नियोजन को आमंत्रित करती हैं। प्रतिष्ठानों को इस तरह आकर्षित किया जाता है कि वे स्थानीय निवासित जनसंख्या से भी अधिक लोगों की आवश्यकताओं की पूर्ति कर सकें। पर्यटक की उच्च मांग के समय वे दिन, सप्ताह व महीने होते हैं, जिनमें अधिक पानी, गैस व बिजली की पूर्ति करनी होती है। होटलों के आंकड़ों से स्पष्ट होता है कि पर्यटनकालीन मौसम में ही सार्वजनिक सेवाओं का अधिक उपयोग किया जाता है।

पर्यावरण एवं पर्यटन

पर्यटन एक बिना दूषित वातावरण उद्योग है। भारत की प्राकृतिक सम्पदा का आधार व संरक्षण, हमारी पर्यटन की वस्तु में योगदान करती है। यदि इसे भली-भांति संगठित किया जाए (अर्थात् यदि पर्यटन का कच्चा माल जैसे प्राकृतिक व सांस्कृतिक आकर्षण का ह्रास व बदलाव नहीं किया जाए) तो राष्ट्रीय सम्पदा का पर्यटन के माध्यम से ह्रास होने का प्रश्न ही नहीं उठता है। प्राकृतिक दृश्य व स्थानीय संस्कृति आगन्तुकों के मुख्य आकर्षण हैं। पर्यटन का विकास पर्यावरण को अन्य

आर्थिक व औद्योगिक विकास की अपेक्षा कम नुकसान पहुंचाता है। भारत की जनसंख्या व क्षेत्र एवं यहाँ के गहरे सामाजिक रीति-रिवाजों को देखते हुए कुछ पर्यटकों का सामाजिक प्रभाव नहीं के बराबर है। साथ ही यह सही है कि कुछ धनी राष्ट्रों में पर्यावरण को दूषित करने, मुख्यतः समुद्र तट व पर्यटन क्षेत्रों को कोलाहित करने का कारण पर्यटन ही रहा है। यदि बुद्धिमता तथा पसन्द के साथ इसे इस्तेमाल किया जाए तो पर्यटन पर्यावरण में सौन्दर्यता व सुधार लाने का एक माध्यम बन सकता है।

राजस्व सर्जन और पर्यटन

विश्व के अधिकतर विकासशील राष्ट्र थोड़ा ऋण में हैं। उनका विकास या तो अन्तर्राष्ट्रीय फण्ड पर आश्रित रहा है अथवा उन्हें भूतपूर्व उपनिवेश राष्ट्रों से ऋण के रूप में मदद प्राप्त होती है। उन राष्ट्रों का घाटे का बजट रहा जिनका विदेशों से प्राप्त आर्थिक मदद से पूर्ण विकास नहीं हो सकता था। विदेशों से प्राप्त मदद व आर्थिक मदद को, पर्यटन के विकास से प्राप्त किए गये करों के माध्यम से आय में वृद्धि करने से कम किया जा सकता है। इस तरह पर्यटन के वित्तीय योगदान का सार्वजनिक राजस्व वित्त के संतुलन को बरकरार रखने में बहुत ही महत्वपूर्ण स्थान है।

विदेश-विनिमय और पर्यटन

किसी भी उत्पादित वस्तु के लाभ के लिए विदेश-विनिमय का होना आवश्यक है। यह उपभोक्ता को असीम सीमा तक जाने के लिए बहुत अनिवार्य है। विशेषकर उन सामग्रियों के संबन्ध में जो दूसरे राष्ट्रों में विकसित के रूप में मौजूद हैं, तथा उनके मंगवाने के अलावा और कोई चारा नहीं है। विकासशील देशों में पर्यटन उद्योग में पूंजी की सामर्थ्यता सम्भवतः कुछ अधिक है; विशेषतः उन अवस्थाओं में जबकि अनिवार्य आधारिक संरचनाओं की लागतों को शामिल किया जाता है। यदि पर्यटन विकसित होता है, तो विदेशी पर्यटक आकर्षित होते हैं, जो कि अपनी पूंजी के लाभकारी विनियोजनों, अच्छे अलीशान होटल, आधुनिक खेल के मैदान व मण्डप स्थल, समुद्र तट आदि की तलाश में लगाते हैं। इसके लिए प्राप्तकर्ता राष्ट्र के पास न तो निर्माणक होते हैं न ही तकनीकी प्रशिक्षित अधिकारी। राष्ट्र के उद्योगों को उत्साहित करने हेतु वे आधुनिकीकरण व नई तकनीकियों को अपनाते हैं, वे पेटेन्ट

हेतु प्रार्थना करते हैं एवं विदेशी लाइसेन्स का उपयोग करने हेतु बाध्य होते हैं। अतएव विदेशी मदद की जरूरत होती है, इसके अतिरिक्त बहुत से अन्य निर्यात पैदा करने वाले उद्योगों की तुलना में यातायात उद्योग विदेशी उद्यमियों के माध्यम से स्वागतकर्ता राष्ट्र की सरकार व निजी उद्यमियों के सहयोग से अधिक अच्छा विकसित हो सकता है।

विदेशीय विनियमन या विनियोजन से पर्यटन उद्योग को बल मिलता है– असली विनियोजन तकनीकी मदद से परमिट व पेटेन्ट आदि के तकनीकी स्तर में वृद्धि होती है। आयातित पूंजी से प्राप्त आय को परिवर्तित करने हेतु संचित सोने की उपलब्धता पूर्ण गारन्टी प्रदान करती है। विदेशी पूंजी प्रबन्ध की दक्षता के इस्तेमाल करने से जो पर्यटन में विकास होता है वह आर्थिक योगदान की दृष्टि से स्थानीय क्षेत्रों में स्थानीय लोगों के माध्यम से किये गये विनियोजन की अपेक्षा कई मात्रा में श्रेष्ठ होता है।

पर्यटन उद्योग के विकास हेतु विदेश विनियोजन का फायदा व उनका ज्ञान होना ही पर्याप्त नहीं है। विकासशील प्राप्तकर्ता राष्ट्र को प्रत्येक क्षेत्रों के जरूरी आधारिक संरचना का विकास करना है। उदाहरणार्थ–पर्यटकों की सुरक्षा हेतु उचित प्रबन्ध करने होते हैं एवं विकसित पर्यटन केन्द्रों तक पहुंचने हेतु परिवहन के सुविधाजनक प्रबन्ध करने चाहिए। विदेशी उद्यमी फायदे को प्रायः स्वदेश ले जाने की स्वतंत्रता के परिणामस्वरूप निर्माण व संचालन के उद्देश्य से विदेशों में निर्मित माल का आयात करना चाहते हैं। ये फायदे व कर की दर, कर-दर उपभोग की आवास वस्तुओं, सामान व पर्यटन सेवाओं पर उत्पादन-कर, दक्षता को इस्तेमाल की स्वतंत्रता, विदेशी कारीगरों को रोजगारों पर लगाने की स्वतंत्रता एवं इन कारीगरों के माध्यम से संचित धन को स्वदेश ले जाने एवं इसी तरह की अन्य सुविधाओं के बारे में सरकारों में दीर्घकालीन सहयोग की आकांक्षा रखते हैं।

विदेशी विनियोजन पर पर्यटन का प्रभाव

विश्व के अधिकांश विकासशील देशों का लक्ष्य मूल रूप से विदेशी मुद्राओं को हासिल करना होता है। बहुत से विकासशील राष्ट्र मूल रूप से विदेशी पूंजी को अर्जित करने के इच्छुक होते हैं। पर्यटन में विनियोजन करने से अर्जन की दर अन्य क्षेत्रों में विनियोजन करने की तुलना में कम होने पर भी फायदा प्रदान करती है। रोजगार के स्तर में तीव्र वृद्धि हेतु यह जरूरी है कि पर्यटन में विनियोजन करते वक्त इनसे पैदा रोजगारों के प्रभावों की तुलना अन्य विनियोजन की

सम्भावनाओं को ध्यान में रखकर की जाए। पर्यटन में विनियोजन के माध्यम से सृजित किया गया कच्चा माल पर्यटन के महत्व को आंकने का द्योतक है। भारत जैसे विकासशील देश में टपकनें (छीजन) व श्रम की अर्धबेरोजगारी के कारण आर्थिक प्रभावों की कमी हो जाती है। किसी भी देश में विदेशी यात्रा के प्रभावों को नापने हेतु गुणित को यदि मापदण्ड समझा जाए तो इसमें व्यय के अतिरिक्त धारा में मांग व इसमें विद्यमान टपकनों (छीजन) के प्रभावों को शामिल किया जाता है।

पूंजी की पर्याप्तता पर पर्यटन का प्रभाव

पर्यटन उद्योग में पूंजी की अधिक आवश्यकता होती है। परंतु यह पूंजी अन्य औद्योगिक विकासों की तुलना में कम होती है। परन्तु इस सम्बन्ध में नई सड़कों, रेलवे बन्दरगाह, हवाई अड्डों आदि जोखिम के निर्माण के परिणामस्वरूप जन्मे आर्थिक प्रभावों को भूलना नहीं चाहिए। इसके अलावा, अच्छे संचालित होटल व केटरिंग प्रत्येक विदेशी पर्यटन के आधार हैं। इन क्षेत्रों के कार्य को पूर्णतः प्राथमिकता मिलनी चाहिये। यह अतिरिक्त धन तभी आकर्षित होता है जब यह अधिक उत्पादित विनियोजन प्रदान करता है। पर्यटन के क्षेत्र में विनियोजन से आय की प्राप्ति निम्नलिखित तरीके से हो सकती है–

(1) दीर्घकालीन विनियोजन जैसे रेस्टोरेन्ट का खोलना।

(2) दीर्घकालीन विनियोजन जैसे शहरी विकास की दृष्टि से एक वन-मण्डल को राष्ट्रीय उद्यान में परिवर्तित करना।

(3) मध्यकालीन विनियोजन जैसे एक सुन्दर स्थान में स्थित होटल अथवा महासम्मेलन होटल।

पर्यटन उद्योग की सबसे बड़ी विशेषता यह है कि इस उद्योग में एक बार पूंजी लगाने के बाद दीर्घ काल तक लाभ प्राप्त होता है। पर्यटन उद्योग की एक विशेषता यह है कि इससे दीर्घकालीन फायदे प्राप्त होते हैं। तुलनात्मक दृष्टि से इस उद्योग में फायदा प्राप्ति का वक्त बहुत ही कम होता है तथा फायदे का अंश अन्य उद्योगों के मुकाबले निश्चित रूप से अधिक होता है। यह उद्योग जल्दी व अधिक फायदे की वजह से बहुत ही प्रिय बनने योग्य है। अतएव विनियोजक अपना धन आय प्राप्ति के पूर्व विचार करके लगाने को तत्पर होते हैं। इसके अतिरिक्त पर्यटन के आर्थिक पहलू का मूल सिद्धांत यह है कि अन्य उद्योगों की तुलना में यह मध्यम तरह का विनियोजन चाहता है। किसी भी राष्ट्र के पर्यटन के तहत वहां की जलवायु,

दर्शनीयता, इतिहास, पुरानी इमारतें, कला, रीति-रिवाज, त्योहार एवं वहां के लोग शामिल हैं। पर्यटक जो कि अपनी बचत की गयी आय का इतना अधिक व्यय करते हैं एवं साथ ही अवकाश व्यतीत करने की धारणा बनाते हैं, वे अवकाश हेतु नये वातावरण के सुहावने बदलाव से कुछ अधिक आकांक्षा रखते हैं। वे अपने साथ केवल स्मरणीय छाप एवं स्मारिकाओं को ही ले जाते हैं।

विकासशील क्षेत्रों में पर्यटन की उत्पत्ति

प्रायः पर्यटन स्थल मुख्य रूप से उन क्षेत्रों में स्थित हैं जो आर्थिक रूप से अविकसित हैं तथा वहां रोजगार का अभाव होता है। पर्यटन का विकास अधिकतर बिना उद्योग के क्षेत्रों, सामान्यतः ग्रामीण, पर्वतीय एवं समुद्रतटीय क्षेत्रों में होता है। शक्ति के साधनों व कच्चे धातु से भरपूर क्षेत्रों एवं भौगोलिक दृष्टि से अनुकूल स्थान अपनी आर्थिक क्रियाओं औद्योगीकरण से ही आरम्भ करते हैं। ये स्थान बहुत ही विकसित होते हैं एवं साथ ही बड़े मार्केट नजदीक होने एवं समानान्तर उत्पादन वृद्धि व तृतीय क्षेत्रों में सेवाओं के परिणामस्वरूप कृषि उत्पादन में भी वृद्धि होती है। जो क्षेत्र अधिक अनुकूल नहीं हैं वे अर्द्धविकसित रहते हैं। उनकी क्रियाशील जनसंख्या विकसित क्षेत्रों की तरफ आकर्षित होती है, जो औद्योगिक दृष्टि से अधिक विकसित नहीं हैं, हालांकि उनके पास अन्य बहुत से आकर्षण हैं जैसे बर्फ से आच्छादित पहाड़ियां, सुन्दर प्राकृतिक सौन्दर्य, असीमित समुद्र तट व अन्य पर्यटक आकर्षण, उनमें क्षेत्रीय विकास व फैलाव बहुत से तरीकों अथवा सामरिक प्रणाली के माध्यम से किया जा सकता है। जिन क्षेत्रों में मुख्य औद्योगिक पर्यटन विकास है, वहां पर्यटन एक उद्योग है, यह उस भूमि व वस्तुओं का उपयोग करता है जिसका औद्योगिक व कृषि की दृष्टि से कोई मूल्य नहीं है जैसे बर्फ (शीतकालीन क्रीड़ाओं हेतु उपयोग होती है), रेत (समुद्र तट पर), बंजर भूमि व रेगिस्तान, राष्ट्रीय पार्क। ये समस्त तत्व अन्य राष्ट्रीय उद्योगों हेतु उत्पादनविहीन हैं, परन्तु पर्यटन हेतु सम्पदा के स्रोत हैं। बड़े मण्डल जो कि कृषि के विकास हेतु अनुकूल नहीं हैं विलासिता हेतु उपयोग किए जा सकते हैं। ये क्षेत्र बाह्य मनोरंजन हेतु अति उत्तम स्थान बन सकते हैं एवं इस तरह एक लाभकारी पर्यटन उद्योग के अंग बन जाते हैं। यदि कोई क्षेत्र कृषिविहीन व उद्योगविहीन है तो वहां जरूरी आधिकारिक संरचना व श्रम शक्ति की सम्भावना नहीं होती एवं इस क्षेत्र में पर्यटन विकास हेतु अत्यधिक संरचना लागत व श्रम के पुनः स्थानीयकरण की जरूरत होगी। इसके अतिरिक्त इस तरह के क्षेत्रों में पर्यटन

के माध्यम से प्रदान अनुकूल विनियोजन व आय के कारण पड़ोसी क्षेत्र भी विकास की क्रिया से प्रवाहित होंगे।

आमदनी के वितरण में भी पर्यटन मददगार साबित होता है। एक कमर्शियल इलाके से जुड़े प्रतिबंधित इलाकों में वित्त का नियोजितकरण हो जाता है, परन्तु पर्यटन क्षेत्र आय के वितरण द्वारा इस प्रभाव को अन्य क्षेत्रों में अग्रसित करते हैं—जैसा कि बहुत से विकासशील राष्ट्रों में पर्यटन के विस्तार से प्रतीत होता है। इन क्षेत्रों में पर्यटन विस्तार से जनसंख्या का निवासित होने, आय वृद्धि, बिजली के घरेलू यन्त्रों के उपयोग व नये आवास का निर्माण व प्रतिष्ठाओं में वृद्धि होती है। यह महसूस किया गया है कि जिन क्षेत्रों में कृषि की क्रियायें सीमित थीं एवं व्यावहारिक रूप में उद्योग नहीं थे एवं इस तरह के विकास की गति बिल्कुल निम्न स्तर पर थी, वे पर्यटन के बढ़ावे से बहुत बड़ी सार्वजनिक व निजी पूंजी के आकर्षण बन चुके हैं। इन क्षेत्रों में विनियोजन व जनसंख्या के उपभोग में वृद्धि होने से स्थानीय आय में वृद्धि हुई है। परिणामस्वरूप स्थानीय जनसंख्या के जीवन स्तर में अनुकूल प्रभाव पड़ता है।

आय का सृजन और पर्यटन

पर्यटन में आय का सृजन आगन्तुकों की संख्या, आगमन के प्रयोजन की अनेक प्रक्रियाएं अथवा कई प्रयोजनों के लिए हर पर्यटन व्यय, रुकने का औसत वक्त, खर्चीले व कम खर्चीले पर्यटकों की संख्या की श्रेणी में बदलाव आदि पर निर्भर करता है। **नेशनल काउन्सिल ऑफ अप्लाइड इकानोमिक्स रिसर्च** के अनुसार, "प्रत्येक पर्यटन क्रिया व यात्रा उद्देश्य हेतु गणना की गयी पर्यटकों से कुल प्राप्तियां, पर्यटकों की कुल संख्या को उनके औसतन ठहरने के समय (दिनों) और प्रति पर्यटक प्रतिदिन व्यय से गुणा करने की राशि के बराबर होती है।" समय-समय पर पर्यटन से प्राप्त आय के सम्बन्ध में बहुत से अध्ययन किए गए हैं। विदेशी पर्यटकों से प्राप्त आय के अनुमान निम्नलिखित माध्यमों से प्राप्त होता है—

1. भारत में पर्यटन का आर्थिक सर्वेक्षण, 1965-66, पर्यटन विभाग की रिपोर्ट।
2. भारत में विदेशी पर्यटकों का सर्वेक्षण, 1968-69 व 1972-73, पर्यटन विभाग के माध्यम से आई. आई. पी. ओ. (1980), नई दिल्ली को स्थापित किया गया।
3. परिवहन व उड्डयन मन्त्रालय, भारत सरकार।
4. इन समस्त रिपोर्टों में प्रति पर्यटक व्यय के आकार व अन्य सम्बन्धित पहलुओं के आँकड़े मिलते हैं।

पर्यटकों के खर्च का वर्गीकरण निम्न तरह मान लिया गया है :

(1) खरीददारी।

(2) खान-पान, रहन-सहन, रेस्टोरेन्ट व मनोरंजन।

(3) आन्तरिक परिवहन।

(4) अन्य।

इन रिपोर्टों में प्रति पर्यटक व्यय को यात्रा के उद्देश्य के अनुरूप निम्नलिखित भागों में बांटा गया है–

(1) अवकाश।

(2) अध्ययन व शिक्षा।

(3) व्यापार।

इन रिपोर्टो में दी गई सूचनायें पर्यटन से कुछ प्राप्तियों को ज्ञात करने हेतु बहुत ही जरूरी हैं।

विदेशी पर्यटक व्यय के माध्यम से उत्पन्न कुल राष्ट्रीय उत्पादन में किए गए शुद्ध योगदान को विदेशी पर्यटक व्यय से प्राप्त शुद्ध आय अथवा प्राप्तियों के माध्यम से प्रदर्शित किया जाता है। कुल प्राप्ति में से पर्यटकों की लागत को घटाने से विदेशी पर्यटकों की शुद्ध प्राप्तियों की गणना की जा सकती है।

पर्यटन के क्षेत्र में विनिवेश प्रक्रिया तथा लाभांश

लागत-लाभ विश्लेषण के सिद्धांत के अनुरूप लाभों को यदि लागत से विभक्त किया जाए तो अनुपात दर की प्राप्ति होती है। परियोजना को लागू करने से जो फायदे होने की सम्भावनाएं होती हैं उन्हें भी फायदे के रूप में परिभाषित किया जाता है। लागत-लाभ में अन्तिम प्राप्ति दर कम-से-कम एक होनी चाहिए अन्यथा वह कुछ भी अर्जित नहीं करेगी। बहुत से विकल्पों का मूल्यांकन करते वक्त उच्च दर वाले विकल्प के सफल होने की अधिक सम्भावना होती है। अतएव लागत-लाभ विश्लेषण को निकालने व अनुमान लगाने में बड़ी सावधानी बरतनी पड़ती है। लागत-लाभ दर तुलना करने की वह प्रणाली है, जिससे यह मालूम किया जाता है कि किये गए व्यय से, सरकार अथवा राष्ट्र को कितनी आय प्राप्त होती है। इसमें राष्ट्र के माध्यम से पर्यटन योजनाओं में संचालन किए गए खर्च व पर्यटकों के माध्यम से उत्पादन व्यापार से अर्जित कर-राजस्व की तुलना की जा सकती है। इससे बजट के समायोजन को स्थिर रखने में मदद मिलती है। कुछ राष्ट्रों में सरकारी पर्यटन संगठनों की लागत का आंशिक भुगतान करने हेतु निजी कोष का भी इस्तेमाल किया

जाता है। परन्तु यह अंश प्रायः इतना कम होता है कि आँकड़ों की दृष्टि से यह खास महत्वपूर्ण नहीं होता है। भारत में राष्ट्रीय पर्यटन के विकास में कार्यक्रमों का संचालन का पूर्ण व्यय का भुगतान सरकार के माध्यम से किया जाता है।

पर्यटन के क्षेत्र में विनिवेश-लाभ की विशेषताएं

पर्यटन के क्षेत्र में किये गये व्यय से लाभ-लागत की सीमा विस्तृत है। निम्नतम सीमा बहुत ही कम व उच्चतम सीमा बहुत ही ऊँची है। यह सीमा 1 से 5 के बीच होती है। बहुत ही औद्योगिक सम्पन्न व आर्थिक दृष्टि से पर्यटन पर कम निर्भर रहने वाले राष्ट्रों में यह सीमा 30 से 10 के मध्य होती है। उदाहरणार्थ–हम यह कह सकते हैं कि भारत में लागत-लाभ दर 3 : 1 है। इसका अर्थ दूसरे शब्दों में यह है कि पर्यटन योजनाओं पर किए गए प्रति एक रुपये के बदले पर्यटकों के माध्यम से किए गए खर्च से उत्पन्न व्यापार से कर-राजस्व के रूप में सरकार 3 रुपये इकट्ठा करती है।

पर्यटन में सबसे अधिक लगी हुई पूंजी या उच्च लागत लाभ दर आगन्तुकों के रचना के अनुक्रय एक दूसरे से जुड़ी होती है। उच्च लागत-लाभ दर वाले राष्ट्रों में अधिक पर्यटन होने की सम्भावना होती है। साधारणतया अत्यधिक पर्यटन व्यापार वाले राष्ट्रों का पर्यटन संचालन बजट के तहत ही होता है।

एक लागत-लाभ दर मार्गदर्शक भी हो सकती है। यह वित्त मन्त्री को बता सकती है कि पर्यटन क्रियाओं पर कितना खर्च हो रहा एवं इससे कितना अर्जित हो रहा है। परन्तु एक उच्च लागत-लाभ दर का कारण सरकार के माध्यम से पर्यटन को वित्तीय मदद प्रदान करने की असफलता के कारण हो सकता है। एक निम्न लागत-लाभ-दर पांच वर्षों तक विद्यमान रह सकती है, जब तक कि बढ़ते हुए कार्यक्रम के परिणामस्वरूप पर्यटन में बढ़ोतरी नहीं होती है।

एक निम्न लागत-लाभ-दर पर्यटन बजट को दक्षता से न खर्च करने का द्योतक है। लाभ-लागत विश्लेषण एक कला है जो कि केवल विनियोजन निर्णय से ही सम्बन्धित है। समस्त विकासशील राष्ट्रों में पर्यटक को प्रोत्साहित करने हेतु नई होटल परियोजनाओं हेतु वित्तीय ऋण व अनुदान हेतु सरकारी सहयोग बढ़ता जा रहा है। सरकारें होटलों में सार्वजनिक विनियोजन हेतु अपने बजट में धन को आवंटित कर रही हैं। इन परिस्थितियों में लागत-लाभ विश्लेषण सम्बन्धित है। जब कभी किसी किस्म का सार्वजनिक विनियोजन किए जाय तो यही प्रणाली अपनानी होगी। सार्वजनिक विनियोजन का मुख्य उद्देश्य यह है कि अधिकांश फायदा अधिक से

अधिक लोगों हेतु प्राप्त हो सके। लागत-लाभ का विश्लेषण समस्त पक्षों के फायदे के विशद पहलू को ध्यान में रखकर किया जाता है, जबकि निजी विनियोजन के निर्णय में इस बात का स्वाभाविक रूप से बिल्कुल ही ध्यान नहीं दिया जाता है। समुदाय के समस्त लागत-लाभ इस विश्लेषण में अध्ययन किये जाते हैं। इसमें कुछ विशेष कठिनाइयां भी हैं जो निम्न प्रकार हैं—

प्रथम—समुदाय के किसी वर्ग को जो लागत के रूप में है, वही दूसरे समुदाय के फायदे हो सकते हैं।

द्वितीय—कितने ही लागत व लाभों को मापना एवं उनको उनकी सम्बन्धित महत्ता के आधार पर सन्तुलित करना बड़ा ही कठिन कार्य है।

तृतीय—यह अभी भी एक उभरती हुई कला है जो कि बृहत् विकास की प्रतीक्षा कर रही है।

चतुर्थ—समस्त लागत व लाभों की जानकारी हासिल करने हेतु तथा उनको पारस्परिक माप हेतु बहुत से परीक्षणों का पूर्व विचार करने हेतु एक प्रणाली निर्मित की जा सकती है।

पूंजी लाभ विश्लेषण के अन्तर्गत देखने योग्य व न देखने योग्य दोनों प्रकार की पूंजी को सम्मिलित किया जाता है। इसके अन्तर्गत उद्भव व विकास उसी स्थल में मौजूद समान सुविधाओं की पूर्ति के प्रभावों का विश्लेषण किया जाता है। किसी भी क्षेत्र में किसी विशेष परियोजना के विकास के प्रभावों से यातायात में भीड़ अथवा दुर्घटनाओं में वृद्धि होती है एवं सौन्दर्य से परिपूर्ण भूमि में एक सुन्दर इमारत के प्रावधान के परिणामस्वरूप समस्त समुदाय को मनोरंजन उपलब्ध होता है।

पर्यटन के क्षेत्र में मुख्य व्यय

पर्यटन के क्षेत्र में मुख्य व्यय निम्नलिखित हैं—

1. पर्यटन उद्योग में विदेशी विनियोजन पर अर्जित फायदे को विदेशों में प्रेक्षित होने वाली राशि।
2. विदेशी ऋणों का ब्याज।
3. विदेशी ट्रेवल एजेन्ट्स व टूर आपरेटर्स के कमीशन का भुगतान।
4. राष्ट्रीय पर्यटक संगठनों (पर्यटन विभाग व राष्ट्रीय विकास निगम) और ट्रेवल एजेन्सी एवं राज्य व शहरी स्तर के होटलों आदि का प्रशासन, संवर्धन व प्रचार पर किया गया व्यय।

5. निश्चित लागत व मरम्मत की चालू लागत में विदेशी पूंजी के अंग। इसके अतिरिक्त कमरों की बिक्री, खाद्य बिक्री केन्द्रों अथवा सम्मेलन कक्षों का किराया आदि एवं खाद्य व आवास की बिक्री की अलग-अलग विदेशी विनिमय प्राप्ति सम्बन्धी आय के आँकड़े।
6. सरकार के माध्यम से पर्यटन योजनाओं के विनियोजन पर की गयी धनराशि के ब्याज से हुई हानि अथवा नुकसान जैसे कर रियायतें, सीमा कर से छूट व कम ब्याज दर के ऋण आदि।
7. भूमि तथा इमारत की लागत।
8. होटलों में विनियोजित पूंजी की विदेशी विनिमय लागत।
9. सरकार की आधारिक संरचना पर पूंजी व्यय जो कि विशेषतः पर्यटन के विकास हेतु आवश्यक है।
10. आवास व्यवस्था में विनियोजन जैसे—होटल, पर्यटक लधु आवास, अतिथिगृह आदि।
11. संचालित चालू लागतों का वर्गीकरण।
12. परिवहन सेवाओं में विनियोजन।
13. विशेष रूप में पर्यटकों द्वारा इस्तेमाल किए गए स्वदेश में उत्पादित माल के आयात किए गए कल-पुर्जे सहित आगन्तुकों द्वारा इस्तेमाल की गयी वस्तुओं व सेवाओं की लागत आय।
14. अन्य सुविधाओं की लागत जैसे सम्मेलन कक्ष, तरण-ताल आदि।

पर्यटन संगठन के संवर्धन पर विदेशीय विनियोजन पूंजी अथवा दूसरे देशों में निर्माणित विभागों के दुरुस्तीकरण की पूंजी बहुत ही ज्यादा है। यह अनुमान है कि कुल विदेशी विनिमय लागत पूर्ण प्राप्ति के 5 प्रतिशत से अधिक नहीं है। यह प्रतिशत अन्य विकासशील राष्ट्रों की तुलना में अनुकूल ही है। यह तर्क है कि सार्वजनिक सेवाओं जैसे—बिजली व पानी पूर्ति, परिवहन व दूर संचार सेवाओं, पर्यटन स्थल के विकास हेतु आयात किए गए यन्त्र आदि के विदशी विनिमय लागत का कुछ अंश पर्यटकों द्वारा इस्तेमाल की गयी सुविधाओं हेतु निश्चित करना चाहिए। आस्ट्रेलिया, वहामाज, बरमूडा व हवाई आदि में निवासित जनसंख्या बहुत ही कम है, परन्तु लाखों आगन्तुक वहां आते हैं। भारत में हमारे पास करोड़ों की जनसंख्या हैं, परन्तु तुलनात्मक रूप में यहाँ बहुत ही कम आगन्तुक आते हैं। सार्वजनिक सुविधाओं एवं सड़क व रेल परिवहन से सम्बन्धित अधिकांश यन्त्रों का स्वदेश में ही निर्माण होता है। पर्यटकों द्वारा इस्तेमाल की गई सेवाओं में केवल विमान सेवायें

ही ऐसी हैं जिनमें बहुमूल्य लागत के विदेशी आयातित यन्त्रों की जरूरत होती है, परन्तु भारत में इण्डियन एअर लाइन्स सेवा के माध्यम से यात्रा करने वाले यात्रियों की संख्या स्वदेशी हवाई यातायात के आकार की तुलना में बहुत ही कम है। भारत में, सरकार के माध्यम से न केवल पर्यटन विकास की आधारिक संरचना में बल्कि साथ ही व्यावसायिक योजनाओं जैसे होटल व पर्यटन स्थल में भी पूंजी का विनियोजन किया जा रहा है।

केन्द्र में औद्योगिक वित्त निगम व राज्यों में राजकीय वित्त व विकास निगम होटल परियोजनाओं हेतु ऋण प्रदान करते हैं। पर्यटन उद्योग, होटल व अन्य पर्यटन उद्यमों में विनियोजन का अनुमान नहीं लगाया गया है। केन्द्र में पर्यटन विभाग बजट का व्यय व इसी भांति राज्य सरकारों का व्यय सरकार की महत्वपूर्ण लागतें हैं। कुछ पर्यटन लागतें जो कि ध्यान में रखनी जरूरी हैं निम्न तरह की हैं–

विदेशों में किये गए विज्ञापन व उन्नति सम्बन्धी खर्चे ताकि विदेशी पर्यटक आकर्षित हो सकें एवं पर्यटन क्षेत्र में क्षमता बढ़ाने हेतु विदेशों में कार्मिकों के प्रशिक्षण के खर्चे।

रेस्टोरेन्ट व्यवसाय व उद्योग के जरिए रहने हेतु आवस की व्यवस्था, खान-पान की व्यवस्था और मनोरंजन के लिए लाइसेंसों की आवश्यकता होती है। इसके तहत उपभोग के सामान का आयात जैसे पेय पदार्थ, खाद्य सम्बन्धी कच्चे माल की भोजन सामग्री, मछलियां, पनीर, चटनी, कांच का सामान आदि हैं और सम्पत्ति सम्बन्धी सामान व कलपुर्जे आदि पूंजी माल के बारे में प्रतिवर्ष किए गये भुगतानों को हिसाब में लिया जा सकता है। विदेशी पहलुओं को विशेषतः होटल क्षेत्र में सहयोग हेतु विदेशी विनिमय में किए गए भुगतान।

8

पर्यटन उत्पाद तथा उनका वाणिज्यीकरण

जिस प्रकार पर्यटन को विशेष स्थलों की जरूरत होती है उसी प्रकार पर्यटन उत्पाद के लिए बाजार की आवश्यकता होती है परंतु पर्यटन स्थल पर पर्यटक खुद आता है लेकिन पर्यटन उत्पाद में यह बात लागू नहीं होती है क्योंकि पर्यटक के पास भ्रमण के लिए अनेक विकल्प मौजूद होते हैं। इसलिए उत्पादक अपने उत्पाद को बेचने के लिए विशिष्ट बाजार ढूंढता है जिससे पर्यटक वहाँ पर आए तथा उत्पादित वस्तुओं का क्रय करे।

इस प्रकार जो उत्पाद आपकी प्रसन्नताओं, व्यापार की आवश्यकताओं और अवकाश के समय को, आपके नियमित निवास स्थान के अतिरिक्त अन्य स्थानों पर संतुष्ट कर देता है, वही साधारणतया पर्यटक उत्पाद कहलाता है।

सामान्य अर्थों में उत्पाद, एक दृश्य, एक वस्तु, एक स्थान, एक व्यक्ति, एक घटना या एक ऐसी संस्था की आवश्यकताओं को संतुष्ट कर सकता है, हो सकता है, जो उत्पाद प्रस्तुत किया जाता है वह दूसरों की आवश्यकताओं को संतुष्ट करने

वाला होता। यह उत्पाद अन्य कीमतों से विनिमय होना चाहिए। अतः उत्पाद को प्रस्तुत करने वालों एवं उसको प्राप्त करने वालों दोनों को ही परस्पर संतुष्ट करने वाला होना चाहिए।

पर्यटन उत्पाद को गुण-दोषों के आधार पर तीन प्रकार से विभाजित किया जा सकता है। वे तीन प्रकार निम्न हैं–

(1) क्रेता की जरूरतों को संतुष्ट करना चाहिए।

(2) उसे वस्तु विनिमय या धन विनिमय में होना चाहिए।

(3) उत्पाद हमेशा प्रस्तुत किया जाना चाहिए।

सामान्यतः यह जरूरी नहीं है कि उत्पाद हर जगह आपके मन-मुताबिक होगा। उत्पाद कुछ भी हो सकता है, जैसे–फिरोजाबाद की कांच की चूड़ियां, बनारस की रेशमी साड़ियां, मुम्बई से उत्पादित सूत्री वस्त्र, आंध्र प्रदेश से उत्पादित पंचमल्ली साड़ी या मुरादाबाद में बनी पीतल की कलात्मक वस्तुएं, अलीगढ़ के ताले अथवा कुछ और वस्तुएं भी हैं। एक से अधिक स्थानों से भी उत्पाद हो सकता है, जैसे–लंदन, काशी, हरिद्वार, बम्बई, गोवा, हवाई, इत्यादि। उत्पाद कोई संस्था भी हो सकती है, जैसे–वर्ल्ड लाइफ फेडरेशन फोरेक्स डिवीजन ऑफ ए ट्रेवल एजेंसी इत्यादि। उत्पाद एक मानव विशेष भी हो सकता है, जो अपनी कार्य शैली से उत्पादन करता है, उदाहरणार्थ–सपेरा, जोकर, टूरिस्ट गाइड, नर्तक या नर्तकी इत्यादि। घटनायें भी उत्पाद का रूप ग्रहण कर लेती है, यथा–केरल व नैनीताल की नौका दौड़, जयपुर एवं मैसूर के हाथियों का उत्सव, अहमदाबाद की पतंग उड़ाने की प्रतिस्पर्द्धाएं आदि। कुछ कार्य प्रणालियां भी उत्पाद का रूप ग्रहण करती हैं।

उत्पाद के उपर्युक्त सभी उदाहरण मूलतः मूर्त ही हैं। इनका सेवा द्वारा उत्पादन होता है या उन सेवाओं से उत्पाद होता है, जो गोचर या लौकिक होती हैं, जैसे–कार, टेलीविजन, माइक्रोफोन, एयर कंडीशनर इत्यादि।

उपर्युक्त उत्पाद निम्न दो भागों में विभाजित हो जाते हैं, गोचर या लौकिक एवं अमूर्त।

सेवाओं द्वारा उत्पादित उत्पाद जैसे–बैंकिंग सेवा, होटल सेवा, हवाई जहाज अथवा रेल सेवा, स्वास्थ्य सेवायें, यात्रा कन्सलटेन्सीज इत्यादि हैं। अतः पर्यटन उत्पाद भी अमूर्त उत्पाद के अंतर्गत आता है या उन सेवाओं के अंतर्गत आता है जो निम्न विशिष्टताओं से परिपूर्ण होती हैं–

1. परिवर्तनशीलता या अस्थिर
2. ग्राहक द्वारा योगदान

3. नश्वरता
4. अविभेद्य
5. अमूर्त अथवा अस्पर्श
6. स्वामित्व की कमी

उपभोक्ता का सहयोग

जब गोचर लौकिक वस्तु उत्पादित होती है, तो वह क्रय की जाती है और उसके उपरांत उसका इस्तेमाल कर लिया जाता है। लेकिन सेवाओं में उसका उत्पाद उपयोगी होता है। भिन्नता केवल इतनी है कि जब लौकिक वस्तु उत्पादित होती है, तो उसके लिए उस समय क्रेता के उपस्थित रहने की आवश्यकता नहीं है। उदाहरण के लिए किसी कम्पनी का क्रय अधिकारी कंपनी के निदेशक के उपयोग के लिए कार खरीद सकता है, लेकिन सेवाओं के सिस्टम में क्रेता का होना आवश्यक है। यदि आप किसी स्थान की यात्रा कर रहे हैं, तो यह तो संभव है कि आपका सचिव आपके लिए टिकट ले आये, लेकिन सेवाओं के उपयोग के लिए यात्रा के लिए आपको स्वयं ही जाना होगा। इस यात्रा सेवा में आपके यात्रा का साधन रेल, जहाज इत्यादि इसके कर्मचारी, कई सुविधाएं सम्मिलित होती हैं। यह सेवा क्रेता के उपस्थित हुए बिना नहीं हो सकती।

स्थायित्व

इसमें सेवाओं को सेवा देने वाले मनुष्य से पृथक् नहीं किया जा सकता। मार्ग दर्शक को किसी स्थान के वर्णन के लिए स्वयं उपस्थित रहना होगा। अध्यापक को पढ़ाने के लिए, चिकित्सक को आपरेशन करने के लिए, निजी पाइलट को वायुयान उड़ाने के लिए निजी रूप से उपस्थित रहना होगा। जो गाइड अनुपस्थित रहेगा, जो अध्यापक कक्षा में नहीं जायेगा, जो चिकित्सक अस्पताल में नहीं जायेगा, जो पायलट जहाज पर नहीं जायेगा, वह अपनी सेवायें नहीं दे सकता। इसलिए सेवा देने वाले को निजी तौर पर उपस्थिति रहकर ही सेवाओं का उत्पादन कर दूसरों को सेवायें दी जा सकती हैं।

अस्थायित्व

व्यक्ति द्वारा उत्पादित सेवाएं मानव द्वारा उत्पन्न उत्पाद से सम्बन्धित होती हैं। इसी वजह से अलग-अलग व्यक्तियों की सेवाओं में अंतर होता है।

उदाहरणार्थ—वायुयान में हर एयर-होस्टेस एक-सी सेवा नहीं कर सकती। इसी प्रकार एक ही एयर होस्टेस प्रातः और संध्यकालीन पालियों में समान सेवा नहीं कर सकती। उसकी सेवाओं पर उसके दैहिक, मानसिक तनाव, मूड और वातावरण इत्यादि का असर पड़ता है, अथवा सेवाओं का मानक समान नहीं हो सकता। सेवाओं के अन्तर का दूसरा कारण है अतिथि अथवा क्रेता का सम्मिलित हो जाना। एक गायक अथवा गायिका किसी उत्सव में अपनी कला का प्रदर्शन तो करता है, पर उसकी कला के प्रदर्शन पर उसके दर्शकों द्वारा किये गये गुण-दोष विवेचन का भी असर पड़ता है। साथ ही कलाकार के मनोभाव, वातावरण एवं उसके साथी कलाकारों का भी अत्यधिक असर पड़ता है। इस प्रकार ये सेवायें समयानुसार एवं व्यक्ति के अनुसार एवं जनभावनाओं के अनुसार ही अच्छी या खराब होती हैं।

शीघ्र नष्ट होने वाली सेवाएं

कुछ सेवाएं ऐसी होती हैं जिनका सिर्फ प्रयोग किया जा सकता है। जैसे—अध्यापन, अभियंता, मार्गदर्शक, गाइड आदि। अगर इन सेवाओं का इस्तेमाल नहीं किया जाता है, तो ये विनष्ट हो जाती हैं। इसके विपरीत उत्पाद विक्रय हेड प्रस्तुत किये जाते है। यदि उन्हें प्रयोग नहीं किया जाता तो उसे भविष्य के विक्रय के लिए स्टोर किया जा सकता है। दूध, फल, ब्रैड आदि उत्पाद नष्ट होने वाले होते हैं, जो कुछ समय के लिए ही स्टोर किये जा सकते हैं अन्यथा वे नष्ट हो जाते हैं। उदाहरणार्थ—एक वायुयान में कुछ खाली सीटें हैं परिणामस्वरूप उन खाली सीटों की क्षति होगी। एक सिनेमाघर में यदि कतिपय लोग नहीं पहुंचते तो सिनेमाघर की सेवायें पूर्ण रूपेण प्रयोग नहीं मानी जातीं।

स्वामित्वाधिकार का आशय

जब हम एक कार या कोई अन्य वस्तु क्रय करने जाते हैं तब उस कार अथवा वस्तु का स्वामित्व वहाँ से हटकर हमारा हो जाता है। लेकिन हम यदि एक टैक्सी किराये पर लेते हैं तो उसका स्वामित्व नहीं बदलता अपितु हमें एक स्थान से दूसरे स्थान तक जाने का अधिकार ही हासिल होता है हम न तो टैक्सी के स्वामी होते हैं, एवं न ही टैक्सी ड्राइवर के। इसी प्रकार होटल में जब हम जाते हैं तो उसके कमरों के स्वामी नहीं होते केवल उनका उपयोग ही कर सकते हैं।

अदृश्यता तथा अछूता

कुछ सेवाएं ऐसी होती हैं जिनका न कोई स्वरूप होता है और न ही उसे छुआ जा सकता है और न उन्हें स्पर्श किया जा सकता है। उदाहरण के लिए किसी स्थान अथवा वस्तु को दिखाने वाले गाइड की टिम्पणियाँ सुनी जा सकती हैं, जबकि यात्रा-अभिकर्ता एक स्थान से दूसरे स्थान के लिए टिकट देता है जो कागज का टुकड़ा मात्र होता है एवं सेवाओं की प्राप्ति के लिए वह प्रवेश पत्र होता है। एक एयर लाइन आवागमन तथा सामान लाने-ले जाने के लिए सेवायें प्रदान करती है।

हम सचमुच में रेलगाड़ी को देखते हैं, जो प्रस्थान स्थान से गन्तव्य स्थान तक ले जाती है तथा कतिपय सुविधाएं एवं आत्म भी देती हैं। दूसरी ओर गोचर अथवा लौकिक उत्पाद विक्रेता बेचता है और क्रेता उसे प्राप्त करता है, जबकि अमूर्त अथवा अदृश्य में भौतिक रूप से कुछ नहीं लेना, जिसे क्रेता और विक्रेता के मध्य रखा जा सके।

कतिपय सेवाएं जैसे डॉक्टर, इन्जीनियर, शिक्षक, गाइड आदि, इन सेवाओं का कोई रूप नहीं होता है। जबकि एयरलांइस, होटल अथवा रेस्टोरेंट, रेल सेवा, बस सेवा इत्यादि सेवायें लौकिक वस्तुओं की सहायता से सेवायें प्रदान करती हैं।

अमूर्त विशेषता वाली सेवाओं में उन सेवाओं को समझने में कुछ मुश्किलें आती हैं। इनमें सेवाओं का वायदा किया जाता है, जो उनके प्रयोग के उपरांत ही आंकी जा सकती हैं। उत्पाद पहले उत्पादित किये जाते हैं एवं तब उनका उपयोग होता है सेवायें पहले बेची जाती हैं, तब उत्पादित होती हैं, और उसके उपरांत ही उन्हें उपयोग में लाया जाता है। उदाहरणार्थ हम दिल्ली से बम्बई का एक एयर टिकट खरीदतें हैं। टिकट खरीदने के बाद वायुयान पर चढ़े यात्रा में और यात्रा के बाद हम सेवा की कुशलता अथवा गुण को आंक सकते हैं कि कितना धन देने से हमें किसकी और कैसी सेवायें उपलब्ध हुईं।

पर्यटन उत्पाद के भेद

अगर देखा जाए तो सेवा उत्पाद या सेवायें ही पर्यटन उत्पाद कहलाती हैं। इस पर्यटन उत्पाद की कई विशेषताएं होती हैं—बड़े-बड़े होटलों, रिसोर्टों को पर्यटन उद्योग का प्रमुख अंग माना जाता है। यह उद्योग पर्यटक सेवा के लिए निर्मित होता है। इस उद्योग के द्वारा कान्फ्रेंस प्लानिंग आदि की सेवा दी जाती है। त्योहार, मेले, इत्यादि घटनायें हैं जिनका मजा लिया जा सकता है पर ये नश्वर हैं। नृत्य, गायन,

वादन आदि देखे जा सकते हैं और मजा लिया जा सकता है। वन्य जीवों का जीवन प्राकृतिक उत्पाद है जो सेवा प्रदान करता है। बीचों, मरुस्थलों आदि का सौन्दर्य कुछ अलग ही होता है। यहाँ लोग विश्राम करने व आनन्द मनाने आते हैं। इसे इकट्ठा नहीं किया जा सकता एवं न ही इनका स्वामित्व प्राप्त किया जा सकता है।

धार्मिक तीर्थ आदि शांति प्रदान करने वाले स्थान हैं जहाँ पारलौकिक शांति मिलती है। अतएव पर्यटन सेवा उत्पाद है, लेकिन इन स्थानों पर जाकर खरीद की जा सकती है। इसकी विशिष्टता एक उत्पाद की भांति हैं, सेवा की भांति नहीं। ये उत्पादन लौकिक हैं एवं उत्पादककर्ता से अलग किये जा सकते हैं। इसी प्रकार रेस्टोरेंट, पैक एण्ड कैरी संस्थान इत्यादि लौकिक उत्पादन जैसे–भोजन, ड्रिंक्स इत्यादि का उत्पादन करते हैं जो कुछ समय के लिए स्टोर किये जा सकते हैं तथा ये अति अल्प अवधि में ही खराब हो जाते हैं।

इन सभी उत्पादों के उत्पादन के समय ग्राहक का रहना आवश्यक नहीं है। लेकिन उसको क्रय करते वक्त तथा उपयोग में लाते वक्त उसका रहना आवश्यक है।

इस तरह पर्यटन उत्पाद को दो भागों में बाँट सकते हैं जो निम्न प्रकार से हैं– 1. सेवा उत्पाद 2. फिजीकल उत्पाद।

पर्यटन उत्पादों का वर्गीकरण

सामान्य रूप पर्यटन उत्पादों का वर्गीकरण इस प्रकार किया जा सकता है–

(1) प्राकृतिक एवं मनुष्यकृत उत्पाद–प्राकृतिक एवं मानवकृत स्थान, या वस्तु या घटना विशेष रूप से पर्यटकों को आकर्षित करती हैं। प्राकृतिक पर्यटन उत्पादों में पर्वतीय स्थान, जलवायु, वनस्पति युक्त बीच, मरुस्थल, द्वीप इत्यादि विशेष तौर पर आते हैं। मनुष्यकृत उत्पादों में मेले, उत्सव, कन्वेंशन, पेंटिंग आर्कीटेक्चर, बाजार, मनोरंजन गृह इत्यादि आते हैं। मुंबई का एस्सेल वर्ल्ड एवं अमेरिका की डिजनीलैंड मानवकृत पर्यटन उत्पादों में आते हैं। प्राकृतिक उत्पाद मानव के लिए विस्मयकारक होते हैं और ईश्वर की ओर से उपहार स्वरूप मिलते हैं, जबकि मनुष्य द्वारा निर्मित उत्पाद मनोरंजन एवं व्यापारिक उद्देश्यों की पूर्ति के लिए होते हैं।

(2) प्रतीकात्मक उत्पाद–प्रतीकात्मक उत्पाद वे होते हैं, जो प्रकृति अथवा मानव दोनों के ही प्रयासों से होते हैं। वन्य जन्तु शरण स्थली, मैरीन पार्क, क्रीड़ाएं, पुष्प उत्सव और प्रदर्शनी इसी प्रकार के उत्पाद हैं। प्रकृति ने स्रोत अथवा साधन दिए और मानव ने उसमें कुछ और जोड़कर उसे उत्पाद के रूप में परिवर्तित किया।

(3) कान्टीनम उत्पादन—कतिपय उत्पाद जोकि मानव निर्मित अथवा प्राकृतिक निर्मित होते हैं, को किसी भी तरह की श्रेणी में वर्गीकृत नहीं किया जा सकता है।

(4) घटना और स्थान पर आधारित उत्पाद—जहाँ कोई घटना आकर्षण का कारण होती है, जहाँ कोई स्थल आकर्षण का कारण होता है, वह स्थान मूल उत्पाद माना जाता है। घटना प्रधान उत्पादों में कार्नीवाल, गोवा का फूड उत्सव, खजुराहो का नृत्योत्सव, स्पोर्टस इत्यादि एवं ताजमहल, स्कीइंग, नेशनल पार्क, मुंबई इत्यादि स्थान मूल उत्पाद माने जाते हैं।

(अ) पैकेज एण्ड प्रोग्राम

(ब) स्थान

(स) पर्यटन

(द) उद्देशीय या गन्तव्य स्थान

(य) घटना

(र) सक्रियता या तीव्रता

(अ) पैकेज एण्ड प्रोग्राम—यह सुविधा नियोजित एवं नियंत्रित होती है। इस सुविधा में एक निश्चित मूल्य पर पर्यटकों को कई तरह की सुविधाएं मिल जाती हैं। इस प्रकार के पर्यटक कार्यक्रमों में रहना, खाना-पीना, घूमना आदि सभी का मूल्य ले लिया जाता है। इन सुविधाओं के अतिरिक्त परिपूरक कैसिनों कूपन आदि भी मिल जायें तो पर्यटन उत्पाद के आकर्षण बढ़ जाते हैं। इसके कार्यक्रम और ज्यादा आकर्षक तब हो जाते हैं जब उनके साथ केन्द्रीय अथवा अंतर्तम आकर्षण और जुड़ जायें। इस तरह की स्थिति में पर्यटक कुछ अधिक धन भी व्यय करते हैं। कान्फ्रेंस आर्गनाइजर्स कान्फ्रेंसों को आयोजित करके पैकेज एण्ड प्रोग्राम उत्पाद सर्वाधिक आकर्षक हो सकते हैं।

(ब) स्थान—कुछ स्थान इस प्रकार के होते हैं, जो अपनी प्राचीनता, सांस्कृतिकता, कलात्मकता, वास्तुकला से पूर्ण विशेषताओं के कारण अत्यंत आकर्षक होते हैं, जैसे—खजुराहो का नृत्योत्सव, महाबलीपुरम के मंदिर, पुरी के मंदिर, कन्याकुमारी के बीच, आगरा का ताजमहल, मिस्र के पिरामिड इत्यादि।

(स) पर्यटन—वर्तमान समय में पर्यटन विशिष्ट स्थलों जैसे—धार्मिक या तीर्थ या ऐतिहासिक जगहों पर जाकर मन को हर्षित करने को एक उत्तम साधन है, जिसके द्वारा पर्यटकों को अलग तरह की सुविधाएं आदि प्रदान की जाती हैं।

अलग-अलग पर्यटन स्थल उन्हें विभिन्न प्रकार से आकर्षित करते हैं। उदाहरणार्थ कोई भी ऐसी संस्था अपने देश के विभिन्न स्थानों पर बस, ट्रेन, वायुयान या पानी के जहाजों के माध्यम से भ्रमण करा सकती है। यही पर्यटक के रहने, खाने-पीने आदि का भी प्रबन्ध कर देती है। यदि पैकेज टूर के कार्यक्रम एवं सर्किट टूर दोनों मिलाकर पर्यटकों के कार्यक्रम बनाये जायें तो वे बड़े ही आकर्षक एवं फायदेमंद हो सकते हैं। पैलेस आन व्हील्स जो पर्यटन विभाग की अति सुव्यवस्थित एवं सुविधापूर्ण रेलगाड़ी है, जो इसी प्रकार के पैकेज और सर्किट कार्यक्रमों को परिपूर्ण करने वाली है और पर्यटकों की सभी आवश्यकताओं और सुख-सुविधाओं से परिपूर्ण है।

(द) उद्देशीय या गन्तव्य स्थान—कुछ स्थान स्वयं में पर्यटन आकर्षणों से परिपूर्ण हैं। वहाँ की घटनाओं, शहरों, विशिष्ट स्थलों का कोई आकर्षण पर्यटकों को नहीं होता अपितु उस शहर का स्थान ही आकर्षण का पूर्ण केन्द्र होता है, जैसे—दिल्ली, चेन्नई, मुंबई, लंदन, न्यूयार्क, आदि पर्यटकों के लिए विशेष आकर्षण हैं। इन स्थानों पर जाकर पर्यटन तो कर ही लेते हैं साथ अपने दूसरे उद्देश्यों को भी पूर्ण कर लेते हैं।

(य) घटना—कतिपय महत्वपूर्ण घटनाएं पर्यटन के विकास में महत्वपूर्ण योगदान देती हैं।

(र) सक्रियता या तीव्रता—जो उत्पाद सक्रियता या तीव्रता प्रधान होते हैं उनमें पर्यटकों को बड़ा सक्रिय रहना पड़ता है, उदाहरण के लिए दुकानदारी, साहसिक स्पोर्टस जैसे—ग्लाईडिंग पेरासलिंग इत्यादि। पर्यटक दुकानों पर सामान अथवा भेंट की वस्तुएं क्रय करने आते हैं। पर्यटक जब तक स्वयं नहीं चाहेगा तब तक वह साहसिक कार्यों में भाग नहीं ले सकता। सक्रिय पर्यटन उद्योग प्रकृति एवं मानव दोनों का ही सहारा ले सकता है।

उपर्युक्त घटनाओं के अतिरिक्त कतिपय घटनायें ऐसी हैं, जो पर्यटक को विशष रूप से आकर्षित करती हैं, जैसे—बोट रेस, ओलम्पिक खेल आदि।

पर्यटन उत्पादों द्वारा आवश्यकताओं की संतुष्टि

पर्यटकों की जरूरतों की पूर्ति अधिक-से-अधिक पर्यटन उत्पाद से संभव है। अब हमें यह देखना है कि पर्यटकों की वे आवश्यकताऐं कौन-सी हैं? पर्यटकों को ऐसा कौन-सा आकर्षण मिलता है जिसके लिए वे दूर-दूर तक यात्रा करते हैं, आवश्यकता के मुताबिक धन व्यय करते हैं। हम आसानी से इन आकर्षण युक्त

आवश्यकताओं को निम्न प्रकार से सात भागों में विभाजित कर सकते हैं जिनके द्वारा पर्यटकों को प्रोत्साहन व प्रेरणा मिल सकती है–

1. विश्राम एवं अवकाश
2. शिक्षा
3. मित्रों-परिचितों तथा रिश्तेदारों से भेंट
4. मनोरंजन एवं आनन्द
5. व्यवसाय
6. विशिष्ट हित
7. स्वास्थ्य

प्रेरक घटक एवं पर्यटन उत्पाद

कई ऐसे तत्व हैं जिनके द्वारा पर्यटन एक उत्पाद होता है। ऐसी आवश्यकताएं एवं आकर्षण हैं जिनसे स्वतः ही किसी न किसी रूप में पर्यटन उत्पाद होता रहता है। दोनों ही एक दूसरे के पूरक कहे जा सकते हैं। कुछ विशिष्ट प्रेरक घटक एवं उनसे उत्पादित पर्यटन उत्पाद निम्नलिखित हैं–

प्रेरक घटक	*पर्यटन उत्पाद*
1. विश्राम एवं अवकाश	योग, खेल-कूद, समुद्र तटीय बीच
2. मित्रों, परिचितों एवं रिश्तेदारों	खरीदारी करना एवं स्मृति भेंट
3. विशिष्ट हित	साहसिक कार्य जैसे पर्वतारोहण, तैराकी इत्यादि
4. मनोरंजन एवं आनन्द	पर्व, उत्सव, वन्य जीवन, मेले इत्यादि।
5. व्यवसाय	सम्मेलन एवं सभायें
6. शिक्षा-दीक्षा	संस्कृति एवं विशिष्ट परम्परायें।
7. स्वास्थ्य	साहसिक खेलकूद

भारतीय पर्यटन सिंहावलोकन

भारत एक धर्म निरपेक्ष देश है। यहां विभिन्न धर्म के लोग स्वतंत्र रूप से रहते हैं। अतः यहां प्रत्येक दिन पर्व अथवा उत्सव होता रहता है। पर्यटन के नजरिये से भी यह देश श्रेष्ठतम राष्ट्रों में से विश्व में एक है। यहाँ के विविध पर्यटन उत्पाद देश को व्यापक प्रगति की ओर प्रेरित करने में काफी सफल रहे हैं। निम्नांकित तत्वों से यह स्पष्ट हो जाता है कि पर्यटन उत्पादों की स्थिति कैसी है।

वन्य जीवन—भारत वन्य जीवन के रूप में भी जाना जाता है। विभिन्न वन्य पशु यहाँ के वनों में स्वतंत्र रूप से विचरण करते रहते हैं। राजस्थान में स्थित रणथंभोर एक ऐसी जगह है, जो अपने बाघों के लिए विख्यात है। कश्मीर की सुप्रसिद्ध शरणस्थली अपने हिरन और बारहसिंगाओं के लिए विश्व प्रसिद्ध है। राजाजी नेशनल पार्क केवल अपने जंगली हाथियों के झुंडों के लिए विख्यात है। इसी तरह स्नेक पार्क विभिन्न प्रकार के सर्पों के लिए और सुन्दर वन का मगरमच्छ पार्क अपने तरह-तरह के मगरमच्छों के लिए विख्यात है। इन सभी से इन वन्य पशुओं के निजी जीवन के विषय में जाना जा सकता है।

(1) सांस्कृतिक पर्यटन उत्पाद—भारत भले ही आर्थिक क्षेत्र में कई राष्ट्रों से पीछे है लेकिन सभ्यता और संस्कृति के क्षेत्र में यह विकासशील देश विश्व के सभी राष्ट्रों से आगे है। यहां की संस्कृति की ख्याति सम्पूर्ण विश्व में है, जो विदेशी पर्यटकों को अपनी ओर आकर्षित करती है। पर्यटन उत्पाद के तौर में भारतीय संस्कृति के नृत्य-गायन, पेंटिंग, मूर्तिकला, वास्तुकला इत्यादि अति प्रसिद्ध हैं। सम्पूर्ण भारत की

यों तो एक ही तरह की संस्कृति रही है, पर जलवायु, स्थान आदि के कारण वह परिवर्तित भी हो जाती है।

भारतीय नृत्य, विशेष तौर पर दक्षिण भारत का भारत नाट्यम, मणिपुर का कुचिपुड़ी नृत्य, उत्तर भारत का कत्थक नृत्य, केरल का कथकली नृत्य, उड़ीसा का ओडिसी नृत्य इत्यादि विश्व प्रसिद्ध हैं। भारतीय चित्रकला में भी विविधता है। राजस्थान की मधुवनी कला, उत्तर प्रदेश की विचित्र कला के अतिरिक्त भी विभिन्न व्यक्तियों द्वारा चित्रकला का अद्भुत प्रदर्शन किया गया है।

इसी प्रकार भारतीय शास्त्रीय संगीत भी सर्वविदित एवं प्रसिद्ध है। तमिलनाडू के मृदंग एवं कश्मीर के जलतरंग संगीत को कौन नहीं जानता। यहाँ संगीतज्ञों ने अपने संगीत के जादू से विश्व को सम्मोहित कर रखा है।

अजंता एवं एलोरा की गुफायें, बौद्ध धर्म की कला और वास्तु शास्त्र के ज्वलंत उदाहरण हैं। पुरी के हिन्दू धर्म के मंदिर विशिष्ट प्रकार की संस्कृति के परिचायक हैं। इसी तरह जैन वास्तुशास्त्र के गुजरात और मुस्लिम वास्तुशास्त्र के फतेहपुर सीकरी (आगरा) तथा अन्य स्थानों पर दर्शनीय हैं।

भारतीय वास्तुशास्त्र कला खजुराहो के मंदिरों में तथा दूसरी इसी प्रकार की अन्य विशेषताएं मंदिरों व स्मारकों में देखी जा सकती हैं। इसकी तुलना किसी दूसरे से करना शोभनीय नहीं कहा जा सकता।

इसी प्रकार भारतीय संस्कृति की अन्य विधायें भी पर्यटन उत्पाद में विभिन्न तरह से सहायक होती हैं।

धर्म—भारत आदि काल से ही धर्म प्रधान देश रहा है और यहाँ की अधिसंख्य जनसंख्या हिन्दुओं की है—मुस्लिम, ईसाई, सिक्ख, जैन, बौद्ध, ज्यूज, इत्यादि भी इस देश की जनसंख्या के ही अंग हैं तथा प्रेमपूर्वक निवास करते हैं। ये सभी धर्मावलम्बी अपने-अपने धर्मानुसार धार्मिक कृत्य पूजा-पाठ आदि स्वतंत्रापूर्वक करते रहते हैं।

साहसिक खेलकूद एवं अन्य कार्य—समुद्र एवं बड़ी नदियों में ये साहसिक खेल-कूद होते रहते हैं। इन खेलकूदों से मानसिक प्रगति के साथ-साथ शारीरिक प्रगति भी होती है।

सम्मेलन एवं सभा—यह बड़ी तीव्रता से पर्यटन उत्पाद के रूप में बढ़ रहा है। ये अधिसंख्य लोगों को आकर्षित करने में समर्थ है। इसमें आराम एवं काम दोनों को सम्मिलित किया जा सकता है। दिल्ली, बंगलौर, कोलकाता, मुंबई इत्यादि में कान्फ्रेंस आदि आयोजित की जाती है।

खरीददारी—पर्यटक भ्रमण के दौरान मेले तथा बड़े बाजारों से विभिन्न तरह की वस्तुएं यादगार के रूप में खरीदता है। सहारनपुर और मैसूर में चंदन की लकड़ी, हाथी दांत, मुरादाबाद में पीतल का सामान, आगरा एवं कानपुर में चमड़े आदि का सामान निर्मित होता है। उन्हें क्रय करने के लिए लोग देश के कोने-कोने से पहुंचते हैं। बनारस की रेशमी साड़ियाँ, कश्मीर के गलीचे विश्व प्रसिद्ध हैं।

पर्व-उत्सव एवं त्योहार—मानव जीवन में पर्व-उत्सव एवं त्योहार निरन्तर होते रहते हैं। त्योहार, मेले आदि मौसम के मुताबिक होते हैं। दीपावली सम्पूर्ण देश में बड़े उत्साह से मनायी जाती है। होली पूरे देश में मिलन का त्योहार होता है। कुंभ,

मेरठ का नौचंदी मेला, अलीगढ़ की नुमाइश आदि भी प्राचीन मेले हैं, जो सदियों से एक विशेष पर्व के तौर पर मनाये जाते हैं, जिनमें ग्रामीण तौर पर सभी शहरी नागरिक सम्मिलित होते हैं और अपनी आवश्यकताओं की पूर्ति करते हैं।

बीच—भारत में कई समुद्र तटीय पर्यटक स्थल हैं, जहां लाखों की संख्या में पर्यटक मनोरंजन करने तथा अवकाश बिताने के लिए पहुंचते हैं। भारत के पास लगभग छः हजार मील लंबा समुद्र तट है। समुद्र के तटीय भागों में नारियल, पाम आदि के असंख्य कुंज हैं, जहाँ पर्यटक पहुंचते हैं। विभिन्न प्रकार के पशु-पक्षी वहाँ दिखाई देते हैं और पर्यटकों का मनोरंजन करते हैं। उड़ीसा, अंडमान निकोबार, गोवा, मुंबई, केरल इत्यादि में भिन्न-भिन्न प्रकार के बीच हैं और वहाँ विभिन्न राज्यों के लोगों का निवास है।

इस प्रकार भारतीय पर्यटन उत्पाद के उत्तम विविध तत्वों के आधार पर कहा जा सकता है कि हमारा यह पर्यटक उत्पाद एक विशिष्ट प्रकार एवं उपयोगी उत्पाद है। इन उत्पादों से आर्थिक, मानसिक, शारीरिक और अन्य सभी प्रकार के फायदे होते हैं।

बाजारीकरण की अवधारणा

वर्तमान समय में पर्यटन के क्षेत्र में बाजारीकरण की विचारधारा उत्पन्न हो रही है। इसका प्रमुख कारण है, पर्यटन की तरफ लोगों की मानसिकता बनना। पर्यटन के क्षेत्र में बड़े-बड़े उद्योगपति मुनाफा कमाने के लिए अधिक से अधिक पूंजी विनियोग करना चाहते हैं। अक्सर विपणन शब्द का प्रयोग प्रबन्धकों द्वारा अपनी कम्पनियों को आधुनिक, गतिशील एवं भविष्य द्रष्टा के रूप में प्रस्तुत करने के लिए किया जाता रहा है। विपणन के सही अर्थ को समझने के लिए हमें थोड़ा पिछले समय में जाना होगा।

बाजारीकरण या विपणन विचारधारा एवं इसका दर्शन--बाजारीकरण विचारधारा का उद्भव सर्वप्रथम संयुक्त राज्य अमरीका में हुआ। इसके पश्चात् यह विचारधारा शनैः-शनैः पश्चिमी यूरोप एवं जापान तक पहुँच गई। जापान ने इस विचारधारा को वास्तविक रूप में विकसित करके अन्य विश्व के राष्ट्रों तक पहुंचाया। जापान ने सबसे समस्त उत्पादों को उपभोक्ताओं की अभिलाषाओं एवं आवश्यकताओं पर आधारित करके एवं इस कार्य में सफलता प्राप्त करके विपणन को एक नया आयाम दिया। वास्तव में जब तक देश का उत्पादन, इसकी मांग से कम रहा, तब तक विपणन की कोई सतत् वृद्धि नहीं हुई। परन्तु बीसवीं शताब्दी के छठे दशक

में उत्पादन पद्धतियों में सुधार, संचार एवं यातायात के क्षेत्रों में नये-नये आविष्कारों के फलस्वरूप उत्पादन की मात्रा निरन्तर बढ़ती गयी। ऐसे समय एवं परिस्थितियों में विपणन के विचार की वास्तविक रूप में आवश्यकता अनुभव हुई।

बाजारीकरण व विपणन की विचारधारा का आलंवन

जब उत्पादन मांग से कम होता था, तब उत्पाद बेचने में कोई वक्त नहीं लगता था एवं विक्रय क्रेता की इच्छाओं पर चलती थी। परन्तु परिवर्तित परिस्थितियों में विपणन के अंतर्गत विक्रेताओं को अपने उत्पादों में क्रेताओं की इच्छानुसार परिवर्तन करना एक आवश्यकता बन गयी, जो विपणन का मूल आधार बनी।

पर्यटन और विपणन की विचारधारा

पर्यटन की समस्त क्रियाओं को बाजारीकरण विशेष रूप से प्रभावित करता है। वर्तमान पर्यटन में पूंजी के रूप में भारी निवेश करना पड़ता है एवं इस क्षेत्र में पर्याप्त आधारभूत संरचना स्थापित करने में वर्षों लग जाते हैं। इस समस्याओं को एक सोची-समझी विपणन नीति के द्वारा ही हल किया जा सकता है।

भारतीय संदर्भ में बाजारीकरण विचारधारा

भारत के पर्यटन क्षेत्र में आगमन अधिक प्राचीन घटना नहीं है। स्वतन्त्रता प्राप्ति के उपरांत भारत ने अपनी अर्थव्यवस्था के लिए संरक्षित एवं प्रतिबन्धित व्यापार नीति अपनायी। उत्पादों की नयी इकाइयों को प्रारंभ करने, पुरानी इकाइयों का विस्तार करने एवं यहाँ तक कि उत्पादों के वितरण हेतु भी सरकारी अनुमति लेने की आवश्यकता होती थी। यही स्थिति पर्यटन के क्षेत्र में भी थी। वायु सेवाओं, होटल व्यवस्थापन और पर्यटन से सम्बन्धित अन्य क्रियाओं पर भी सरकारी नियंत्रण था। निजी कम्पनियों को इस क्षेत्र में आने की कोई अनुमति नहीं थी। परिणामस्वरूप विपणन विचारों के क्षेत्र में कोई महत्व नहीं मिल सका। यह स्थिति सन् 1980 के बीच तक बनी रही।

भारतीय अर्थव्यवस्था के परम्परागत ढांचे में सुधार लाने का प्रयास सर्वप्रथम स्व. राजीव गांधी ने किया। परन्तु दुर्भाग्यवश उनकी असामयिक मृत्यु के कारण कोई ठोस योजना नहीं बन सकी। इसके उपरांत तत्कालीन प्रधानमंत्री नरसिम्हा राव ने 1994 में उद्योग व्यापार को सरकारी नियंत्रण से मुक्त करने एवं भारतीय अर्थव्यवस्था को विश्व अर्थव्यवस्था के समकक्ष लाने के लिए नयी आर्थिक नीति की घोषणा की,

जिसके अंतर्गत अर्थव्यवस्था के उदारीकरण के लिए कार्यक्रमों की घोषणायें की गयीं। इस नीति का सार इस प्रकार था।

इस प्रकार हम देखते हैं कि पर्यटन के क्षेत्र में विपणन की विचारधारा का प्रारम्भ उपर्युक्त ऐतिहासिक घटना के फलस्वरूप हुआ था। बाजारीकरण विचारधारा के फलस्वरूप ही पर्यटन के क्षेत्र में अधिक विकास हो रहा है। 1995 के सर्वेक्षण के अनुसार भारत में पर्यटन आगमन का आंकड़ा 2 मिलियन को पार कर गया। पर्यटन से सम्बन्धित सेवाओं की प्रकृति, वर्गीकरण, विशेषतायें एवं पर्यटन योगदान।

किसी देश के पर्यटन को विकसित करने में वहां की विचारधारा की विशेष भूमिका रहती है। बाजारीकरण की विचारधारा से पर्यटन का विकास ही नहीं होता बल्कि देश की आर्थिक स्थिति भी मजबूत होती है, साथ में विदेशी मुद्रा अधिक-से-अधिक आने की संभावना रहती है। राष्ट्र में पर्यटन की संभावनाओं में सकारात्मक वृद्धि करने के लिए हमें उन जरूरी सेवाओं की प्रकृति व विशेषताओं को जानना जरूरी है, जो पर्यटन से प्रत्यक्ष तौर से सम्बद्ध हैं, तभी हम एक सकारात्मक विपणन विचारधारा को अपनाने में सफल हो सकते हैं।

पर्यटन विकास की दृष्टि से जरूरी सेवाओं के अन्तर्गत निम्नलिखित आवश्यक तत्वों को रखा जा सकता है–

पर्यटन उत्पाद–पर्यटक उत्पाद विपणन में एक प्रमुख एवं महत्वपूर्ण भूमिका अदा करते हैं। वास्तव में पर्यटन का मूल आधार ही पर्यटक उत्पाद है जिससे आकर्षित तथा वशीभूत होकर राष्ट्रीय पर्यटकों का आगमन संभव होता है। पर्यटकों की दृष्टि से इनका विशेष महत्व होता है। सर्वप्रथम पर्यटक उत्पाद की प्रकृति, वर्गीकरण एवं विशेषताओं को जानना बहुत महत्वपूर्ण होता है।

पर्यटकों की दृष्टि में पर्यटन उत्पाद के महत्व को इन शब्दों में प्रकट किया जा सकता है–

भारत एक विकासशील देश है तथा इसकी अर्थव्यवस्था मिश्रित है। यहां की अर्थव्यवस्था में पर्यटन उत्पादन को विशेष रूप से शामिल किया गया है। क्योंकि पर्यटन उत्पादन से विदेशी मुद्रा का अधिक अर्जन होता है। इस उत्पाद में आकर्षण, परिवहन, आवास और मनोरंजन आदि महत्वपूर्ण तत्व सम्मिलित रहते हैं। कोई भी देश अपने पर्यटक उत्पादों को निम्न रूप में प्रकट कर सकता है–

(1) संस्कृति

(2) संग्रहालय एवं स्मारक

(3) शिल्प एवं कलात्मक संसाधन

(4) प्राकृतिक संसाधन

पर्यटकों की इच्छाओं एवं आवश्यकताओं के अनुसार ही इन उत्पादों को प्रस्तुत किया जाता है। विपणन की दृष्टि से मुख्य रूप से निम्नांकित तीन आधारों पर उत्पादों को विकसित करने की जरूरत होती है--

(i) आकर्षण

(ii) सुविधायें

(iii) पहुंच या उपलब्धता

पर्यटन प्रोत्साहन

पर्यटन प्रोत्साहन के द्वारा राष्ट्रीय तथा अन्तर्राष्ट्रीय स्तर के पर्यटकों को ट्रेवल एजेन्टों के द्वारा पर्यटन से सम्बन्धित विभिन्न तरह की सुविधाओं से अवगत कराया जाता है। उदाहरण के लिए परिवहन, आवास, खान-पान इत्यादि की सुविधाएं।

पर्यटन प्रोत्साहन के पीछे मूल धारणा यह है कि उत्पाद से सम्बन्धित नवीनतम सूचनायें पर्यटकों को आगमन के लिए प्रेरित करती हैं।

पर्यटन प्रोत्साहन का कार्य निम्न माध्यमों द्वारा किया जा सकता है:–

– सांस्कृतिक आदान-प्रदान

– पोस्टर

– संचार माध्यम

– दृश्य श्रव्य सामग्री

– मेले एवं प्रदर्शनी

– विज्ञापन

पर्यटन प्रोत्साहन की विशेषता यह है कि इसकी प्रकृति पर्यटकों में पर्यटक स्थान के प्रति जागरूकता लाती है जिससे पर्यटकों से जुड़े उत्पादों की बिक्री की संभावना बढ़ जाती है।

पर्यटन उत्पाद की बाजारी कीमत

पर्यटन उत्पाद की बाजारी कीमत बहुत हद तक पर्यटन बाजार को प्रोत्साहित करती हैं। उत्पाद चाहे कितना ही अच्छा हो, ठोस एवं स्थायी हो, यदि उत्पाद की कीमत प्रति स्पर्धात्मक कीमत से ज्यादा है, तो आय का आकार छोटा हो जायेगा।

हर उपभोक्ता का सामान्य व्यवहार यही होता है कि वह दूसरे वैकल्पिक उत्पादों की कीमतों को ध्यान में रखकर ही विनिमय कार्य करता है। दुर्भाग्यवश हमारे प्राथमिक बाजार भौगोलिक दृष्टि से इतनी अधिक दूर हैं कि पर्यटकों द्वारा दी जाने वाली कीमत बहुत अधिक हो जाती है।

पर्यटन बाजार

पर्यटक भ्रमण के दौरान बाजार में जिन-जिन वस्तुओं का क्रय करता है, वे सभी पर्यटन बाजार के अन्तर्गत आते हैं। बाजारों को विविध वर्गों में विभक्त करके हम पर्यटक आगमन के महत्वपूर्ण क्षेत्रों का आंकलन करके अपनी विपणन नीति में परिवर्तन कर सकते हैं तथा उसे और भी ज्यादा प्रभावशाली बना सकते हैं। कुल पर्यटन बाजार को मुख्य रूप से तीन भागों में विभाजित कर सकते हैं–

पर्यटन बाजार पर अधिकार

जिन पर्यटकों का उद्देश्य पर्यटन क्षेत्रों में जाकर अपना छुट्टी बिताना होता है, उन पर्यटकों को प्रायः पर्यटन बाजार पर बोझ समझा जाता है। क्योंकि ये पर्यटक किसी भी वस्तु के क्रय-विक्रय से ज्यादा घूमने-फिरने पर अधिक ध्यान देते हैं।

पर्यटन बाजार का व्यवसाय

इसमें अपनी व्यावसायिक आवश्यकताओं की पूर्ति के लिए आने वाले पर्यटकों को सम्मिलित किया जाता है।

सामान्य रुचि का पर्यटन बाजार

उपर्युक्त दोनों उद्देश्यों के अतिरिक्त अन्य सभी उद्देश्यों की पूर्ति के लिए आने वाले पर्यटकों को इस बाजार में सम्मिलित किया जाता है।

पर्यटन और परिवहन

पर्यटन सेवाओं में परिवहन की महत्त्वपूर्ण भूमिका होती है। यदि पर्यटन में परिवहन की अच्छी व्यवस्था न हो तो पर्यटन सफल नहीं हो सकता। यह पर्यटकों को विभिन्न पर्यटन स्थलों पर पहुंचाने का काम करती हैं। चूंकि पर्यटन का यह एक अनिवार्य अंग है। अतः इन सेवाओं का वजूद व्यवहार, परिचालन, पर्यटन की दृष्टि से प्रमुख हो जाता है।

दुर्भाग्यवश भारत में सड़क परिवहन संख्या के आधार पर तो काफी बढ़ा है, पर गुणवत्ता के आधार पर इसमें आज भी कमी है। (Indian Railways, Air India, Indian Airlines) अभी सरकारी क्षेत्र में होने के कारण पूर्णतः व्यावसायिक धारणा रखते हैं। अतः बाजार पक्ष दुर्बल पड़ जाता है। देखा गया है कि सरकारी वाहनों की तुलना में निजी कम्पनियाँ बाजार नीति को बेहतर तरीका अपनाकर परिवर्तित लाभ प्राप्त कर रही हैं। इसके पीछे उपभोक्ताओं की रुचियों के अनुसार अपनी सेवाओं में परिवर्तन करने की व्यवस्था है। सरकारी सेवाओं को भी अपनी परम्परागत सोच को बदलकर वास्तविक रूप में पर्यटक परिवहन के क्षेत्र में विपणन को अपनाने की आवश्यकता है।

बाजार की योजनागत संक्रियाएं तथा विपणन योजनाएं

पर्यटन उत्पाद में बाजार प्रक्रिया योजना का बड़ा ही महत्व होता है। प्रायः पर्यटन उत्पाद से सम्बन्धित सभी व्यापारिक संस्था को एक विपणन कार्य योजना की आवश्यकता होती है। यह योजना अल्पकालीन या दीर्घकालीन हो सकती है। अल्पकालीन योजना अपने निर्धारित समय में आय-प्रवाह पर्यटन, उत्पादों का चयन और लाभदायकता के तात्कालिक मुद्दों से सम्बद्ध रहती है, जबकि दीर्घकालीन कार्य योजना में विकास के लक्ष्यों तथा रणनीति, विस्तार तथा लाभदायकता निर्धारित की जाती है।

बाजार प्रक्रिया पर आने से पहले हमें इस क्षेत्र के विशेषज्ञों द्वारा दिए गये दो मूल मन्त्रों को आत्मसात करना चाहिए।

विपणन प्रक्रिया की कार्य योजना को निम्नलिखित प्रकार से स्पष्ट किया जा सकता है—

संयुक्त विश्लेषण

कोई भी पर्यटन उद्योग अपने लक्ष्यों को निर्धारित करने के लिए एकजुट होकर भिन्न-भिन्न विषयों पर अध्ययन करता है। लक्ष्यों का निर्धारण चालू आर्थिक एवं बाजार के वातावरण के अध्ययन के उपरांत किया जाता है। कम्पनियाँ उपस्थित प्रतियोगता के आधार पर सुदृढ़ता एवं दुर्बलता (Strength and Weaknesses) की परख करती हैं। साथ ही साथ चुनौतियों व प्राथमिकताओं (Threats and Opportunities) के अंतर्गत की जाती है। लक्ष्यों के अंतर्गत एक दिये गये समय में बाजार के खण्डों के तहत पर्यटकों के आगमन की भी जांच पड़ताल होती है।

बाजार अन्वेषण

परिभ्रमण उत्पाद सभी वर्गों द्वारा उपयोग नहीं किया जाता है क्योंकि बहुत से लोग इस पर आने वाले व्यय को सहन नहीं कर सकते। अतः एक बाजारकर्ता को इसके लाभ-हानि, आकर्षण, प्रतियोगात्मक स्थिति इत्यादि बातों को जानना आवश्यक है। इन सभी बातों की जानाकारी शोध द्वारा ही सम्भव है।

निम्नलिखित आवश्यक प्रश्नों के उत्तर बाजार अन्वेषण में ढूंढने चाहिए।

1. वर्तमान एवं भूत में बाजार की क्या प्रवृत्तियाँ रही हैं?
2. कीमतों का कौन-सा स्तर वे स्वीकार करते हैं?
3. भविष्य के ग्राहक कौन हो सकते हैं?
4. इन प्रवृत्तियों को कौन-कौन से कारक प्रभावित कर रहे हैं?
5. पर्यटन बाजार का कुल आकार क्या है?
6. प्रतियोगिता की तुलना की दृष्टि से हमारी मजबूती तथा कमजोरियां कौन-कौन सी हैं?
7. वे कहाँ रहते हैं?
8. वे हमारे उत्पादों में क्या-क्या देखना चाहते हैं?
9. वर्तमान ग्राहक कौन एवं क्या है?

बाजार शोध आंकड़ों के प्रकार पर निर्भर करती है। ये आंकड़े प्राथमिक अथवा द्वितीयक हो सकते हैं।

प्राथमिक आंकड़े मुख्यतया पहली बार क्षेत्र-सर्वेक्षणों द्वारा जमा किये जाते हैं। सूचनाओं को प्रश्नावली के आधार पर तैयार किया जाता है। इसके द्वारा हम व्यापार की स्थिति, पूर्व की संख्या, पर्यटकों का व्यय, व्यापार के स्रोत, लागत, लाभ इत्यादि के सम्बंध में सूचनायें प्राप्त कर सकते हैं।

द्वितीय आंकड़े वे होते हैं जो पूर्व से ही सुलभ रहते हैं। इनके एकत्रीकरण की प्रक्रिया को एक रिसर्च भी कहा जाता है। द्वितीय आंकड़ों के स्रोत पत्र-पत्रिकाएं, जनगणना रिपोर्ट, वार्षिक कम्पनी रिपोर्ट, पर्यटन संस्थाओं के सर्वेक्षण, अन्तर्राष्ट्रीय संस्थाओं के सर्वेक्षण इत्यादि होते हैं।

बाजार विभाजन

बाजार विभाजन एक महत्वपूर्ण प्रक्रिया है। इस प्रक्रिया का महत्व अधिकांश उत्पादकर्ता देते हैं। इस प्रक्रिया के तहत उत्पादकर्ता ये जान पाते हैं कि उनके उत्पाद

को बाजार के किन वर्गों द्वारा अधिक पसंद किया जा रहा है। कौन वर्ग उसके उत्पाद में अधिक दिलचस्पी रखता है। हालांकि यह प्रक्रिया भी बाजार शोध के द्वारा पूरी हो सकती है। हम निम्न आधार पर बाजार खण्डीकरण का कार्य कर सकते हैं—

भौगोलिक आधार पर

1. प्राथमिक बाजार
2. द्वितीयक बाजार
3. बाह्य बाजार

विश्राम यात्रा के आधार पर

1. वरिष्ठ नागरिक बाजार
2. युवा बाजार
3. विशेष रुचि बाजार

मानव सम्पदा का पर्यटन में महत्व

पर्यटन मुख्य रूप से मानवीय संपदा पर ही निर्भर है। यदि पर्यटन में मानवीय संपदा को नजरअंदाज कर दिया जाए तो उसका महत्त्व नष्ट हो जाता है। वर्तमान आर्थिक युग में मानव को आर्थिक संसाधनों का प्रमुख अंग माना जा रहा है। इसे पूंजी का एक हिस्सा माना जा रहा है। यह विचार किसी भी देश में तब से शुरू हुआ जब से कल्याणकारी एवं लोकतंत्रात्मक सरकारों की स्थापना हुई। सर्वेक्षणों द्वारा स्पष्ट भी हो चुका है कि जिस क्षेत्र में भी मानव संसाधन पर्याप्त तौर पर विकसित है, उन क्षेत्रों का विकास तुलनात्मक तेजी के साथ सम्पन्न हुआ है। मानव संसाधनों के अभिप्राय को स्पष्ट करते हुए सर आर्थर लुईस ने इस प्रकार वर्णित किया हैः—

मानव संसाधन के रूप में तभी परिवर्तित हो सकता है, जबकि उसे किसी विशेष क्षेत्र में ज्ञान, बुद्धि प्रशिक्षण इत्यादि आधारों पर विशेषज्ञता प्राप्त हो चुकी हो। बौद्धिक परिपक्वता के साथ-साथ शारीरिक एवं मानसिक स्वास्थ्य को भी मानव संसाधन के अन्तर्गत रखा गया है।

पर्यटन की दृष्टि से जब मानव संसाधन को देखा जाता है, तो पर्यटन के विभिन्न क्षेत्रों जैसे—गाइड सेवाओं, प्रकाशनों, होटल, पोषाहार, परिवहन सुविधाओं इत्यादि में मनुष्य के महारत हासिल करने से है। सचमुच में मानव संसाधन के रूप में पर्यटक का विकास तभी कर सकता है, जबकि उन्हें पर्यटन से सम्बन्धित प्रशिक्षण

दिया जाए। उदाहरण के लिए हम होटल व्यवसाय को पर्यटन के क्षेत्र में तभी लाभकारी बना सकते हैं, जब हमारे पास होटल प्रबन्धकों की पर्याप्त कुशलता हो।

मानव संसाधन किसी राष्ट्र के लिए अनमोल खजाना होता है। किसी राष्ट्र का विकास मानव संसाधन पर ही निर्भर करता है। राष्ट्र के विकास में इसका योगदान प्रत्यक्ष या परोक्ष दोनों रूप में रहता है। खासकर पर्यटन क्षेत्र से सम्बन्धित सभी क्रिया-कलाप मानव संसाधन पर ही निर्भर करते हैं। अतः भारत सरकार ने पर्यटन क्षेत्रों में मानव संसाधनों को और अधिक विकसित करने के लिए अनेक योजनाएँ तथा कार्यक्रम बनाये हैं। इनका अध्ययन निम्नांकित तरीके से किया जा सकता है।

(क) होटल प्रबन्ध कैटरिंग तकनीकी—यद्यपि सरकार ने होटल प्रबन्ध कैटरिंग तथा पोषाहार के क्षेत्र में प्रशिक्षण कार्यक्रम 1962 से ही शुरू कर दिया था, परन्तु व्यावसायिक आधार एवं पर्यटन की दृष्टि से इसका विधिवत् श्रीगणेश 1984 से हुआ, जब पर्यटन मंत्रालय ने राष्ट्रीय होटल प्रबन्ध एवं कैटरिंग प्रौद्योगिकी परिषद् की स्थापना की। इस परिषद् के मुख्य लक्ष्य निम्न हैं—

1. इस क्षेत्र में अन्तर्राष्ट्रीय स्तर की नवीनतम क्रियाओं का लेखा-जोखा रखना।
2. राष्ट्रीय स्तर पर प्रमाणन तथा सत्यापन करना।
3. उपयुक्त प्रशिक्षण कार्यक्रमों के निर्माण के लिए एक तकनीकी भंडार ग्रह के रूप में काम करना।
4. सम्बद्ध संस्थाओं में स्टाफ के सदस्यों के लिए शैक्षिक एवं अन्य अहर्तायें निर्धारित करना।
5. पाठ्यक्रम एवं सुविधाओं की आवश्यकताओं का मानकीकरण करना।
6. होटल प्रबन्ध केटरिंग प्रौद्योगिकी से सम्बन्धित विकास पर सरकार को राय-मशविरा देना।

इस प्रकार होटल प्रबन्धन एवं कैटरिंग के क्षेत्र में हुए अत्याधुनिक परिवर्तनों के परिणामस्वरूप इन संस्थानों में छात्रों की रुचि एवं दृष्टिकोण में व्यापक बदलाव हुए हैं। वर्तमान समय में देश में 18 होटल प्रबन्धन संस्थान तथा 15 भोजन कला संस्थान स्थापित किये जा चुके हैं। इसके अतिरिक्त पूर्वोतर राज्यों में डिल्लोमा स्तर के पाठ्क्रमों को प्रारंभ किया गया है।

उपर्युक्त संस्थानों के वित्त पोषण का कार्य केन्द्र सरकार द्वारा अनुदान के माध्यम से होता है।

(ख) भारतीय स्कीइंग एवं पर्वतारोहण संस्थान—इस संस्थान की स्थापना 1968 में हुई थी। इस संस्थान की स्थापना का प्रमुख उद्देश्य पर्यटकों को निम्न

क्रीड़ाओं के स्थान के विषय में जानकारी प्रदान करना है। ये जानकारियां निम्न हैं–

1. गुलमर्ग हिम क्रीड़ा परियोजना
2. पैराग्लाइडिंग
3. जल क्रीड़ा
4. स्कीइंग
5. रिबर रनिंग स्कीइंग
6. पर्वतारोहण एवं चट्टारोहण

वर्तमान वित्तीय वर्ष में इस संस्थान ने औली एवं पटनीटाप में स्कीइंग, हिमाचल में ट्रेकिंग, ऋषिकेश में रंनिग इत्यादि में प्रशिक्षण उपलब्ध कराया है।

राष्ट्रीय जल क्रीड़ा संस्थान

पर्यटन विभाग ने पर्यटन की लोकप्रियता को बढ़ाने के लिए 1990 में प्रायः पर्यटक स्थलों पर एक राष्ट्रीय जल क्रीड़ा संस्थान की स्थापना की है। इस संस्थान की प्रमुख बातें अग्रलिखित हैं–

1. गोवा सरकार ने इस संस्थान के लिए 18 एकड़ अतिरेक भूमि एवं 60 प्रतिशत उपकरण प्राप्त कराये हैं।
2. वर्तमान में इसे भारतीय पर्यटन तथा यात्रा प्रबन्ध संस्थान को हस्तान्तरित किया जा चुका है।
3. यह संस्थान सेलिंग, विंड सर्फिंग, जल स्कीइंग, आउटबोर्ड मोटर रख-रखाव, जीव रक्षा तकनीकों एवं जलक्रीड़ा के सम्बन्ध में प्रशिक्षण प्रदान करता है। संस्थान ने अपने प्रशिक्षण को गोवा से बाहर 11 राज्यों तक पहुँचाया है।

मार्गदर्शक प्रशिक्षण

इस प्रशिक्षण में विशेष रूप से पर्यटन की आधारभूत संरचना से सम्बन्धित बातें प्रशिक्षक को अवगत कराई जाती हैं। देश में विदेशी पर्यटकों के आवागमन में तेजी से वृद्धि हुई है और यह संख्या 3.2 मिलियन के आँकड़े को पार कर गयी हैं। फलस्वरूप गाइडों की मांग में भी बेतहाशा वृद्धि हुई है और गाइडों की बढ़ती हुई मांग को पूरा करने के लिए 1996 से 3 टियर गाइड प्रणाली प्रारंभ की गयी है जिसके अंतर्गत क्षेत्रीय स्तर, राज्य स्तर तथा स्थानीय स्तर पर गाइड प्रशिक्षण कार्यक्रम चलाये जा रहे हैं। पर्यटन से सम्बन्धित प्रशिक्षण में देशी भाषाओं के साथ-साथ अन्तर्राष्ट्रीय

भाषाओं का भी प्रशिक्षण विशेष रूप से दिया जाता है, क्योंकि पर्यटन में राष्ट्रीय और अन्तर्राष्ट्रीय स्तर के पर्यटक उपस्थित रहते हैं।

भारतीय पर्यटन और ट्रेवल प्रबन्ध संस्थान

पर्यटन मंत्रालय ने 1983 ई. में एक निबंधित सोसायटी के रूप में भारतीय पर्यटन एवं यात्रा प्रबन्धक संस्थान की स्थापना की थी। इस संस्थान का मुख्य लक्ष्य मानव संसाधन विकास की क्रियाओं में प्रशिक्षण कार्यक्रमों, पर्यटन तथा यात्रा उद्योग के हित से सम्बन्धी अनुसंधान कार्यक्रमों को प्रारंभ करना था।

भारत सरकार के पर्यटन विभाग द्वारा इस प्रबंधन को आर्थिक सहायता दी जाती है। इसकी स्थापना उच्च शिक्षा की आवश्यकताओं को पूर्ण करने, यात्रा एवं पर्यटन उद्योग से सम्बन्धित प्रशिक्षण एवं इस क्षेत्र में उत्कृष्टता के एक केन्द्र के रूप में की गयी है। 1992 में इसे ग्वालियर में विपणन कर दिया गया।

इस प्रबंधन द्वारा पर्यटन प्रबन्ध में डिप्लोमा, गन्तव्य स्थल प्रबन्धन में डिप्लोमा, यात्रा प्रचालन सेवा प्रबंधन, पर्यटन विपणन भाषा पाठ्यक्रम, वन्य जीव पर्यटन प्रबन्धन, आदि पाठ्यक्रम चलाए जाते हैं। विश्वविद्यालयों के द्वारा भी पाठ्यक्रम संचालित किये जाते हैं। इस प्रबंधन द्वारा प्रबन्ध विकास कार्यक्रमों में मुख्य रूप से पाठ्यक्रमों को शामिल किया जाता है।

(1) यात्री सेवाओं का प्रबन्ध
(2) पर्यटन एवं यात्रा में प्रबन्ध समस्यायें
(3) प्रभावी पर्यटन विकास
(4) पर्यटन में कम्प्यूटर तकनीकी
(5) पर्यटन विकास का पर्यावरणीय क्षेत्र।

9

वर्तमान पर्यटन क्षेत्र में आर्थिक सिद्धांतों की भूमिका

पर्यटन अब पूरे विश्व में उद्योग का रूप धारण कर चुका है। इसके द्वारा विश्व के कई विकसित या विकाशील राष्ट्रों को अपनी आर्थिक अर्थव्यवस्था सुदृढ़ करने में आर्थिक रूप से सहायता मिलती है तथा कुछ राष्ट्र ऐसे हैं जो पर्यटन उद्योग पर ज्यादा निर्भर हैं। सिद्धांतों की महत्त्वपूर्ण भूमिका होती है। पर्यटन ही एक ऐसा माध्यम है जो विकसित राष्ट्रों को प्राथमिक कार्यों में व्यस्त कर सकता है एवं जिन राष्ट्रों में जिन्होंने यह अवस्था पार कर ली है वे अपनी अर्थव्यवस्था की परिपक्वता में इसे प्रेरक की जरूरत के रूप में महसूस करते हैं। पर्यटन की कुछ विशेषताएं जैसे–अपना पारम्परिक अस्तित्व, अर्थव्यवस्था की स्वाधीनता, राष्ट्र के प्रत्येक क्षेत्र का प्रभावित होना, तीव्रगति से विकास के लिये पूंजी के उत्पादन की कम दरें आदि के परिणामस्वरूप यह राष्ट्रीय विकास का बहुत ही महत्वपूर्ण पहलू है।

युनाइटेड चैम्बर्स ऑफ कॉमर्स के मतानुसार, "किसी भी क्षेत्रीय, प्रान्तीय अथवा सामुदायिक विकास कार्यक्रम के लिये पर्यटन की उन्नति एक मुख्य संचालन

है। नये व्यापार व नये धन को एक नये क्षेत्र में लाने के मूलतः तीन तरीके हैं, ये हैं—कृषि विकास, औद्योगिक विकास एवं पर्यटन विकास। संभवतः तीनों में पर्यटन विकास एक शीघ्रगामी एवं कम से कम कठिन तरीका है। सम्प्रदायों के विकास की यह नयी धारणा और अवतन विकसित क्षेत्रों से पर्यटन के विकास की विचित्रता से वेतन व व्यापार की रुचि में बढ़ोतरी सम्भव है।

अन्य आर्थिक प्रक्रियाओं जैसी पर्यटन भी एक प्रक्रिया है। हालांकि पर्यटन स्वयं में संगठित उद्योग है परन्तु इसकी सीमाएँ बहुत विशाल हैं। विभिन्न स्थानों पर बिखरी हुई हैं एवं इसके फायदे राष्ट्र की समस्त जनता को उपलब्ध होते हैं। वे सभी व्यक्ति जो पर्यटन स्थलों में निवास करते हैं, पर्यटन क्रिया में सक्रिय होते हैं। समस्त पर्यटक चाहे वे स्वदेशी हों अथवा विदेशी, क्षेत्रीय अर्थव्यवस्था के विकास में लोगों को रोजगार प्रदान करवाने में सहायक होते हैं एवं साथ-ही-साथ अच्छी संस्कृति को कायम करने का अवसर प्रदान करते हैं। हर नागरिक को अपने शहर, संस्कृति, रिति-रिवाज व इतिहास पर गर्व होता है।

श्री पीटर (Peter) के अनुसार पर्यटन के मुख्य रूप से पांच कार्य हैं जो निम्न प्रकार हैं—

1. बहु प्रभाव के माध्यम से सामान्य आर्थिक क्रियाओं में सहयोग,
2. विश्व के कार्यों में लोगों का उत्साह बढ़ाकर सामाजिक फायदा प्रदान करना एवं विदेशी व्यक्ति व विदेशी वस्तुओं का ज्ञान हासिल (प्राप्त) करना,
3. रोजगार के अवसर में वृद्धि करना,
4. जिन क्षेत्रों में औद्योगिक विकास नहीं है वहाँ पर विकास को बढ़ावा देन,।
5. दुर्लभ मुद्रा अर्जित करके व्यापार संतुलन में योगदान करना।

पर्यटन की अर्थव्यवस्था तथा गुणक का असर

आर्थिक लाभ—चेकोस्लोवाकिया, इटली, जर्मनी, अमेरिका में प्रारंभ से ही राष्ट्र की अर्थव्यवस्था में किए गए पर्यटन के अध्ययन सांख्यकीय मापदण्ड की समस्याओं से सम्बन्धित रहे जो कि किसी भी पर्यटन के सांख्यकीय अध्ययनों में पैदा होती है। इन अध्ययनों ने पर्यटन के क्षेत्र में बड़ा योगदान दिया है। वर्तमान समय के अध्ययनों में पर्यटन का अन्तर्राष्ट्रीय स्तर पर निरीक्षण किया गया है। आधुनिक युग में प्रकाशित श्री माइकल पीटर्स (Peters 1969) का कार्य उल्लेखनीय है। जिन्होंने ये सूचनायें IOTO (अब WTO) स्रोतों से प्राप्त की थी। उनके अनुसार स्वदेशी व विदेशी पर्यटकों का कुल व्यय विश्व की कुल उपभोक्ता सम्बन्धी खर्चों का करीब

5% है। राष्ट्रीय आय में वृद्धि व अवकाश के वक्त में वृद्धि के कारण बढ़ोतरी दर में वृद्धि हुई है। यह भी देखा गया है कि विकसित राष्ट्रों में आय सुविधाओं की आय सम्बन्धी लचकता की मांग बहुत ही अधिक है। अन्तर्राष्ट्रीय पर्यटन की मांग तो आय की लचकता से अत्यधिक भरपूर है।

पर्यटन में आर्थिक स्रोतों का विकास होता है। अतः लाभ प्राप्ति का प्रश्न बहुत ही महत्वपूर्ण है। ये लाभ प्राप्ति किसी भी उपक्रम के उद्देश्यों के रूप में प्राप्त की जा सकती है। लाभ प्राप्ति चाहे वे सकारात्मक हों या नकारात्मक एवं जिनका उद्देश्यों से कोई सम्बन्ध न होने पर भी कुछ मिले-जुले कारणों से महसूस हो सकते हैं।

भारत में मुख्य रूप से पर्यटन को आर्थिक दृष्टिकोण से देखा गया लेकिन अंतर्राष्ट्रीय सद्भावना पर ध्यान नहीं दिया गया। इस तरह मुख्य उद्देश्य रहा है पर्यटन के आर्थिक फायदे जैसे पर्यटन का राष्ट्र की अर्थव्यवस्था में प्रभाव, रोजगार के साधनों का सृजन, विदेशी मुद्रा का अर्जन तथा राष्ट्र की आय का साधन आदि मालूम करना। अदृष्टिगत व्यापार के क्षेत्र का अधिकतर अनुपेक्षण नहीं किया गया है। विकसित अर्थव्यवस्था में विकास कार्यों हेतु विदेशी मुद्रा की अति जरूरत होती है, जोकि अदृष्टिगत व्यापार के माध्यम से प्राप्त करने का एक अच्छा अवसर था। यह भी उदाहरणीय है कि यूगोस्लाविया एवं बुलगारिया जैसे साम्यवादी राष्ट्रों ने भी पर्यटन की महत्ता समझ ली है एवं एक महत्वपूर्ण पर्यटक मार्किट स्थापित कर ली है। ब्लाडीमीर अन्कुडिनोभ विदेशी पर्यटन बोर्ड के विभागाध्यक्ष के अनुसार–

"बिजली विज्ञान प्लास्टिक संगणक एवं बहुत से अन्य उद्योग प्राचीन उच्च कोटि के उद्योगों की अपेक्षा बड़ी तीव्र गति से विकसित हो रहे हैं। परन्तु यहाँ एक उद्योग है जो कि वृद्धि दर में किसी नये आने वाले उद्योग से प्रतिस्पर्धा कर सकता है। यह उद्योग है–पर्यटन।"

पर्यटन के आर्थिक लाभों का वास्तविक अनुमान हम उत्पादन रोजगार पर पड़े प्रभावों का अध्ययन किए बिना नहीं कर सकते। आज यह विश्व का उद्योग है, जिसमें बहुत ही ज्यादा वित्त निवेशित है। आवास परिवहन पर्यटन स्थल क्षेत्र एवं सहायक सुविधाएं व सेवाएं, जो कि पर्यटन के मुख्य पहलू हैं, पर किए गए नियोजन से पर्यटन-प्रिय राष्ट्रों में बहुत ही उच्च आय प्राप्त होती है। रोबर्ट इरवेज ने विकासशील अर्थव्यवस्था में पर्यटन के योगदान के सन्दर्भ में बताया कि उत्पादन के साधनों का पर्यटन में उपयोग करने से कितनी अधिक आय प्राप्त होती है। यह तभी मालूम हो सकता है जबकि पर्यटन व अन्य वैकल्पिक क्षेत्रों में उत्पादन के कार्य का सही अनुमान लगाया जाए।

गुणित प्रभाव–पर्यटन एक उद्योग है जो कि अर्थशास्त्र के सिद्धांतों के माध्यम से प्रभावित होता है। पिछले 40 वर्षों में पर्यटन से सम्बन्धित अर्थशास्त्रियों ने एक सिद्धांत की खोज की जिसे पर्यटन के गुणक प्रभाव के नाम से पुकारा जा सकता है। 1959 में प्रोफेसर पाल ए. साइमूलशन, अध्यक्ष अर्थशास्त्र विभाग, मोसेचेसेस्ट इन्स्टि्यूट ऑफ टेक्नोलोजी ने एक गुणित सिद्धांत की खोज की थी। इस सिद्धांत का सूत्र इस तरह है–

$$K = \frac{1}{C}$$

1–इस सिद्धांत के अनुरूप 'K' गुणित 'C' पर आश्रित है जो कि क्रमशः उपभोक्ता व आय के पर्यटन के सम्बन्ध के द्योतक हैं, इस प्रकार,

K= गुणक

C= उपभोग में परिवर्तन

Y= आय में बदलाव

यह सिद्धांत पर्यटन पर भी लागू होता है। पर्यटकों के माध्यम से किया गया व्यय (पर्यटक खर्च), व्यय करने के पश्चात् रुकता नहीं है। यह राष्ट्र की अर्थव्यवस्था में तब तक विचलन भ्रमण करता रहता है जब तक पर्यटक किसी स्थान की यात्रा करता रहता है व उस क्षेत्र में धन व्यय करता है। यही धन स्थानीय लोगों की आय का साधन बन जाता है। फिर से यह आय खर्च की जाती है और यह सिलसिला चलता रहता है, इस तरह यह धन कई हाथों से निकलता रहता है एवं खर्च व पुनः खर्च किया जाता है। राष्ट्रीय आय पर उस तरह के खर्च का गुणित प्रभाव पड़ता रहता है एवं धन विभिन्न हाथों से होकर गुजरता रहता है। राष्ट्रीय मुद्रा जो कि विदेशी मुद्रा हेतु विनिमय की जाती है देश के अन्दर प्रवेश करके सारे मार्केट में फैल जाती है। इस तरह गुणित का तेज प्रभाव पर्यटन व आय के अनुपात पर निर्भर करता है जो कि समस्त राष्ट्रीय अर्थव्यवस्था की विभिन्न शाखाओं में पुनः वितरित होती रहती है।

विदेशी पर्यटकों के माध्यम से व्यय किए गये धन का राष्ट्र की आय पर बहुत ही अच्छा प्रभाव पड़ता है। जैसे जब कोई यात्री अपने होटल के बिल का भुगतान करता है तो होटल प्रबंध के माध्यम से प्राप्त किया गया धन भिन्न-भिन्न ऋणों के भुगतान अथवा विदेशी पर्यटनों के दूरभाष, कर्मचारी सेवाओं, वस्तुओं तथा अन्य सेवाओं पर व्यय किया जायेगा। जो लोग इस तरह होटलों से धन प्राप्त करते हैं वे अपनी निजी सेवाओं अथवा यात्राओं हेतु भुगतान करते हैं। यात्रियों के माध्यम से

प्राप्त यह धन दुबारा खर्च किया जाता है एवं अर्थव्यवस्था के विभिन्न भागों में प्रवेश करता रहता है। अतएव प्रत्येक वक्त नयी आय का सृजन होता है जब तक कि इसमें से टपकन न दी जाये। जैसे आय का भुगतान, विदेशी में निवेशित धन संचय आदि जो कि धन को विचलन से पृथक कर ले।

''आगन्तुकों के धन की अटूट बदलाव शृंखला गुणक के प्रभाव को निर्मित करती है। जितनी इन हाथों की संख्या अधिक होगी जिनसे यह धन गुजरता है उतनी ही कुल राष्ट्रीय आय वस्तु और राष्ट्रीय खाते पर इसके लाभों का प्रभाव पड़ता है।''

पर्यटकों के खर्चे का पलटाव अथवा खर्चे का गुणित प्रभाव प्रत्येक अर्थव्यवस्था में भिन्न होता है। व्यय की इस धारणा में टपकन की वजह से आने वाली खर्च की धाराओं से क्षेत्र की आय घटती रहती है। इस तरह का अनुभव केवल अनुमान लगाकर ही किया जा सकता है। इन सब मामलों में इस बहाव का प्रभाव केवल एक वर्ष तक ही देखा जा सकता है। हालांकि एक वर्ष के पश्चात् भी धन की राशि विभिन्न हाथों में परिवर्तित होती रहती है। यह अनुमान है कि 12 महीनों के पश्चात् इस धन का अर्थव्यवस्था पर खास प्रभाव नहीं रहता है। इस तरह के टपकन के कई मुख्य कारण हैं जैसे—आयात, बचत (संचय), देश के बाहर पर्यटन (विदेश यात्रा), विदेशों के निवेश इत्यादि। जितना ज्यादा टपकन होगा उतना ही गुणित कम होगा। इस टपकन के कारण आने वाले खर्च की धारायें कुछ व्यय के चक्र को पूर्ण करने पर शून्य की सीमा तक पहुंच जाती हैं।

पर्यटकों के माध्यम से किया गया व्यय दो तरह के प्रभाव प्रदर्शित करता है—प्रत्यक्ष और अप्रत्यक्ष। पर्यटक के माध्यम से किए गए क्षेत्रीय खर्च से पैदा प्राथमिक आय को प्रायः प्रत्यक्ष प्रभाव कहा जाता है। तत्पश्चात् अर्जित की गयी आय को अप्रत्यक्ष प्रभाव कहते हैं। उदाहरणस्वरूप एक पर्यटक किसी भी पर्यटन इलाके में 800 रुपए खर्च करता है, उसमें से श्रमिकों और मालिकों के लिए 400 रुपये की आमदनी होती है एवं आयात करने के परिणामस्वरूप आय 800 रुपये खर्च व्यय चक्र से टपकन के रूप में बाहर हो जाती है। दूसरे चक्र में, 500 रु. में से 40% टपकन होने से 200 रु. रह जाता है। तीसरे चक्र में 200 रु. घटकर 800 रु. रह जाते हैं एवं यह चक्र तब तक चलता रहेगा जब तक कि यह राशि शून्य तक नहीं पहुंचती है, यदि हम समस्त अप्रत्यक्ष आय को प्रत्यक्ष आय में जोड़ दें तो कुल रकम 1000 रु. बनती है, क्योंकि 1000 रु. प्रत्यक्ष आय (500 रु.) का दुगुना हुआ है। गुणक निकालने का यह बहुत ही सीधा व सरल ढंग है, परन्तु व्यावहारिक दृष्टि से प्रत्येक चक्र में टपकन से सम्बन्धित आँकड़े मालूम करना बहुत ही कठिन कार्य है।

हम एक और उदाहरण लेते हैं, एक नवनिर्मित होटल इकाई ले लीजिए जो कि विदेशी कम्पनी के अधिकार में है। कुछ मान्यताएँ इस तरह हैं :

होटल की आय	10,00,000 रु.
स्वदेशी माल के क्रय का मूल्य (खान-पान, गोस्त आदि)	3,00,000 रु.
आयात किए गए विदेशी माल का मूल्य (शराब, पेय पदार्थ, साबुन आदि)	2,00,000 रु.
राष्ट्र के कर्मचारियों को भुगतान किया गया वेतन	3,00,000 रु.
	8,00,000 रु.
	2,00,000 रु.

कुल फायदा 2,00,000 रु. में से, 1,00,000 रु. निर्यात किया गया एवं 1,00,000 रु. उसी समय देश के उद्योगों में विनियोग/निवेशित किया गया है, 3,00,000 रु. के राष्ट्रीय माल के तहत कृषि तथा अन्य स्वदेशी उद्योगों से क्रय किया गया माल है। इस 3,00,000 रु. में से विभिन्न औद्योगिक कर्मचारियों को मजदूरी दी जाती है। यह मजदूरी पुनः वितरित हो जाती है एवं अन्य क्रय करने में योगदान करती है इस तरह प्राप्ति का अंश गुणक के प्रभाव में और भी बढ़ोतरी कर देता है। 2,00,000 रु. आयात हेतु खर्च किया जाता है व 1,00,000 रु. का फायदा देश के बाहर विदेशों में चला जाता है। इस तरह 3,00,000 रु. राष्ट्र की हानि है जिसको टपकन के नाम से पुकारा जाता है।

इस प्रकार प्रारंभ की 10,00,000 रु. की रकम में से 7,00,000 रु. (3,00,000 + 3,00,000 + 1,00,000) ही राष्ट्रीय चक्र की क्रिया में रहता है एवं सक्रिय रूप से चलता रहता है। यदि यह मान लिया जाए कि यह अनुपात आने वाली अवस्था में भी यही रहता है, तो हम इस धन के चलन के प्रभाव की निम्न तरह गणना कर सकते हैं–

अवस्था	चलन का प्रभाव
1. आरंभिक खर्च	दस लाख
2. द्वितीय चक्र	$1 \times 0.7 = 0.7$
3. तृतीय चक्र	$1 \times 0.7 \times 0.7 = 0.7^2$
4. चतुर्थ चक्र	$1 \times 0.7 \times 0.7 \times 0.7 = 0.7^3$

अवस्था	चलन का प्रभाव
5. पंचम चक्र	$1 \times 0.7 \times 0.7 \times 0.7 \times 0.7 = 0.7^4$
6. NCR चक्र	$1 \times 0.7 \times 0.7 \times 0.7 \times 0.7 \times 0.7 = 0.7^5$

गुणित करने पर प्रतीत होता है कि धन का प्रति 'N' चलन निम्न के बराबर है–

N = 3.33

3.33 को इस उदाहरण में, गुणक का प्रभाव कहा जा सकता है। दूसरे शब्दों में, प्रारम्भ में किए गये पर्यटकों के व्यय का प्रभाव आरंभ होता है, जिससे विभिन्न स्रोतों पर तीन गुना ज्यादा प्रभाव पड़ता है।

आगन्तुकों द्वारा किया गया खर्च किसी भी निर्धारित इलाकों की आमदनी में अपने आप बढ़ता रहता है। यह व्यय की गयी रकम बढ़ती रहती है एवं गुणित का गुणक कही जाती है। यह प्रदर्शित करती है कि पर्यटक खर्चों के परिणामस्वरूप आय में कितनी वृद्धि होती है। यदि पर्यटकों का व्यय 10 लाख रु. है व गुणित का मूल्य 1.9 है तो आय में 19 लाख रु. की वृद्धि हो जायेगी। जैसे ही होटल मालिक स्थानीय परिवहन व रेस्टोरेन्ट मालिक व अन्य मालिकों से आय प्राप्त करते हैं, इस आय में वृद्धि होना आरंभ हो जाती है। दूसरी तरफ कर्मचारी अपना वेतन भुगतान करते हैं। मकान मालिक व अन्य प्राप्तकर्ता अपनी आय व्यय करते हैं। इस तरह कुल आय का भाग प्रत्यक्ष रूप में पर्यटकों के माध्यम से किए मूल खर्चे के योग से ज्यादा होता है। कितना अधिक यह गुणित होगा, यह अनुमान इस बात पर निर्भर करता है कि हर अवस्था में प्रारम्भिक आय में से कितना पुनः व्यय की गयी अर्थात् वह राशि जो बचत नहीं है एवं कितने टपकन थे जो कि उस धन से पैदा होती है और जो देश के आयात करने के परिणामस्वरूप बाहर चला जाता है। उदाहरणार्थ–पर्यटकों के माध्यम से आयात की गई वस्तुओं पर किए व्यय आय से बहुत ही कम बढ़ोतरी करते हैं। पर्यटन आय गुणित के माध्यम से अनुमान लगाया जा सकता है कि विदेशी पर्यटकों के व्यय का स्वदेशी अर्थव्यवस्था पर क्या प्रभाव पड़ता है। जब कभी गुणित एक से अधिक होगा (जो प्रायः अधिकांश दशाओं में होता है) तो पर्यटन से प्राप्त के लाभों को पूर्णतः प्रदर्शित नहीं कर सकती है। गुणित के मान व समुदायों की आय व रोजगार के क्षेत्रों में प्राप्त फायदे अर्थव्यवस्था में टपकन की वर्तमानता (मौजूदी) के अनुरूप घटते रहते हैं।

गुणक प्रक्रिया

प्रथम स्थिति—गुणित की दर का अनुमान लगाने हेतु आवश्यक है कि पर्यटक व्यय को राष्ट्रीयता तथा आय के अनुरूप निश्चित किया जाए, प्रत्येक व्यक्ति खर्च का बंटवारा विभिन्न प्रकार के द्वारा किया जा सकता है। उदाहरणस्वरूप खान-पान, आवास, स्थानीय परिवहन, मनोविनोद, खरीदारी आदि तथा अन्य छोटी मदें जैसे—समाचार पत्र, डाक टिकट आदि। समूह समारोह अथवा पैकेज यात्रा की वजह से आवास व खान-पान के व्यय का अलग-अलग विभाजन करना बड़ा ही जटिल है। पैकेज चात्रा में तो होटल के आवास के खर्च में ही समस्त अथवा कुछ खाने-पीने के खर्चे भी शामिल होते हैं। इसी तरह पर्यटन स्थलों में स्थित होटल तो यूरोपियन प्लान की अपेक्षा अमेरिकन प्लान अथवा संशोधित अमेरिकन प्लान ही प्रदान करते हैं। व्यय करने के ढंग से पर्यटन के रोजगार सम्बन्धी स्रोत एवं इसका राज्य की आय में योगदान का निश्चय किया जा सकता है, क्योंकि प्रति वर्ग में अलग-अलग श्रमिक गुणक अनुपात है तथा हर श्रेणी में कर सीमा शुल्क एवं उत्पादन शुल्क के रूप में अलग-अलग होते हैं।

दूसरी स्थिति—द्वितीय परिक्रमा में टैक्सी चालक, बैरा, दुकान मालिक जो कि प्रथम परिक्रमा से लाभान्वित हैं अपना फायदा व आय स्थानीय उत्पादित वस्तुओं व सेवाओं में व्यय करते हैं।

तीसरी स्थिति—तृतीय परिक्रमा में, द्वितीय परिक्रमा के लाभान्वित वर्ग अपने वेतन का कुछ अंश वस्तुओं व सेवाओं में व्यय करते हैं।

चौथी व परवर्ती स्थिति—द्वितीय, तृतीय व परवर्ती परिक्रमाओं का व्यय उन वस्तुओं पर होता है जो कि कच्चे माल, खान-पान से भिन्न होता है। अप्रत्यक्ष रूप में इससे देश की राष्ट्रीय आय में वृद्धि होगी। पर्यटकों का व्यय वस्तुओं व सेवाओं की पूर्ति करने वालों को आय प्रदान करता है और अन्य उद्योगों में भी आय सृजन करता है। व्यय के द्वितीय, तृतीय व परवर्ती परिक्रमा को एकमुश्त करके राष्ट्रीय आय, राजकीय आय व रोजगार में पर्यटन के योगदान का एक स्पष्ट अनुमान लगाया जा सकता है।

मूल घटक—आयात व संचय आदि के माध्यम से की गयी टपकन को गुणक के माध्यम से परिदर्शित किया जा सकता है। निर्यात अथवा पर्यटकों के व्यय के माध्यम से की गयी आर्थिक क्रियाओं के प्रत्येक रुपये में एक नयी सम्पत्ति

का सृजन होता है। कुछ प्राकृतिक लाभों से यदि एक निर्यात उद्योग कम गुणित के कारण भी निर्यात वृद्धि कर सकता है, तो इस तरह के उद्योग को अन्य उद्योगों जिनमें गुणित की अधिकतम मात्रा है परन्तु निर्यात बिक्री योग्यता प्राप्ति की सीमितता है, की अपेक्षा अधिक मान्यता प्राप्त होती है। जैसे बहुत से कृषि उद्योगों में बहुत ही उच्चतम गुणित है क्योंकि कृषि उत्पादन में कम से कम आयात की गयी वस्तुओं की जरूरत होती है। प्रायः इन कृषि क्षेत्रों में रहन-सहन का स्तर कम होता है एवं उनकी आय में बहुत तीव्र उतार-चढ़ाव होते हैं जो कि मार्केट की परिस्थितियों सम्पत्ति व अन्य बाह्य शक्तियों पर निर्भर करती है।

दूसरी तरफ, कुछ उत्पादन में लगे हुए उद्योगों में ऊँची मजदूरी प्रदान करने से रहन-सहन के स्तर को ऊँचा करने में मदद मिलती है। परन्तु ये उद्योग अधिकतर कल-पुर्जे व कच्चा माल आयात करते हैं। इसीलिए इनमें गुणित कम ही होता है।

पर्यटन उद्योग में तो गुणित का प्रभाव बड़ी ही नाटकीयता से मालूम किया जा सकता है। मूल रूप से यह दो प्रमुख बातों (तत्वों) पर निर्भर करता है :

प्रत्येक चलन परिपथ में हानि का प्रभाव होगा, यह भी सम्भव है कि पर्यटन के माध्यम से छोड़े हुए धन का 1/10 अथवा 1/3 निर्यात किया जाए, यदि प्राप्त किया गया धन निर्यात किया जाए तो परिस्थितियां कुछ आश्चर्यजनक रूप ग्रहण कर लेती हैं। इस दशा में पर्यटन उद्योग पूर्णतः सक्रिय रहता है एवं दूसरे स्वागतकर्ता राष्ट्र को कोई फायदा प्राप्त नहीं होता है।

यह वित्त चलन में अनेक दफा एक व्यक्ति के हाथ से दूसरे व्यक्ति के हाथ में परिवर्धित होता रहता है। एक वर्ष के भीतर इन चलनों की संख्या में चार-पांच से दस पन्द्रह तक वृद्धि हो सकती है। जितना जल्दी इनका चक्कर परिभ्रमण होगा उतना ही अधिक गुणक का प्रभाव होगा, अतएव यह अति जरूरी है कि राष्ट्र के नागरिक पर्यटकों से प्राप्त धन को व्यय करने के इच्छुक होंगे। उपभोग व व्यय की क्रिया तीव्र होनी चाहिए। यदि लोग इस धन को संचय करने में जुट जाएं अथवा अनुत्पादक बचत में लगा दें (जैसे सोना, चाँदी, जेवर आदि) तो गुणक के प्रभाव का बिल्कुल लोप हो जायेगा। आधुनिक अर्थशास्त्री इसी विचार के प्रेरक हैं कि धन यथा शीघ्र खर्च होना चाहिए।

गुणक के असरों व प्रभावों के प्रक्रिया को पूरा होने के लिए पर्यटक खर्च को छः उप-क्षेत्रों या विभागों में विभाजित किया गया है जो निम्न प्रकार हैं—

खर्च का ब्यौरा	*खर्च का प्रतिशत*
खान-पान	32
खरीददारी	25
आवास	25
स्थानीय परिवहन व्यवस्था	5
हृदयावलोकन एवं मनोविनोद	15
अन्य	3
योग	**100**

यह वर्गीकरण विदेशीय यात्रियों के किसी भी राष्ट्र या क्षेत्र में जाने के पश्चात् किये हुए खर्च से जुड़ा है। कुछ अन्य प्रकार का वर्गीकरण जो उपरोक्त प्रक्रिया को सर्वमान्य है, जो अद्योलिखित हैः

व्यय की मदें	*प्रतिशत*
आवास	30%
खान-पान	35%
स्थानीय परिवहन5%	
खरीददारी	25%
अन्य	5%

पर्यटन के गुणक के प्रभावों का पता करने के लिए निम्नलिखित बातों पर ध्यान दिया गया है जो इस प्रकार हैं–

स्थानीय लोगों के माध्यम से क्रय की गयी वस्तुओं जैसे कपड़ा अथवा उपभोग किए जाने वाले शीघ्रकालीन सामान का अधिकतम भाग या तो पूर्णतः अथवा आंशिक में आयात किया जाना।

पर्यटकों के माध्यम से देश के भ्रमण में इस्तेमाल किए जाने वाले यन्त्रों व ईंधनों का आयात किया जाना।

पर्यटकों के माध्यम से उपभोग किए गए खान-पान व सामग्री की खरीददारी का अधिकतर प्रतिशत जो आयात किया गया।

अन्य किस्म की टपकनें जैसे विदेशों में पुनः विनियोजन, संचय एवं लेन-देनों में बिलम्ब का अधिपत्य रहना। अन्य शब्दों में, इन टपकनों की वजह से अर्थव्यवस्था की उपभोग सम्पन्नता व क्षमता कम रहेगी जोकि बचत का एक अन्य तरीका है एवं

साथ ही पलटाव की क्रिया भी कम रहेगी। तुलनात्मक रूप में श्रमिक कम महँगा रहेगा एवं यह कुल प्रदान की गयी सेवाओं एवं उत्पादित माल की लागत का 1/3 से अधिक नहीं होगा।

वस्तुतः सम्पूर्ण वर्गों या श्रेणी के सालाना लेन-देन का क्रम ज्यादा नहीं होगा। इसलिए यह आवश्यक है कि एक साल के भीतर खर्च के पांच चक्रक होंगे। कुछ अर्थव्यवस्था में लेन-देन संख्या, पलटाव की क्रिया का पूर्णतः लोप होने से पहले, 13 तक पहुंच सकती है परन्तु फिर भी ये अतिरिक्त लेन-देन क्षेत्र की कुल आय में विशेष वृद्धि नहीं करते हैं।

व्यावहारिक रूप से, बहुत-सी परिस्थितियों में चौथी व पाँचवीं लेन-देन के पश्चात् किसी दी गयी श्रेणी में धन का विचलन फिर से नहीं हो पाता है। इस तरह का अनुमान आंशिक रूप में बहुत से लेन-देने की छोटी संख्या को छोड़ने हेतु तथा अंशतः पूर्व स्थिति कायम रखने हेतु लिया गया है। हालांकि व्यक्तिगत रूप में ये लेन-देन छोटे होते हैं परन्तु वे सामूहिक रूप में पर्यटकों के व्यय के पलटन की क्रिया को होटलों से प्राप्त मदों की अपेक्षा ज्यादा ऊपर उठा देते हैं।

पर्यटन में विभिन्न निर्यात समाग्रियों की तरह खर्च के ब्योरे की प्रक्रिया के अनुसार एक नवीन आमदनी का निर्माण व विकास होता है। कृषि, खान, साज-सज्जा तथा अन्य उद्योग, जिनमें आयात की बहुत ही कम जरूरत होती है, की अपेक्षा पर्यटन में गुणित का अंश बहुत कम होता है। अतएव पर्यटन के गुणित के प्रभावों की प्रशंसा करने के बजाय यह उचित होगा कि पर्यटन के उन मूल लाभों की तरफ जोर दिया जाए जिनकी वजह से यह कृषि व अन्य उत्पादक वस्तुएं जो कि निर्यात की जाती हैं, उनकी अपेक्षा अपना पृथक अस्तित्व रखते हैं। उदाहरण के तौर पर पर्यटन या यात्रा का निर्यात व्यापार के रूप में विशिष्ट लाभ यह होता है कि जो सामग्री इसके अन्तर्गत पैदा की जाती है वह कुछ अजीब व अनोखी होती है। कोई भी अन्य उद्योग, पर्यटन के माध्यम से प्रदान की गयी वस्तु के समान वस्तु प्रदान नहीं कर सकता है। जितना अधिक वह गुणित होगा उतना ही अधिक परियोजनाओं में विनियोजन होगा, उनमें यात्रा उद्योग के विकास से अधिकतम फायदा होगा, गुणित के फायदे भिन्न-भिन्न हो सकते हैं, परन्तु यह सम्भव है कि इसका कारण केवल यह नहीं कि गुणित के वास्तविक लाभों में भिन्न-भिन्न लेन-देनों में विभिन्नता है अपितु गुणित को निकालने की प्रणाली भी भिन्न हो सकती है। गुणित की दर शुद्ध अर्जित विदेशी मुद्रा के माध्यम से निकाली जाती है। जितनी अधिक टपकनें होंगी उतनी यह कम होगी। 1957-59 में पूर्वी एशिया व प्रशान्त राष्ट्रों में सर्वेक्षण किया गया एवं इन देशों की निम्न औद्योगिक प्रगति को देखा जाए तो प्रतीत होता है कि भारत की समस्त

पर्यटन सम्बन्धित आवश्यकताएं स्वदेश में ही उपलब्ध हैं एवं टपकन कुल प्राप्ति के 5 प्रतिशत से अधिक नहीं है। यह उचित होगा कि गुणक को कुछ ऊँचा ही अनुमानित किया जाए।

पर्यटन गुणित की गणना की पद्धतियां

यहाँ पर पर्यटन गुणित की गणना हेतु निम्न मान्यताएं दी गयी हैं :

एक पर्यटक के माध्यम से किया गया खर्च 1000 रु. है।

प्रथम वर्ष में पर्यटक के माध्यम से धन व्यय करने के बाद सम्भवतः 5 अथवा 6 लेन-देन होंगे।

करीबन 13-14 लेन-देनों के बाद धन स्वयं लोप हो जाएगा।

व्यय करने की मदें तथा उनका प्रतिशत निम्नलिखित है–

मद	*प्रतिशत व्यय*
आवास	25%
खान-पान	32%
खरीददारी	23%
स्थानीय परिवहन	7%
ह्दयावलोकन तथा मनोविनोद	10%
अन्य	5%

यह अनुमान है कि खरीददारी की श्रेणी में पर्यटकों के क्रय का 40 प्रतिशत पूर्ण उत्पादित आयात की मदें हैं जो कि पर्यटक अपने देश को ले जाते हैं और इस क्रय की 30 प्रतिशत ऐसी वस्तुयें हैं जो कि स्थानीय प्रक्रिया पर निर्भर करती हैं एवं स्थानीय कच्चे माल के माध्यम से निर्मित की जाती हैं। यह एक मानवीय तथ्य है कि जितना आयात का प्रतिशत ज्यादा होगा एवं आयात पर निर्भरता होगी उतना ही पर्यटक रुपये के पलटाव की क्रिया कम होगी। जब आयात पर धन व्यय किया जाता है, तो धन राष्ट्रीय अर्थव्यवस्था से बाहर चला जाता है जिससे राष्ट्र को हानि होती है।

उपभोग की आदतें

उपभोग की विभिन्न विधियाँ होती हैं जिसे भिन्न-भिन्न प्रयोग में लाया जाता है। गुणित के सिद्धांत में 'उपभोग में परिवर्तन' को कुल व्यय व प्रथम खाने के उत्तर

को माना जा सकता है, इस व्यय को 'आय में परिवर्तन' मानकर प्रतिशत प्रति रुपये पलटाव अथवा गुणित के रूप में निकाला जा सकता है।

कुल व्यय अथवा आय में 3272/50 रु. व उपभोग में 2272/50 रु. बदलाव है एवं उपभोग की प्रवृत्ति से गुणक की संख्या दो=रु. 2272/50÷3272/50=69. 4 प्रतिशत है। 50 प्रतिशत उपभोग की प्रवृत्ति से गुणक की संख्या 2 होगी, गुणक एक साधारण तरह की क्रिया में विद्यमान होता है, यह किसी भी विशेष स्थान पर विद्यमान रहता है। दूसरी तरफ, पर्यटन एक फैला हुआ व विनम्र तरह का संचालन है जिसमें पर्यटक अपने धन को विभिन्न तरह के सामान व सेवाओं पर व्यय करते हैं, जिनकी खरीददारी का सम्पूर्ण अर्थव्यवस्था पर बहुत ही तीव्र प्रभाव पड़ता है।

एक साधारण गुणित बचत की सीमान्त प्रवृत्ति का व्युत्क्रम है। मान लीजिये कि उपभोग की प्रवृत्ति = 4/5 व बचत की सीमान्त प्रवृत्ति = 1/5, तो पर्यटन का गुणित = 5 होगा। टपकनें व बचत से व्यय का प्रभाव (धारा) मन्द हो जाती है। बचत की सीमान्त प्रवृत्ति टपकनें व संचित धन का ही मिश्रण है। जितनी ज्यादा बचत की सीमान्त प्रवृत्ति होगी उतना ही गुणित कम होगा, यदि बचत की सीमान्त प्रवृत्ति 1/3 है तो गुणक 3 होगा।

पीटर का गुणक सिद्धांत

पीटर ने एक रूप में औसतन गुणित का विकास किया जो इस तरह है

$$Kt = \frac{1}{S - ea}$$

जिसमें,

Kt = पर्यटन का गुणक

S = बचत की प्रवृत्ति

ea = नागरिकों के माध्यम से विदेशों में पर्यटन व्यय की सीमान्त प्रवृत्ति

'ea' गुणक आयात की अन्तिम आयात प्रवृत्ति को दर्शाती है। वस्तुएं व सेवायें, जो कि देश के नागरिक विदेशी पर्यटन के वक्त उपभोग करते हैं।

पीटर का गुणित पर्यटन अर्थव्यवस्था में एक औसत गुणित है। यह पर्यटन खर्चों के गुणित के प्रभावों का प्रदर्शन नहीं है।

केन्ज की व्यय की गुणक अवधारणा

श्री केन्ज़ के अनुसार

$$K=\frac{1}{S+M}+\frac{1}{1-C+M}$$

K = गुणक

S = बचत की सीमान्त प्रवृत्ति

M = आयात की सीमान्त प्रवृत्ति

C = उपभोग की सीमान्त प्रवृत्ति

इसी प्रकार क्लेमेन्ट के अध्ययन के अनुरूप जिन राष्ट्रों में कम पलटाव हेतु ज्यादा टपकनें हैं, पर्यटकों का किया गया व्यय अदृश्य होने से पहले करीब प्रतिवर्ष 3.2 अथवा 3.5 बार पलटाव होता है, आर्थिक दृष्टि से विकसित राष्ट्रों में जहां टपकनें कम हैं यह सम्भवतः उच्च पक्ष में 4.3 तक हो सकता है, 3.2 गुणक बहुत कम है एवं किसी भी सामान्य अवस्था में पैदा हो सकता है।

यह विभिन्न उद्योगों के भिन्न-भिन्न कच्चे माल के आकार को ध्यान में नहीं रख सकता है। इसके अनुसार स्वतंत्र भाग में वृद्धि होने से सम्पूर्ण अर्थव्यवस्था में गुणित के एक ही प्रभाव होंगे।

पर्यटक उद्योग के विशेष आकार (बनावट) के कारण एक धृष्ट प्रणाली की जरूरत होती है तथा इसके लिए अन्तर्उद्योगीय लेन-देनों के मिश्रण की आवश्यकता होती है।

अतः,

$$Kt=\frac{\Delta T}{\Delta V}$$

जहाँ पर कि,

ΔT= पर्यटकों से प्राप्त आय में बदलाव (निर्यात)

ΔV= राष्ट्र की आय में प्रवाहित बदलाव जिसके कारण पर्यटकों से प्राप्त काम के परिवर्तन, जिसमें कि लेन-देन मिश्रण में एक नियत गुणक दिया है। लेन-देन का मिश्रण नियत गुणक व विभिन्न क्षेत्रों के अर्जित एक नियत पर्यटकों से प्राप्त आय है। इस गुणक का अर्थ उद्योगों के तहत स्पष्ट है बशर्ते कि हम इस तरह की मान्यताओं को अपने विश्लेषण में स्वीकार करें।

इस प्रकार, यदि $(I-A)^{-1}$ जो कि लियोनटिफ का पाद है तो इसमें I = परिचित ढांचा है व A = तकनीकी गुण का ढांचा तो $\Delta Y = \Delta T\ (I-A)^{-1}$ जहाँ पर कि T = पर्यटक प्राप्ति के वृत्त खण्ड का संकेतक है।

उपभोग को भी यदि इस ढाँचे के अन्दर लें तो लियोनटिफ के एक बन्द प्रणाली में $Kt = (I-A)^{-1}$।

केन्ज़ का समस्त अर्थव्यवस्था हेतु औसतन भारित गुणक समस्त व्यक्तिगत क्षेत्रों/वर्गों हेतु है। क्योंकि पर्यटकों से प्राप्त राशि बहुत से वर्गों के माध्यम से अर्जित की जाती है, इसलिए पर्यटन का गुणक हर वर्ग हेतु वही होगा जो कि लियोनटिफ के पाद में है एवं जिन वर्गों के माध्यम से पर्यटक प्राप्ति आय के वितरण के माध्यम से भारित है।

अमेरिकी सरकार के जरिए स्थापित एक अध्ययन दल ने प्रशान्त टीम में पर्यटक व्यय के प्रभाव के अध्ययन हेतु प्रशान्त द्वीप व सुदूर पूर्व के 17 राष्ट्रों का सर्वेक्षण किया, इस अध्ययन के अनुरूप प्रारम्भ में किया गया व्यय एवं साथ में उन लोगों के माध्यम से प्राप्त पर्यटक मुद्रा पुनः व्यय को ध्यान में रखकर एक आर्थिक क्रिया का सार जन्म लेती है तथा सिद्ध होती है, जोकि 3.2 से 4 बार ऊँची थी। न्यूनतम गुणित, उन राष्ट्रों में मान्य है, जहाँ पर विदेशी मुद्रा में टपकनों का प्रतिशत तुलनात्मक रूप में अधिक है एवं उच्चतम गुणित उन देशों पर लागू होता है जिनमें टपकनें कम हैं अथवा बिल्कुल नहीं हैं।

गुणक के प्रभावी तत्व

गुणक को प्रभावित करने वाले मुख्य तत्व निम्नलिखित हैं–

1. पर्यटन व्यापार पर विनियोजित विदेशी पूँजी पर अर्जित फायदे को बाहर प्रेक्षित करना।
2. पर्यटकों को भिन्न-भिन्न सुविधायें प्रदान करने हेतु आयात की गयी पूर्तियाँ।
3. इसके तहत पर्यटन के उद्देश्य से किए गए आयात व्यय जिसमें वे वस्तुएं, सामान व यंत्र आते हैं जो होटल निर्माण व अन्य पर्यटक सुविधाओं हेतु आयात की गई हों
4. विदेशों में मार्केट से सम्बन्धित क्रियाओं पर की गई विदेशी मुद्रा की लागत।
5. विदेशी पर्यटकों के उपभोग हेतु किया गया आयात।

इन समस्त टपकनों का शुद्ध प्रभाव यह है कि इनसे निर्यात किए गए राष्ट्रों में आय अर्जित होती है तथा रोजगार के स्रोत बढ़ते हैं। अतः जितनी अधिक टपकनें होंगी उतना ही कम गुणित फायदा होगा। परिणामस्वरूप स्वदेशी अर्थव्यवस्था को पर्यटन का फायदा कम प्राप्त होगा।

जहाँ तक गुणक का संबंध है यह अनुमान लगाया जा सकता है कि भारत में आयात पर कठोर प्रतिबन्ध है इसलिए गुणित उच्च स्तर पर ही रहेंगे।

पर्यटन उद्योग का पूर्ण रूप से विकास नहीं हो पाया है। यह अभी विकास की दिशा में अग्रसर है इसलिए इस उद्योग को विकसित करने के लिए पर्यटन स्थलों पर यातायात, होटल, स्वास्थ्य सेवाएं, संचार आदि की व्यवस्था अच्छी होनी चाहिए। यह सत्य है कि पर्यटक अन्य सुविधाओं जैसे जन सुविधाओं पर थोड़ा-सा ही बोझ डालते हैं परन्तु अन्य क्षेत्रों जैसे अलीशान होटल, उच्च श्रेणी के रेस्टोरेन्ट आदि में विनियोजन का मुख्य कारण पर्यटन के माध्यम से अर्जित आय की आकांक्षाओं पर आधारित होते हैं। यह सुझाव दिया जाता है कि अधिकांश विकासशील राष्ट्रों में पर्यटन पर विनियोजन का एक मुख्य आकर्षण यह है कि विनियोजन के परिणाम प्राप्त करने से पूर्व का फायदा अर्जन समय बहुत ही कम होता है।

विशेष तौर पर गुणित के फायदे भिन्न-भिन्न होते हैं। परन्तु इस विभिन्नता का मुख्य कारण प्रणाली में मतभेद है न कि विभिन्न राष्ट्रों में प्रयोग किए गये गुणित के असली गुण। ये अनुमान विभिन्न क्षेत्रों में अनुमानित पर्यटक गुणित की विभिन्नता को प्रदर्शित करते हैं, जो निम्नलिखित प्रकार से हो सकते हैं–

1. यह विभिन्नता गलत तरीके इस्तेमाल करने के कारण भी हो सकती है। श्री जिन्दर के अनुमान जो कि पूर्वी कैरिवियन के सम्बन्ध में थे, में मुख्य समस्या गलत प्रणाली का प्रयोग था, जिसके परिणामस्वरूप गुणित स्वयं गुणन हो गया।
2. यह विभिन्नता पर्यटन उद्योग में पर्यटन प्राप्ति के बनावट में अन्तर होने के बजाय अर्थव्यवस्था के आकार में विभिन्नता अथवा उपभोग के आकार के कारण हो सकती है।
3. कुछ औसत गलत स्थानों पर इस्तेमाल हो जाने से प्रणाली से संबंधित संदिग्ध तरह की गलतियाँ भी हो सकती हैं।
4. आँकड़ों की कमी के कारण भी यह हो सकता है।
5. प्रत्येक क्षेत्र से प्राप्त कुल लाभ, कुल मजदूरी व वेतन, किराया, ब्याज के अनुमान से देश की उत्पादन वस्तुओं को निकाला जा सकता है; जिसके माध्यम से आय के तत्वों (Y) को गुणित किया जा सकता है। अतएव आँकड़ों की कमी एवं अन्य आर्थिक सन्दर्भों से सिद्धांतों को उधार लेने से राष्ट्र के लोगों की आय मालूम करने का यह एक साधारण तरीका नहीं है।
6. इनके गुणित का प्रभाव नहीं पड़ेगा।

7. यह भी जरूरी है कि प्रवासी जो कि विदेशी भी हैं, के माध्यम से आर्जित आय एवं विदेशी व देशवासी के माध्यम से प्रेक्षित आय के हर क्षेत्र में अन्तर किया जाए। क्योंकि यह आय अर्थव्यवस्था में पुनः खर्च नहीं की जायेगी।

इन कुल देश के उत्पादन के तत्वों के मुख्य ढाँचे से अलग करने से अप्रत्यक्ष आय के प्रभाव पर बहुत ही तेज असर पड़ने से विभिन्न क्षेत्रों तथा सम्पूर्ण अर्थव्यवस्था के गुणित के आधार पर प्रभाव पड़ता है। आमतौर पर देखा गया है कि विदेशी प्रवासियों द्वारा प्राप्त होने वाली आय को जनहित के कार्यों पर व्यय नहीं किया जाता। जबकि आजकल हर राष्ट्र कल्याणकारी कार्यों के बढ़ावे पर जोर दे रहा है। पर्यटन उद्योग में विदेशी मालिकों के प्रभुत्व व विदेशी का दक्ष रोजगारों में होने के परिणामस्वरूप यह सम्भव है कि राष्ट्र की आय का राष्ट्रीय गुणित कुल राष्ट्रीय उत्पादन के गुणित के मुकाबले में कम होगा।

मान्यताएँ—जिन लघुकालीन नीतियों के संबन्ध में यह कहा जाता है कि सभी स्रोत और मजदूरों तथा पूंजी का सही प्रयोग नहीं हो पा रहा है। ऐसी परिस्थितियों में प्रयोग हेतु केन्ज ने मूल गुणित सिद्धान्त बनाया। इन परिस्थितियों में एक बन्द अर्थव्यवस्था में वास्तविक आय व रोजगार में वृद्धि हो सकती है। यदि अन्तिम मांग के किसी भी तत्व में स्वतंत्र वृद्धि हो तो वास्तविक आय में वृद्धि गुणित के आकार से निर्धारित की जा सकती है।

एक स्वतंत्र अर्थव्यवस्था में आय सृजन की क्रिया, बचत की सीमांत प्रवृत्ति से प्रतिबन्धित होती है। परन्तु जब तक $S+m<1$ है, गुणक भी इकाई से अधिक रहेगा। ये समस्त समान मान्यताएं जो कि समाज हेतु समस्त स्रोतों के शून्य अवसर लागत के समान समझी जाती हैं, लागू की जा सकती हैं। जहाँ पर किसी वर्गीय गुणित को उस वर्ग की अन्तिम उत्पादन वृद्धि के परिणामस्वरूप असली आय में वृद्धि को प्रदर्शित करने हेतु प्रयोग किया जाता है। परन्तु एक विकासशील अर्थव्यवस्था में, विशेषतौर पर न प्रयोग की जाने वाली अंश पूंजी को इस अर्थव्यवस्था की विशेषता नहीं कहा जा सकता है। वर्तमान परिपेक्ष्य में यह सैद्धान्तिक और अनुभव प्राप्त करने में कोई मदद नहीं पहुंचता है। अतः इसके अनुसार श्रम मूल्य विहीन है।

अतः सामान्य रूप से कहा जा सकता है कि स्रोतों की विकासशील राष्ट्रों में अवसर लागत स्थिति होगी, इस तरह की परिस्थितियों में कम वक्त में अन्तिम मांग में वृद्धि होने से कीमतों में वृद्धि हो सकती है। हालांकि मूल्यों में वृद्धि होने से आय के वितरण, उपभोग की प्रवृत्ति तथा उपभोग के आकार में भी बदलाव की सम्भावना होगी।

मुख्य रूप से देखने में आया है कि अधिक आय वाले समाज में विदेशों से आयातित वस्तुओं के इस्तेमाल करने की प्रवृत्ति इस कदर हावी है कि इस आय वर्ग को दोबारा वितरण से लाभ प्राप्त होंगे। आयात की क्रिया में विचलन की प्रवृत्ति रहेगी ताकि आयात की औसत प्रवृत्ति में वृद्धि हो, इसके अतिरिक्त, यदि विनिमय दर नियत है तो स्वदेशी मूल्य में वृद्धि की प्रवृत्ति रहेगी, जिससे उत्पादित वस्तुओं के विकल्प के रूप में आयात अधिक होगा, यह मानने का कारण यह है कि उद्योगों के मध्य के लेन-देन के ढाँचे के गुणक (ढाँचे में उद्योग सहित) में आश्रित होकर ही बदलाव आयेंगे एवं ये बदलाव आयात गुणक पर प्रभाव डालकर गुणित में भी कमी लायेंगे। परिणामस्वरूप इस तरह पीटर का यह सुझाव गलत होगा कि सामान्य नियम के समान विकासशील राष्ट्रों ने यदि पर्यटन गुणक को माना है, तो पर्यटक व्यय में बदलाव होने से आय में भी चौगुनी वृद्धि हो जायेगी।

इसके अतिरिक्त जब तक अंश पूंजी व मजदूर-शक्ति के प्रयोग बावत अतिरिक्त सूचनाएं उपलब्ध नहीं हैं, गुणित जो प्रदर्शित करना चाहता है वह ठीक स्पष्ट नहीं होगा—भले ही वह उद्योगों के मध्य के लेन-देन के ढांचे से निकाला गया है।

हालांकि अंश पूंजी व मजदूर-शक्ति के भाग जहाँ पूर्णतः प्रयोग में नहीं हैं, उन दशाओं में पर्यटक गुणक हमें इस समस्या के समाधान हेतु कोई सूक्ष्मकालीन ढंग नहीं बतलाता है। यदि अन्य निर्यात उद्योगों में अत्यधिक क्षमता है, तो पर्यटन का विस्तार इस क्षेत्र में कोई खास सहायता नहीं कर पायेगा। क्योंकि इन उद्योगों से स्रोतों का विचलन नहीं हो पाता है। केवल दीर्घकाल में ही यह विचलन होता है, परन्तु इन स्रोतों का विवरण विभिन्न समस्याएं पैदा करता है। दूसरी तरफ यदि अधिकतम क्षमता स्वदेशी उद्योग में है तो यह महसूस होगा कि लघु काल में घाटे के बजट के माध्यम से सरकारी व्यय में वृद्धि से यह अधिक प्रभावशाली होगा क्योंकि कम-से-कम सरकारी व्यय की हिदायतें प्रभाव डालने हेतु कुछ विशेष किस्म की बनायी जा सकती हैं।

पर्यटक गुणित की आलोचना के आधार पर यह तथ्य दृष्टिगत होता है कि पर्यटक गुणित का जिस तरह से प्रयोग किया जा रहा है उसमें पर्यटन के विस्तार से प्राप्त होने वाले दीर्घगामी फायदों का ही आकलन किया जाता है। वो परिस्थितियां जिनमें यह हो सकता है बिल्कुल ही अवास्तविक हैं, जैसे पर्यटन के माध्यम से प्रत्यक्ष व अप्रत्यक्ष रूप में इस्तेमाल किए गए समस्त स्रोतों की अवास्तविक शून्य अवसर लागत।

पर्यटन में गुणित के प्रभावों को नियोजन हेतु प्रयोग किया जाता है। पर्यटन के बावत इस तरह के तर्क की कभी आलोचना नहीं हुई है। अमेरिका के वाणिज्य विभाग के माध्यम से गठित सर्वेक्षण सन् 1560 में प्रशान्त द्वीप व पूर्वी एशिया के राष्ट्रों हेतु स्थापित किया गया था। इसके अनुसार गुणित प्रभाव कई राष्ट्रों में 3.2 से 4.5 तक रहता है। नेशनल काउन्सिल ऑफ इकोनोमिक्स रिसर्च ने भारत हेतु गुणक का अनुमान 3.6 और 3.2 लगाया है। विनियोजन, निर्यात और राज्य उपभोग के व्यय में वृद्धि के समान इसका भी गुणित प्रभाव पड़ता है। कहने का आशय यह है कि आगन्तुकों के प्रत्यक्ष व्यय की अपेक्षा आर्थिक क्रियाओं के जरिये राष्ट्र की आर्थिक व्यवस्था पर अत्यधिक प्रभाव पड़ता है।

गुणित का अनुमान तभी उचित होगा जब हम उससे सम्पूर्ण राष्ट्र की विभिन्न आर्थिक क्रियाओं में महत्वपूर्ण रूप में अलग-अलग महसूस करें।

भारत में पर्यटन उद्योग में अधिक लाभ प्राप्त करने के लिए यह आवश्यक है कि इस उद्योग में प्रयोग आने वाली प्रमुख आयातित वस्तुओं पर गहरी निगाह रखी जाए तथा यह भी ध्यान देने की जरूरत है कि जहाँ तक संभव हो इन वस्तुओं की जगह पर स्थानीय वस्तुओं का प्रयोग हो। इससे पर्यटक गुणित की वृद्धि में महत्वपूर्ण प्रभाव पड़ेगा और इस तरह राष्ट्र की सम्पूर्ण अर्थव्यवस्था को फायदा हो सकेगा।

पर्यटन का आर्थिक विकास में योगदान

पर्यटन का आर्थिक विकास में महत्त्वपूर्ण योगदान होता है इसलिए अधिकांश विकासशील राष्ट्रों में आर्थिक विकास की अधिकतम दर प्राप्त करने हेतु पर्यटन ही एक मूल्यवान विशेषता है। यह आर्थिक विकास की गाड़ी की एक धुरी है।

सम्पूर्ण विश्व में पयर्टन में जो धन सामान्य निवास स्थान से बाहर अर्जित किया जाता है वह पर्यटन के स्थलों में व्यय होता है। आगन्तुकों के व्यय से प्राप्त होने वाले क्षेत्रों में जो क्रय शक्ति में वृद्धि होती है यही पर्यटन का सबसे आश्चर्यजनक आर्थिक प्रभाव है। पर्यटन व्यय के कारण जो धन के प्रभाव का सृजन होता है, वह गुणक के प्रभावों के माध्यम से समस्त अर्थव्यवस्था में अपना स्थान बना लेता है।

राष्ट्रीय आय में पर्यटन का योगदान

विश्व के अधिकांश विकसित या विकासशील राष्ट्रों के राष्ट्रीय आय की वृद्धि में पर्यटन उद्योग का महत्वपूर्ण योगदान होता है। क्योंकि वर्तमान समय में पर्यटन

के स्वरूप में पर्याप्त बदलाव आये हैं और यह एक व्यय साध्य बनाता जा रहा है। राष्ट्रीय और अन्तर्राष्ट्रीय दोनों ही स्तर पर होने वाली आय से राष्ट्रीय आय में वृद्धि होती है। हालांकि स्वदेशी पर्यटन के क्षेत्र में यह तथ्य सत्य नहीं उतरता है, क्योंकि जहाँ तक खान-पान के व्यय का प्रश्न है यह अर्थव्यवस्था में एक स्थान से दूसरे स्थान में परिवर्तित होता रहता है। पर्यटन के तहत सेवाओं व उपभोक्ताओं की वस्तुओं की मांग सम्मिलित है, अन्यथा इनका उत्पादन ही नहीं होता। इससे उत्पादन के क्षेत्रों को बढ़ावा मिलेगा, जिससे स्वयं राष्ट्रीय उपभोग में वृद्धि होगी। क्योंकि पर्यटन व्यय के आकार का प्रभाव उत्पादन की वृद्धि में परिलक्षित होता है, अतः विभिन्न क्षेत्रीय कच्चे माल के माध्यम से उस क्षेत्र व अन्य क्षेत्रों में आय वृद्धि की क्रिया से देखा जा सकता है। यह विकास उपभोग से बिल्कुल स्वतंत्र है एवं इससे समस्त अर्थव्यवस्था पर धारावाहिक प्रतिक्रिया के रूप में प्रभाव पड़ता है। आवास का निर्माण, पर्यटकों हेतु शहरीकरण, आधारिक संरचना की संस्थापना, बढ़ते हुए पर्यटन के परिवहन व यन्त्र आदि कुछ उदाहरण हैं, जिनसे जाहिर होता है कि पर्यटकों की माँग से किस तरह विनियोजन को बढ़ावा मिलता है। इस तरह यहाँ एक वेग वृद्धि सिद्धान्त लागू होता है। साधारण अर्थव्यवस्था में जब कभी उपभोग में संचित दर से वृद्धि होती है, तो यह सिद्धान्त लागू होता है परन्तु यदि प्रति समय यह कम दर से बढ़ रही हो तो यह बिल्कुल लागू नहीं होगा एवं समान रूप में पर्यटन में भी लागू होता है। पर्यटन के क्षेत्र में उन्नति का द्विमुखी विकास व संचित विकास कायम रहना चाहिए।

निम्नलिखित मांगों का विनियोजन की वेग वृद्धि पर असर भिन्न-भिन्न क्षेत्रों में दिखाई देता है जो इस प्रकार है–

1. होटल व अन्य सहायक आवास का निर्माण।
2. संस्थापना तथा आधारित संरचना के कार्य, सड़कों का जाल, स्वछता, जल व बिजली की पूर्ति, हवाई अड्डे, शहरीकरण की योजनाएँ।
3. खाद्य सामग्री, मांस, पेय पदार्थ तथा अन्य कृषि सम्बन्धी वस्तुओं का उत्पादन।
4. परिवहन के साधन व सह-कल पुर्जों आदि का उत्पादन।
5. पर्यटकों के माध्यम से क्रय की जाने वाली वस्तुओं के उत्पादन में वृद्धि।
6. वित्तीय क्षेत्र, बैंकिंग, बीमा व अन्य व्यावसायिक सेवाओं का पुनः निर्माण।
7. नये व्यापारिक संस्थानों का निर्माण।

पहले दो मांगों का राष्ट्र की अर्थव्यवस्था पर बहुत ही अधिक प्रतिकूल प्रभाव पड़ता है।

अधिकांश राष्ट्रों में पर्यटन का उसकी अर्थव्यवस्था में प्रमुख स्थान होता है। देश वित्तीय व्यवस्था में पर्यटन के योगदान को ज्ञात करने के लिए आवश्यक हो जाता है कि स्वदेशी और विदेशी दोनों ही स्तर पर प्राप्त होने वाली आय की सही-सही जानकारी प्राप्त हो। अतएव अन्तर्राष्ट्रीय यात्रा से प्राप्ति व राष्ट्रीय आय की तुलना इस महत्ता की सूचक हो सकती है। इन आँकड़ों से प्रतीत होता है कि बारवडोज में राष्ट्रीय आय का एक बहुत ही अच्छा अंश पर्यटकों के माध्यम से योगदान किया जाता है। इसके अतिरक्त मैक्सिको, जोर्डन, पनामा व स्पेन में पर्यटन एकमात्र महत्वपूर्ण निर्यात रहा है एवं कनाडा में अखबारी कागज के पश्चात द्वितीय स्थान रखता था। ब्रिटेन में यह बहुत बड़ी रकम स्टर्लिंग में अर्जित करता है। पर्यटन 24 से भी अधिक राष्ट्रों में निर्यात की जाने वाली तीन मुख्य मदों में से एक है। पर्यटकों से प्राप्त आय का अदृष्टिगत निर्यात के क्षेत्र में आस्ट्रेलिया, कनाडा, कॉस्टारीका, डोमिन रिपब्लिक, इक्वेडर, इल सलबडोर, हन्डूराज, आयरलैंड, इटली, मैक्सिको, पुर्तगाल, स्पेन व स्विट्जरलैंड का प्रथम स्थान है। बहुत से राष्ट्रों में राष्ट्रीय आय के साथ पर्यटन से प्राप्त आय का एक अच्छा अनुपात है। भारत में पर्यटन आय, राष्ट्रीय आय में बहुत ही कम योगदान करती है। 1971-72 के संयुक्त राष्ट्र संघ के विकास कार्यक्रमों व इण्डियन मार्केट रिसर्च ब्यूरो के सर्वेक्षणों के अनुसार, पर्यटन प्राप्ति के सरकारी आँकड़े असली पर्यटन अर्जन के प्रत्यक्ष व अप्रत्यक्ष स्रोतों से 50 प्रतिशत से अधिक नहीं हैं।

पर्यटन के महत्त्व तभी पता चल सकता है जब छुट्टियाँ बिताने वाले तथा व्यापार यात्रा इत्यादि करने वाले यात्रियों के कुल खर्च को अनुमानित किया जाए और इस खर्च की तुलना अन्य राष्ट्रीय आय से की जाए। जिस तरह एक व्यक्तिगत परिवार हेतु पूरे एक वर्ष हेतु कुल आय-व्यय के बजट की समस्या है, यही समस्त राष्ट्र पर लागू होती है। पारिवारिक खर्च के अतिरिक्त केन्द्रीय, राज्य व स्थानीय अधिकारी व व्यापारिक संगठनों का भी ऐसा होता है। इन सबके हिसाब रखने के सिद्धांत व अर्थशास्त्र के नियम एक समान हैं। प्रति वर्ष राष्ट्र के कुल व्यय व उत्पादित माल व सेवाओं का मूल्य आँका जाता है। राष्ट्र में व्यक्तिगत व संघीय संस्थाओं की आय के जोड़ को ही राष्ट्रीय आय कहते हैं। यह एक वर्ष के अन्दर उत्पादित वस्तुओं तथा संस्थाओं का मूल्य है इसकी गणना करने के ढंग को उत्पादन प्रणाली, कहते हैं। इसकी आय प्रणाली जिसमें व्यक्तिगत व संघीय आय का जोड़ करती हैं, एवं व्यय प्रणाली जिसमें लोगों के क्रम के मूल्य, जिसमें जो क्रय व संचय किया गया, के योग से तुलना की जाती है। इस तरह ये तीन प्रणालियाँ, उत्पादन

प्रणाली आय प्रणाली व व्यय प्रणाली एक-दूसरे पर नियंत्रण का कार्य करते हैं। इन तीनों प्रणालियों के अनुमान का सार एक ही है और वह है कुल राष्ट्रीय उत्पादन। कुल राष्ट्रीय उत्पादन में से कुछ राशि पूंजी को अटल रखने अथवा इसके ह्रास हेतु रखी जाती है, ताकि पर्यटन के माध्यम से निर्मित राष्ट्रीय आय ज्ञात हो सके। यही एक वर्ष की न्यूनतम राशि है जो कि पर्यटन व्यय के माध्यम से उत्पादित होती है। ज्यों ही पर्यटक व्यय करना आरंभ करते हैं राष्ट्रीय आय में वृद्धि होने लगती है। वह होटल मालिकों, रेस्टोरेन्ट, यात्रा संचालकों आदि को धन का भुगतान करते हैं। यह धन प्राप्त करने वालों की आय में जुड़ जाता है एवं राष्ट्रीय आय का अंश बन जाता है। इसी तरह होटल मालिक भी रेस्टोरेन्ट, टूर आपरेटर आदि को धन भुगतान करते हैं। यह धन प्राप्तकर्ताओं की आय में जुड़ जाता है एवं राष्ट्रीय आय का अंश बन जाता है। ठीक इसी भांति रेस्टोरेन्ट मालिक व टूर संचालक इस धन को वेतन, मजदूरी, ब्याज पूर्ति व किराये के रूप में भुगतान करते हैं और इससे फिर राष्ट्रीय आय में वृद्धि हो जाती है। यह प्रक्रिया तब तक कार्यान्वित रहती है जब तक कि वित्त व्यवस्था में संचालित होती रहती है लेकिन इसकी औसत क्रमशः अल्प होती रहती है।

प्रतिबिम्बित दर

आन्तरिक आयात व निर्यात की दर को ही प्रतिबिम्बित दर कहते हैं। इसको किसी एक राष्ट्र की बढ़ोतरी अथवा किसी एक वस्तु अथवा सेवा की बढ़ोतरी को आयात के रूप में व्यक्त किया जा सकता है। किसी भी राष्ट्र की निर्यात की गयी बिक्री को यदि उस राष्ट्र में हमारे नागरिकों के माध्यम से किये गये पर्यटन खर्च से विभक्त किया जाए तो उसे प्रतिबिम्बित दर कहते हैं। श्री पीटर ग्रे के अनुसार, "यह मानते हुए कि विनिमय दर स्थिर है व किसी किस्म के प्रलोभन मूल्य बदलाव नहीं है तो विदेशी माल व सेवाओं के बढ़ते हुए क्रय क्रेता राष्ट्र के विदेशी मुद्रा भण्डार में एक अन्तिम बोझ के रूप में साबित होंगे। सीमान्त प्रवृत्ति के ऊपर आनुपातिक बोझ के परिणामस्वरूप विक्रेता राष्ट्र को स्वदेशी वस्तुओं पर खर्च, क्रेता राष्ट्रों से प्रत्यक्ष आयात, अन्य राष्ट्रों से आयात पर खर्च करना होता है और साथ ही इसका प्रभाव तृतीय राष्ट्रों की सीमान्त प्रवृत्ति पर पड़ने से इन राष्ट्रों को आयात आदि पर व्यय करना होता है। यह आयात में वृद्धि करने के परिणामस्वरूप हुए आनुपातिक बोझ एकक (इकाई) व प्रति आयातित इकाई हेतु गारंटी किये हुए वृद्धि निर्यात की दर के अन्तर के बराबर होगा। अतिरिक्त आयात की दर को ही प्रतिबिम्बित दर के

नाम से पुकारा जा सकता है। प्रतिबिम्बित दर को किसी राष्ट्र की आयात की बढ़ोतरी अथवा किसी एक वस्तु अथवा सेवा के आयात की बढ़ोतरी के रूप में भी व्यक्त किया जा सकता है। पिछली दशा में यह दर समस्त पूर्ति करने वाले देशों की प्रतिबिम्बित दर की औसत है जोकि इन राष्ट्रों में वितरित आयात वृद्धि के माध्यम से भारित करके प्राप्त की गई है।

10

पर्यटन के आर्थिक प्रभाव एवं विश्व पर्यटन की नई दिशा

परंपरागत रूप से यह माना जाता है कि पर्यटन देशों के बीच और राष्ट्र के भीतर आपसी समझ और तालमेल स्थापित करता है एवं भावनात्मक एकीकरण को बढ़ावा देता है। परंतु इसके आर्थिक महत्व से लोग साधारणतया परिचित नहीं हैं। हाल में और वह भी बीसवीं सदी के उत्तरार्द्ध से पर्यटन को आर्थिक विकास के एक महत्त्वपूर्ण उपकरण के रूप में पहचाना जाने लगा है। रोजगारोन्मुखता, विदेशी मुद्रा अर्जन, आय में वृद्धि और उत्पादन वृद्धि में पर्यटन के योगदान को देखते हुए यह कहा जा सकता है कि इस उद्योग की हमें जरूरत है। इसके इसी महत्व को देखते हुए सातवीं पंचवर्षीय योजना (1985-90) में पर्यटन को उद्योग का दर्जा प्राप्त हो गया है। आर्थिक क्षेत्र के तहत रोजगार, आय, विदेशी मुद्रा अर्जन एवं विकास इत्यादि आते हैं।

रोजगार—भारत सरकार की राष्ट्रीय पर्यटन कार्य योजना के अनुसार 1989-90 के दौरान भारत में पर्यटन उद्योग ने प्रत्यक्षतः 55 लाख व्यक्तियों के लिए रोजगार उत्पन्न किया। यह अनुमान है कि 2000 ई. तक यह रोजगार बढ़कर दोगुना हो जाएगा।

पर्यटन उद्योग में प्रत्यक्ष एवं परोक्ष रोजगार करने की अद्‌भुत सामर्थ्य है। होटल से लेकर भिन्न-भिन्न टूर-ऑपरेटरों, पर्यटन कार्यालय, परिवहन संचालकों, टूरिस्ट गाइडों इत्यादि के रूप में लोगों को इससे प्रत्यक्ष रोजगार उपलब्ध होता है। पर्यटन के, विविध व्यवसाय से सम्बद्ध लोगों को अप्रत्यक्ष रोजगार हासिल होता है। होटलों में सब्जी, मांस, मछली, मुर्गी, अनाज इत्यादि आपूर्ति करने वालों, बिजली मिस्त्री प्लम्बर, बढ़ई एवं बिस्तर, पर्दा आदि देने वालों एवं अन्य सभी प्रकार की सहायता के अलावा आपूर्तिकर्ताओं एवं टैक्सी चालकों आदि को अप्रत्यक्ष रोजगार मिलता है।

वस्तुतः पर्यटन संबंधी कार्यकलापों से अप्रत्यक्ष रूप में दूर-दूर तक लोग सम्बद्ध होते हैं तथा इसकी एक शृंखला बन जाती है। इसके द्वारा हस्तशिल्प और हस्तकलाएं बेचने वाली दुकानों और इम्पोरियमों के द्वारा हजारों शिल्पियों, कारीगरों और बुनकरों को रोजगार उपलब्ध होता है। वस्तुतः उत्पादन की अपेक्षा पर्यटन से सम्बद्ध गतिविधियों में रोजगार एवं निवेश का अनुपात कहीं अधिक है। यहाँ यह तथ्य स्पष्ट कर देना जरूरी है कि पर्यटन ने उन क्षेत्रों में रोजगार उत्पन्न किए जहां रोजगार के वैकल्पिक अवसर अत्यल्प मौजूद थे। उदाहरणार्थ हिल स्टेशनों को ही लीजिए इन पर्यटन स्थलों पर कई परिवार अपने जीविकोपार्जन के लिए मुख्य तौर पर पर्यटन पर ही निर्भर हैं। स्थानीय होटलों, रेस्तरां, परिवहन, दुकान आदि में स्थानीय जनता को रोजगार उपलब्ध होता है। पर्यटन उद्योग में लगे लोगों और उनके परिवार वालों की दैनिक उपयोग के कुछ सामानों, सेवाओं एवं शिक्षा आदि की जरूरत पड़ती है जिससे दुकानों, अस्पतालों एवं स्कूलों आदि में अप्रत्यक्ष तौर पर रोजगार उत्पन्न होता है। इस तरह पर्यटन से रोजगार का व्यापक अवसर उत्पन्न होता है इससे पांच सितारा होटलों में काम करने वाले उच्च प्रशिक्षित प्रबंधकों से लेकर कमरे में काम करने वाले सेवाकर्मियों, परिवहन कर्मचारियों, शिल्पियों आदि तक को रोजगार उपलब्ध होता है।

पर्यटन के विकास से बेरोजगार और अल्प रोजगार प्राप्त युवकों और युवतियों को रोजगार प्राप्त करने का सुनहरा अवसर प्राप्त होता है। रोजगार संबंधी इन सकारात्मक पक्षों के बावजूद स्थानीय जनता की दृष्टि से कुछ मुद्‌दों पर गौर करना आवश्यक है।

1. पर्यटन द्वारा पैदा किया गया ज्यादातर रोजगार मौसमी होता है। परिणामस्वरूप रोजगार नियोक्ता अस्थायी रोजगार को प्रधानता देते हैं अथवा मौसम के दौरान बाहर से मजदूर मंगा लेते हैं। इससे सही अर्थों में स्थानीय जनता को रोजगार नहीं मिलता।

2. कतिपय स्थानों पर स्थानीय लोगों की शिकायत होती है कि उन्हें छोटी-छोटी नौकरियां ही मिल पाती हैं, प्रबंधक जैसी नौकरिया बाह्य लोगों को ही मिलती हैं। होटल, रेस्तरां, ट्रेवल एजेंसियों, दुकानों इत्यादि पर बाहरी व्यक्तियों का स्वामित्व होता है। अतः पर्यटन से हुई आय का फायदा स्थानीय लोगों को नहीं मिल पाता है।
3. मौसम के दौरान पर्यटक के कारण अनेक जरूरी वस्तुओं के मूल्य बहुत बढ़ जाते हैं। इससे स्थानीय जनता बुरी तरह प्रभावित होती है।

अतएव पर्यटन विकास की योजना बनाते वक्त इन मुद्दों पर भी विचार करना चाहिए। इन्हें नजरअंदाज करने से पर्यटन स्थलों पर तनाव पैदा होगा।

आय—पर्यटन का रोजगार एवं आय एक दूसरे से गहन रूप में सबद्ध हैं और ये एक ही स्रोत अर्थात् पर्यटक व्यय पर आधारित होते हैं। पर मजदूरी और वेतन, ब्याज, किराया और फायदे के रूप में आय होती है। पर्यटन से कई प्रकार की वस्तुओं और सेवाओं की मांग होती है। उदाहरणार्थ—आवास, खाद्य एवं पेय, लंबी दूरी के और स्थानीय परिवहन, मनोरंजन, खरीददारी स्थल, दर्शनार्थ गाइड आदि। पर्यटकों द्वारा किए गए खर्चे से इन सेवाओं को स्पष्ट आय ही नहीं होती है वरन् इससे होटलों, रेस्तरों, परिवहन संचालकों और कामगारों, ट्रेवल अभिकर्ताओं और उनके कर्मचारियों, मनोरंजन गृहों के मालिकों और वहाँ के कर्मचारियों, कलाकारों, शिल्पियों तथा कला की वस्तुएं बनाने वाले लोगों को भी रोजगार उपलब्ध होता है। इन सभी सेवाओं से जुड़े रोजगार एवं इनसे होने वाली आय निम्नांकित पर आश्रित होती है—

1. उनके खर्च करने का ढंग
2. पर्यटकों की संख्या
3. पर्यटकों द्वारा किया गया व्यय

इस प्रकार पर्यटन से काफी मात्रा में अप्रत्यक्ष तौर पर रोजगार पैदा होता है और आय अर्जित होती है। पर्यटक जो व्यय करता है वह एक आदमी से दूसरे आदमी के हाथों में चला जाता है। परंतु एक व्यक्ति से दूसरे व्यक्ति के पास जाने में राशि घटती चली जाती है, क्योंकि हर व्यक्ति अपना भाग अपने पास रखता जाता है। अब देखना यह है कि पर्यटन और आपूर्तिकर्ताओं और इनसे बेहतर रोजगार प्राप्त करने वाले व्यक्तियों की कितनी जरूरत है। इससे एक नया आयाम सामने आता है। पर्यटन से उत्पन्न मांग से और ज्यादा रोजगार की जरूरत होगी और आय में भी वृद्धि होगी। परिणामस्वरूप इसमें एक के उपरांत एक श्रृंखलाबद्ध तरीके से वृद्धि होती जाती है। पर्यटन उद्योग अपने स्थानीय बाजार क्षेत्रों से जितना

अधिक सेवाएं एवं वस्तुएं खरीदेगा उसका गुणित असर उतना ही अधिक होगा। सेवाओं और वस्तुओं के आयात की प्रवृत्ति अधिक होगी गुणित प्रभाव उतना घटेगा। भिन्न-भिन्न प्रकार के करों, चुंगी और प्रवेश शुल्क आदि से केंद्र और राज्य सरकार की भी आय होती है।

विदेशी मुद्रा—हाल के वर्षों में भारत की विदेशी मुद्रा अर्जन संबंधी आर्थिक गतिविधि में पर्यटक की अहम् भूमिका रही है। 1985 में भारत में 8,36,908 पर्यटकों का आगमन हुआ जो 1992 में बढ़कर 18,67,651 हो गया। इसी तरह विदेशी मुद्रा अर्जन 1985 में 1800 करोड़ रुपये था जो 1992 में बढ़कर 3916 करोड़ रुपये हो गया। भारत पर पर्यटन के आर्थिक प्रभाव संबंधी संयुक्त राष्ट्र की रिपोर्ट के अनुसार पर्यटक आगमन में वृद्धि होने पर भी डॉलर की दृष्टि से पर्यटन की आय में कमी देखी गई है। इस परिघटना के कुछ अनुमानित कारण इस प्रकार हैं—

1. बाजार में विदेशी मुद्रा की कालाबाजारी के कारण रिसाव,
2. विगत वर्षों में रुपये की अवमूल्यन की ऊँची दर
3. बजट पर्यटकों के अनुपात में वृद्धि आदि।

पर्यटन की आवश्यकताओं को पूर्ण करने के लिए वस्तुओं या सेवाओं के आयात मूल्य को रिसाव कहते हैं। रिसाव जितना अधिक होगा, विदेशी मुद्रा का अर्जन उतना कम होगा और कम रिसाव का अभिप्राय है अधिक विदेशी मुद्रा अर्जन। इस प्रकार रिसाव को कम करके विदेशी मुद्रा अर्जन को बढ़ाया जा सकता है। यह सब पर्यटन के प्रकार, स्वरूप और पर्यटकों की श्रेणी पर निर्भर करता है। तीसरी दुनिया के राष्ट्रों की सरकारों को इस रिसाव को कम करने की रणनीति बनानी होगी, जो इस प्रकार हो सकती है—

1. बहुराष्ट्रीय कंपनियों के नियंत्रण को रोकना।
2. पर्यटकों पर कुछ प्रतिबंध लगाना, जैसे—उन्हें होटल के बिलों का भुगतान और खरीददारी इत्यादि में विदेशी मुद्रा ही देनी होगी।
3. स्थानीय संसाधनों का प्रयोग कर आयात विकल्प तैयार करना।

इस प्रकार विदेशी पर्यटन से अंतर्राष्ट्रीय व्यापार को बढ़ावा मिलता है। विदेशी पर्यटक आभूषण एवं कीमती पत्थर, कालीन, कपड़े, हस्तशिल्प और चमड़े के बने सामान खरीदते हैं।

ये विदेशी पर्यटक अपने-अपने राष्ट्रों में इन पदार्थों के प्रचार के माध्यम भी बनते हैं। इससे विदेशों में इन चीजों की मांग बढ़ती है और इनके निर्यात में बढ़ोतरी होती है। भारतीय हस्तशिल्प की श्रेष्ठ गुणवत्ता और पर्यटकों के द्वारा

विश्वभर में इनके प्रचार होने से इस क्षेत्र की निर्यात संभावनाएं अत्यधिक बढ़ गई हैं।

पर्यटन को उचित प्रकार से समझकर एवं सुनियोजित तरीके से नीति निर्माण कर इसका उपयोग आर्थिक विकास और उत्थान के लिए किया जा सकता है। नए रोजगार आय के मौके प्रदान कर विदेशी मुद्रा अर्जन, पिछड़े क्षेत्रों में अधिसंरचनात्मक विकास, प्रतिबंधित क्षेत्रों को कुछ सीमा तक अंतर्राष्ट्रीय पर्यटकों के लिए खोलकर अर्थव्यवस्था को एक नई ऊँचाई दी जा सकती है। अतएव अब पर्यटन को एक हाशिए पर खड़े विलासितापूर्ण क्रियाकलाप के तौर पर नहीं देखा जा सकता। निर्यात उद्योग के समान यह भी आर्थिक विकास का एक महत्त्वपूर्ण यन्त्र है। तथापि विकास की प्रक्रिया में मेजबान जनता के हितों का ध्यान अवश्य रखा जाना चाहिए।

विश्व पर्यटन की नई दिशाएं

पर्वतीय क्षेत्रों में कुछ आधारभूत सुविधाओं का अभाव ही क्षेत्रीय विकास में मुख्य रुकावट का कारण रहा है। यही कारण है कि यहां के रास्तों में न ही विदेशी एवं न ही स्वदेशी पर्यटक दिखाई देते हैं। किसी भी क्षेत्र विशेष की संपदा है वहां के प्राकृतिक स्रोत, जनसंपदा एवं मानवकृत विभिन्न प्राकृतिक संपदा का सदुपयोग, पर्वतीय क्षेत्र गढ़वाल एक अद्वितीय एवं संपन्न प्राकृतिक क्षेत्र है। तुलनात्मक दृष्टिकोण से यह विश्व प्रसिद्ध स्विट्ज़रलैंड के स्तर का क्षेत्र है। सम्पूर्ण क्षेत्र हिमालय के गोद में बसा हुआ है तथा एक सुंदर एवं रमणीक क्षेत्र है। साल में छः किस्म के मौसम, हर रोज नये-नय त्यौहार, जन एवं वन, वनस्पतियां, निराले पर्वत व झरने, नदी व झीलें, वन क्षेत्र, जानवरों की लुप्त प्राय प्रजातियों वाले अभयारण्य, शस्य श्यामला वनस्पति की अथाह संपदा, जंगलों से घिरी हुई मनमोहक ग्रामीण जीवन क्रिया, अपनी भाषा, वेशभूषा, एक प्राकृतिक पर्यावरण एवं परिस्थिति परिवर्तन की विचित्र जीवनचर्या, आदि-आदि, समस्त क्षेत्र की प्राकृतिक छटा में चार चांद लगा देती है। स्वयं गढ़वाल अपने बदौलत प्राचीन एवं प्राकृतिक सौन्दर्यता का ऐसा उदाहरण प्रस्तुत करता है जो विश्व स्तरीय स्पेन जैसे प्रतिष्ठित पर्यटन के क्षेत्र में उपलब्ध हो सकती है। प्रकृति ने इस क्षेत्र को इतना विविध संपन्न बनाया है कि कोई भी व्यक्ति इसे सही-सही तौर पर समझ या समझ पाने का दावा नहीं कर सकता है। क्षेत्र को भली प्रकार समझने का सबसे उत्तम तरीका यही है कि आप खुद यहां भ्रमण एवं खुद ही क्षेत्र विशेष की संपदा का अनुभव उठायें। गढ़वाल की गौरवशाली संपदा अपनी भव्यता से आपको अनायास ही स्तब्ध कर देगी।

विकास के पहलू

किसी भी क्षेत्र के सर्वांगीण विकास और उन्नति के लिए तीन विशेष पहलुओं पर अधिक ध्यान देना चाहिए :–

1. कृषि विकास
2. औद्योगिक विकास एवं
3. पर्यटन विकास।

समस्त अर्थव्यवस्था के आधार-स्थल होने से ये तीन पहलू विकास की विभिन्न गतिविधियों को जन्म देते हैं। किस क्षेत्र विशेष में किस पहलू का कितना महत्त्व है, यह निर्भर करता है, वहां की प्राकृतिक संपदा एवं दूसरी सम्बन्धित आवश्यकताओं पर। पहले एवं दूसरे पहलुओं के अलावा एवं विश्व की बदलती हुई परिस्थितियों में इन पर्वतीय क्षेत्रों में पर्यटन विकास संभवतः सबसे तीव्र गति से समस्त विकास की क्रिया का सृजनकर्ता है। इतिहास की ओर नजर डालें तो मालूम पड़ता है कि सम्पूर्ण हिमालय क्षेत्र के लोग ज्ञान के आदान-प्रदान, एक दूसरे को जानने-समझने, सामान खरीदने एवं बेचने, व्यापार मेले एवं प्रदर्शनियों में भाग लेने तथा शोध-अनुसंधान आदि गतिविधियों को जानने के लिए सदियों से दूर-दराज के क्षेत्रों की यात्रा करने आये हैं। इन गतिविधियों ने व्यापार को बढ़ावा दिया है। परिणामस्वरूप व्यापारिक क्रियाओं का पर्यटन विकास पर जबरदस्त असर पड़ा है। आर्थिक दृष्टिकोण से विश्लेषण करें तो पर्यटन विकास की परिधि में लगभग सभी प्रकार की आर्थिक क्रियाओं का समावेश है। हालांकि कुछ अन्य पहलुओं का भी योगदान रहता है। खासकर पर्वतीय क्षेत्रों के विकास पहलुओं का योगदान रहता है। विशेषकर पर्वतीय क्षेत्रों के विकास से सम्बन्धित आर्थिक मुद्दों का अपना विशेष दबाव हमेशा ही रहा है।

उचित समय

समय आ गया है कि हम अपनी पर्यटन संबंधी नीतियों की तरफ ध्यान दें, इनमें आवश्यक परिवर्तन लायें तथा क्षेत्र की आवश्यकतानुसार पर्यटन के नये आयामों की खोज करें। अब इस क्षेत्र की हिमाच्छादित सुरम्य पहाड़ियों के आकर्षण को विश्व की नजरों में अनदेखी नहीं किया जा सकता है। ब्रह्मपुत्र एवं सिन्धु घाटियों के बीच स्थित एवं लद्दाख से अरुणाचल तक फैली हुई हिमालय की 2500 कि.मी. लम्बी ऊंची पर्वत श्रृंखलाएं विश्व पर्यटन के मानचित्र में एक मुख्य वर्णनीय स्थान

रखती हैं। इस क्षेत्र की हिमाच्छादित सुरम्य पहाड़ियां दर्शकों को अपने जादुई आकर्षण से मंत्रमुग्ध कर देती हैं। लेखक और कवियों ने इसकी सौन्दर्यता को अपनी लेखनी द्वारा वर्णित करने के प्रयास किए। ऋषि मुनियों ने इनकी गुफाओं में तपस्या के लिए अनुकूल वातावरण पाया है। भ्रमणकर्ताओं, यात्रियों ने इस क्षेत्र की सार गर्भित महिमा तक पहुंचने के प्रयास किये हैं। भारत के दूर-दराज के स्थान से ग्राम निवासी सदियों से बद्रीनाथ, केदारनाथ जैसे ऊंचे और दुर्गम तीर्थस्थलों से यात्रा करते आये हैं।

परिवर्तन

आज समय के साथ-साथ परिस्थिति परिवर्तन ने क्षेत्रीय मुद्दों की परिभाषा में एक गहन बदलाव ला दिया है। यही उचित समय होगा कि हम प्रगति एवं विकास शब्दों पर बुद्धिमता एवं धैर्यता से फिर विचार करें। वर्षों से हम पर्वतीय क्षेत्रों को विभिन्न अर्थों में पेश करते आये हैं। अब उचित होगा कि क्षेत्रीय विकास के लिए एक संसाधन के रूप में इसके महत्व को समझें व स्वीकार करें। क्षेत्र का विशिष्टतायें वर्णित करते हुए एक विस्तृत नीति निर्धारण की ओर ध्यान देना आवश्यक है। नीति निर्धारण में पर्वतीय क्षेत्रों के असीम आकर्षणों, योजनाबद्धता, सीमांकन एवं एकीकृत क्षेत्रीय कार्यक्रम के जरिये से वास्तविकता को बदलने की ओर ध्यान देना जरूरी है। साथ ही सार्वजनिक एवं निजी क्षेत्र के उपक्रमों, एअर लाइन्स, रेलवे एवं सड़क परिवहन प्रणालियों, नगरपालिका और स्थानीय निगमों के साथ-साथ सांस्कृतिक एवं शैक्षणिक संस्थाओं सहित अंतर्राष्ट्रीय, राजकीय एवं क्षेत्रीय स्तर पर पर्यटन के साथ सक्रिय रूप में सभी संबंधित एजेन्सियों से एक सामूहिक प्रयास के रूप में पर्यटन विकास के लिये उत्तरदायित्व का विशेष तौर से उल्लेख किया जाना भी आवश्यक है।

इसके अलावा पर्यटन के हितों को उच्चतम सीमा तक बढ़ाने के लिए यात्रा परिपथ की धारणा पर आधारित एक कार्यक्रम योजना का तैयार होना अति आवश्यक है। योजना में चुनिन्दा परिपथों में गहन विकास प्राप्त करने, कुछ ही क्षेत्रों में केन्द्रीयकरण की प्रवृत्ति को दूर करने, पर्यटन आकर्षणों की विविधता को बढ़ावा देने और आर्थिक रूप से पिछड़े क्षेत्रों का उद्घाटन करने का प्रस्ताव होना भी आवश्यक है। पर्यटन सुविधाओं की सफल सेवाओं और प्रभावी प्रबन्ध सुनिश्चित करने के लिए मानव शक्ति विकास और प्रशिक्षण पर भी उचित जोर दिया जाना चाहिए।

पर्यटन को एक उद्योग के रूप में मान्यता देकर तथा कर-प्रोत्साहन, आर्थिक सहायता, राष्ट्रीय और राजकीय वित्तीय संस्थाओं के द्वारा ऋणों की मंजूरी में

प्राथमिकता, बिजली एवं पानी की व्यवस्था में प्राथमिकता आदि प्रोत्साहनों और सुविधाओं की व्यवस्था होना आवश्यक है।

पर्वतीय कार्य योजना

कार्य योजना में पर्यटकों के आगमन में वृद्धि, पर्यटन के माध्यम से विदेशी मुद्रा अर्जित करना और रोजगार प्राप्ति के लिये नीतियों का गठन निम्न मुद्दों पर आधारित हो :–

- पर्वतीय पर्यटन संरचना का सुधार।
- विद्यमान संरचना का सर्वोत्तम उपयोग सुनिश्चित करने के लिए गन्तव्यों के विवरण के साथ-साथ एकीकृत वृद्धि के लिए चयनात्मक आधार पर क्षेत्रों का विकास।
- मानव शक्ति संसाधनों के विकास के लिए संस्थाओं की फिर संरचना करना एवं उन्हें शक्तिशाली बनाना।
- विदेशी पर्यटकों की आवागमन और विदेशी मुद्रा अर्जन में वृद्धि करने के लिए उपयुक्त नीति तैयार करना।

पर्यटन का विकास के लिये निम्न गतिविधियों पर खास ध्यान दिया जाना आवश्यक है :

- चुनिन्दा पर्यटन क्षेत्रों/स्थलों का विकास।
- पर्यटन की विविधता में अवकाश और सवकाश पर्यटन को समर्पित करना।
- ट्रेकिंग, हिम-क्रीड़ा, वन्य-जीवन एवं जल क्रीड़ा जैसी साहसिक पर्यटन क्रियाओं का सुसंगठित और सुनियोजित विकास।
- पर्वतीय संस्कृति के सदृश संस्कृतिवाले क्षेत्रों तथा नई स्रोत बाजारों की खोज।
- क्षेत्रीय विरासत परियोजनाओं की पुनर्स्थापना।
- क्षेत्रीय छवि का निर्माण कार्य एवं मुख्य बाजारों में विपणन योजनाएं आरम्भ करना।
- अलग-अलग पर्यटन केन्द्रों पर सस्ता आवास उपलब्ध कराना।
- सार्वजनिक क्षेत्रों के नियमों में सेवा-स्तरों में सुधार लाना।

समन्वय एवं संयोजन

जिस प्रकार एक संसाधन प्रणाली में कच्चा माल विभिन्न मशीनों से गुजरते हुये तैयार उत्पादन में परिवर्तित हो जाता है ठीक उसी प्रकार पर्यटन

सप्लाई के अनेक घटकों को भी इस तरह सुसंगठित किया जाना चाहिए कि यात्रा की योजना बनाते हुये पर्यटकों की प्रत्याशा उनकी वापसी पर पूर्ण संतोष और चिरस्मरणीय याद में परिवर्तित हो जाए। इसलिए पर्यटन संरचना में सभी घटकों की उपलब्धता एवं सम्पादन एक सही समन्वय एवं संयोजन द्वारा आवश्यक होना चाहिए।

पर्यटक स्वयं में विकास की एक मिश्रित प्रक्रिया है। पर्यटन विकास का अर्थ केवल पर्यटन सुविधाओं जैसे होटल व्यवस्था, रेस्तरां, मनोरंजन गतिविधियों से ही नहीं है, इसके अन्दर सड़क मार्ग, पानी, बिजली, सीवरेज, दूर-संचार सुविधाएं आदि भी सम्मिलित हैं। हरेक पर्यटक अपनी यात्रा पर एक सुखद एवं आनन्दकारी अनुभव पाने की तलाश में रहता है। अतः पर्यटन उत्पादन में निम्न तथ्यों का सम्मिलित होना आवश्यक है :–

- पूर्ण क्षेत्र में शान्ति एवं स्थिरता का वातावरण।
- सुरक्षा एवं सुनिश्चितता का आश्वासन।
- व्यावहारिक एवं मेजबान समाज।
- एक ऐसा उद्योग एवं व्यवस्था जो मुस्कराहट के साथ अपेक्षित सेवाएं उपलब्ध करा सके।

ट्रेकिंग या पैदल यात्रा

पर्वतीय क्षेत्रों में सैर करने का सबसे अच्छा एवं लुभावना तरीका है पैदल यात्रा करना। पैदल यात्रा करनेवाले सैलानियों के लिए समस्त हिमालय क्षेत्र स्वर्गमय है। ट्रैकिंग एक कम खर्च करने वाली साहसिक गतिविधि है। इसमें भारी-भरकम उपकरणों की आवश्यकता नहीं होती है। अगर जरूरत है तो बस कुछ दूर तक चलने-बढ़ने एवं उतरने की क्षमता और प्रकृति को सराहनेवाली दृष्टि।

ट्रेकिंग कोई भी व्यक्ति अकेले भी कर सकता है और 10 या 20 व्यक्तियों के समूह में भी। ट्रेकिंग के लिए बस थोड़े से अनुभव की जरूरत है और शुरू में यह आसान रास्तों से प्रारम्भ की जा सकती है। इसके लिए सबसे कम साज-सामान की जरूरत होती है। एक स्लीपिंग बैग और रकसैक–ये दो ही मुख्य चीजें हैं। ट्रेकिंग करने वाला अपनी इच्छानुसार खर्च कर सकता है। ट्रेकिंग किसी भी समय, किसी के साथ और किसी भी स्थल से अपनी सुविधानुसार आरम्भ की जा सकती है। व्यक्ति अपनी ट्रेकिंग की योजना खुद बना सकते हैं और यही कारण है, कि सम्पूर्ण हिमालय के लोग ट्रेकिंग के प्रति दिन-प्रतिदिन आकर्षित हो रहे हैं। बुनियादी सेवाओं की कमी

में ट्रेकिंग करनेवाला वास्तविक मनोरम स्थलों तक पहुंच ही नहीं पाता है। ट्रेकिंग के रास्तों पर सुरक्षित एवं अच्छे आवास सुविधाओं की भी भारी कमी है।

पिंडारी और हर की दून आदि कुछ प्रमुख मार्गों पर तो पी. डब्ल्यू. डी. अथवा वन-विभाग के विश्रामगृह हर आठ से दस कि.मी. की दूरी पर स्थित हैं। परन्तु इनमें 6-8 व्यक्तियों के रहने की सुविधा है। यह बात ध्यान देने योग्य है कि पहाड़ों पर चलने वाले ट्रेकर की इन ट्रेकर हट्स या टेन्टों में रहने की व्यवस्था की जा सकती है। इस साहसिक क्रिया के लिए बड़े टूरिस्ट लॉज की आवश्यकता बिल्कुल नहीं है। एक हाल जैसे कमरे में 10-20 व्यक्तियों के रहने का प्रबन्ध एक के ऊपर एक बड़े-बड़े बंकर बनाकर किया जा सकता है। शौच की व्यवस्था खड्ड के एक सिरे पर की जा सकती है और पानी की कमी की समस्याएं कुछ हद तक टिसू पेपर के इस्तेमाल से सुलझायी जा सकती हैं। इन आवास स्थलों के बाहर कूड़ा डालने के लिए एक बर्तन रखा जा सकता है और समय-समय पर कूड़े को नष्ट करने का कार्य यहां नियुक्त निरीक्षक के हवाले किया जा सकता है। स्थानीय लोगों को सही प्रशिक्षण देकर ट्रेकिंग के लिए तैयार किया जा सकता है। आवश्यकता बड़े-बड़े होटलों की नहीं है, बल्कि बहुत से ट्रेकर्स हट्स के निर्माण करने की है, ताकि ट्रेकिंग करनेवाला व्यक्ति केवल एक स्लीपिंग बैग और जेब में कुछ पैसे लेकर अधिक स्वतंत्रता से गढ़वाल क्षेत्र का भ्रमण कर सके। उसे भारी-भरकम साज-सामान साथ में ले जाने की जरूरत न हो और वह बेरोक-टोक पहाड़ों का आनन्द ले सकता है।

इस सुव्यवस्थित कार्यक्रम द्वारा क्षेत्रीय लोगों की बेरोजगारी की समस्या का भी काफी हद तक समाधान हो सकेगा। साथ ही अधिक-से-अधिक लोग ट्रेकिंग के लिये गढ़वाल हिमालय की ओर आकर्षित होंगे।

पूरे गढ़वाल एवं कुमाऊँ को खास क्षेत्र घोषित करके प्रदेश के लिये परमिट देने अथवा प्रतिबन्धित करने में भी लचीला रुख अपनाना चाहिए ताकि अत्यधिक देशी एवं विदेशी पर्यटकों को यहां आने के लिये प्रेरित किया जा सके। साथ ही ट्रेकिंग के प्रत्येक मार्ग पर प्रवेश शुल्क लगाया जा सकता है और पदयात्रियों को वातावरण को साफ रखने की उनकी जिम्मेदारी के विषय में अवगत कराया जा सकता है। ट्रेकिंग मार्गों की पूर्ण जानकारी, आवास स्थलों का व्यवस्था और उनके बीच की दूरी आदि बातों की जानकारी भी ट्रेकर्स को दी जा सकती है। यात्रियों को यह भी बतलाना चाहिए कि नदी, झरने, झीलें व तालाब कहां-कहां पर स्थित हैं तथा हिमस्खलन का खतरा किन-किन स्थलों पर है। वहां उन्हें या तो सतर्कतापूर्वक जाना होगा अथवा जाने से बचना होगा।

सुरक्षा एवं सूचना के दृष्टिकोण से सभी प्रवेश मार्गों को वायरलैस द्वारा जिला मुख्यालयों से जोड़ा जाना चाहिए जिससे आपातकालीन स्थिति में आवश्यक सहायता अतिशीघ्र प्रभावित लोगों तक पहुंचाई जा सके।

ट्रेकिंग को लोकप्रिय बनाना अति आवश्यक है। प्रारम्भ में कुछ बाधाएं हो सकती हैं इन पर तत्काल ध्यान देना आवश्यक है। ध्यान देने योग्य मुख्य बातें हैं :–

- प्रमुख ट्रेकिंग मार्गों को भली-भांति निर्देशित किया जाए।
- ट्रेकिंग मार्गों पर स्थित पेड़-पौधों, वन सम्पदा, प्राकृतिक आकर्षणों, पशु-पक्षियों, ग्रामीण जीवन आदि के बारे में जानकारियां उपलब्ध हों।
- हर की दून, फूलों की घाटी आदि प्रमुख ट्रेकिंग मार्गों पर ठहरने के लिये ट्रेकर्स हट या कॉटेजेज तथा मूलभूत सुविधाओं की उपलब्धि।
- प्रचलित ट्रेक मार्गों पर कूड़े-कचरे की सुव्यवस्था।

पूरा गढ़वाल एवं कुमाऊँ क्षेत्र साहसिक पर्वतीय ट्रेकर एवं पर्यटकों का स्वर्ग है। आज के पर्यटक जोखिम पसंद हैं, उनके लिये गढ़वाल की प्राकृतिक संपदा एवं संपन्न वातावरण स्वागत द्वार पर खड़ी है। हमें केवल बुनियादी परिवर्तन करने हैं। पर्यटन सम्बन्धी योजनाएं, विचार एवं भावनाएं कठिन प्रतीत होती हैं परन्तु आज समय एवं क्षेत्र की पुकार है तथा भविष्य में विकास की नई-नई दशाएं हैं। आज नहीं तो कल गढ़वाल एवं कुमाऊँ का विकास अवश्यंभावी है।

पर्वतारोहण

पर्वतारोहण का सीधा सम्बन्ध पर्वतों से है। साहसिक पर्वतारोहण से तात्पर्य उस खेल-क्रीड़ा से है जिसमें पर्वत के अति दुर्गम स्थल तक चढ़ाई करके निर्देशित स्थान तक पहुंचने के प्रयत्न किए जाते हैं। गढ़वाल एवं कुमाऊँ क्षेत्र में इस साहसिक खेल-क्रीड़ा की असीमित संभावनाएं हैं। स्वतंत्रता के पूर्व यूरोपियनों के लिए पर्वतारोहण एक मनोरंजन प्राप्ति का साधन था। सन् 1930 के लगभग कई अंग्रेज समूह-दल नन्दा देवी (25654 फीट), कैमट (25473 फीट) एवं अन्य चोटियों पर चढ़ने में सफल रहे। सन् 1936 की जर्मन की सिनोल च (22597 फीट) एवं 1939 की पोलैण्ड की नन्दादेवी के पूर्व की सफलताएं उल्लेखनीय हैं।

पर्वतारोहण को एक खेल-क्रीड़ा के रूप में बीज उगाने का श्रेय 1940 में दून स्कूल के दो अध्यापकों श्री आर.एल. होल्डवर्थ एवं जी.ए.के. मार्टिन को है। ये भारतीय विद्यार्थियों को हिमालय में ट्रेकिंग एवं चढ़ाने के लिये गये थे। भारत का

सर्वप्रथम अभियान सन् 1951 में प्रारम्भ हुआ। इस वर्ष दून स्कूल के एक अन्य अध्यापक गुरुदयाल सिंह ने त्रिशूल चोटी के अभियान दल का नेतृत्व करके चोटी पर पहुंचने का साहसिक कार्य सम्पन्न किया। पांच वर्ष बाद श्री एन. डी. नयाल ने श्री जे.जे.एम. गिवसन के साथ मिलकर बन्दरपूंछ अभियान सम्पन्न किया था। सन् 1993 में एवरेस्ट पर्वत पर चढ़ने का प्रयास किया गया तथा गढ़वाली बछेन्द्रीपाल, प्रथम भारतीय महिला पर्वतारोही बनी।

पर्वतारोहण एक रोमांचकारी एवं साहसपूर्ण क्रिया बन जाने के परिणामस्वरूप उत्तरकाशी के हिमालय पर्वतारोहण स्टेशन की स्थापना की गई। यह संस्थान पर्वतारोहण के अनेक पहलुओं में प्रशिक्षण देने के लिये अति उत्तम कार्य कर रहा है। इसके अन्तर्गत ऊंचाई पर चढ़ने के प्रशिक्षण, शारीरिक बनावट, पर्वतारोहण का इतिहास, पर्वतीय जीव-जन्तु एवं वनस्पतियां, जियो-मारफोलॉजी, वन्य प्राणी व पर्वतारोहण की विभिन्न प्रणालियां एवं पद्धतियां प्रचलित हैं। यहां 500 मीटर से 600 मीटर की ऊंचाई पर चढ़ाई करने के लिये क्रियात्मक प्रशिक्षण भी दिया जाता है। साथ ही यह संस्थान उच्च प्रशिक्षण भी प्रदान करते हैं। पर्वत चोटियों पर चढ़ाई करना पर्वतारोहण के लिये अति आवश्यक है एवं यह संस्थान सुप्रसिद्ध चोटियों जैसे कैमेट व त्रिशूल के अभियान का प्रशिक्षण भी उपलब्ध कराता है। युवक एवं युवतियों को पर्वत श्रृंखलाओं की चोटी पर चढ़ने के पाठ्यक्रम के लिए विशेष कोर्स भी संचालित किए जाते हैं।

उत्तरकाशी में संस्थान की स्थापना के बाद पर्वतारोहण को एक साहसिक खेल के रूप में बढ़ावा मिला। भारतीय पर्वतारोही संघ की स्थापना के बाद इस जोखिम भरी मनोरंजक क्रीड़ा में लोगों की रुचि और भी बढ़ी। इस संघ द्वारा शौकीन पर्वतारोहियों के लिये सब तरह का आवश्यक सामान जुटाया गया और साथ ही आर्थिक सहायता की भी व्यवस्था की गयी। परिणामतः देश के विभिन्न शहरों में पर्वतारोहण क्लब बनने लगे। पर्वतों का खेल होने के नाते पर्वतारोहण को बढ़ावा देना चाहिये तथा समस्त देश में जगह-जगह क्लब खोलने की व्यवस्था होनी चाहिए।

स्कीइंग

स्कीइंग साहसिक खेलों में सबसे ज्यादा प्रचलित खेल है। बड़े हर्ष की बात है कि स्कीइंग होने की संभावनाओं को ध्यान में रखते हुए गढ़वाल क्षेत्र में औली स्की रिजार्ट को चुना गया है। परिणामस्वरूप इस परियोजना में सरकार ने करोड़ों रुपये खर्च किए जिसके परिणामस्वरूप आज औली में अत्यधिक

टेक्नॉलोजी के साथ-साथ अनुभवी प्रशिक्षक भी उपलब्ध हैं। औली को जोशीमठ से जोड़नेवाली केबल कार एशिया में अपनी तरफ की सबसे लम्बी है। औली में चियर लिफ्ट, स्की लिफ्ट, आरामदेह लकड़ी के केबिन और रेस्तरां हैं। औली में हर किस्म के स्की ढलान भी उपलब्ध हैं। दुर्भाग्यवश हर तरह की सुविधाओं से सम्पन्न होने के बावजूद औली अधिक लोकप्रिय नहीं हो सका है। इसका प्रमुख कारण असुगम यात्रा, अकुशल प्रबन्ध एवं अन्य समस्याएं हैं। कुछ समस्याएं निम्न हैं :–

- गरम पानी का अभाव
- कमरों को गर्म करने के लिए पुराने तरीकों का इस्तेमाल
- उपकरणों, विशेषकर स्की जूतों की आधुनिकीकरण की आवश्यकता
- कैफ्टीरिया का अकुशल प्रबन्ध
- समुचित चिकित्सा की अपर्याप्ता

औली को अधिक लोकप्रिय बनाने के लिये उपर्युक्त तथ्यों पर विशेष ध्यान देना जरूरी है। निकट की विमान सेवा उपलब्ध कराकर, आवश्यक वस्तुओं की खरीददारी की सुविधाएं और प्रबन्ध व्यवस्था को चुस्त-दुरुस्त बनाकर औली को भविष्य का सबसे अधिक आकर्षक पर्यटन स्थल बनाया जा सकता है।

गत् कुछ वर्षों से हिमालय स्कीइंग लोगों को अपनी ओर आकर्षित करने में सफल रहा है तथा इस साहसिक खेल की ओर रुचि बढ़ती जा रही है। नये-नये प्रशिक्षुओं की अलग-अलग ग्रुपों में प्रशिक्षण की व्यवस्था होनी चाहिए ताकि वे कम ढलान वाले और छोटे बर्फीले टीलों पर इस खेल का आनन्द प्राप्त कर सकें। अभ्यस्त तथा प्रशिक्षित खिलाड़ियों के लिये तो यह स्थल बहुत ही अच्छा है। औली में केवल कार द्वारा कई हजार फुट लम्बे बर्फ के ढलानों पर लम्बी दौड़ का आनन्द लिया जा सकता है। निम्न स्तरीय स्कीइंग के लिये गढ़वाल एवं कुमाऊँ क्षेत्र के कई अन्य स्थलों की खोज की जा सकती है।

पैराग्लाडिंग एवं हैंग ग्लाइडिंग

मुक्त आकाश में चील की तरह उड़ने एवं विचरण करने का रोमांचकारी अनुभव लेने के लिए पैराग्लाइडिंग एवं हैंग ग्लाइडिंग एक बेहतरीन खेल है। पैराग्लाइडिंग के लिए इस्तेमाल किए जानेवाले एयरो फाइल डैन साहसिक लोगों के बीच अधिक लोकप्रिय होती जा रही है, जबकि हैंग ग्लाइडिंग केवल प्रतियोगी स्पर्द्धा तक ही सीमित होती जा रही है।

इन साहसिक खेलों को बढ़ावा देने हेतु गढ़वाल एवं कुमाऊँ क्षेत्र में अत्यधिक संभवनायें हैं। लोकप्रियता बढ़ाने के लिये तथा स्थानीय पर्यटकों को उत्साहित करने के लिये इन खेलों का आयोजन अवकाश के दिनों में आबादी क्षेत्रों में किया जा सकता है। अन्य साहसिक खेलों की तरह इनका विकास प्रशिक्षकों के अभाव तथा उपकरणों के बहुत महंगे होने के कारण बाधक हो रहा है।

जल-क्रीड़ाएं

समस्त गढ़वाल एवं कुमाऊँ क्षेत्र में जलक्रीड़ा की असीम संभावनाएं हैं। सम्पूर्ण क्षेत्र में अनगिनत झीलों एवं नदियों को देखते हुए पानी के रोमांचक खेलों के लिये कई प्राकृतिक सुविधाएं हैं। गंगा, यमुना तथा अन्य नदियों की तरंगों पर नौकायन जैसी जलक्रीड़ाओं का आनन्द लिया जा सकता है। झीलों एवं तालों के किनारों को आकर्षक केन्द्र के रूप में विकसित किया जाना चाहिए। महंगे उपकरण एवं प्रशिक्षित कर्मियों की समुचित व्यवस्था की जरूरत है।

बाहर स्कीइंग भी एक मनोरंजक खेल है। मोटर बोट व चालक सहित स्कीइंग का सामान किराये पर उपलब्ध कराया जाता सकता है। नये प्रशिक्षुओं को यह खेल सिखाने के लिये प्रशिक्षण की व्यवस्था आवश्यक है।

रिवर राफ्टिंग ऐसी रोमांचक क्रीड़ा है जिसके विकास की सम्पूर्ण हिमालय क्षेत्र में असीम संभावनाएं हैं। गढ़वाल क्षेत्र में यह खेल केवल ऋषिकेश के निकट गंगा नदी तक ही सीमित है। इस खेल को क्षेत्र के अन्य भागों में प्रचलित करने के प्रयास किये जाने चाहिए। अन्य नदियां भी हैं जहां इस खेल की अनुकूल स्थितियां मौजूद हैं। प्रमुख समस्या उपकरणों के महंगे होने की है। आमतौर पर इसमें समस्त उपकरणों को आयात करना पड़ता है। इसके अलावा प्रशिक्षकों का उपलब्ध न होना एक बड़ी समस्या है। खेल के लिये पथ-प्रदर्शक अथवा गाइड बाहर से आमंत्रित करने होते हैं। आवश्यक उपकरणों का देश में ही निर्माण करके और प्रशिक्षित गाइडों की संख्या में वृद्धि करके ये कमियां दूर की जा सकती हैं। इस खेल के लोकप्रिय होने की बड़ी सम्भावनाएं हैं। भारतीय सेना की टीमों ने तो यह साबित कर दिया है कि लद्दाख, सिन्धु घाटी से लेकर अरुणाचल में ब्रह्मपुत्र नदी में रिवर राफ्टिंग संभव है। देवप्रयाग से लेकर ऋषिकेश तक अनेक एजेन्सियों द्वारा नियमित रूप से रिवर राफ्टिंग के समारोह आयोजित किये जाते हैं।

वन्य प्राणी दर्शन

पेड़-पौधों, फूल-फल, पशु-पक्षियों और वनस्पति प्रेमियों के लिए सम्पूर्ण हिमालय क्षेत्र एक स्वर्ग के समान है, यहां के जंगलों में ऐसे पशु एवं पक्षी रहते हैं जो अत्यन्त ही दुर्लभ जाति के हैं। वन संपदा इतनी घनी है कि असंख्य किस्म की वनस्पति, जड़ी-बूटियां आदि केवल इसी क्षेत्र में उपलब्ध हैं, इस संपदा का संरक्षण करके बेहतरीन बाजार के माध्यम से हम वन्य जीव प्रेमियों को पर्वतीय क्षेत्र की ओर आकर्षित करने में सफल हो सकते हैं। साथ ही समस्त वन संपदा को सही मायने में प्रस्तुत करने हेतु निम्न कदम उठाये जा सकते हैं :—

- प्राकृतिक वातावरण की समुचित जानकारी रखने वाले प्रशिक्षित पथ-प्रदर्शकों की व्यवस्था।
- ठहरने हेतु उचित स्थानों की सुविधा एवं आरक्षण की व्यवस्था।
- उद्यानों एवं वनों के भीतर पर्यावरण के पहलुओं के मद्देनजर अनुकूल परिवहन सुविधाओं की व्यवस्था।
- समस्त वन्य प्राणी क्षेत्रों तक पर्यटकों के पहुंचने की सुविधाएं।
- अधिकारियों के लिये उचित कार्यक्रम एवं अध्ययन की व्यवस्था हो।

अर्थव्यवस्था के दूसरे विकास कार्यों के साथ-साथ पर्यटन विकास सम्भव है। पर्यटन विकास का लाभ अन्य लोगों के अलावा व्यापारी, व्यवसायी, कृषक, दुकानदार, आवास और खान-पान इकाइयों, वित्तीय संस्थाओं, परिवहन उद्योग मालिकों, कलाकारों, शिल्पियों और दूर-दराज के रमणीय इलाकों तक अवश्य पहुंचेगा। साथ ही पर्यटन बढ़ावे से रोजगार की असीम सम्भावनाएं बढ़ जाती हैं, क्योंकि पर्यटन श्रम-प्रधान उद्योग है। इसके अलावा पर्यटन के अप्रत्यक्ष लाभ भी हैं जैसे विभिन्न क्षेत्रों के लोगों के बीच बेहतर आपसी सम्बन्ध स्थापित करना, राष्ट्रीय एकीकरण, सामुदायिक विकास, जीवन स्तर में सुधार, सामंजस्य की भावना का सृजन इत्यादि एक सुव्यवस्थित एवं नियंत्रित पर्यटन से किसी भी क्षेत्र की अर्थव्यवस्था आत्मनिर्भरता की ओर अग्रसर होती है।

11

भारत में पर्यटन नीति और नियोजन

पर्यटन एक ऐसी गतिविधि है जो मेहमान (Guest) और (Host) दोनों ही समुदायों को प्रभावित करती है। यह अर्थव्यवस्था के समस्त क्षेत्रों, सामान्य तौर पर समाज और पर्यावरण को प्रभावित करती है। अनियमित पर्यटन की बढ़ोतरी कभी-कभी इन संसाधनों को समाप्त कर देती है, जिन पर पर्यटन विकास को अकेले बाजार की शक्तियों पर नहीं छोड़ दिया जाना चाहिए। इसके लिए एक सावधानीपूर्वक तैयार की गई नीति आवश्यक है जिसकी रचना समाज और आर्थिक कारकों में सम्पूर्ण तौर पर अनुकूल सम्बन्धों को ध्यान में रख कर की जानी चाहिए। नीति संबंधी उद्देश्यों की पूर्ति हेतु पर्यटन नीति बनाना और योजनाएं तैयार करना एक मुख्य कार्य है। वर्षों के अनुभव से यह साबित हो चुका है कि पर्यटन विकास में नीति निर्माण और नियोजन का काफी महत्व है। पर्यटन विश्व अर्थव्यवस्था के बहुत महत्वपूर्ण अंश के रूप में उभरा है। यह न सिर्फ विकासशील देशों के लिए ही एक प्रमुख आर्थिक गतिविधि है अपितु बहुत से विकसित देशों के लिए भी उतना ही

महत्वपूर्ण है। आज पर्यटन विश्व का सबसे वृहद उद्योग है। 1994 में पर्यटन से आमदनी 321 अरब अमरीकी डॉलर थी। इसमें 12 करोड़ 70 लाख लोगों को सम्पूर्ण विश्व में रोजगार मिला। 330 खरब डॉलर के निवेश पर आधारित यह विश्व का सबसे बड़ा उद्योग है और इसने तेल की खोज एवं मोटर गाड़ियों के उत्पादन आदि को भी पीछे छोड़ दिया है। हालांकि कतिपय देशों के पास तकरीबन एक जैसे संसाधन हैं, फिर भी हम देखते हैं कि उनमें से कुछ देशों में पर्यटन की दृष्टि से दूसरे देशों की तुलना में सैलानियों की अधिक संख्या आती है और ऐसे देश दूसरे देशों की तुलना में तकरीबन विदेशी मुद्रा विनिमय अर्जित करते हैं। सम्बद्ध दशा की सरकारों की पर्यटन नीति में इसके कारण मौजूद हैं।

किसी भी देश या पर्यटन स्थल में पर्यटन के विकास हेतु एक निश्चित योजना और लक्ष्य होना चाहिए। ज्यादा से ज्यादा मुनाफा कमाने, अनियंत्रित जन पर्यटन के निषेधात्मक प्रभाव को खत्म करने अथवा रोकने हेतु और सतत् विकास के लिए ऐसा करना आवश्यक है। इस प्रकार निम्नलिखित कारणों से पर्यटन नियोजन की जरूरत पड़ती है–

1. विकास के क्षेत्रों को पहचानने के लिए।
2. सार्वजनिक एवं निजी क्षेत्रों की भूमिका का संयोजन
3. विकास हेतु पर्यटन के विविध प्रकार के चुनाव के लिए
4. विकास को समुचित निर्देशन देने हेतु
5. नियमन करने एवं कानून पढ़ाने के लिए
6. निषेधात्मक प्रभाव रोकने हेतु
7. किसी स्थल का विपणन करने के लिए
8. सेवाओं की गुणवत्ता बढ़ाने के लिए
9. मानव संसाधनों को प्रशिक्षित करने हेतु।

ज्यादातर देशों की सरकारें पर्यटन नीति निर्माण और लक्ष्य निर्धारण को प्राथमिकता देने लगी हैं। लक्ष्य की पूर्ति के लिए समुचित नियोजन एवं योजना की निगरानी की जरूरत होती है। इस नियोजन प्रक्रिया में निम्नांकित चरणों से गुजरना पड़ता है–

1. वर्तमान स्थिति आदि के संदर्भ में विकास क्षेत्र की तालिका प्राप्त करना।
2. परियोजना का अध्ययन एवं इसकी शर्तों को पहचानना।
3. सर्वेक्षण के परिणामों का विश्लेषण, संश्लेषण एवं इसके प्रति अपनाया जाने वाला रवैया।

4. परियोजना का लक्ष्य और उद्देश्य का निर्धारण।
5. योजना एवं सुझावों को व्यवहार में लाना।
6. व्यावहारिक परिस्थिति को देखते हुए विकास की नीति का निरूपण
7. आगे बढ़ने से संबंधित सुझाव।
8. सुधार एवं पुनर्मूल्यांकन की जानकारी प्राप्त करना।
9. कार्यान्वयन के परिणाम की निरंतर सूचना प्राप्ति।
10. मूल्यांकन की जानकारी की प्राप्ति।

भारतीय पर्यटन नीति का प्रारंभ वर्ष 1982

भारत में एक विषय के तौर पर पर्यटन हमारे संविधान में शामिल नहीं है। 1950 में परिवहन मंत्रालय के एक हिस्से के रूप में इसे मान्यता प्राप्त हुई। पर्यटन विभाग की स्थापना 1965 में हुई एवं उसी समय राज्य सरकारों ने भी अपने पर्यटन विभागों तथा पर्यटन विकास निगमों की स्थापना की। नवंबर 1982 में भारत सरकार ने प्रथम बार अपनी पर्यटन नीति की घोषणा की एवं इसका मुख्य उद्देश्य विदेशी पर्यटकों के सम्मुख भारत को एक आकर्षक पर्यटन स्थल के तौर पर प्रस्तुत करना था। इस नीति में निम्नांकित उपाए सुझाए गए।

1. पर्यटन को निर्यात उद्योग का दर्जा देना,
2. यात्रा-कार्यक्रमों के विकास के लिए विशेष दृष्टिकोण अपनाना,
3. सैलानियों को आकृष्ट करने हेतु राष्ट्रीय विरासत का उपयोग,
4. निजी उद्योग की हिस्सेदारी के लिए आमंत्रण आदि।

पर्यटन के विकास को ध्यान में रखते हुए इस नीति में निम्नांकित उदेश्यों को सम्मिलित किया गया–

1. इसमें समाज, राज्य दोनों को सामाजिक आर्थिक लाभ होता है। लोगों को रोजगार मिलता है, आमदनी बढ़ती है और राज्य को राजस्व और विदेशी मुदा प्राप्त होती है। सामान्य रूप से सभी लोगों के स्तर में बढ़ोतरी होती है।
2. यह राष्ट्रीय अतंर्राष्ट्रीय स्तर पर एक जोड़ने वाली ताकत बनेगी और भ्रमण के द्वारा आपसी समझ विकसित होगी।
3. इससे देश के युवाओं को न सिर्फ रोजगार मिलता है अपितु खेलकूद, रोमांच और इसी तरह के राष्ट्रीय चरित्र निर्माण संबंधी गतिविधियों में सम्मिलित होने का मौका भी मिलता है। देश के युवाओं को सही दिशा दिखाने में पर्यटन एक कारगार हथियार बन सकता है।

4. इससे दुनिया की दृष्टि में हमारी नई तस्वीर बनेगी और हमारी जीवन शैली, सांस्कृतिक अभिव्यक्ति और विरासत के नूतन आयाम सामने आ सकेंगे। परम्परा को अक्षुण्ण रखने के साथ इसके नए आयामों से परिचित कराना पर्यटन का प्रयोजन होगा। पर्यटन समृद्धि लाता है पर इसके साथ-साथ हमें यह भी ख्याल रखना होगा कि इससे हमारी सामाजिक एवं सांस्कृतिक पहचान और मूल्य नष्ट न हों और न ही हमारे प्राकृतिक संसाधनों को नुकसान पहुंचे। इसके प्रतिकूल इन्हें इस रूप में विकसित करना चाहिए कि हमारी सामाजिक एवं सांस्कृतिक शक्ति में संवृद्धि हो। पर्यटन में भारत को अपनी शर्तों पर अपने को प्रस्तुत करना चाहिए तथा दूसरे देशों की संस्कृतियों तथा जीवन शैलियों की नकल कभी नहीं करनी चाहिए।
5. घरेलू तथा अन्तर्राष्ट्रीय पर्यटन के द्वारा देश के युवाओं को एक दूसरे को समझने में सहायता मिलती है। इससे उन्हें एक दिशा मिलती है, राष्ट्रीय एकता एवं सामंजस्य की राह पर चलने का एक सकारात्मक मौका मिलता है।

इस नीति में अन्तर्राष्ट्रीय पर्यटन को विदेशी मुद्रा अर्जन का एक मुख्य स्रोत माना गया है और इसके विकास को प्राथमिकता दी गई है। निम्न तालिका 2002-2003 में पर्यटन पर हुई व्यय राशि को दर्शाती है।

वर्ष 2002-03 में पर्यटन के लिए राज्यवार अनुमोदित धनराशि

1. उतर प्रदेश	2	50.75	19.50
2. आसाम	9	76.81	61.88
3. अरुणाचल प्रदेश	5	4.13	3.25
4. बिहार	8	50.50	50.50
5. छतीसगढ़	9	30.80	9.85
6. गोवा	1	0.05	0.05
7. गुजरात	2	19.71	5.91
8. हरियाणा	8	33.22	31.10
9. हिमाचल प्रदेश	30	77.93	76.04
10. जम्मू कश्मीर	3	9.44	8.95
11. झारखण्ड	0	0	0
12. कर्नाटक	6	90.25	62.55

13. केरल	11	86.14	82.99
14. मध्य प्रदेश	18	71.12	57.48
15. महाराष्ट्र	8	2.38	54.62
16. मणिपुर	2	0.52	0.26
17. मेघालय	3	7.03	2.12
18. मिजोरम	6	14.12	4.85
19. नागालैण्ड	5	36.05	32.34
20. उड़ीसा	2	4.75	1.57
21. पंजाब	3	2.30	1.46
22. राजस्थान	13	109.97	109.62
23. सिक्किम	13	34.62	26.98
24. तमिलनाडु	5	55.90	31.61
25. त्रिपुरा	5	21.61	6.78
26. उतरांचल	3	54.80	41.80
27. उतर प्रदेश	3	29.50	29.50
28. पश्चिमी बंगाल	5	20.11	6.00
29. अण्डमान निकोबार	0	0	0
30. चण्डीगढ़	3	0.77	0.66
31. दादर और नागर हवेली	2	0.81	065
32. दिल्ली	14	50.40	44.90
33. दमन दीयु	3	4.95	1.69
34. लक्ष्यद्वीप	0	0	0
35. पाण्डिचेरी	2	0.79	0.63
कुल	212	112.11	868.09

पर्यटन के संदर्भ में कहा गया है कि सैलानियों में सबसे ज्यादा संख्या घरेलू सैलानियों की होती है। हालांकि अखिल भारतीय स्तर पर घरेलू सैलानियों की सांख्यिकी सुलभ नहीं है पर रेलवे और बसों में यात्रा कर रहे यात्रियों का अनुमान लगाया जा सकता है। हमारे यहाँ लाखों लोग प्रतिदिन तीर्थयात्रा हेतु अपने आप प्रेरित होकर घूमने के लिए निकलते हैं। पर इन लोगों के लिए समुचित सुविधा उपलब्ध नहीं है। अतः इस तरह की सुविधाएं उपलब्ध कराना आखिर सरकार का ही कर्त्तव्य है।

सांस्कृतिक पर्यटन की प्रासंगिकता पर वाद-विवाद करते हुए यह स्वीकार किया जाता है कि भारत में ज्यादातर विदेशी पर्यटक हमारी पुरातात्विक और स्मारकीय निधि को देखने के लिए आते हैं। यह कई सर्वेक्षणों और अध्ययनों से भी सामने आया है। हालांकि इन स्मारकों की देखरेख एवं रख-रखाव का जिम्मा भारतीय पुरातत्व सर्वेक्षण विभाग पर है परंतु यह परामर्श दिया गया है कि इन स्थलों पर योजनाबद्ध ढंग से पर्याप्त पर्यटन सुविधाओं का प्रबंध किया जाना चाहिए। इसके लिए राज्य सरकार के साथ-साथ संबद्ध एजेंसियों की भी मदद ली जानी चाहिए इसका उद्देश्य पुरातात्विक स्मारकों की समृद्ध विरासत का पूरा उपयोग करना है। धार्मिक स्थलों, हमारे लोक नृत्य-संगीत आदि को देखने हेतु पूरे विश्व से पर्यटक भारत की ओर खींचे चले आते हैं।

यहाँ इस बात का स्मरण रखना चाहिए कि पर्यटन सिर्फ केंद्र या राज्य सरकार का नहीं बल्कि निजी क्षेत्रों एवं शैक्षिक तथा सांस्कृतिक संगठनों का भी दखल है। पर्यटन एक व्यापक गतिविधि है, यह कोई मूर्त उपभोक्ता वस्तु नहीं है, अतएव राष्ट्रीय मापदण्ड पर इसके संपूर्ण प्रभाव को आंकना कठिन हो सकता है। परंतु जहां मापदण्ड स्पष्ट है और किसी खास परिस्थिति के संदर्भ में पर्यटन को आंकना है, वहां ऐसा संभव होता है, उदाहरणार्थ गुलमर्ग शीतकालीन खेलकूद परियोजना में लागत या मुनाफे के अध्ययन से पता चलता है कि 25 से 30 सालों की अवधि में शीतकालीन खेलकूद विकास से तकरीबन 600 करोड़ रुपए मूल्य का कुल सामाजिक लाभ हुआ है। अंतर्राष्ट्रीय पर्यटन क्षेत्र में 1979-80 में विदेशी मुद्रा अर्जन में 20.4 प्रतिशत की बढ़ोतरी हुई जबकि भारत में यह बढ़ोतरी दर 12.8 प्रतिशत रही। निर्यात रणनीति समिति ने 80 के दशक की अपनी रिपोर्ट में यह बात की है। 1979 में कुल विश्व निर्यात में भारत का भाग 10.42 फीसदी था जबकि विश्व पर्यटन आमदनी में इसका हिस्सा 0.52 प्रतिशत तक सीमित रहा। स्मरण रखना चाहिए कि भारत के लिए पर्यटन की दृष्टि से 1979 सबसे खराब वर्ष रहा था। यह सिद्ध तथ्य है कि यदि पर्यटन को पूरा मौका मिले तो यह अन्तर्राष्ट्रीय व्यापार में अपने घाटे को कम कर सकता है। इसके साथ-साथ पर्यटन की दृष्टि से समर्थ क्षेत्रों को इससे सामाजिक आर्थिक फायदा भी होता है।

पर्यटन नीति की कार्य योजना के तहत पर्यटन के फायदे युवकों में राष्ट्रीय एकता की भावना जाग्रत करना, युवा छात्रावास गतिविधियों का बढ़ावा एवं पूरे देश में समन्वय और सामंजस्य की भावना फैलाना सम्मिलित है। ऐसा महसूस किया गया है कि इसमें देश के विविध भागों के युवा जन एक दूसरे को जान पहचान पांएगे,

समझ पाएंगे और देश के विविध क्षेत्रों से परिचित हो सकेंगे। इस तरह भारत के युवा एक सूत्र में पिरोए जा सकेंगे। इसके अतिरिक्त पर्यटन के माध्यम से इनमें एक अंतर्राष्ट्रीय समझ भी पैदा की जा सकेगी।

यह भी परामर्श दिया गया है कि पर्यटन उद्योग के निवेश को पर्याप्त बनाने हेतु भारत सरकार के सभी संबद्ध मंत्रालयों से वित्तीय एवं अन्य प्रकार की मदद प्राप्त करने का हर संभव प्रयास किया जाएगा। इस प्रक्रिया में इस तथ्य को सामने रखने की कोशिश की जाएगी, कि हमारे देश को अपने विकास कार्यों हेतु विदेशी मुद्रा की सख्त जरूरत है। पर्यटन उद्योग विदेशी मुद्रा के अर्जन में मददगार हो सकता है। ऐसी आशा की जाती है कि इस विदेशी मुद्रा के अर्जन में पर्यटन कई तरह से सहायता कर सकता है। इंजीनियरिंग के सामान, हथकरघा और हस्तशिल्प उद्योग एवं अन्य निर्यातोन्मुख उद्योगों द्वारा अर्जित विदेशी मुद्रा में भी इससे वृद्धि होगी। इस संदर्भ में पर्यटन उद्योग को निर्यातोन्मुख उद्योग का दर्जा दिलाने का भी प्रयास किया जाएगा। इस नीति के अंतर्गत संरक्षण और विकास के बीच संतुलन स्थापित किया जाएगा। ताकि–

1. पुरातात्विक स्मारकों आदि की लूट खसोट न हो, और
2. पर्यावरण की दृष्टि से संवेदनशील प्रदेशों की पारिस्थितिक व्यवस्था में कोई असंतुलन कायम न हो।

परिभ्रमण हेतु निर्मित योजनाओं का महत्व

केन्द्र एवं राज्य सरकार पर्यटन की प्रसारणशील दुनिया में उचित योजना पर ज्यादा से ज्यादा जोर देती है। यह पर्यटन प्रोत्साहन और गन्तव्य क्षेत्र के भौतिक, सामाजिक तथा सांस्कृतिक पर्यावरण को सुरक्षित रखने में सरकारों के बीच संतुलन सुनिश्चित करने हेतु किया जाता है। पर्यटन विकास से संबंधित क्रियाओं में योजना के महत्व का उद्देश्य पर्यटन संबंधी क्रियाओं एवं स्थानीय पर्यावरण के बीच संतुलन बनाए रखना है। ऐसा मुख्य रूप से विकासशील देशों में है। यहाँ यह प्रभाव सबसे सबल है क्योंकि उन्हें ऐसी संरचनात्मक सुविधाएं विकसित करनी चाहिए जो किसी भी पर्यटन क्षेत्र की सांस्कृतिक विशेषताओं को सुरक्षित रख सकें एवं हर तरह के सामाजिक, आर्थिक विकास को प्रोत्साहित भी करें। साथ ही साथ पर्यटन को प्रोत्साहित करने हेतु अपने भौतिक पर्यावरण की सुरक्षा भी करनी चाहिए एवं बढ़ानी चाहिए।

पर्यटन संबंधी गतिविधियों तथा पर्यटन प्रोत्साहन और अधिसरंचनात्मक सुविधा का विकास किसी भी पर्यटन क्षेत्र पर सामाजिक, सांस्कृतिक, आर्थिक, भौतिक पर्यावरणीय प्रभाव डालेगा। विकासशील निर्णय लेने से पूर्व वैज्ञानिक

तकनीकियों का प्रयोग करते हुए योजनाबद्ध मूल्यांकन एवं विश्लेषण करना चाहिए। ये पर्वतीय पर्यावरण के संवेदनशील प्रतिवेशों, वन्य जीवों एवं तटीय रिसॉर्ट एवं दूसरे परम्परागत सांस्कृतिक दृष्टि से संवेदी गतव्यों पर ज्यादा लागू होते हैं। इस तरह पर्यटन विकास के प्रत्येक कार्यक्रम का मौलिक उद्देश्य सकारात्मक प्रभावों को बढ़ावा देना और गंतव्य क्षेत्रों के सामाजिक, आर्थिक एवं भौतिक पर्यावरण पर पड़ने वाले नकारात्मक प्रभावों को घटाने वाला होना चाहिए।

परिभ्रमण नियोजन : विस्तृत परिप्रेक्ष्य में

राष्ट्रीय पंचवर्षीय योजनाओं में पर्यटन विकास हेतु महत्व दिए जाने वाले मुख्य क्षेत्र निम्नलिखित हैं जो स्थानीय और राष्ट्रीय स्तर पर पर्यटन नियोजन से संबंधित हैं।

1. सांस्कृतिक और ऐतिहासिक महत्व की राष्ट्रीय विरासत की परियोजनाओं का पुनरुद्धार और संतुलित विकास हो जिससे भारत एक सांस्कृतिक पर्यटन स्थल के रूप में स्थापित हो सके। साथ ही पर्यटन का राष्ट्रीय विरासत के संरक्षण के मुख्य तौर पर उपयोग हो सके।
2. सीमित संसाधनों को बहुत ज्यादा क्षेत्रों, केन्द्रों में फैलाने के बजाए कुछ ऐसे चुने हुए पर्यटक क्षेत्रों एवं केंद्रों को विकसित किया जाए जो सैलानियों में लोकप्रिय हों।
3. अधिक से अधिक सैलानियों को आकर्षित करने और उनके ठहरने की अवधि को बढ़ाने हेतु हिमालय, रेतीले तटों वाले सुनहरी धूप से विशाल समुद्र तटीय क्षेत्रों एवं वन्य जीव जैसे पर्यटन संसाधनों का इस्तेमाल करके ट्रेकिंग, शीतकालीन क्रीड़ा, वन्य जीव पर्यटन और तटीय रिसॉर्ट जैसे परम्परागत क्षेत्रों को विकसित करना चाहिए।

वृहत, मध्य और लघु आयामी पर्यटन कॉम्पलेक्स की योजना विकास पर्यटन संसाधनों और पर्यटन गतिविधियों को संघटित कर सकेंगे। इससे एक तरफ तो सामाजिक, आर्थिक एवं पर्यावरणात्मक उद्देश्यों की दृष्टि से और दूसरी तरफ पर्यटकों की ढांचागत सुविधाओं की जरूरतों की दृष्टि से अधिकतम लाभ प्राप्त हो सकेगा। पर्यावरणीय संरक्षण को ध्यान में रखते हुए विविध गंतव्य क्षेत्रों के लिए प्रायोजित अनेक पर्यटन गतिविधियों की ढांचागत सुविधाओं की मांग की आपूर्ति का प्रबंध पर्यटन विकास योजनाओं के द्वारा किया जाना चाहिए। पर्यटन गतिविधि क्षेत्र अपने अंदर विविध प्रकार के उद्देश्यों को समेटे हुए है—

1. अनुकूल जलवायु वाले सुखदायी क्षेत्र यथा पर्वतीय रिसॉर्ट।

2. सहनशील मनोरंजन हेतु प्राकृतिक सौंदर्य के विस्तृत क्षेत्र।
3. ऐतिहासिक स्थल, स्मारकीय कॉम्पलेक्स, मेले एवं उत्सवों वाले केन्द्रों सहित सांस्कृतिक पर्यटन के क्षेत्र।
4. मनोरंजन और अवकाश व्यतीत करने हेतु क्षेत्र जैसे समुद्र तट के रिसॉर्ट।
5. विशेष दिलचस्पी के क्षेत्र अर्थात् वन्य जीव क्षेत्र, आकर्षक वनस्पति और प्राणी समूह के क्षेत्र, पक्षी विहार आदि।
6. धार्मिक पर्यटन क्षेत्र जैसे तीर्थ स्थल और मंदिरों के शहर।
7. साहसिक पर्यटन क्षेत्र अर्थात् वन्य जीव क्षेत्र, आकर्षक वनस्पति और प्राणी समूह के क्षेत्र, पक्षी विहार आदि।

अधिसंरचना सुविधा का विकास

अगर पर्यटन को किसी क्षेत्र के विकास कार्यों के अभिन्न घटक के रूप में प्रभावी ढंग से कार्य करना है तो पर्यटन से संबंधित भौतिक अधिसरंचनात्मक सुविधाओं का विकास आवश्यक है। अधिसंरचनात्मक सुविधाओं की कल्पना भी क्षेत्र एवं मेजबान जनसंख्या की सम्पूर्ण जरूरतों को ध्यान में रखते हुए करनी चाहिए। किसी रिसॉर्ट को पर्याप्त पैमाने पर आकर्षण प्रदान करना पर्यटक केंद्र के विकास हेतु एक अत्यावश्यक पहलू है। मोटे तौर पर कहा जा सकता है कि पर्यटन केंद्र को सामूहिक तौर पर गुण प्रदान करने वाले तीन प्रमुख घटक निम्नांकित हैं–

(1) पहुंचने की सुविधा : पर्यटन की दृष्टि से प्रमुख नगर के विशिष्ट स्थानों हेतु निकटवर्ती हवाई अड्डे, रेलवे स्टेशन, सड़क मार्ग, टर्मिनल अथवा फिर अन्य प्रवेश बिन्दुओं से शहर की ओर जाने की सुविधा परिवहन के तीन परम्परागत साधनों अर्थात् सड़क, रेल और वायु द्वारा देश के समीपस्थ स्थल से पर्यटन केंद्र और पर्यटन गंतव्य क्षेत्र तक पहुंचने की सुविधा और क्षेत्र के भीतर सैलानी की रुचि के अन्य महत्वपूर्ण केंद्रों के साथ संयोजन।

(2) आवासीय सुविधाएं और सेवाएँ : पर्यटन केन्द्र एवं गंतव्य क्षेत्रों पर मात्रात्मक एवं गुणात्मक रूप से सैलानियों की जरूरतों को संतुष्ट करने वाले पर्याप्त आवास की व्यवस्था महत्वपूर्ण कारक है। यदा-कदा अच्छा आवास पर्यटकों को ज्यादा समय तक विश्राम करने की प्रेरणा देता है। भारत के मौजूदा संदर्भ में इस मौलिक सुख-सुविधा का अभाव सैलानियों के आगमन के लिए सबसे बड़ा बाधक है। विभिन्न केंद्रों पर दिए जाने वाले आवास की मात्रा एवं स्वरूप पर्यटकों की जरूरतों पर निर्भर करेगा। किंतु कुल मिलाकर पर्यटन हेतु आवास आरामदेह समग्र सुविधाओं

तथा सेवाओं सहित और आर्थिक तथा भौतिक दृष्टिकोण से स्थिति के अनुसार होना चाहिए। साथ ही आवासीय विकास हेतु जरूरी अन्तर्निर्मित सुविधा होनी चाहिए।

पर्यटक कॉम्पलेक्स आकर्षित बनाने हेतु निरंतर विद्युत एवं सुरक्षित जल आपूर्ति, मल निकास, अपवहन और स्वच्छता प्रबंध आदि मौलिक सुविधाएं एवं सेवाएं ऐसे आवश्यक तत्व हैं जिन्हें सुनिश्चित कर लेना चाहिए इसके अलावा नागरिक सेवाएं जैसे चिकित्सा सुविधा, डाक तार और बैंकिंग की सुविधाओं को आवश्यक भाग समझा जाता है। फिर भी इन सभी पक्षों में पर्यटन के प्रकार एवं गंतव्य की स्थिति के अनुसार बदलाव होता रहता है।

(3) मनोरंजन पक्ष : यदि आवास उपलब्धता प्रमुख रूप से पर्यटन के ठहरने की अवधि को बढ़ाने में प्रभावी है तो किसी सैलानी केंद्र पर पर्याप्त मनोरंजन के तत्वों की व्यवस्था कुल मिलाकर पर्यटन स्थल के समय बढ़ा सकती है। विकास का यह पक्ष विश्राम एवं मन बहलाव हेतु अनिवार्य रूप से महत्वपूर्ण है। पर्यटन के संदर्भ में मनोरंजन की बहुत-सी अभिव्यक्तियां हैं जिसमें बाहरी सक्रिय मनोरंजन के आयोजन के अलावा हर प्रकार के व्यावसायिक मनोरंजन भी सम्मिलित हैं। महानगरों और प्रमुख शहरों के संदर्भ में यह समस्या किसी हद तक हल हो चुकी है, जो स्थानीय शहरों की भी आवश्यकताओं में शामिल है। फिर भी छोटे स्थानों में, जो संभवतः ज्यादा पर्यटन रुचि और आकर्षण के हो सकते हैं, ऐसे मनोरंजनात्मक सुख साधन पर्यटकों हेतु दिए जाने चाहिए। किंतु ऐसा सामाजिक, सांस्कृतिक परिवेश और गंतव्य को ध्यान में रखते हुए ही करना चाहिए।

सैलानी अधिसंरचनात्मक सुविधा के प्राथमिक तत्वों के अलावा सहायक अधिसंरचनात्मक सुविधाओं और सेवाओं को पर्यटक केंद्र के व्यापक विकास हेतु समझा जाता है। ये अधिकतर पारम्परिक और देश कला एवं कौशल पर्यटक रुचि के कुटीर उद्योगों के विकास के साथ-साथ सहायक सुविधाओं जैसे इसी हेतु के कार्यों में लगे शिल्पियों एवं कारीगरों हेतु आश्रम और इसी प्रकार के उपयोग के लिए भूमि निर्धारण पर आधारित है। पर्यटन सेवाओं के फलस्वरूप सुख साधनों के प्रबंधन के लिए सेवा करने वाले लोगों की आवश्यकता होती है। उनके लिए भी घर और सहायक आवश्यकताएं जरूरी हैं। इस पक्ष की उपेक्षा के फलस्वरूप जीवन की अमानुषिक परिस्थितियाँ सहित मलिन बस्तियों की उत्पति भी होगी।

परिभ्रमण मास्टर प्लान

विश्व के सारे देशों में जहां पर्यटन विकास के प्रति सुव्यवस्थित राष्ट्रीय नीति का एक भाग है, वहां पर्यटन अधिसरंचनात्मक सुविधा के विकास के लिए पर्यटन

मास्टर प्लान के प्रतिपादन को महत्व दिया गया है। यह कई कारकों से प्रभावित होता है जिनमें से कुछ निम्नांकित हैं—

1. पहुंच एवं अधिसरंचनात्मक सुविधा की तैयारी से संबंधित वित प्रबंधन की कठिनाइयां।
2. सकारात्मक कारक जिन्होंने विशेष संसाधन के उचित उपयोग को प्रभावित किया है अथवा प्रभावित कर सकते हैं।
3. पर्यटन विकास हेतु किसी विशेष संसाधन की वहन क्षमता से सम्बद्ध पर्यावरणीय और पारिस्थितिक प्रभाव एवं पर्यटन विकास कार्यक्रम के वृहद एवं सूक्ष्म स्तर पर सामाजिक, सांस्कृतिक प्रभाव पड़ता है।

निश्चित तौर पर वृहद एवं सूक्ष्म दोनों स्तरों पर सम्पूर्ण पर्यटन मास्टर प्लान का भौतिक विकास योजना के मूल घटक के तौर पर विशेष महत्व का है। वृहद स्तर पर पर्यटन विकास हेतु व्यापक मास्टर प्लान में निम्नलिखित बातें सम्मिलित होंगी—

1. विकसित, संरक्षित आदि किए जाने वाले क्षेत्रों का विस्तृत ब्यौरा देने वाली भौतिक योजना।
2. पर्यटन और मनोरंजन हेतु नीतियाँ और प्राथमिकताएं।
3. अधिसंरचनात्मक सुविधा एवं विकास की योजना।
4. कार्रवाई की योजना एवं परिवर्तनों तथा उनके प्रभावों के प्रबोधन की क्रियाविधि।
5. कार्यान्वयन, समन्वयन एवं वित प्रबंधन हेतु नीति।
6. पारिस्थितिक और सामाजिक, आर्थिक कारकों और उनसे संबंधित कठिनाइयों का हल।

असंगत उपयोगों के जोखिम में डाले जाने वाले सीमित क्षमता के संसाधन जैसे कोई लोकप्रिय पर्यटन केंद्र, पर्वतीय रिसॉर्ट अथवा तटीय कॉम्पलेक्स पर से दबाव कम करके सीमित करने या नियंत्रित करने हेतु मौलिक विकास नीति में निम्नलिखित को शामिल होना चाहिए :-

1. सुविधाओं को सीमित करना।
2. विभिन्न गतिविधियों का स्थानिक दृष्टि से विभक्तिकरण।
3. वैकल्पिक गंतव्यों का विकास करना।
4. प्रवेश पर प्रतिबंध लगाना।

मास्टर प्लान के प्रतिपादन की तरफ पहले कदम के रूप में पर्यटन संसाधनों की विशेषताओं एवं संभावनाओं का सर्वेक्षण करना जरूरी होगा। इन्हें गतव्य क्षेत्र के

भौतिक, सामाजिक, आर्थिक एवं पर्यावरणीय गुणों के सम्बंध में विश्लेषित करना चाहिए। पर्यटन गंतव्य क्षेत्रों की योजना, विकास और प्रबंधन के पर्यटन मास्टर प्लान के विकास की मूलभूत रूपरेखा निम्नलिखित द्वारा स्थापित होगी।

1 प्रक्षेपित सैलानी आगमन का विश्लेषण।

2. संसाधनों का संश्लेषण और पर्यटन उद्देश्यों से उनका अनुकुलतम प्रयोग।

मास्टर प्लान के भौतिक घटक पर्यटक सुसाधयता एवं ढांचागत सुविधा पर्यटन गतिविधि क्षेत्र के विविध भागों के आन्तरिक एवं बाहर प्रवेश्यता एवं संयोजन और प्रमुख मनोरंजनात्मक तथा खुले स्थानों का एक संगठित खुले स्थान के तंत्र के तौर पर विभाजन की व्यवस्था की रूपरेखा स्थापित करेंगे।

पर्यावरणीय सुरक्षा, प्राकृतिक स्थलों का विकास, स्थल विकसित करना आदि से संबंधित उपाय और कार्यक्रम ऐसे मास्टर प्लान के महत्वपूर्ण घटक हैं। इसके अलावा प्रतिवेश की पृष्ठभूमि से कॉम्पलेक्स का संयोजन और परिसर के विकास को नियंत्रण में रखने के उपाय को भी सुनिश्चित करना चाहिए। परिभ्रमण परिसर के विकास को प्रतिवेश की ऐसी छोटी बस्तियों से जोड़ने की उचित कार्रवाई की योजना पर भी विचार करना आवश्यक है जिनकी सामाजिक एवं आर्थिक गतिविधि कॉम्पलेक्स में हो रही पर्यटन गतिविधियों से जुड़ी हुई है। इन नीतियों एवं कार्यक्रमों का सविस्तार ब्यौरा दिया जाना चाहिए तथा कॉम्पलेक्स तथा पर्यटन अधिसरंचनात्मक सुविधा को समयबद्ध विकास की संगठित भौतिक विकास योजना द्वारा मजबूती प्रदान करनी चाहिए।

पर्यटन केंद्र एवं गंतव्य क्षेत्र बहुत अधिक लोगों के एकत्र होने के केन्द्र हैं और उनके लिए सुरक्षित और सुविधाजनक प्रवेश, पार्किंग की सुविधा, पैदल यात्रियों के चलने, एकत्र होने के लिए स्थान, भोजन व्यवस्था और आवास हेतु पर्यटक अधिसंरचनात्मक सुविधा, नागरिक सेवाएं, पर्यटक क्रय-विक्रय, सैलानियों की मनोरंजनात्मक गतिविधियों को स्थिति के अनुसार आकर्षण और ऐसी सुविधा कि सैलानियों द्वारा प्रक्षेपित मांग को सुलभ कराना चाहिए। स्पष्ट रूप से व्यक्तिगत सुख साधनों की उपलब्धता और पर्यटक अधिसंरचनात्मक सुविधा का माप स्थान-स्थान पर बदलेगा। फिर भी इस दशा में इन समस्त अधिसरंचनात्मक सुविधाओं को पर्यटक स्थल के संबंध में स्थानिक दृष्टि से आयोजित करना चाहिए और इनके बीच उचित क्रियात्मक संबंध की उचित भौतिक योजना और रूपरेखा की जरूरत है।

पर्यटन से संबंधित क्रियाएं व्यावहारिक रूप से भू-विस्तीर्ण के अनुरूप हैं। चाहे पर्यटक केंद्र नगर विकास क्षेत्र का एक हिस्सा हो या अलग कॉम्पलेक्स के रूप में स्थित हो या फिर पर्यटक रिसॉर्ट के रूप में स्वयं शहर क्षेत्र हो। पर्यटन क्षेत्र के

सम्पूर्ण संघटित विकास पर जोर दिया जाना चाहिए और विभिन्न पर्यटन केंद्रों के लिए प्रत्येक विकास योजना को वृहत नगर योजना अथवा क्षेत्र नगर योजना के रूप में कल्पना करनी चाहिए।

इस संदर्भ में इस बात पर भी जोर दिया जाना चाहिए कि महानगर और प्रमुख नगरों के मामले में सैलानियों की जरूरतों विशेषकर अधिसंरचनात्मक सुविधा के क्षेत्र में इन शहरों की आवश्यकताएं उन नगरों में चल रही योजनाओं से ही पूरी हो सकती हैं। किंतु छोटे नगरों अथवा अलग-अलग स्थानों पर पर्यटकों के लिए दी जाने वाली अधिसंरचनात्मक सुविधाओं और सेवाओं को इस तरह से व्यवस्थित करना चाहिए कि स्थानीय जनता भी उसका प्रयोग उचित विस्तार और आवर्धन के साथ कर सके। फलस्वरूप इन स्थितियों में पर्यटन पर निवेश से अधिकतर ऐसे छोटे नगरों और केंद्रों के विकास को हर तरह का लाभ पहुंचाना चाहिए जहां पर्यटन क्षेत्र स्थित हों। इसके अलावा विकास योजना को उपयुक्त चरणों में कार्यान्वित किया जाना चाहिए। इसकी रूपरेखा इस तरह तैयार की जानी चाहिए कि भविष्य में संवर्धन और विस्तार किया जा सके।

शहरी पर्यटन स्थल

किसी पर्यटन क्षेत्र की स्थापना का महत्वपूर्ण कदम जहां पर्यटक स्थल नगर में स्थित हो तो उसे इस प्रकार विभाजित करना चाहिए कि चारों ओर पर्याप्त क्षेत्र पर्यटन गतिविधि के लिए विशेष क्षेत्र के रूप में बचा रहे। किसी नगर में जहां ऐसे पर्यटक स्थल हों जैसे ऐतिहासिक नगरों तथा तीर्थ स्थानों में वहां इस बात का ध्यान रखना चाहिए कि ऐसे विस्तृत स्थानों की पहचान की जा सके जो बहुत से स्थलों को एक कॉम्लेक्स के रूप में घेरे हुए हो। सैलानियों को भौतिक विकास का एक संगठित उपागम संभव हो सके। ऐसे मामलों में सैलानी स्थल की अधिसरंचनात्मक सुविधा की योजना एवं भौतिक विकास की कल्पना सम्पूर्ण पर्यटन विकास योजना की सीमा के अंदर होनी चाहिए। पर्यटन विकास योजना में निम्नलिखित को शामिल होना चाहिए।

पर्यटक अधिसंरचनात्मक सुविधाओं की स्थापना हेतु प्रत्येक स्थल के चारों ओर स्थल निर्धारण और इसके साथ ही जहां तक संभव हो उचित नियमों द्वारा यह सुनिश्चित करना कि प्रतिवेश में अत्यधिक विकास न हो। एक प्रमुख मौलिक धारणा यह होनी चाहिए कि अपने समीपवर्ती प्रतिवेश में सहनशील उपयोगों की स्थापना द्वारा पर्यटन स्थान की पहचान और उपयोगिता को बनाए रखा जाए।

पर्यटन अधिसंरचनात्मक सुविधा के विकास हेतु निर्धारित स्थान के भीतर ध्यान देने योग्य महत्वपूर्ण बात यह है कि वास्तविक सैलानी स्थल अधिसरंचनात्मक सुविधा में निम्न न हो अपितु उसे सबसे प्रमुख होना चाहिए। ऐसा तभी संभव होगा जब निर्धारित अपितु के आंतरिक स्थानिक संघटन की योजना ऐसी हो कि अधिसरंचनात्मक पर्यटन कॉम्पलेक्स स्थान प्रवेश सड़क से प्रत्यक्ष रूप से जुड़ा हो और वहां तक सुगमता से पहुंचा जा सके। अधिसंरचनात्मक सुविधा तथा पर्यटक स्थल पर ज्यादा से ज्यादा विस्तृत हरा-भरा क्षेत्र होना जरूरी है। इसके लिए आवश्यक है कि धीरे-धीरे क्षेत्र को विकसित किया जाए। इसकी जरूरत नहीं है कि अधिसरंचनात्मक सुविधा को लम्बे चौड़े क्षेत्र में फैला कर विकसित किया जाए अपितु इस सुविधा को सुनियोजित ढंग से विकसित किया जाना चाहिए जिससे निर्धारित स्थल का भाग हरा-भरा रह सके। कार्यशीलता की दृष्टि से अधिक बेहतर सुविधाएं प्रस्तुत की जा सकें।

योजना के माध्यम से पर्यटन

सैलानी केंद्रों के संदर्भ के संरक्षण में केंद्रों की पर्यटन विशेषता की सुरक्षा भी सैलानियों की होती है और नगरों के मामले में यह पक्ष विशेष रूप से सैलानियों का होता है क्योंकि वह सब मिलकर विरासत और ऐतिहासिक महत्व के कारण सैलानियों के लिए आकर्षण के केंद्र बनते हैं। ऐसे शहरों की आर्थिक विभिन्नता के समान ही भौतिक विकास को इस तरह नियोजित करना चाहिए कि उसकी विरासत और ऐतिहासिक पवित्रता को भौतिक रूप से बनाए रखा जा सके। ऐसा उनके प्रतिवेश में पर्याप्त हरा-भरा मध्यवर्ती क्षेत्र रख कर और पर्यटन की प्रगति के हित में नया विकास करके किया जा सकता है।

सैलानी स्थल में भौतिक संरक्षण को सामान्य सौन्दर्यीकरण तथा नवीकरण करके अभिव्यक्त किया जा सकता है जिससे अच्छे उद्यानों, फर्श और पर्यटक सामग्री जैसे सैलानी साहित्य, गाइडें, कलाकृतियों की दुकानें आदि सहित उसे अधिक सुंदर और आकर्षक बनाया जा सके। तथापि यह ऐसा मामला है जो अत्यधिक नियंत्रण उद्देश्य, अभिज्ञता एवं अति संवेदनशीलता के योग्य है। फिर भी बड़े पैमाने पर नवीनीकरण और सौंदर्यीकरण से भी ज्यादा सुरक्षा और संरक्षण पर जोर देना चाहिए।

स्थल विकास में, विशेषकर पर्यटन गतिविधि क्षेत्रों में वर्तमान प्राकृतिक विशेषताओं का संरक्षण सैलानियों के आकर्षण को बनाए रखने हेतु महत्वपूर्ण है। इसके अलावा उचित रूप से निर्धारित योजना के अनुसार प्रचुर वृक्षारोपण को स्थल

विकास कार्यक्रम का आवश्यक भाग होना चाहिए। अधिसरंचनात्मक सुविधा हेतु निर्धारित स्थानों के भवनों की मौलिक योजना के साथ एक मंद गति का विकास जुड़ा होना चाहिए।

किसी परिभ्रमण स्थल की भौतिक विकास योजना का प्राथमिक उद्देश्य सैलानियों को आकर्षित करना है क्योंकि इस दृश्य योजना में आकृति एवं सौंदर्य उनके प्रतिपादन और भूदृश्य निर्माण, नवीकरण और सामान्य सौंदर्यीकरण अथवा पर्यटकों के लिए आवासीय सुविधा सुलभ कराने से संबंधित है। अतः सैलानी स्थलों और परिसरों की भौतिक योजना को शहर की दशा और भूदृश्य योजना सहित भौतिक धरातल और अधिसंरचनात्मक सुविधाओं के संश्लेषण का प्रयास करना चाहिए जिससे उभर कर सामने आने वाला विकास कार्यात्मक एवं पर्यावरणीय रूप से अनुकूल हो और साथ ही सौंदर्य की दृष्टि से देखने में आकर्षक हो।

निजी या सामूहिक पर्यटकों की रुचि को ध्यान में रखते हुए पर्यटक स्थान के विकास की योजना बनाई जानी चाहिए। इसे परिभ्रमण स्थल और परिवेश की भौतिक विशिष्टताओं से संबंधित होना चाहिए। ऐसा प्रयत्न स्थल के विवेकपूर्ण विकास और पर्यटक संसाधनों के अनुकूलतम, उपयोग के लायक बना देता है। समस्त अधिसंरचनात्मक सुविधाओं का विकास स्मारक के इर्द-गिर्द करना चाहिए। बड़े कॉम्पलेक्सों के विषय में भी ऐसे ही सिद्धांत अपनाए जा सकते हैं। एक सम्पूर्ण स्थानिक परिरूप की रचना पर विचार करके ऐसा किया जा सकता है। इसमें ही सैलानी रुचि के निजी केंद्र होंगे जिन्हें समग्र परिरूप का एक हिस्सा समझा जा सकता है। सुंदर प्राकृतिक स्थलों में भी इस तरह की योजना की जा सकती है।

योजना की आवश्यकताएं

1. पर्यटक संसाधन और अधिसंरचनात्मक सुविधा की जरूरत का मूल्यांकन।
2. क्षेत्र और अन्य पक्षों की वहन क्षमता के समनुरूप विकास की सीमा तय करना।
3. स्थानीय पर्यावरण पहाड़ी आवास पर विकास का प्रभाव।

इस तरह सारे विकासीय कार्यक्रमों में विशेष तौर पर पर्यटन संबंधी विकास में निम्न बातों पर मुख्य ध्यान रखना चाहिए।

1. उनकी सामाजिक सांस्कृतिक पहचान की सुरक्षा करना एवं कायम रखना।
2. नगर की अर्थव्यवस्था में समग्र रूप से सुधार।
3. जीवन की गुणवत्ता का विकास।

योजना आयोग द्वारा पर्यटन को प्रोत्साहन

भारत सरकार के योजना आयोग ने जून 1982 में पर्यटन को एक उद्योग का दर्जा प्रदान किया। सातवीं एवं आठवीं योजना में पर्यटन को काफी महत्व दिया गया। आने वाले वर्षों हेतु योजना बनाने के लिए 1986 में योजना आयोग ने राष्ट्रीय पर्यटन समिति की स्थापना की। समिति ने 1988 में अपनी रिपोर्ट पेश की जिसके मुख्य सुझाव निम्नांकित हैं।

1. पर्यटन क्षेत्र की विकास रणनीति निम्न लागत अर्थव्यवस्था, उत्पादकता का उच्च स्तर, अधिसरंचना की कार्यकुशलता में सुधार एवं प्रतियोगिता को प्रोत्साहन देना आदि के सिद्धांतों पर आधारित होनी चाहिए।

2. विविध पर्यटन विकास परियोजना को राशि आवंटित करते समय पूंजी श्रम अनुपात एवं प्रति डॉलर अर्जित पूंजी संसाधन लागत और प्रति पर्यटक सेवा की जानकारी प्राप्त करना उपयोगी होगा। इस हेतु पूंजी की कमी को देखते हुए पहले देश के कतिपय पर्यटन स्थलों को विकास हेतु चुनना चाहिए।

3. पर्यटन उद्योग को ऋण देने हेतु और विशेष जरूरतों को पूरा करने के लिए 100 करोड़ रुपए की आंरभिक लागत से पर्यटन वित्त निगम की स्थापना का परामर्श दिया गया। फलस्वरूप भारतीय पर्यटन वित्त निगम की स्थापना की गई।

आवास के क्षेत्र में निम्नांकित सुझाव दिए गए–

1. अधिक से अधिक युवा छात्रावासों का निर्माण।
2. व्यक्तिगत अतिथि गृहों, पर्यटन आवास, वन-आवास आदि जैसी अनुपूरक आवास व्यवस्था की स्थापना को बढ़ावा एवं प्रोत्साहन देना।
3. महलों, पुरानी हवेलियों जैसी राष्ट्रीय संपदाओं की सुरक्षा, रखरखाव तथा आवास गृहों में रूपांतरण।
4. आवसीय क्षेत्रों को प्रोत्साहित करते हेतु ऋण लेने और चुकाने की सुविधा और वित्तीय प्रोत्साहन देना और संचालन तथा प्रबंध में कार्यक्षमता बढ़ाने हेतु कार्यात्मक लागत को प्रोत्साहन देना।

परिवहन के क्षेत्र में सुझाव दिए गए–

1. एयर चार्टर का उदारता से अनुमोदन करना चाहिए।
2. भारत आने और यहाँ से जाने वाले सैलानियों की सुविधा विदेशी वायुसेवाओं के प्रति उदार नीति अपनानी चाहिए।
3. नदी मार्गों की संभावनाओं का पूर्ण रूप से विकास किया जाए।

4. रेल मंत्रालय में पृथक से एक पर्यटन निदेशालय होना चाहिए।
5. सड़क परिवहन के सुधार का विशेष प्रबंध किया जाना चाहिए। चार राजमार्गों एवं मोटर मार्गों का विकास किया जाना चाहिए, जिनके किनारे समुचित दूरी पर ठहरने एवं रुकने का इंतजाम हो।
6. ट्रेवल एजेंसियों की वायुसेवाओं और रेलवे आरक्षण से जुड़े टर्मिनल सुलभ ,कराए जाने चाहिए।
7. पैलेस ऑन व्हील्स की तर्ज पर दूसरे यात्रा मार्गों पर भी विशेष पर्यटन रेलगाड़ियाँ चलाई जानी चाहिए।
8. इंडियन एयर लाइन्स का आंशिक निजीकरण होना चाहिए।

एयर इंडिया को संयुक्त स्टॉक कंपनी में बदल दिया जाना चाहिए। इससे इस वायुसेवा में अपेक्षित पेशेवर रूझान आ सकेगा एवं यात्रियों और सुविधाओं का बेहतर इंतजाम हो सकेगा।

पर्यावरण के नुकसान को नियंत्रित करने के विषय में कोई गंभीर कदम नहीं उठाया गया है। इस संदर्भ में निम्नांकित सुझाव दिए गए–

1. किसी खास तरह के पर्यटन की अनुमति जैसे–ट्रेकिंग।
2. पारिस्थितिकी और रोमांच समूह से जुड़े गैर-सरकारी संगठनों का मदद लेना।
3. स्थानीय परिस्थिति के अनुसार पर्यटक सुविधाओं का विकास करना। जैसे–पहाड़ी इलाकों में विस्तृत पैमाने पर सड़क बनाने के स्थान पर पगडंडियों, साईकिल मार्ग, रज्जु मार्ग आदि की संभावनाओं का विकास।
4. जिस क्षेत्र में परिभ्रमण का विकास करना हो उसकी वहन क्षमता का मूल्यांकन करना।
5. जिन क्षेत्रों में पर्यावरण को खतरा हो उन इलाकों में सैलानियों को पूरी तरह से बंद करना। उदाहरणस्वरूप अंडमान निकोबार के अधिकतर द्वीप समूह।
6. विविध क्षेत्रों में विशेष पर्यटन इलाकों का निर्माण जहां पर्यटन गतिविधि को केंद्रित किया जा सके।
7. क्षेत्रों को बारी-बारी से सैलानियों के लिए खोलना और बंद करना ताकि पर्यावरण को सुरक्षित रखा जा सके।
8. एक समय पर सीमित संख्या में सैलानियों के प्रवेश की अनुमति।

अंतर्राष्ट्रीय पर्यटन में भारत की आमदनी बढ़ाने हेतु निम्नांकित उपाय सुझाए गए–

1. पर्यटन के विकासात्मक और विपणन प्रयत्नों को प्रतियोगितापूर्ण धार देने के लिए एक बाजार विकास पद्धति की स्थापना की जानी चाहिए।
2. पर्यटन सूचना नेटवर्क की स्थापना हेतु संचार और कम्प्यूटर के क्षेत्र में हुए तकनीकी विकासों का प्रयोग किया जाना चाहिए।
3. विदेशों में पर्यटन कार्यालयों के चुनाव और स्थापना के पर्याप्त संसाधन जुटाये जाने चाहिए एवं योग्य, पेशेवर और विपणन में माहिर अधिकारी की नियुक्ति की जानी चाहिए।
4. सार्वजनिक और निजी अभिकरणों से संसाधन ग्रहण कर मुख्य बाजारों में एक राष्ट्रीय छवि निर्माण एवं विपणन (Marketing) योजना बनाई जानी चाहिए।
5. गैर परपंरागत सैलानियों के ढंग से पर्यटन करने वाले पर्यटकों को भारत अपनी और खींच सकता है और इस दिशा में प्रयत्न करना चाहिये।
6. यूरोप एवं उत्तरी अमेरिका में पर्यटन अपनी चरम सीमा पर पहुंच गया है, वहां से सैलानियों को आकर्षित किया जा सकता है।

आठवीं पंचवर्षीय योजना (1992-97)

मई 1992 में भारत सरकार ने संसद में पर्यटन कार्य योजना रखी जो आठवीं योजना के दस्तावेज का आधार बनी। आठवीं योजना के कतिपय प्रमुख सुझाव इस प्रकार हैं—

पर्यटन के विकास में राज्य के विकास की संपूर्ण रणनीति का नियोजन, निजी क्षेत्र में निवेश को बढ़ावा देने हेतु वित्तीय मदद, उद्योग, उपभोक्ता और पर्यावरण के हितों की सुरक्षा हेतु प्रभावी नियमन को विकसित करना आदि की भूमिका होगी। आठवीं योजना के दौरान मुख्य ध्यान यूरोप, अमेरिका एवं जापान जैसे राष्ट्रों में आने वाले सैलानियों पर केंद्रित किया गया जो ज्यादा पैसा खर्च कर सकते हैं। पर्यटन परिवहन आदि के विकास में निजी क्षेत्र के हस्तक्षेप को प्रोत्साहित किया गया।

वस्तुतः आठवीं योजना के दस्तावेज का मुख्य जोर निजी उद्यम की भागीदारी एवं पैसा खर्च करने वाले सैलानियों को आकृष्ट करने पर केंद्रित है।

राष्ट्रीय कार्य योजना 1992

उद्देश्य—भारत सरकार की राष्ट्रीय पर्यटन कार्य योजना के उद्देश्य इस तरह हैं—

(1) पर्यटन उत्पाद की विविधता : भारत में परंपरागत सांस्कृतिक पर्यटन को सुरक्षित रखते हुए परिवर्तित उत्पादों में विभिन्नता लाई जानी चाहिए। उपभोक्ता की बदलती मांग को देखते हुए आराम, रोमांच, सम्मेलन एवं प्रोत्साहन पर्यटन के क्षेत्र का भी समुचित विकास किया जाना चाहिए।

(2) अंतर्राष्ट्रीय पर्यटन का विकास और विदेशी मुद्रा अर्जन : वैश्विक पर्यटन से विदेशी मुद्रा अर्जन में काफी सहायता मिलती है और देश की आवश्यकताओं को देखते हुए पर्यटन का विकास करना चाहिए ताकि सदी के अंत तक विदेशी मुद्रा अर्जन 2,440 करोड़ रुपये से बढ़कर 10,000 करोड़ हो जाए।

(3) राष्ट्रीय विरासत और पर्यावरण का संरक्षण : पर्यटन का विकास करते समय सांस्कृतिक अभिव्यक्ति एवं विरासत की सुरक्षा की जाएगी तथा कला और शिल्प को संरक्षण प्रदान किया जाएगा। पर्यावरण का संरक्षण और संवर्धन पर्यटन विकास का अभिन्न अंग होना चाहिए।

(4) विश्व पर्यटन में भारत के योगदान में बढ़ोतरी : अभी विश्व के कुल विदेशी सैलानियों के आवागमन में भारत का सिर्फ 0.4 प्रतिशत अंशदान है। कार्य योंजना का एक मुख्य उद्देश्य अगले पांच वर्षों में इसे बढ़ाकर 1 फीसदी करना है।

(5) घरेलू पर्यटन खासकर कम आय वाली श्रेणी का विकास करना : पूरेवर्ष स्थानीय सैलानियों की संख्या सबसे अधिक होती है। भारत में भी घरेलू सुविधाओं का सुधार और प्रचार किया जाएगा। इसमें खासकर कम आमदनी वाले पर्यटकों का ध्यान रखा जाएगा ताकि वे भी पर्यटन और छुट्टी का आनंद उठा सकें।

(6) रोजगार के अवसर बढ़ाना : पर्यटन उद्योग प्रत्यक्ष एवं अप्रत्यक्ष रूप में तकरीबन 1.3 से 1.4 करोड़ लोगों को रोजगार मुहैया करता है। सदी के अंत तक आज की तुलना में ये रोजगार के अवसर कम से कम दोगुने हो जाने चाहिए।

(7) क्षेत्रों का सामाजिक आर्थिक विकास : पर्यटन के क्षेत्र विशेष के व्यक्तियों को सामाजिक आर्थिक लाभ होता है और उनका जीवन स्तर ऊँचा उठता है। यह उन क्षेत्रों में भी विकास का मार्ग प्रशस्त करता है जहां अन्य आर्थिक गतिविधियों की संभावना नहीं होती है।

पर्यटन विकास की व्यूह रचना

इन उद्देश्यों की पूर्ति हेतु निम्नलिखित रणनीति अपनाई गई है–

1. मानव संसाधनों के विकास हेतु संस्थाओं की पुनर्रचना और दृढ़ीकरण।
2. विदेशी सैलानियों के आगमन और विदेशी मुद्रा अर्जन बढ़ाने के लिए एक सुचारू नीति का निर्माण।

3. वर्तमान अधिसरंचना के अधिकतम उपयोग हेतु स्थल का विपणन करने के साथ-साथ समन्वित विकास के लिए चयनित आधार पर क्षेत्रों का विकास।

पर्यटक अधिसरंचना में सुधार : अब तक राज्य ने सेवा क्षेत्र में अत्यधिक योगदान दिया है अब इसे निजी क्षेत्र की पहल और निवेश के लिए छोड़ देना चाहिए। विकास की गति को बढ़ाने हेतु अब घरेलू और विदेशी दोनों ही क्षेत्रों में निजी निवेश को प्रोत्साहन देने की नीति अपनानी होगी। पर्यटन मंत्रालय की भूमिका नीति निर्माण तक सीमित होगी।

आवास : हमारे देश में होटल, पर्यटन अधिसरंचना का एक प्रमुख और जरूरी अंग है। यह आंकलन लगाया गया है कि भारत में अनुशासित श्रेणी में 44,405 होटल सुलभ हैं, अगले तीन वर्षों में इन्हें दोगुना करने का लक्ष्य है। किसी होटल द्वारा अर्जित 50 फीसदी विदेशी मुद्रा आयकर से मुक्त होगी और अगर 50 प्रतिशत विदेशी मुद्रा पर्यटन उद्योग में पुनर्निवेशित की जाती है तो यह भी आयकर से मुक्त होगी। पर्यटन क्षेत्रों में भी इस विकास से गति लाने हेतु सरकार ने घोषणा की कि ग्रामीण इलाकों, हिल स्टेशनों, तीर्थ स्थलों एवं विशेषीकृत पर्यटन स्थलों में होटल खोलने पर व्यय कर नहीं लगेगा और आयकर में 50 प्रतिशत की छूट दी जाएगी।

सरकार होटलों को बनाने हेतु सस्ती दर पर ऋण भी मुहैया करा रही है। 4,5 स्टार होटलों के लिए 7.5 लाख तक 1 प्रतिशत ब्याज की मदद एवं अन्य अनुशासित होटलों के लिए 3 प्रतिशत ब्याज की दर पर सहायता ऋण दिया जाता है। बड़े शहरों में खुलने वाले 4-5 स्टार होटलों और अन्य होटलों के लिए अब यह रियायत खत्म कर दी जाएगी। विशेष क्षेत्रों और विशेषीकृत स्थलों में विकास को गति देने हेतु सभी 1, 2, 4 स्टार के होटलों के लिए ब्याज सहायता बढ़ाकर 5 प्रतिशत कर दी जाएगी।

कई छोटे एवं बड़े महत्व की हवेलियां ऐसी हैं जिन्हें आराम से होटलों में रूपांतरित किया जा सकता है। इनमें से अनेक की हालत खस्ता है एवं इन्हें होटल के तौर पर बदल कर राष्ट्रीय विरासत होटलों के विकास को प्रोत्साहित करने का निर्णय लिया गया है। अपने नियमों को उदार बनाते हुए भारतीय पर्यटन वित्त निगम अब विरासत होटल परियोजनाओं हेतु 50 लाख से अधिक अग्रिम ऋण मुहैया कराएगा। एक नई योजना लागू की जाएगी जिसके अन्तर्गत इस श्रेणी के होटलों को 5 अथवा कुल लागत की 25 प्रतिशत की छूट दी जाएगी। ग्रामीण क्षेत्रों में स्थित हवेलियों पर व्यय कर नहीं लगेगा और आयकर में 50 प्रतिशत की छूट होगी। आशा की जाती है कि इन रियासतों से भारत में विरासत होटलों में तेजी से बढ़ोतरी होगी।

स्थानीय पर्यटन को प्रोत्साहन

घरेलू पर्यटन को बढ़ावा देने के उपाय भी किए जा रहे हैं। अल्प आय वाले सैलानियों हेतु खास तौर पर तंबूनुमा आवासों का इंतजाम किया जा रहा है। रोमांचकारी कैंपों की स्थापना हेतु तंबू आवासों और अन्य सुविधाओं को जुटाने हेतु केंद्र सरकार राज्य पर्यटन विकास निगमों को वित्तीय सहायता मुहैया कराएगी। घरेलू और विदेशी दोनों तरह के सैलानियों को सस्ता और स्वास्थ्यकर आवास उपलब्ध कराने के लिए अतिथि भुगतान आवास को भी प्रोत्साहित किया जा रहा है। इसी तरह अपार्टमेंट होटलों को मान्यता प्रदान करने हेतु वर्गीकरण नियमों में ढील दी जा रही है ताकि घरेलू सैलानियों को सस्ते आवास प्रदान किए जा सकें। इसके अतिरिक्त घरेलू पर्यटक में लगे टूर-ऑपरेटर, ट्रेवल ऐजेंटों को मान्यता प्रदान करने हेतु नियम में सुधार लाया जा रहा है।

वस्तुतः समस्त मान्यताप्राप्त टूर-ऑपरेटरों और ट्रेवल एजेंटों को यह सलाह दी गयी है कि वे स्थानीय पर्यटन के लिए भी कार्य करें।

तीर्थयात्रा पर्यटन : पर्यटन में तीर्थयात्रियों की संख्या अत्यधिक होती है। तीर्थस्थलों में सैलानियों की अधिसंरचनात्मक सुविधाओं का अपेक्षित विकास आवश्यक है। इस उद्देश्य से केंद्र सरकार हर वर्ष केंद्रीय सहायता के रूप में 5 करोड़ रुपये राज्य पर्यटन विकास निगम/निजी उद्यमियों को प्रदान करेगी। इसका प्रयोग प्रमुख तीर्थ स्थलों में आवासीय सुविधाओं के विकास, यात्रा के दौरान आवश्यक सुविधाओं के विकास एवं परिवहन व्यवस्थाओं के विकास के लिए किया जाएगा।

शिल्प ग्राम :- सूरजकण्ड मेले और शिल्प ग्राम, उदयपुर की तरफ बड़ी संख्या में सैलानी आकृष्ट होते हैं। इन केंद्रों की सफलता को देखते हुए पर्यटन विभाग देश के विविध भागों में शिल्प ग्राम स्थापित करने और इन प्रदेशों के परंपरागत हस्तशिल्पों को प्रोत्साहित करने की दिशा में विचार कर रहा है।

मेले और पर्व त्योहार—मेले एवं पर्व सिर्फ भारत की सांस्कृतिक विरासत का ही प्रतिनिधित्व नहीं करते अपितु ये सैलानियों के आकर्षण के प्रमुख केंद्र भी बन गए हैं। इनकी तरफ पर्यटन मंत्रालय का ध्यान गया है और विदेशी सैलानियों को आकर्षित करने हेतु उन्हें विकसित और प्रचारित करने की ओर ध्यान दिया जा रहा है। पुष्कर मेला, सोनपुर पशु मेला, पंतग उत्सव, एल्लेप्पी नौका दौड़, मैसूर एवं कुल्लू का दशहरा पर्व आदि सैलानियों के आकर्षण के केंद्र बन गए हैं। देश में इन परंपरागत मेलों और पर्वों के विकास के लिए मंत्रालय उदारता से वित्तीय मदद देगा और अंतर्राष्ट्रीय बाजार में जोर-शोर से इनका प्रचार करेगा।

सड़क परिवहन : सड़क परिवहन पर्यटन अधिसंरचना का एक अन्य प्रमुख अंग है। अगले दो वर्षों में सरकार उत्तर प्रदेश और बिहार के बौद्ध स्थलों हेतु वातानुकुलित बस की व्यवस्था करेगी यात्रा व्यापार के लिए रियायती कीमत पर 300 वातानुकूलित इकाइयों को आयात करेगी। यह स्मरण रखना होगा कि इन पर्यटन बसों में पर्यटकों की सभी तरह की सुविधाओं का इंजताम हो। यह महसूस किया गया है कि देश के सैलानियों को उच्च स्तरीय आरामदायक बसों, कारों की आवश्यकता है। सरकार कुछ ऐसी योजना सामने लाएगी जिसके द्वारा पर्यटन के क्षेत्र की इस महत्वपूर्ण आवश्यकता को पूरा किया जा सके।

भ्रमण व्यापार : ट्रेवल एजेंट एवं टूर ऑपरेटर पर्यटन उद्योग के एक प्रमुख अंग हैं। देश के वृहद आकार को देखते हुए भ्रमण व्यापार को इस तरह प्रोत्साहित किया जाना चाहिए कि अधिक से अधिक लोग भ्रमण व्यापार की सीमा में आ जाएं। ट्रेवल एंजेट और टूर-ऑपरेटर को मान्यता प्रदान करने हेतु तत्सम्बन्धी नियमों और कायदे-कानूनों को व्यवस्थित करने के लिए सरकार को प्रयत्न करना होगा।

पर्यटक आवागमन को बढ़ावा : सरकार ने उदारवादी चार्टर नीति की घोषणा की है और नियम के अन्तर्गत नागरिक उ़डयन के महानिदेशक ने स्वयं चार्टर फ्लाइट हेतु शीघ्र अनुमति दे दी है। चार्टर फ्लाइट के आगमन से पर्यटन यातायात में तेजी से बढ़ोतरी होगी हाल में लागू की गई एयर टैक्सी योजना से राष्ट्रीय वाहकों की कार्यक्षमता में वृद्धि होगी और देश में सैलानियों का आवागमन सुगम हो जाएगा। सरकार ने हाल में ही आने वाले पर्यटन मौसम में गोवा और मैसूर के मध्य पैलेस ऑन व्हील्स की तर्ज पर एक पर्यटक रेलगाड़ी चलाने की घोषणा की है। जल परिवहन द्वारा भी सैलानियों के आवागमन की व्यवस्था की जाएगी। नदियों और समुद्रों में चलने वाले जल-वाहनों का प्रचलन शुरू करने हेतु तेजी से इस क्षेत्र को बढ़ावा दिया जाएगा।

विदेशी निवेश : नई औद्योगिक नीति के परिशिष्ट 111 में न. 4 में होटल पर्यटन से संबद्ध उद्योग को सूचीबद्ध किया गया है, जिसके अन्तर्गत कुल निवेश का 51 फीसदी विदेशी निवेश स्वयंमेव स्वीकृत हो जाएगा। इस तरह की उदार नीति से यह स्पष्ट है कि पर्यटन क्षेत्र में विदेशी निवेश को आकृष्ट करने हेतु विशेष प्रयास किए जा रहे हैं। सरकार ने प्रधानमंत्री के प्रधान सचिव की अध्यक्षता में एक समीति की स्थापना की है जो हमारे देश में निवेश करने वाले विदेशी निकायों से वार्ता करेगी। पर्यटन क्षेत्र में विदेशी निवेश को प्राप्त करने एवं उसे ठोस स्वरूप प्रदान करने के लिए पर्यटन सचिव की अध्यक्षता में मंत्रालय में एक समिति गठित की है जो मुख्य समिति की मदद करेगी।

2001-2003 तक विदेशी मुद्रा का वितरण

वर्ष	रुपये करोड़ में	परिवर्तन प्रतिशत	यू.एस. डॉलर लाखों में	परिवर्तन प्रतिशत
2001 (जन. दिस.)	14344,00	—	3042	—
2002 (जन. दिस.)	14195,00	-1,0	2923.00	3.9
2003 (जन. दिस.)	17049,41	20.1	3602,88	23,3

स्रोत : बाजार सर्वेक्षण विभाग, पर्यटन विभाग, भारत सरकार, नई दिल्ली।

सम्मेलन पर्यटन : आज सम्मेलन एवं गोष्ठी का पर्यटन में महत्व अत्यधिक बढ़ गया है। सम्मेलन पर्यटन को प्रोत्साहित करने हेतु देश में सर्वसुविधा सम्पन्न सम्मेलन परिसरों का निर्माण करना होगा ताकि भारत में अधिक-से-अधिक अंतर्राष्ट्रीय सम्मेलनों को आयोजित किया जा सके। अंतर्राष्ट्रीय स्तर की सभी तरह की सुविधाओं से युक्त सम्मेलन नगर बनाने की ओर सरकार को ध्यान देना होगा।

एकीकृत विकास के लिए चयनित आधार पर क्षेत्रों का विकास : केंद्र सरकार ने पर्यटन सुविधाओं के एकीकृत विकास हेतु राज्य की कुछ सरकारों के साथ मिलकर कतिपय विशेष पर्यटन क्षेत्रों के निर्माण का प्रस्ताव रखा है। प्रस्तावित विशेष पर्यटन क्षेत्र की योजना के अंतर्गत निर्यातक सामान उत्पादित करने वाले उद्योगों के विकास हेतु बनाई गई अधिसूचित निर्यात संसाधन क्षेत्र योजना के समान ही है। मास्टर प्लान तैयार करते समय विशेष पर्यटन क्षेत्रों की सीमा निर्धारित कर दी जाएगी और इस क्षेत्र में निवेशकों को होटल एवं पर्यटन से संबंधित उद्योगों हेतु रियायती दर पर जमीन आवंटित की जाएगी। विशेष पर्यटन क्षेत्र के प्रशासन हेतु राज्य सरकारों द्वारा विशेष क्षेत्र विकास प्राधिकरण की स्थापना की जाएगी जिसे राज्य सरकार के विविध अधिनियमों द्वारा पूर्ण शक्ति प्रदान की जाएगी। विशेष क्षेत्र विकास प्राधिकरण इस क्षेत्र का मास्टर प्लान बनाएगा एवं निवेशकों को एक ही स्थान पर सारे कार्य निपटाने की सुविधा प्रदान करेगा। सड़क परिवहन, बस अड्डा, सड़क के किनारे आवश्यक सुविधाएं, बिजली आपूर्ति, जल आपूर्ति, कानून और व्यवस्था एवं नगर निगम सेवा जैसी आधारभूत अधिसंरचना सुलभ कराने का जिम्मा राज्य सरकार का होगा। केंद्र सरकार हवाई अड्डा और वायु सेवा, रेलवे स्टेशन, संचार सेवा और डाक/तार, बैंकिंग सेवा जैसी सुविधाएं जुटाएगी। निवेशकों को आकर्षित करने हेतु केंद्र और राज्य सरकार दोनों विशेषीकृत केंद्रीय और राज्य करों में छूट देंगी और कम से कम दस वर्षों तक पानी एवं बिजली की दर में बढ़ोतरी नहीं करेंगी।

विदेशी पर्यटकों का क्षेत्रीय आधार पर वितरण

क्र.सं क्षेत्रीय आधार के निवासी	*कुल*	*(% कुल पर्यटक)*
1. दक्षिणी एशिया	4,29,707	19.1
2. उत्तरी यूरोप	3,52,568	15.7
3. पश्चिमी एशिया	2,75,202	12.2
4. दक्षिण-पूर्वी एशिया	2,26,733	10.1
5. उत्तरी-अमेरिका	1,97,427	8.8
6. पश्चिमी यूरोप	1,96,542	8.7
7. पूर्वी यूरोप	1,84,287	8.2
8. माईक्रोनेशिया	80,319	3.6
9. पूर्वी एशिया	77,711	3.5
10. दक्षिणी यूरोप	74,370	3.3
11. आस्ट्रेलिया एवं न्यूजीलैण्ड	63,035	2.8
12. पूर्वी अफ्रीका	29,547	1.3
13. दक्षिणी अफ्रीका	19,418	0.9
14. पश्चिमी अफ्रीका	6,084	0.3
15. दक्षिणी अमेरिका	7,621	0.3
16. उत्तरी अफ्रीका	5,495	0.2
17. केरेबियन	4,150	0.2
18. मध्य अफ्रीका	2,568	0.1
19. मध्य अमेरिका	2,568	0.6
20. अन्य	14,452	0.6
कुल	22,49,138	100.0

विशेष पर्यटन क्षेत्रों का चयन केंद्र सरकार करेगी। इस तरह के विशेष पर्यटन क्षेत्रों के सफलतापूर्वक कार्यान्वयन हेतु जो अन्य अधिकतम सुविधाएं प्रदान करने के इच्छुक होंगे उनसे केंद्र सरकार राय परामर्श देगी। इन क्षेत्रों में विदेशी निवेश को आकर्षित करने का विशेष प्रयास किया जाएगा।

पूंजी जुटाने एवं पर्यटन से जुड़ी गतिविधियों में सहयोग देने हेतु पर्यटन क्षेत्र सहयोग योजना की शुरुआत की जाएगी। पर्यटन से जुड़ी इन गतिविधियों का निर्धारण सरकार करेगी। अण्डमान और लक्ष्यद्वीप जैसे स्थानों में पर्यटन हेतु निर्णय

लिया गया है। ऐसा करते समय उस स्थान की वहन क्षमता और पर्यावरण के संरक्षण का विशेष ध्यान रखा जाएगा। इसमें अमीर सैलानियों पर ध्यान केंद्रित किया जाएगा और यह ध्यान रखा जाएगा कि अल्प संख्या में लोग यहाँ आएंगे। दूसरे शब्दों में कम लोग अधिक पैसा खर्च करें। सैलानियों को दी जाने वाली सुविधाएं द्वीप के प्राकृतिक साहचर्य के अनुरूप होंगी।

मानव संसाधन विकास के लिए संस्थाओं की पुनर्रचना और दृढ़ीकरण : हमारे देश में होटल प्रबंधन खान-पान तकनीक की 15 एवं फूड क्राफ्ट की 16 संस्थाएं कार्यरत हैं। आवासीय तथा आवागमन क्षेत्र को मिलाकर अभी अनुमानतः कुल 8,500 प्रशिक्षित कार्मिकों का कारोबार होता है जबकि हर वर्ष 28,000 प्रशिक्षित कार्मिकों की आवश्यकता पड़ती है। अतः इस कमी को दूर करने हेतु सरकार प्रशिक्षण कार्यक्रम को मजबूत बनाएगी। फूड क्राफ्ट इंस्टीट्यूट विद्यार्थियों को सामान्य प्रशिक्षण प्रदान करता है। शुरुआती वर्षों में यह काफी उपयोगी साबित हुआ। शेफ प्रशिक्षण संस्थानों जैसी कुछ विशेषीकृत संस्थाओं की स्थापना कर इन गतिविधियों को आगे बढ़ाने की आवश्यकता है। पूरे देश और क्षेत्र विशेष में प्रशिक्षित मानव संसाधन की आवश्यकता को पूरा करने हेतु एक राष्ट्रीय पाक संस्थान की स्थापना की जाएगी।

होटल प्रबंधन एवं खान-पान तकनीक संस्थानों ने इन आवश्यकताओं को पूरा करने में विशेष योगदान दिया है और इन संस्थानों के छात्रों को होटल उद्योग में आसानी से रोजगार मिल जाता है। इसके अतिरिक्त क्षेत्रीय आधार पर विद्यार्थियों का नामांकन किया जाएगा। देश के भ्रमण व्यापार को प्रशिक्षित मानव संसाधन प्रदान करने हेतु भारतीय पर्यटन और भ्रमण प्रबंधन संस्थान को श्रमिक एवं उपकरण देकर मजबूत बनाया जाएगा। राष्ट्रीय विश्वविद्यालयों को भी इस उद्यम में सम्मिलित किया जाएगा और देश में पर्यटन पाठ्यक्रम चलाने के लिए वित्तीय और अन्य मदद दी जाएगी।

विपणन और विदेश में प्रचार : अधिकतम मुनाफा कमाने हेतु विपणन और विदेशी प्रचार के क्षेत्र में पर्यटन विभाग एयर इंडिया एवं विदेशों में कार्यरत अन्य एजेंसियों से संपर्क स्थापित करेगा। प्रत्यक्ष रूप से ग्राहक के पास पहुंचकर अब सीधे प्रचार पर बल दिया जाएगा। प्रचार-प्रसार की रणनीति के अन्तर्गत परिचय भ्रमण का आयोजन किया जाएगा। डिपार्टमेंटल स्टोर एवं अन्य संगठनों के साथ मिलकर संयुक्त प्रचार का कार्यक्रम चलाया जाएगा और भारत के प्रचार के लिए टूर-ऑपरेटरों को मदद प्रदान की जाएगी। सरकार ने अपने स्थानीय कार्यालयों और कुछ देशों में पर्यटन सूचना केंद्र टूरनेट (Tournet) की स्थापना

की है। भारत सरकार के पर्यटन कार्यालयों के अलावा हवाई अड्डों, होटलों और मित्र देशों में कम्प्यूटरीकृत बहुभाषीय पर्यटन सूचना उपलब्ध करने का भी प्रस्ताव है। इससे हमारे देश में आने के इच्छुक सैलानियों को विविध भाषाओं में परिवहन, साधन, भोजन, आवास, रुचि के क्षेत्र और सेवा की आगत की तैयार सूचना मिल जाएगी।

कार्य योजना का सार

1. विशेष लक्ष्य समूह को स्मरण में रखकर विदेशी कार्यालयों का पुनर्गठन।
2. सूचना क्रांति, सभी प्रमुख बाजारों में भारत की सकारात्मक छवि प्रस्तुत करने हेतु सूचना व्यवस्था का पुनर्गठन।
3. पैलेस ऑन व्हील्स की सफलता को देखते हुए प्रमुख पर्यटन मार्गों पर पर्यटन रेलगाड़ी की शुरुआत।
4. विशेष मार्गों पर नदी-यातायात का इंतजाम।
5. विरासत, होटल, स्वास्थ्य, आराम स्थलों की विशेष श्रेणी का निर्माण एवं कार्यात्मक विशेषज्ञता आदि सुविधाएं उपलब्ध कराना।
6. निवेश बढ़ाने एवं अपेक्षित विकास हेतु अधिसूचित क्षेत्रों के तौर पर विशेष पर्यटन क्षेत्रों का निर्माण।
7. जिन मार्गों पर एयर एंडिया उड़ान नहीं भरता अथवा कम भरता है उन मार्गों के लिए मुक्त आकाश नीति।
8. घोषित नीति स्वयंमेव अनुमति के अन्तर्गत उदारवादी चार्टर।
9. मुख्य अंतर्राष्ट्रीय हवाई अड्डों पर वायु सेवाओं, रेलगाड़ियों, होटलों एवं पर्यटन सूचना की व्यवस्था।
10. चयनित पर्यटन स्थलों हेतु विशेष वायु सेवा, होटल पैकेज।
11. विशेषकृत क्षेत्रों अथवा यात्रा-कार्यक्रमों में पर्यटन और संबंधित उद्योग को वित्तीय मदद प्रदान करने हेतु विशेषीकृत पर्यटन क्षेत्र मेंदद योजना की शुरुआत करना।

गहन विकास हेतु चुने हुए भ्रमण मार्ग एवं स्थल :

1. बागडोगरा-सिक्किम-दार्जिलिंग-केसिंपोंग
2. भुवनेश्वर-पुरी-कोणार्क
3. हैदराबाद-नागार्जुन सागर तिरुपति
4. ग्वालियर-शिवपुरी ओरछा खजुराहो

5. कुल्लू-मनाली लेह
6. जैसलमेर-जोधपुर बीकानेर बाड़मेर
7. इंदौर-उज्जैन माहेश्वर ओंकारेश्वर मांडू
8. ऋषिकेश, नरेन्द्र नगर, गंगोत्री, बद्रीनाथ
9. चेन्नई, ममल्लपुरम पांडिचेरी।

स्थल :

1. लक्षद्वीप द्वीपसमूह
2. मनाली सोलन नाला
3. अण्डमान द्वीप समूह
4. कांगड़ा पंग बांध।

पर्यटन योजना : महत्वपूर्ण तथ्य :

1. भारत पर्यटन विकास निगम की स्थापना एक सार्वजनिक क्षेत्र के उपक्रम के तौर पर 1 अक्तूबर, 1966 को की गई थी।
2. पर्यटन विभाग को एक स्वतंत्र मंत्रालय का दर्जा 23 मई, 1998 को दिया गया, जिसका नाम पर्यटन मंत्रालय रखा गया।
3. विश्व पर्यटन में भारत का हिस्सा वर्ष 2001 की अवधि में 0.37 फीसदी रहा।
4. चीनी पर्यटकों हेतु भोजन, भ्रमण को सुविधाजनक बनाने हेतु चीन द्वारा भारत में वर्ष 2003 तक पर्यटन कार्यालय स्थापित किया जाना है।
5. पर्यटन विभाग प्रचालनरत होटलों को स्टार प्रणाली की विभक्तिकरण कर 1-5 स्टार तक एवं हाल ही में चालू की गई हेरिटेज होटल श्रेणी में करता है।
6. होटलों का निर्माण करने हेतु विदेशी पूंजी निवेश की सीमा को 100 फीसदी की अनुमति प्रदत्त कर दी गई है।
7. 1950 से पहले निर्मित हवेलियों, दुर्गों, किलों एवं निवास स्थानों में प्रचालनरत होटलों हेतु हेरिटेज होटल की एक नई वर्गीकरण श्रेणी शुरू की गई है।
8. पर्यटन को प्रोत्साहन देने के उद्देश्य से विदेशी सैलानियों हेतु आगमन वीसा योजना शुरू की गई है।
9. दिल्ली, आगरा एवं जयपुर की दर्ज पर चेन्नई, बंगलौर एवं त्रिवेन्द्रम को पर्यटन की दृष्टि से गोल्डन ट्राएंगल के रूप में विकसित करने का विचार केन्द्र सरकार के विचाराधीन है।
10. वर्ष 2001-2002 के दौरान 24.23 लाख विदेशी सैलानियों ने भारत की यात्रा की। अप्रैल-दिसम्बर, 2002 के दौरान यह संख्या 16.15 लाख थी।

11. विदेशी सैलानियों द्वारा भारत में औसतन 1260 डॉलर प्रति पर्यटक व्यय किए जाते हैं और यह औसत विगत तीन वर्षों में तकरीबन स्थिर बनी हुई है। इसके विपरीत भारतीय सैलानियों को विदेशों में खर्च हेतु उपलब्ध कराई जाने वाली विदेशी मुद्रा में जून 2000 में भारी बढ़ोतरी कर दी गई है।

12. विगत वर्षों में पर्यटन से देश को होने वाली शुद्ध प्राप्तियां जहां धनात्मक रही हैं वहीं 2001-2002 के शुरुआती महीनों में यह प्राप्तियां ऋणात्मक हो गई थीं क्योंकि विदेशी मुद्रा भारतीय सैलानियों द्वारा विदेशों में खर्च की गई।

भारत में विदेशी सैलानियों का आगमन एवं उपार्जित विदेशी मुद्रा

वर्ष (अप्रैल दिसम्बर)	*विदेशी पर्यटक आगमन लाखों में*	*अनुमानित विदेशी मुद्रा, आय मिलियन अमरीकी डालर*
1995-96	21.90	2713
1996-97	23.34	2878
1997-98	23.71	2914
1998-99	23.97	2993
1999-2000	25.97	2993
2000-2001	26.99	3168
2001-2002	24.23	2960
2002-2003	16.75	2203

स्रोत : बाजार सर्वेक्षण विभाग, पर्यटक विभाग, भारत सरकार, नई दिल्ली।

निष्कर्ष : योजना की प्रक्रिया आवश्यक रूप से लम्बे समय के पर्यटन की संभावनाओं वाले कुछ क्षेत्रों अथवा स्थानों की पहचान करती है। स्थलीय योजना पर्यटन की प्रमुख विशेषताएं हैं। योजना बनाना तकरीबन वैज्ञानिक प्रक्रिया हो चुकी है। नियंत्रित पर्यटन आज की जरूरत है। पर्यावरण की सुरक्षा के लिए तकरीबन पर्यटकों पर पूर्ण प्रतिबंध लगाना समय की मांग है। वास्तव में पर्यटन संरक्षण का दोस्त है उसका, दुश्मन नहीं, वन्य प्राणियों के राष्ट्रीय उद्यानों से यह साबित हो चुका है। यह भी देखने में आया है कि विपरीत पर्यटक न होते तो कुछ स्थानों पर संरक्षण कार्य नहीं होता। इन क्षेत्रों में पर्यटन पर्यावरण के रखवाले के तौर पर कार्य करता है। अतः पर्यावरण के संरक्षण के साधन के तौर पर पर्यटन को प्रभावशाली बनाने हेतु किसी विशेष पर्यटन स्थल को स्थान की प्रकृति और स्थिति के अनुसार योजनाबद्ध करना उपयोगी साबित होता है।

भारतीय हस्तशिल्प एवं पर्यटन

हस्तशिल्प संस्कृति का एक अनिवार्य अंग है जो पर्यटन में भी अपनी प्रमुख भूमिका निभाता है। भारत में कई दूसरे देशों ने भारतीय हस्तशिल्पों को संग्रहित किया गया है, जो सैलानियों के आकर्षण का केन्द्र बने हुए हैं। शिल्पियों द्वारा सदियों से बनायी जाने वाली बेहतरीन कलाकृतियाँ भारतीय शिल्पकारों की कुशलता और अनूठी शिल्पकला की बेहतरीन मिसाल हैं।

औद्योगिक क्रान्ति से पूर्व भारत का हस्तशिल्प उत्पादन विश्व का सबसे वृहद उद्योग था, औद्योगीकरण ने हजारों वर्षों से चले आ रहे इस उद्योग को अत्यधिक कमजोर कर दिया। हस्तशिल्प उद्योग में मोटे तौर पर दो तरह की वस्तुएं बनायी जाती थीं।

1. धनी वर्ग के लोगों हेतु इन वस्तुओं के निर्माण में उच्च कोटि के कौशल की जरूरत होती थी।
2. आम उपयोग की वस्तुएँ जो सस्ती थीं, निर्माण में अपेक्षाकृत कम कौशल की जरूरत होती थी।

ब्रिटिश काल के दौरान औपनिवेशिक नीतियों एवं उद्योगों के आक्रमण के परिणामस्वरूप दोनों तरह के हस्तशिल्पों का तेजी से ह्रास हुआ।

मशीनी वस्तुओं के उत्पादन ने हस्तशिल्प उद्योग को पतन के मार्ग पर ला खड़ा किया।

इतिहास में वर्णित भारतीय हस्तशिल्प

भारत में कला एवं शिल्प की प्राचीन परम्परा रही है। इतिहास के विविध कालखण्डों में भारतीय हस्तशिल्प के उत्थान एवं पतन का सार बड़ा दिलचस्प रहा है।

ब्रिटिश काल से पूर्व भारतीय हस्तशिल्प

परिष्कृत रूप से मिट्टी के बर्तनों का निर्माण तो हड़प्पा काल में ही शुरू हो गया था। भारत का सबसे प्रधान शिल्पीय उत्पाद कपड़ा मुगलकाल में अपने उत्कर्ष पर था। गुप्तकाल में धातु शिल्प, हाथी दांत शिल्प, आभूषण निर्माण आदि शिल्पों की मांग विदेशों में भी थी। इस काल में शिल्पीय उत्पादन का संगठन मुख्य तौर पर चार प्रकार होता था :

1. कतिपय शहर में रहकर स्वतन्त्र रूप से कार्य किया करते थे। यह उत्पादन मुख्य तौर पर बिक्री के लिए किया जाता था। मूसलीपट्टनम्, बनारस और कासिम बाजार में साड़ियों, कश्मीरी कालीनों व शालों का उत्पादन अपने उत्कर्ष पर रहा। परिवार के लोग मिलकर काम करते थे किन्तु शिल्पकारों के पास निजी रूप में पूंजी की कमी होती थी और उत्पादन भी अल्प होता था।
2. ज्यादातर उत्पादन ग्रामीण अर्थ व्यवस्था एवं जजमानी व्यवस्था से जुड़ा था।
3. मध्यकाल के दौरान शासकों हेतु, कुलीनों की जरूरतों को पूरा करने हेतु कारखाने स्थापित किये गये। यह शिल्पीय उत्पादन का एक अन्य रूप था। कारखानों में निपुण कारीगर नियुक्त किये जाते थे, जो कुलीन वर्ग हेतु महंगी विलास की वस्तुएँ बनाते थे।
4. 17वीं सदी में शिल्पीय उत्पादन के लिए एक ऐसी व्यवस्था शुरू हुई जिसमें व्यापारी उत्पादन हेतु कारीगरों को अग्रिम भुगतान किया करते थे।

--कारीगरों को व्यापारियों द्वारा दिये गये निर्देशों के अनुरूप माल उत्पादित करना पड़ता था।

—निश्चित समय में कारीगरों को माल की आपूर्ति करनी होती थी।

फलस्वरूप माल की गुणवत्ता, मात्रा और उत्पादन समय पर शिल्पियों का नियंत्रण हो गया।

परम्परागत तौर पर भारतीय शिल्प किसी जाति विशेष द्वारा बनाये जाते थे जैसे जुलाहे कपड़ा बनाते थे, मोची जूता बनाते थे, सुनार गहना बनाते थे, लोहार लोहे के औजार बनाते थे एवं बढ़ई लकड़ी का काम करते थे।

औपनिवेशिक काल

अंग्रेजों के आ जाने के पश्चात् शिल्पीय उत्पादन तेजी से गिरा। सन् 1757 से पूर्व ईस्ट इंडिया कम्पनी अपने साथ सोना चांदी लाती थी और बदले में भारत से हस्त निर्मित कपड़े और अन्य सामान ले जाती थी, जिससे भारत को खूब मुनाफा हुआ। परन्तु 1757 के बाद इस स्थिति में बदलाव आया और भारतीय शिल्पकारों को अपने उत्पादन को कम मूल्यों पर बेचना पड़ा। फलतः शिल्प उद्योग कमजोर पड़ता गया।

स्वातंत्र्योपरांत भारत

स्वतन्त्रता के पश्चात् भारतीय हस्तशिल्पों के विकास की ओर ध्यान दिया गया। हस्तशिल्पों की समस्याओं का पता लगाने और इनके विकास का मार्ग सुझाने

हेतु नवम्बर 1952 में अखिल भारतीय हस्तशिल्प बोर्ड की स्थापना की गयी। इस्तशिल्प निर्यातों को बढ़ाने हेतु 1958 ई. में भारतीय हस्तशिल्प और हथकरघा निर्यात निगम की स्थापना की गयी। भारतीय शिल्पों को ग्राहकों तक पहुंचाने के लिए स्थान-स्थान इम्पोरियम खोले गये। दिल्ली में 1953 में शिल्प संग्रहालय खोला गया। सहकारी क्षेत्र में दिल्ली स्थित केन्द्रीय कुटीर उद्योग इम्पोरियम एक मुख्य विपणन संस्था है।

इसके अतिरिक्त भारत सरकार ने उदयपुर, राजस्थान स्थित शिल्प ग्राम, अहमदाबाद में कपड़ा संग्रहालय एवं वर्तन संग्रहालय, पुणे में राजा दिनकर केलकर संग्रहालय की स्थापना की।

हस्तशिल्प क्षेत्र की दुर्बलताएं

इन संभावनाओं के बावजूद अंतर्राष्ट्रीय हस्तशिल्प निर्यात में भारत का वास्तविक हिस्सा सन्तोषजनक नहीं है। इसके अधोलिखित कारण हैं :

समुचित अधिसंरचना का अभाव
पुराने जमाने के औजारों का प्रयोग
पूंजी का अभाव
बाजार अनुसंधान का अभाव
बिचौलिये।

शिल्प के विविध प्रकार

यंत्रों के आगमन के बावजूद कारीगरों एवं शिल्पियों द्वारा उत्पादन जारी रहा। इस उत्पादन के दो प्रकार हैं :

(1) रोजमर्रा के उपयोग की वस्तुएँ एवं
(2) सजावट हेतु बनाई गयी वस्तुएँ

मिट्टी के शिल्प और बर्तन : चीनी मिट्टी से निर्मित वस्तुएं पोर्सिलेन एवं सेरामिक।

पत्थर से निर्मित वस्तुएँ: तमिलनाडु में सेलम जिला पत्थर शिल्प हेतु है। ताजमहल का सुन्दर स्थापत्य एवं अलंकरण आगरा के पत्थर शिल्पियों को सदियों से प्रेरित व प्रभावित करता आ रहा है। बिहार में गया पत्थर के काम का प्रधान केन्द्र है। देवी देवताओं की मूर्तियाँ यहां बनाई जाती हैं। जयपुर में पत्थर से देवी-देवताओं की मूर्तियां बनाई जाती हैं।

लकड़ी से निर्मित शिल्प : लकड़ी पर नक्काशी एवं बेंत का फर्नीचर।

धातु शिल्प : उभरी नक्काशी अथवा उभरी छपाई की तकनीक, धातु की सतह पर नक्काशी के द्वारा आकृति बनाना, एक धातु पर दूसरी धातु के तारों से सजावट, सांचे द्वारा आकृतियों बनाना एवं मिश्रित धातुओं से मूर्तियां बनाना।

प्रमुख केन्द्र : लद्‌दाख, श्रीनगर, मुरादाबाद, अलीगढ़, बनारस, बंगाल में वर्द्धमान एवं मिदनापुर, जयपुर की मीनाकारी आदि।